COSMIC IMAGERY

KEY IMAGES IN THE HISTORY OF SCIENCE

BY THE SAME AUTHOR

Theories of Everything

The Left Hand of Creation
(with Joseph Silk)

L'Homme et le Cosmos
(with Frank J. Tipler)

The Anthropic Cosmological Principle
(with Frank J. Tipler)

The World within the World

The Artful Universe

Pi in the Sky

Perché il mondo è matematico?

Impossibility

The Origin of the Universe

Between Inner Space and Outer Space

The Universe that Discovered Itself

The Book of Nothing

The Constants of Nature:
From Alpha to Omega

The Infinite Book:
A Short Guide to the Boundless,
Timeless and Endless

New Theories of Everything

COSMIC IMAGERY

KEY IMAGES IN THE HISTORY OF SCIENCE

JOHN D. BARROW

W. W. NORTON & COMPANY
New York • London

First American Edition 2008

Printed in China

For information about permission to reproduce selections from this book, write to Permissions, W. W. Norton & Company, Inc., 500 Fifth Avenue, New York, NY 10110

For information about special discounts for bulk purchases, please contact W. W. Norton Special Sales at specialsales@wwnorton.com or 800-233-4830

Manufacturing by RR Donnelley Shenzhen
Production manager: Julia Druskin

Library of Congress Cataloging-in-Publication Data

Barrow, John D., 1952–
Cosmic imagery : key images in the history of science / John D. Barrow.
p. cm.—(1st Amer. ed.)
Includes bibliographical references and index.
ISBN 978-0-393-06177-2 (hardcover)
1. Cosmology—History—Pictorial works. 2. Astronomy—History—Pictorial works. 3. Cosmology—History. I. Title.
QB981.B2797 2008
520.22'2—dc22

2008017099

W. W. Norton & Company, Inc.
500 Fifth Avenue, New York, N.Y. 10110
www.wwnorton.com

W. W. Norton & Company Ltd.
Castle House, 75/76 Wells Street, London W1T 3QT

1 2 3 4 5 6 7 8 9 0

To Brenda and Don

I can't think of a single Russian novel in which
one of the characters goes to
a picture gallery

W. Somerset Maugham

CONTENTS

PART 2: **Spatial Prejudice**

PART 3: **Painting by Numbers**

PART 4: **Mind Over Matter**

PREFACE

'What is the use of a book,' thought Alice,
'without pictures or conversations?'
Lewis Carroll

Books, especially books on scientific topics, generally use pictures to illustrate what they have to say. Pictures save words. They change the pace, alter the style, and make things more memorable. They are between a thing and a thought.

This book is not intended to be like that. Its words were wrapped around the pictures and the parts they played in making a lasting and vivid contribution to our scientific understanding of the Universe. Sometimes these images recorded information in a new and impressive way; sometimes they took advantage of new ways of doing or seeing things; and at other times they simply told a story for which words alone were not enough.

Yet this is far from being a 'picture book'. The images each carry a story that is important, unusual, or simply untold. Taken together they create a wide-ranging picture of scientific progress in which there is time for history and space for geography.

The motivation for such a book grew partly out of sociological and technical changes within science itself. In just a few years, the presentation of science at all levels, from technical seminars for fellow experts to popular expositions for the general public, has become extremely visual. The ubiquity of PowerPoint, web-streamed video, digital photography, and artificial computer simulation has meant that images dominate science in a way that would have been technically and financially impossible just 20 years ago. There is a visual culture in science and it is rapidly changing.

Visuality penetrated the practice of science just as deeply as it fashioned its presentation. Small desktop computers revolutionised research and enabled complex and chaotic behaviours to be studied visually for the first time by single individuals and small research groups armed with nothing more than an inexpensive off-the-shelf box of silicon. Small science became bigger. The results of its new forms of experimental mathematics were dramatically visual and started a trend towards the investigation of emergent complexity by means of direct simulation. 'Publication' no longer meant only a paper on paper.

We have witnessed a revolution in the history of science. Not the sort of revolution that philosophers of science once believed in – they don't happen any more – but a revolution brought about by new tools, different ways of seeing, and novel ways of understanding. Nothing old needed to be overthrown to make way for the new.

The future of science will be increasingly dominated by artificial images and simulations. It will be harder for iconic images to last in the face of ever-greater technical facility. So, now is an interesting time to look backwards as well as forwards. I hope that the pictures chosen here will focus attention on the important role they played in facilitating mental pictures and guiding the way science and mathematics developed our understanding of Nature and Nature's laws.

Unfortunately, there are some special forms of complexity where pictures actually make matters worse, and the writing of this book gradually revealed itself to be one of them. Fortunately, I was helped by many people along the way who worked tirelessly to locate pictures, 'firsts', high-quality images, and copyright holders. Will Sulkin, Jörg Hensgen and Drummond Moir at The Bodley Head, Random House, played the major role in sustaining the project and transforming it from black-and-white type into the volume you are now holding. Our children, although no longer children, have maintained an unexpected interest in the final product, perhaps suspecting that there will be an accompanying video game. Elizabeth has learnt that these projects do have an end, and is especially patient on hearing continually that it is nigh. Her vital support in so many ways enabled it all to get done without too many other things remaining undone.

A large number of friends and colleagues helped. For their assistance with discussions, text, pictures, and sources, I would especially like to thank Sarah Airey, Mark Bailey, June Barrow-Green, Nadine Bazar, Alan Beardon, Richard Bright, Rosa Caballero, Alan Chapman, Pamela Contractor, Jim Council, Carl Djerassi, Richard Eden, Kari Enqvist, Gary Evans, Patricia Fara, Ken Ford, Marianne Freiberger, Sandy Geis, Gary Gibbons, Owen Gingerich, Sheldon Glashow, Edward Grant, Peter Hingley, Sharon Holgate, Michael Hoskin, Martin Kemp, Rob Kennicutt, Paul Langacker, Imre Leader, Raimo Lehti, Nick Mee, Simon Conway Morris, Andrew Murray, Dimitri Nanopoulos, Chris Pritchard, Helen Quigg, Stuart Raby, Martin Rees, Simon Rhodes, Adrian Rice, Graham Ross, Martin Rudnick, Chris Stringer, Rose Taylor, Frank Tipler, John Turner, Steven Weinberg, John A. Wheeler, Denys Wilkinson, Robin Wilson, Tracey Winwood and Alison Wright. All of them answered questions, made suggestions, or provided information or images with helpful enthusiasm.

John D. Barrow
Cambridge, 2008.

INTRODUCTION
EVERY PICTURE TELLS A STORY

'Someone who is on his way to
somewhere can hardly keep company
with someone who is going nowhere,'
he said. 'But we can try. You can come
along with me if you carry my suitcase.'

Henning Mankell[1]

WE like pictures. They were the first things we ever saw. Our minds were not made for letters, numbers, double-entry book-keeping, musical scores, or mathematical equations – all of these are postscripts to the human story. Our senses evolved in an environment that was appreciated as something to be understood and remembered as a picture. Sensitivity to a variety of that environment's qualities was engendered in generations of survivors who discovered that they were inclined to act in ways that were safer and more enduring than those of Mr and Mrs Average.

From these humble beginnings we have inherited a liking for pictures. We find them entertaining, we find them educational, we find them memorable, and we find them inspiring. Our earliest relics of cultural anthropology reveal images of unnerving sophistication, such as those in the caves at Lascaux, that would be recognised as works of art even today. Pictures have played a role in binding primitive societies together in life-supporting ways, they have defined entire periods of human history by their style and subjects, and they have maintained tradition and social memory over long periods of time. They have focused religious emotions and contemplation, and they have invited us to ponder the inner workings of minds that use us innocently as subjects. In all these manifestations, pictures seek to represent and encapsulate something of reality in a way that has an instant impact: something that is memorable without needing to be remembered.

Every subject has its iconic images. We all know many of them in the realms of art or functional design. From the *Mona Lisa* and the Alhambra to the map of the London Underground or Tower Bridge, certain images persist and are hugely influential. They shape our memories of the world. The same is true in science. Some images have defined our steps in understanding the Universe; others have proved so effective in communicating the nature of reality that they are part of the process of thinking itself, like numbers or the letters of the alphabet. Others, equally influential, are so familiar that they appear unnoticed in the scientific process, part of the vocabulary of science that we use without thinking.

This book is about scientific imagery. We are going to look at some of the images and pictures that have played a key role in shaping our scientific picture of the world. Some of those images are so subtle that they dominate our way of practising science or describing reality without us even noticing. Others are ubiquitous icons that dominate the presentation of a whole branch of science, or our conception of its history. Others still possess an aesthetic quality but have a scientific subtext that makes them important for our story.

The use of diagrams and pictures in the practice of science and in presenting it to others is no longer an activity with an artistic imperative. The scientists who create a new form of visual representation may draw the pictures for themselves, but more often the final version will be the work of others. A technical artist (or even a computer program) will produce a prettier version from their rough sketches. What true artist would follow such a course? Scientists are trying to present information in an instantly recognisable fashion. But sometimes, as we shall see, their efforts were more enduring and more influential than they ever imagined. At other times, they were unusually controversial.

Some important scientific images are human constructions with a special purpose, but others are records of natural phenomena, the result of new instruments taking the first glimpses of a new world of the ultra small, the vanishingly faint, the stunningly explosive, or the unimaginably remote. They mark our entry into new realms that require maps because evolution never found it a worthwhile investment to equip our unaided senses to penetrate them. As we go there, we shall find that some pictures have come to play a scientific role that rivals that of words and numbers. Carefully constructed families of pictures can act as a calculus all of their own. Like any successful systems of symbols, with an appropriate grammar they enlarge the number of things that we can do without consciously thinking.

In all these considerations we recognise in the background an immediate technological impetus that provides scientists with new ways to use images. In the last twenty years we have witnessed one of the greatest innovations in human history: the Internet (with its World Wide Web) has collectivised human thinking and information retrieval in a revolutionary way and outstripped the biological capability of individuals, allowing the instant retrieval and exchange of pictures. Yet there is also something remarkably visual about the development of the modern science that has accompanied this revolution. The advent of small, inexpensive computers with superb graphics has changed the way many sciences are practised, and the way that all sciences present the results of experiments and calculations.

Long ago, computers were huge and mightily expensive – the preserve of large well-funded research groups studying block-busting problems. They built bombs, predicted the weather, or tried to understand the workings of the stars. But the personal-computer revolution changed all that. It allowed the study of subjects such as chaos and complexity to flourish. Mathematics became an experimental subject. Individuals could follow previously intractable problems by simply watching what happened when they were programmed into a personal computer. The output was, as often as not, a sequence of pictures or movies of how a complex process developed and continued to completion. There is no simple formula that tells how galaxies end up assembling into all the shapes and sizes that we see, or how dramatic turbulence occurs when a fast-flowing river cascades through a series of rapids; these problems are too complicated to be solved exactly using pencil and paper alone. But pictures and film can reveal the crucial points in these processes by transforming the mathematical equations that control their behaviour into simulations. The rapid development of spectacular graphics has enabled almost anyone to present these conclusions in a fashion that is more visually impressive than the achievements of the movie industry just twenty years ago. The PC revolution has made science more visual and more immediate. It has enhanced the intuitive power of the human mind to appreciate the behaviour of complex patterns of behaviour from experience by creating films of imaginary experiences of mathematical worlds.

The last ten years have also seen a growing emphasis upon the visual in many areas of culture. Mass-appeal films are defined increasingly by the presence of special visual effects. Storyline is almost superfluous in the race to string together more and more cataclysmic images, some real, but most largely the products of computer simulation. When they appear on DVD we are given extra visuals and additional incestuous coverage of the making of the film. Experimental theatre introduces challenging new techniques on to the stage that make visual experience more vivid. Popular music is accompanied by the ubiquitous video or DVD: sound is not enough. Animated films have reached new levels of sophistication and even books come with their own websites. Words are no longer enough.

The practice of science has followed a parallel course. Conference presentations and seminars used to rely on blackboard and chalk, handouts, overhead projectors, or 35mm slides. Today, scientists project data via PowerPoint, videos of computer simulations are routine, and audioclips, film, and split-screen graphics are commonplace in scientific presentations to colleagues and the general public. Even the publication of scientific research has been swept up in the technological revolution. It is now possible to publish huge quantities of colour images and video in online journals or websites associated with print journals. This would have been impossible twenty-five years ago. Importing pictures into articles is a relatively simple matter – a few mouse clicks suffice. For a great scientific expositor such as Arthur Eddington, writing in the 1930s, the inclusion of diagrams was an expensive and laborious business and, not surprisingly, you won't find any in most of his books. Today, the opposite is true.

In this climate, it is a challenge to think about the role of images in science, not just today, but over many hundreds of years. There are definitive images, graphs, pictures, and charts that play a key role in widening our appreciation of the world. This book draws together a personal selection of these special pictures. While many such images would figure in almost anyone's gallery of science, others are more subjectively chosen because of an importance that is not immediately obvious, or which involves something that has not been recognised at all before. Scientific pictures are often not just about science. They may be interesting because they are scientific in origin, but yet have an undeniable aesthetic quality. They may even have been primarily works of art that possess a scientific message.

Every one of the pictures we will show has a story. Sometimes that story is about its creator, sometimes it is about the scientific insight that flowed from the picture, sometimes it is about the technique of representation itself, sometimes it is that the simple image was to assume an unforeseen importance which stimulated an entirely new way of thinking, and sometimes it is simply a tale of the unexpected.

Stars in Your Eyes

Astronomy is what we have now instead of theology. The terrors are less, but the comforts are nil.

John Updike[2]

The Globular Star Cluster NGC 6397, one of the closest to Earth, 8,500 light years away.

NO area of science is more picturesque than astronomy. The sky is a natural canvas on which to draw imaginary connections between the stars and our ideas about them. Some of humanity's earliest written records show the phases of the Moon, and tell stories that are fashioned by the appearance of the night sky. The rising and setting stars, the great circles of stars around the Celestial North Pole in the northern sky, constellations of meaning, and the eclipsing of the Moon and Sun were the greatest events that ancient humans ever witnessed. They spoke of the reliability and regularity of the Cosmos: its dark sky full of navigation beacons, their recurring movements a celestial clockwork by which to order one's life and nurture the land. We look first at some of the special pictures of the stars that were drawn and painted with great skill. Some had artistic intent, others were more prosaic. Pictures give way to diagrams. Diagrams inspire generalisations and simplifications. They replace masses of data by a compression of information. Armed with the key images that emerged from early twentieth-century astronomy, we see how the subjects of modern astrophysics and cosmology emerged, supported by early photographs and diagrams that encapsulated the diversity of the Universe and its dramatic runaway expansion. Black and white gave way to colour. The abstractions of Einstein's curved space and time gave rise to analogies and pictures that played a new explanatory role. Space and time gave way to space–time, visible light was augmented by images across the rest of the electromagnetic spectrum, and we realised that we could see back towards the apparent beginnings of time. Cosmology now gives us pictures of other universes, and makes the map of our own ever more complex.

Yet the responses that these celestial images have continually provoked tell a parallel story of equal interest. They have created a perspective of our own place in the Universe. The Universe, big and old, dark and cold, creates feelings about the relationship between life and the impersonal immensities of the cosmos. The patterns of the stars once gave rise to complicated beliefs in astrological influences upon human actions and psychologies. Astrology became astronomy, but the images of the Universe that the Hubble Space Telescope has provided have influenced us in extraordinary ways.[3] They have an aesthetic appeal that resonates with a pioneering spirit of adventure: the discovery of new vistas, spectacular events of stellar birth and death that were once the preserve of science fiction books and films.

Andreas Cellarius's northern hemisphere and its sky, 1660.

MIDNIGHT'S CHILDREN

THE CONSTELLATIONS

Lisa: Remember, Dad, the handle of the Big Dipper points to the North Star.
Homer: That's nice, Lisa, but we're not in astronomy class. We're in the woods.

The Simpsons

ONCE the night sky was dark for everyone, everywhere. There was no artificial light, and the Moon and the stars were visible to the naked eye with a clarity that is no longer possible in modern cities. The stars were also comfortingly familiar: their positions recognised and relied upon for navigation over land and sea; their patterns a portent of disasters; their regularity and predictability the seed of human faith in a Universe that was lawful and regular rather than the plaything of temperamental deities bent upon vengeance and never-ending internecine battles. The stars were important.

The most persistent and vivid picture of the star-spangled sky has remained the map of the constellations. The stars were grouped together into collections that suggested the shapes of animals or everyday objects, perhaps to enshrine some religious or mythological association between those things and their position on the sky, or simply as an aide-memoire for reading the sky. Their slow and steady progress across the darkness, night after night, as the Moon waxed and waned, and the Earth undertook its annual orbit around the Sun, allowed time to be measured in new ways, and provided mariners with a means to navigate after nightfall. If you are travelling on land and night falls, then you can stop and wait for daylight; at sea that possibility is not always open to you.

The slowly evolving map of the constellations has played a multifaceted part in human history. It has fuelled superstition, furthered scientific astronomy, aided navigation, and created a sense of oneness with the Universe. The night sky is the oldest shared human experience. Its re-creation in pictures and illuminated manuscripts has preserved and elevated that experience in many cultures and deservedly finds a place in any gallery of great scientific images.

The annual path of the Sun, viewed from our terrestrial perspective, traces out on the celestial sphere a great circle on the sky. In ancient times this was divided into twelve signs, or 'houses', of the zodiac by the twelve constellations through which the Sun passed in sequence on its (apparent) annual journey around the Earth.[4] These twelve signs are still those used credulously today in the astrology columns of many magazines and newspapers around the world. Actually, the signs of the zodiac (which means literally 'circle of animals') differ from the constellations of the zodiac, even though they share similar names. The constellations are groups of adjacent stars that formed some discernible and suggestive pattern. The signs of the zodiac,[5] by contrast, consist of twelve equal zones of 30 degrees in length, or one hour on a clock face, giving a total of an entire circle of 360 degrees around the sky. Conventionally, the constellations are each taken to be 18 degrees wide in extent on the sky. At first the signs and their synonymous constellations were closely related, but gradually more and more ancient constellations were named and they soon greatly outnumbered the signs. Since the signs were only used for astrological purposes they could remain twelve in number, but the constellations were crucial for navigation and so navigators in different parts of the world needed different markers and lines in the sky, hence the continual additions to the menagerie of constellations to ensure adequate sky coverage. Most people still know their star sign and astrology is still peddled in some quarters as a way of predicting human personality and behaviour by means of the character traits traditionally associated with individuals born under a particular star sign.

Another fascinating feature of the oldest constellation maps has enabled astronomers to speculate about the place and the time of their original creation. When the Earth rotates it wobbles slightly, just like a precessing top, so that its axis of rotation (currently almost, but not quite, the same as the axis through the North and South magnetic Poles) does not always point in the same direction in the sky but traces a circle that takes about 26,000 years to complete.

At present, our northern rotational axis points conveniently towards a star that we have dubbed the 'Pole Star', or 'Polaris', but far in the past or the future the celestial North Pole would have pointed, or will point, in a different direction, either at another star or at no star at all. For example, in 3000 BC the Pole Star would have been Alpha Draconis, but when Shakespeare has Julius Caesar say he is 'constant as the northern star',[6] this is a complete anachronism: there would have been no northern star in Caesar's day.

This precession of the celestial North Pole means that the sky looks different in significant ways to observers located at different latitudes on the Earth and at different times in history. Most interesting of all, an ancient astronomer observing from a latitude of, say, L degrees north would have been unable to see a disc of the sky centred on the south celestial pole and spanning an angle of $2L$ degrees on the sky. By looking at maps of the ancient constellations, several nineteenth-and twentieth-century astronomers have claimed to locate the latitude of the earliest known constellation mappers close to 35 degrees north from the angular extent of the empty zone in their maps of the southern sky.

This 35-degree line of latitude runs through the Mediterranean Sea, intersecting the Minoans, the Phoenicians (modern-day Lebanon) and the Babylonians (modern-day Iraq). Then by locating the centre of that empty zone it is possible to extrapolate the history of the precession backwards to find out when the South Pole was at the centre of the empty zone. This gives an interval of time between about 1800 and 2500 BC for when the ancient constellations were laid down.[7]

The reasons for their suggestive shapes and names are lost. But if we assume that the inventors were Mediterranean navigators then many of the animal shapes make good sense, tracing the rising of stars on the sky and permitting the setting of a sailing course at night. They also took on a wider significance. Maps symbolise a human desire to understand and be in control of our surroundings. To map a territory was tantamount to possessing it. Maps of the heavens offered an ultimate reassurance that all was well in the Universe, that we were at a focal controlling point within it, and had a special part to play in its unfolding story.

Each religious tradition embraced the zodiac and the constellations in its own way. Julius Schiller produced a Christianised picture of the constellations in which the strange pagan symbols were replaced by names from the Old and

New Testaments. But the greatest draughtsman and artist of all those who strove to represent the constellations was the Dutch-German mathematician and cosmologist Andreas Cellarius. His *Celestial Atlas of Universal Harmony*, the *Atlas Coelestis seu Harmonica Macrocosmica* of 1660, is one of the most beautiful books ever created. The hand-coloured engravings are the Sistine Ceiling of the map-maker's art, combining vivid colours and exuberant figures within the geometrical constraints of a sky map of the constellations.[8]

The oldest image of the ancient constellations available to us today is one that is enshrined in stone rather than on paper and associated with a very different type of 'atlas'. In the National Archaeological Museum of Naples, famed for its ancient Egyptian remains, is the Farnese Atlas, a second-century AD Roman statue of the bearded figure of Atlas bearing a white marble globe of the constellations on its shoulders. The statue, standing more than two metres tall, is bending on one knee and is partly covered by a cloak. The celestial globe it bears is 65 cm in diameter and is damaged only by a hole through the top that cuts through the constellations of Ursa Major and Minor. In all, forty-one constellations are shown drawn in positive relief; no single stars are shown. The equator, the two tropics, the colures,[9] and the polar circles are shown as coordinate lines in relief around the sky on the surface. The globe is laid out with remarkable precision and the positions of defining points on the sky are accurate to about 1.5 degrees. In 2005, Bradley Schaefer[10] of the University of Louisiana re-analysed the layout of the constellations shown on the Farnese globe and showed that with high probability they are an image of the long-sought lost star catalogue of Hipparchus of Rhodes, the greatest astronomer of the ancient world. One of Hipparchus' many accomplishments, made possible by the extraordinary precision of his observations, was the discovery of the 26,000-year precession of the Earth's rotation axis that we discussed above. By studying the locations of the constellations and the blind spots in the sky coverage alone, Schaefer dated the constellation map shown on the Farnese globe to 125 BC with an uncertainty of only about fifty-five years either way. Hipparchus created his star catalogue in 129 BC but it vanished in antiquity and has remained unknown until now except for references to it by others.[11] Ironically, Hipparchus' own discovery of the precession was the key tool in dating the globe that bears an image of his map.

The Farnese Atlas.

Yet, the constellations have created one strange mystery of history. The

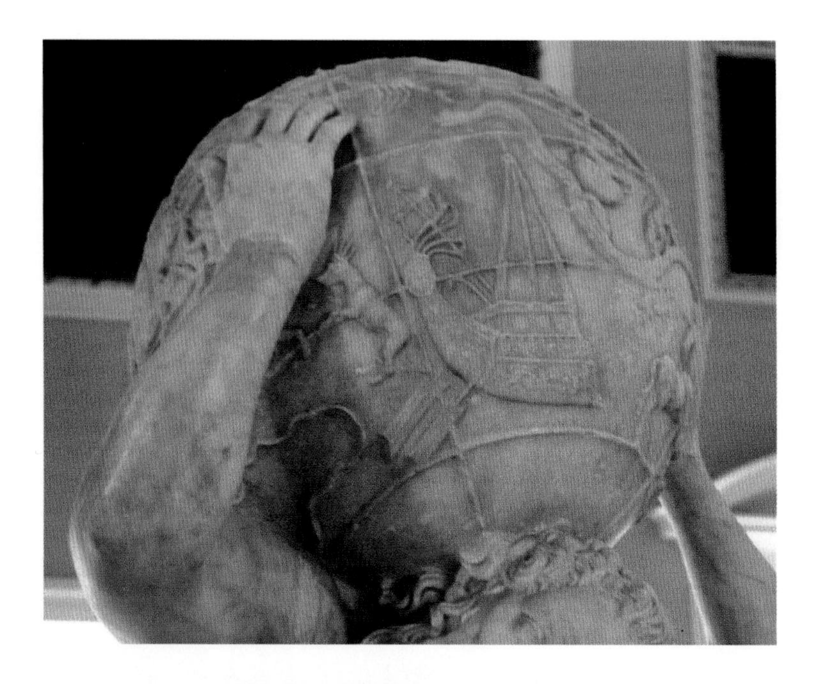

(*Top*) Front view, showing the colure dividing the sky between Canis Major and Argo.

(*Bottom*) Back view, showing the equator, the ecliptic, the equinoctial colure and the two tropics. The rim of the horn of Aries lies right on the colure.

trick of using the Earth's precession history in conjunction with empty regions of ancient sky maps to date and locate the creators of the knowledge those maps contain was first exploited by a little-known Swedish amateur astronomer, Carl Swartz, who in 1807 published a book[12] on the subject in Swedish and French, with a second edition in 1809. From the maps available to him, he deduced that the original constellation map-makers probably lived around 1400 BC near a latitude of 40 degrees north and he suggested the city of Baku, on the coast of Armenia, as their likely home. Subsequently, other astronomers, such as the Astronomer Royal, Edward Maunder,[13] in 1909, and Michael Ovenden[14] in 1965 (neither of whom seemed to know of Swartz's book), narrowed the map-makers' latitude down to around 36 degrees north and their epoch to the range 2500–1800 BC. Ovenden believed (unjustifiably in the minds of many historians)[15] that the Minoans were the prime candidates because they were seafarers with an advanced culture at the target latitude who suddenly disappeared owing to a natural disaster in about 1450 BC. But he noticed a strange historical conundrum. One of the most useful records of ancient astronomical knowledge is contained in a long and detailed prose poem by Aratus of Soli,[16] which is entitled the *Phaenomena* (*The Appearances*) and was published in about 270 BC. He listed forty-eight constellations and their relative positions on the sky in his poetic tribute to the great Greek astronomer and mathematician Eudoxus of Cnidus,[17] who lived between 409 and 356 BC. Ancient literature talks of 'the sphere of Eudoxus' and it is generally believed that he possessed a celestial globe, but nothing more is known about it or about his astronomical writings and maps. Fortunately, Aratus used Eudoxus' now-lost writings to create his poetic description of the sky and so he gives us a constellation-by-constellation guide to Eudoxus' sky. But when Hipparchus studied the poem 150 years later he was puzzled. The sky according to Aratus and Eudoxus was not one they could ever have seen. Taking precession into account, they included constellations that were not visible to them when and where they lived, and they omitted some of the constellations that were visible. Ovenden's analysis of the poem indicates that it describes a sky that would have been visible at latitudes between 34.5 and 37.5 degrees north at a time between about 3400 and 1800 BC – very similar to the time and place we have deduced from the ancient constellation maps as well.[18]

One possible conclusion to draw is that Eudoxus inherited ancient knowledge from another civilisation in another location that he didn't know how to update – probably he didn't realise it needed updating – because he didn't know about precession. If he was in possession of a very ancient celestial globe, he might have reported the sky that was inscribed upon the globe rather than the sky that he saw. Was there a mysterious ancient globe from Egypt or some other Mediterranean civilisation, such as that in Babylonia?

This is a strange story but alas, in this case at least, fact does not seem to be as strange as fiction. A careful scrutiny of the evidence for a void in the southern sky of the size and location claimed by Ovenden and others before him reveals a number of serious errors[19] in their analysis and large uncertainties concerning the practical visibility of stars. In an attempt to distil out the fact from the fiction, Bradley Schaefer has subjected the problem to a modern astronomical analysis,[20] checking all the historical information available, reviewing past analyses, and introducing a new and more reliable technique to isolate when the constellation makers might have worked. He tracks the visibility of stars in six key southern constellations Altar (Ara), Southern Crown (Corona Australis), Southern Fish (Pisces Australis), Water (Aquarius), Centaur (Centaurus), and Argo from 3000 BC onwards, and computes the most northerly latitude[21] at which the creators of each of these constellations could have lived, at any given epoch. The results are interesting because the maximum latitude of visibility has a rather different variation with the passage of time for each of these constellations.[22] Two constellations (Water and Southern Fish) become visible increasingly further north as time passes, while all the others become visible only at less and less northerly latitudes. As a result, the curves of latitude of visibility versus time of the first two have to intersect the four others. Intriguingly, we see that *all* the curves roughly intersect in a fairly narrow range of latitudes and epochs. If all these constellations were created by a single civilisation (rather than being invented separately over a significant period of time) then this crossover indicates when and where it would have to have been done.

The latitude limit coincidence has a spread of only about 6 degrees around 500 BC and only about 2.5 degrees around 300 BC. Schaefer concludes that his realistic treatment of the uncertainties leads to a range from 900 BC to 330 BC for the date of these six constellation makers with 30 to 34 degrees north as

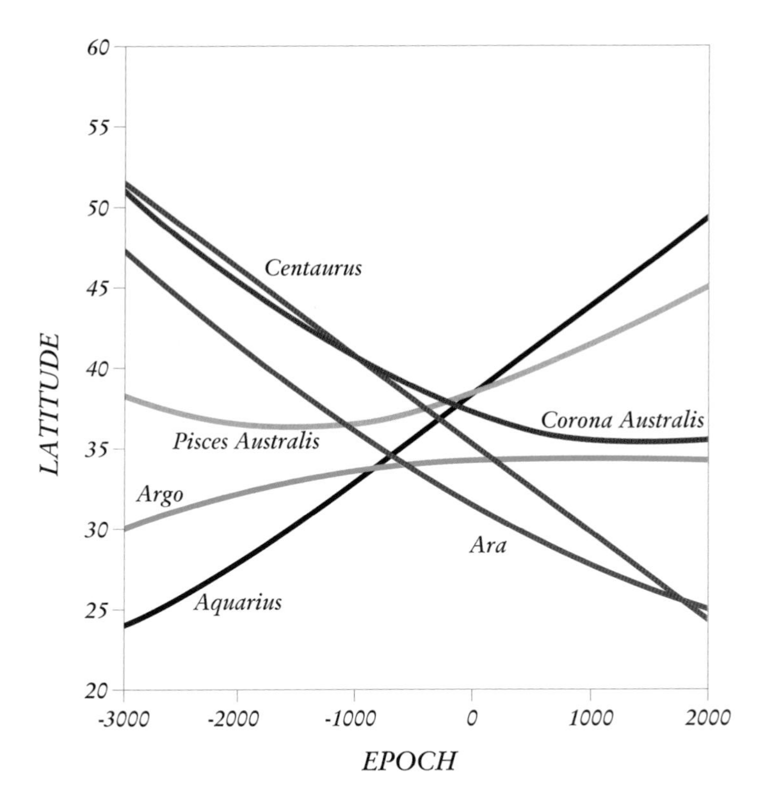

their likely latitude.[23] This is very different to the older estimates and is historically much easier to understand. It rules out the Minoans and the ancient Greeks, but it is a perfect fit for the Babylonians at 32.5 degrees north. This is an appealing conclusion because we already know that the Babylonians contributed about 16 constellations to the celestial picture in about 500 BC, and one of those was the Fish, which was in Schaefer's sample of six.

It is remarkable that the ancient picture of the constellations remains something that we not only admire for its beauty and drama but can use to discover vital facts about those who first framed the story that it tells.

Bradley Schaefer's Realistic latitude limits and epochs, for the makers of six distinctive constellations.

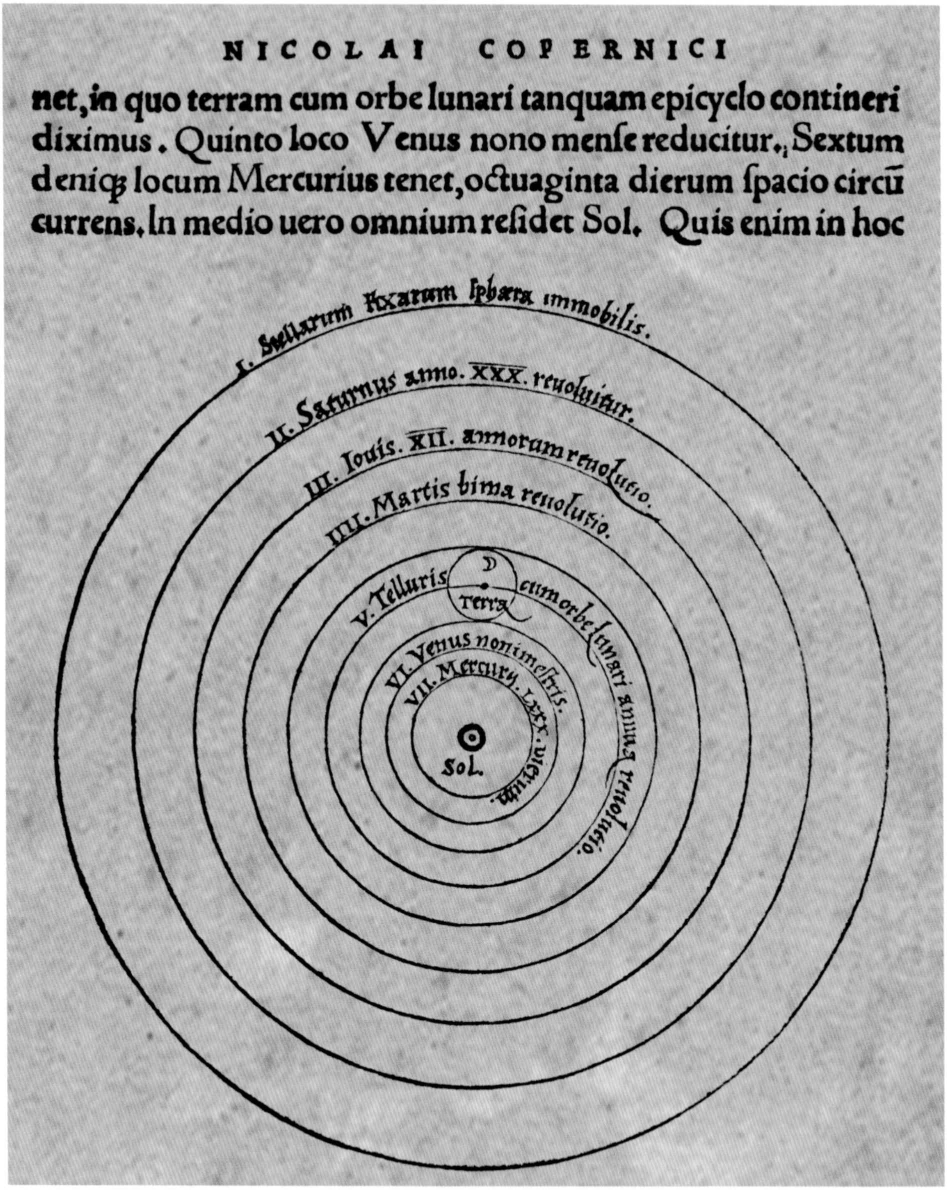

NICOLAI COPERNICI

net, in quo terram cum orbe lunari tanquam epicyclo contineri diximus. Quinto loco Venus nono menſe reducitur. Sextum deniq; locum Mercurius tenet, octuaginta dierum ſpacio circū currens. In medio uero omnium reſidet Sol. Quis enim in hoc

I. Stellarum fixarum ſphæra immobilis.
II. Saturnus anno. XXX. reuoluitur.
III. Iouis. XII. annorum reuolutio.
IIII. Martis bima reuolutio.
V. Telluris cum orbe lunari annua reuolutio.
terra
VI. Venus nonimestris.
VII. Mercurij. LXXX. dierum.
Sol.

Heliocentric model, published by Nicolaus Copernicus (1473–1543) in 1543 in his book *De revolutionibus orbium coelestium* (*On the Revolutions of the Celestial Spheres*). This Sun-centred model revolutionised astronomy, replacing the previous Earth-centred (geocentric) model. The diagram, labelled in Latin, shows concentric spheres around the Sun (centre). The outer, fixed sphere (I, the stars) surrounds rotating spheres (II–VII) for Saturn, Jupiter, Mars, Earth (and the Moon), Venus, and Mercury.

EMPIRE OF THE SUN

THE COPERNICAN WORLD PICTURE

How many students does it take to change a light bulb on the campus of a Scottish university? At Edinburgh, just one – he holds the bulb and the world revolves around him.

Anonymous Scottish academic[24]

NICOLAUS COPERNICUS is generally regarded as a revolutionary – the scientist who dethroned humanity from a central position in the Universe. But, if he was a revolutionary at all, he was certainly a reluctant one and the reality has turned out to be more complicated and far less dramatic.[25] Copernicus' great book, *De revolutionibus orbium coelestium* (*On the Revolutions of the Celestial Spheres*), was delivered to the printers in 1543, shortly before his death, and its impact was muted. Not many copies were printed and few of those ever read. Yet, in time, Copernicus' work became the rallying point for the transformation of our view of the Universe. It began a slow war of attrition that would eventually overthrow the ancient Ptolemaic picture of a Universe with the Earth at its centre in favour of a heliocentric model that is firmly established today.[26] Yet, its impact on our worldview was arguably greater than that on our world model.

The advances in printing during the early sixteenth century meant that Copernicus' illustrated book could be accurately printed with diagrams embedded in the text at the points where they are discussed. The most famous of his diagrams at the beginning of his discussion shows a simple model of the solar system with the Sun located at its centre. The outermost circle marks the boundary of the 'immobile sphere of the fixed stars' beyond our solar system. The other six outer circles are the spheres for the orbits of the six then-known planets. Going from the outside inwards, they denote Saturn, Jupiter, Mars, Earth (with its adjacent crescent Moon), Venus and Mercury, respectively. All follow circular tracks around the central Sun (*Sol*). The Moon was believed to move in a circle around the Earth.

The systems of Copernicus and Ptolemy were not the only pictures of the Universe that were on offer by the sixteenth and seventeenth centuries. The accompanying picture from Giovanni Riccioli's *Almagestum Novum* (*The New Almagest*)[27] of 1651 nicely summarised the world pictures on offer to astronomers in the post-Copernican era. It shows six different models of the solar system (labelled I–VI). Model I is the Ptolemaic system with the Earth located at the centre and the Sun's orbit around it lies outside the orbits of Mercury and Venus; Model VI is the Copernican system we have just seen illustrated by Copernicus himself. Model II is the Platonic system, where the Earth is central. The Sun and all the planets are in orbit around it but the Sun lies inside the orbits of Mercury and Venus. Model III is the so-called Egyptian system, in which Mercury and Venus revolve around the Sun, which, along with the outer planets, revolves around the Earth. Model IV is the Tychonic system (of Tycho Brahe, 1546–1601) in which the Earth is fixed at the centre and the Moon and the Sun revolve around the Earth, but all the other planets revolve around the Sun. The orbits of Mercury and Venus are therefore partly between the Earth and the Sun, while the orbits of Mars, Jupiter, and Saturn encircle both the Earth and the Sun. Model V, called the semi-Tychonic system, was invented by Giovanni Riccioli, the author of *Almagestum Novum* and the creator of this picture. In his model the planets Mars, Venus, and Mercury orbit the Sun which, together with Jupiter and Saturn, orbits the Earth. Riccioli wanted to distinguish Jupiter and Saturn from Mercury, Venus, and Mars because they were known to have moons like the Earth (the two moons of Mars had not yet been discovered) and so their orbits must be centred on the Earth rather than on the Sun.

Copernicus has given his name to an entire philosophy of the world. An 'anti-Copernican' view of anything in science is a pejorative description of a view looked upon as blinkered or biased towards a human perspective or position in some unjustified way. In cosmology the 'Copernican principle' is often used to add prestige to the idea that we should always assume our position in

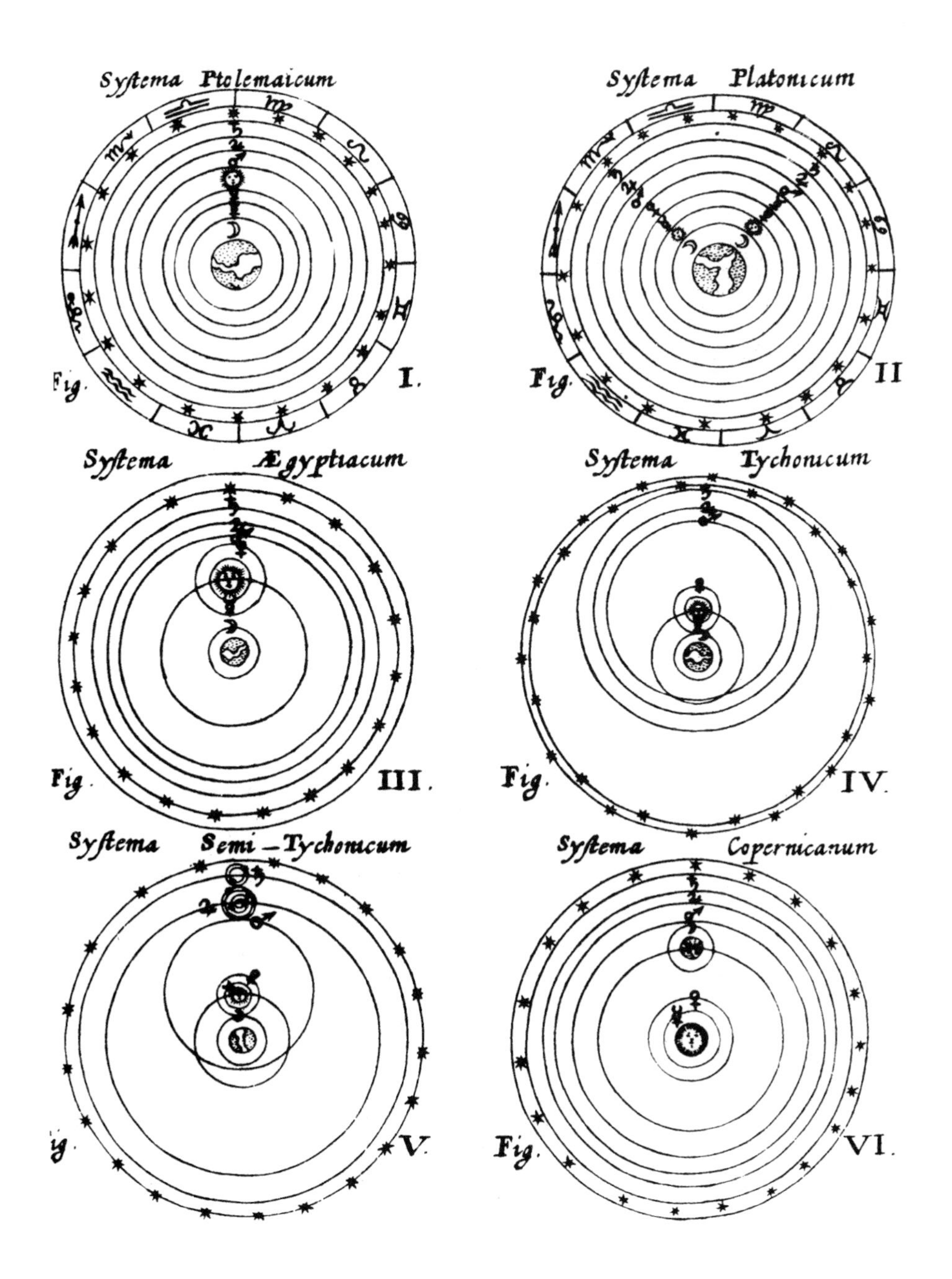

Giovanni Riccioli's summary in 1651 of the six world pictures of the post-Copernican era, from his *Almagestum Novum*.

the Universe is not special – that we have a typical view of the Universe and should build our theories accordingly. Sometimes it is used as a philosophical underpinning for the assumption of a universe model that is on the average the same everywhere throughout space. While this cosmic egalitarianism is an important general lesson for scientists, we have come to appreciate that it contains pitfalls of its own. For while we have no reason to expect that our position in the Universe is special in every way, we would be equally misled were we to assume that it could not be special in *any* way. We now understand that life can only exist in regions of the Universe that have certain features – we could not exist at the centre of a star or in a region of the Universe where the density of matter is too low for stars to form – and at special times – we could not exist when the Universe was too young to have produced elements like carbon in the stars.[28]

If 'typical' places in the Universe do not allow life to develop there then we cannot be in a typical position. This moderation of the Copernican perspective plays a crucial role in the testing of predictions in modern cosmology[29] and draws us back into the issue of what the Universe is like in an unexpected and un-Copernican way. By way of compensation, the development of our scientific descriptions of the Universe have become Copernican in a far deeper way. Newton's famous laws of motion unfortunately only hold for a very special type of observer – one who does not rotate or accelerate relative to the most distant stars. If you rotate, then, contrary to Newton's first law of motion for unaccelerated observers, you will see stars accelerating past your window even though they are acted upon by no forces. Einstein recognised and cured this problem in the way our laws of Nature were formulated. It was scandalous that we could have a description of natural laws that picked on some special collection of observers for whom, by virtue of their motion, the world looks simpler. It is okay that they should find Nature to be simpler in some respects but unacceptable that they should find Nature simpler in all respects. Einstein's great achievement was to find a system of laws of Nature which will be seen to be the same by all observers no matter how they are moving. The true Copernican perspective lives in the realm of the unseen laws of Nature rather than in the complicated world of their outcomes where we find planets, stars, and galaxies. Copernicus' picture did more than picture the solar system correctly: it painted a new world picture.

STARRY, STARRY NIGHT

THE WHIRLPOOL GALAXY

Starry, starry night,
paint your palette blue and grey
Don McLean

ON 13 October 1773, Charles Messier was busy observing a comet that had appeared in the night sky that year. He was rewarded with the discovery of something much more spectacular – a pair of galaxies[30] with brilliant centres. This doublet became the fifty-first entry in Messier's great catalogue of astronomical objects and is known to astronomers today by Messier's catalogue number, M51, or by the more evocative name of the 'Whirlpool Galaxy'.

In the spring of 1845, William Parsons, the third Earl of Rosse, began observing with his great six-foot reflecting telescope – the 'Leviathan of Parsonstown' at Birr Castle, in Ireland's County Offaly. After he had put up with bad weather for a few weeks, conditions became cold and dry and he turned the world's largest telescope towards M51. The Earl was excited by what he was the first human to see: spiral patterns of stars, seemingly swirling in great 'spiral convolutions' about the centre of the galaxy. This was before the availability of photographic plates and so the Earl did what all astronomers then did: he made careful drawings of his observations. These meticulous drawings, first made in April 1845, began to circulate in the scientific world and generated considerable interest when they were first shown at the meeting of the British Association for the Advancement of Science, in Cambridge, that June. And so began the study of the spiral structure of galaxies. These swirling patterns of stars are so familiar to us now, because we have seen countless

The Earl of Rosse's original 1845 drawing of M51, with its counterpart, a modern NASA image, below.

The Starry Night by Vincent van Gogh, 1889. Oil on canvas, 73.7 x 92.1 cm. Acquired through the Lillie P. Bliss Bequest 472.19, Museum of Modern Art, New York.

pictures of beautiful spiral galaxies in magazines, that we have to make some effort to take ourselves back to the nineteenth century and appreciate the impact of these striking pictures, which resolved stars in other galaxies for the first time.[31]

Soon afterwards, at the Paris Observatory, the French astronomer Camille Flammarion (1842–1925), who had also been among the first to have observed the Whirlpool, published an extremely successful and influential book entitled *L'Astronomie populaire* in 1879, of which over 100,000 copies were sold. It was translated into English as *Popular Astronomy* in 1894 by John Ellard Gore. It was the continental *Brief History of Time* of its day and was widely read by educated people of all interests and persuasions. Flammarion included

Rosse's striking drawing of the double galaxy and it was thereby exposed to the whole of the French-reading world and eventually to English readers as well.

I believe that one of the readers of that book must have been the great Post-Impressionist painter Vincent Van Gogh. If you look at his most famous work, *The Starry Night*, with the eyes of an astronomer, there is something familiar about the sky in this small painting that now hangs in the Metropolitan Museum of Art in New York. Despite its title, the dominant impression is one of great spiral swirls of light joining two centres. No one could ever have seen the spiral pattern of stars in a galaxy unless they had looked through Rosse's telescope or seen his drawings. The similarity with the Earl of Rosse's drawings is remarkable and I believe that Van Gogh would have seen those drawings in the press following the publicity attracted by them, or in Flammarion's book during the 1880s when it was big news all over France, and gained his astronomical inspiration from them.[32]

We know that Van Gogh was very interested in the sky – his painting *Moonrise* has been precisely dated by reference to the appearance of the sky, while others, like *Starry Night over the Rhone* and *Café Terrace,* have overtly astronomical content. Some astronomers have even argued about whether the background stars at the top of *The Starry Night* follow closely the pattern of stars in the constellation of Aries, which was Van Gogh's own sign of the zodiac (he was born on 30 March 1853). 'For my part,' Van Gogh once remarked, 'I know nothing with any certainty but the sight of the stars makes me dream.'[33] *The Starry Night* was completed in Saint-Rémy, Provence, in June 1889. Just thirteen months later, driven to despair, Van Gogh shot himself.

Today we can experience the magnificence of the Whirlpool anew thanks to the resolving power of modern telescopes. Although M51 is 37 million light years away, the Spitzer Space Telescope images on the following page show the details of the spiral arms and the companion galaxy in vivid contrast. They also show one of the most important developments in twentieth-century astronomy – multi-wavelength observations. Modern telescopes on mountain tops and in space are able to observe the radiation coming from celestial objects across a wide range of wavelengths. This capability has revealed that stars and galaxies look very different in different wavebands. There is no longer a unique image of a galaxy such as M51 and its companion. Here we see, side by side, an image in optical light, quite close to the sensitivity range of the human eye,[34]

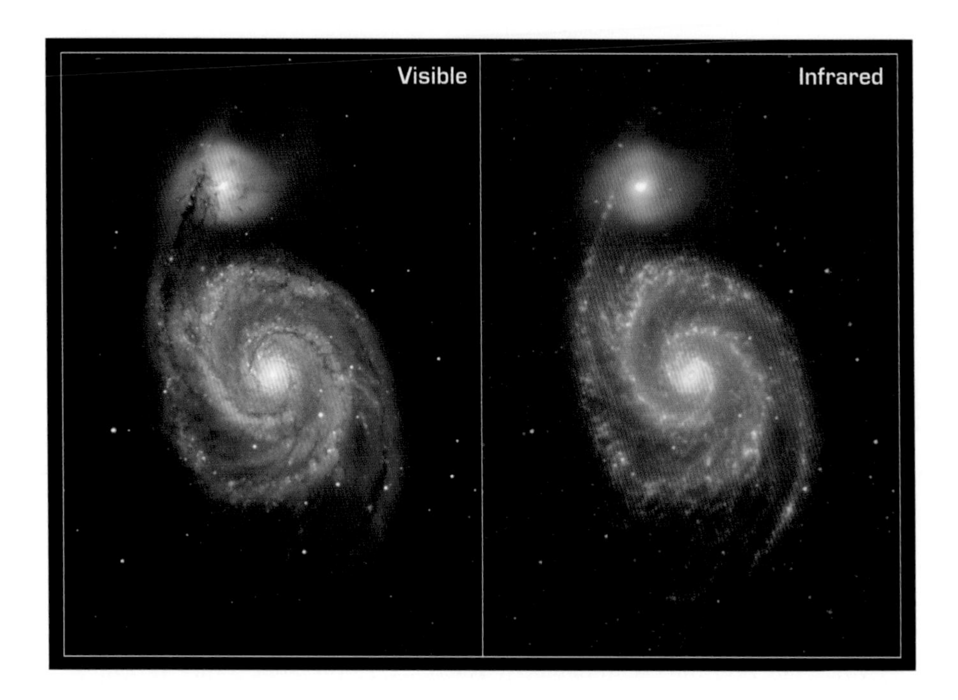

M51 in two wavebands taken by the Spitzer Space Telescope, showing the stars (visible) and the dust (infrared) as two separate components.

alongside one taken in the infra-red, which preferentially 'sees' the clouds of gas and lanes of dust from which the stars are formed. Images taken in the radio, x-ray, ultra-violet or gamma ray wavebands would give us further pictures of the galaxy that are sensitive to other aspects of its make-up and energetics. Just as an x-ray picture of a person will look very different to an optical image taken with the digital camera on your mobile phone, so all cosmic objects have a complicated spectrum of light emission that spans a range of wavelengths.

Astronomers believe that the gravitational influence of the smaller companion played a vital role in creating the spectacular spiral pattern that its larger neighbour displays. The spiral arms are not solid or fixed features. They are places where the gravitational forces are largest and material gets squeezed the most. This facilitates the formation of the bright young stars that illuminate the spiral pattern. As the galaxy rotates, different material passes through these regions and illuminates the spiral pattern. It is like a traffic jam of stars. Look from the air and you will see that cars are more densely packed in a slow-moving traffic jam if a lane has been closed on a section of the motorway, but the jam is always composed of different cars. It is a density wave of cars and the cars flow through it at a different speed to the speed of the wave as a whole.

The same is true in the spiral galaxy. The spiral pattern rotates at a speed that differs from that of the stars themselves as they rotate about the centre of the galaxy. In some places the speeds are the same and there is a longer and stronger gravitational encounter between different stars. This is where stars preferentially form and where there is more starlight. These are the places we notice most.

Van Gogh's famous picture is not a scientific image; it plays no part in the study of galaxies; but over the past century it has served as the impressionistic signature of the stars, impressing itself upon the minds of scientists and art lovers as a point of contact between art and the Universe. Its author combined original astronomical observations with his innovative vision of light and reality in an enduring way that looks as fresh and exciting now as the day he finished it. And today, by virtue of the growth in telescopic and photographic techniques, the stars and galaxies he could only imagine in simple form have indeed assumed the larger significance for our place in the Universe that his painting suggests. Their part in our mental sky is as prominent as the one they played in Van Gogh's starry night.

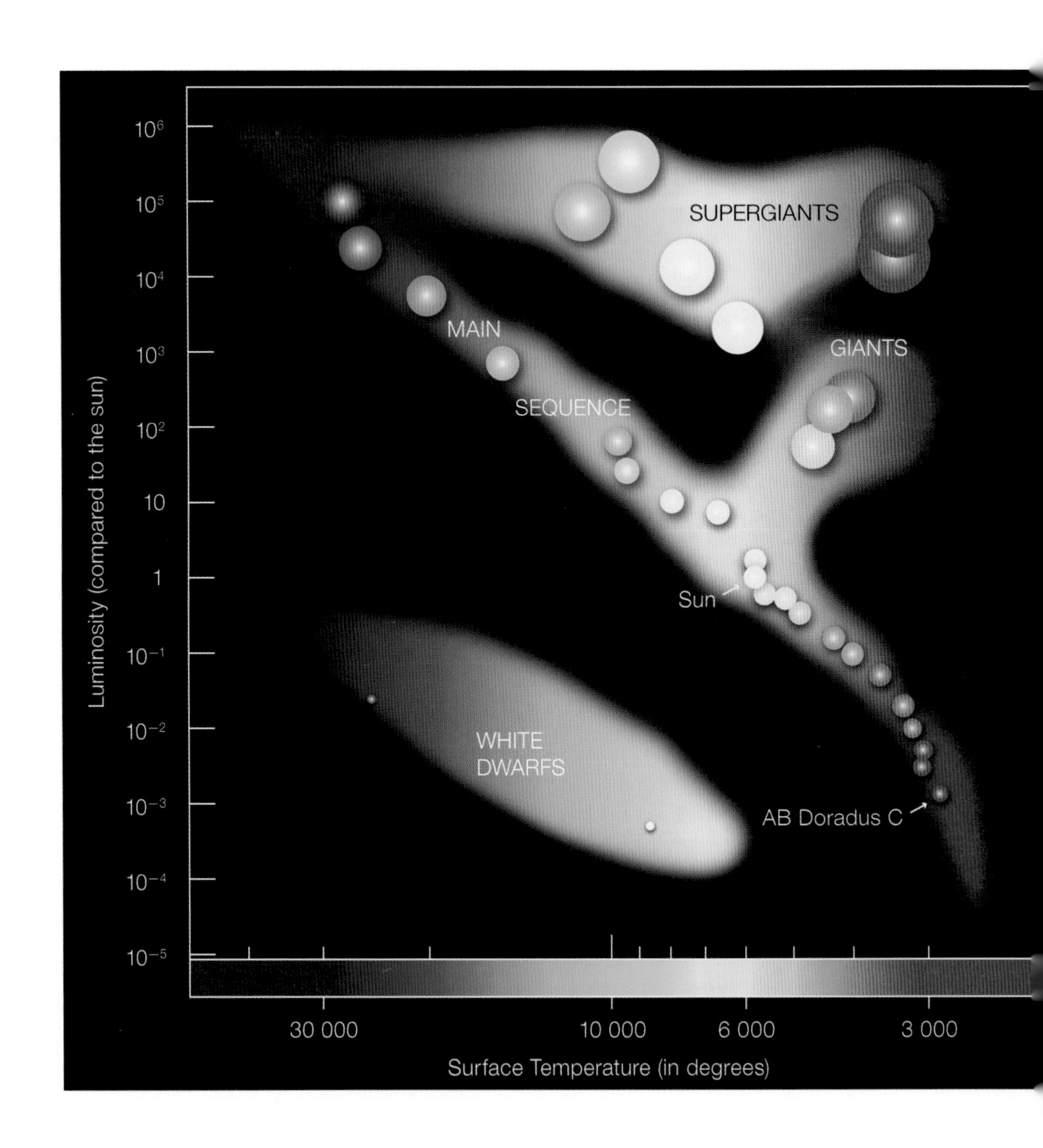

Modern Hertzsprung–Russell diagram showing populations of stars according to their luminosities and temperatures (in degrees Kelvin).

SO, YOU WANNA BE A STAR

THE HERTZSPRUNG–RUSSELL DIAGRAM

Man hath weaved out a net, and this net
throwne upon the Heavens, and
now they are his own.
John Donne

THE most famous diagram in astronomy was devised by the independent efforts of a Danish astronomer, Ejnar Hertzsprung (1873–1967), and an American, Henry Norris Russell (1877–1957). It was first drawn by Hertzsprung in 1911 and then by Russell later that year, using more data, and became known thereafter to astronomers as the 'Hertzsprung–Russell diagram', or 'HR diagram'. Put simply, it tells the life story of stars.

Hertzsprung came from a very wealthy family. His father had wanted to be an astronomer but could not find a job after his studies finished, so he moved into the commercial world and ended up as the director of a successful insurance company. The younger Hertzsprung graduated in chemistry but, perhaps influenced by his father's unfulfilled ambitions and his own strong interest in photography, gradually developed a strong amateur interest in astronomy. He was lucky. His family's wealth enabled him to set up as a private scientist, needing neither salary nor position.[35] It was a good investment on their part. Ejnar became one of the most important astronomers of the twentieth century, producing a prodigious amount of research, even when in his nineties. He soon got a position as well. In 1908 he was made an 'extraordinary' professor at the Potsdam Observatory; in 1919 he became the assistant director of the Leiden Observatory, then in 1935 director, in which post he remained until his retirement ten years later.

Russell was equally successful and became the foremost American astronomer of his day.[36] He was the son of a church minister and graduated from Princeton before working at the Cambridge University Observatory, returning to Princeton as an instructor; he then became professor in 1911, and finally director of the Observatory in 1912. For forty-three years, beginning in 1900, he contributed an astronomical column to *Scientific American*. In many ways Hertzsprung and Russell had careers that appear indistinguishable to the layperson. It is rather fitting that their names are forever linked by the stars they devoted their lives to studying.

Up until the mid-nineteenth century very little was known about stars save for their positions and apparent brightnesses on the sky; only a few nearby stars had their distances from the Earth known with reasonable accuracy. Gradually, though, the application of the new technique of spectral analysis allowed astronomers to determine information about the different colours contributing to the spectrum of starlight and, by using the Doppler effect, to use any systematic shift in their wavelengths[37] to determine how rapidly they were moving, and in which direction, along our line of sight towards them. By the end of the century, stars had been divided into four principal groupings according to their colours – white, yellow (like the Sun), orange, and red – and it was suspected that this colour sequence might be a reflection of temperature variations. Large catalogues of data started to be accumulated, notably by Henry Draper, an accomplished New York medical professor who as a sideline became one of the most brilliant pioneers of astrophotography, and stellar astronomers found themselves in possession of a huge population of objects to make sense of by using statistical methods.

Hertzsprung and Russell both sought correlations between different properties of stars – looking for correlations is what astronomers do because they cannot carry out experiments on the Universe in the way that experimental physicists can. If you think stars get hotter the bigger they are, you search out all the stars that you can and see if there is a correlation between their temperatures and their sizes.

In 1905, Hertzsprung published a paper[38] in a slightly obscure journal of scientific photography where he pointed out that bright red stars must be very large, and the scarcity of large red stars must mean that they evolve very quickly through that phase of their lives, so there must be a connection between the brightness and the colour of a star. He first met Russell, his scientific twin, in 1910 while on a visit to the USA and was able to share his findings with him. In 1911, Hertzsprung finally discovered that if he plotted a graph of the apparent brightnesses of stars in the Pleiades and Hyades star clusters against their colours, then a definite correlation emerged, showing the stars getting redder in colour as they got fainter in brightness. Russell presented a similar graph (rotated by 90 degrees) for all known stars in a talk to the Royal Astronomical Society at their monthly meeting in London in 1913.[39]

They both found that about 90 per cent of all stars lay along a narrow diagonal band in an HR diagram that is now called the 'main sequence'. Until then, astronomers had thought that it was just as likely for a star to be hot but dim, or hot and bright, as it was to be cool and bright. Suddenly it was clear

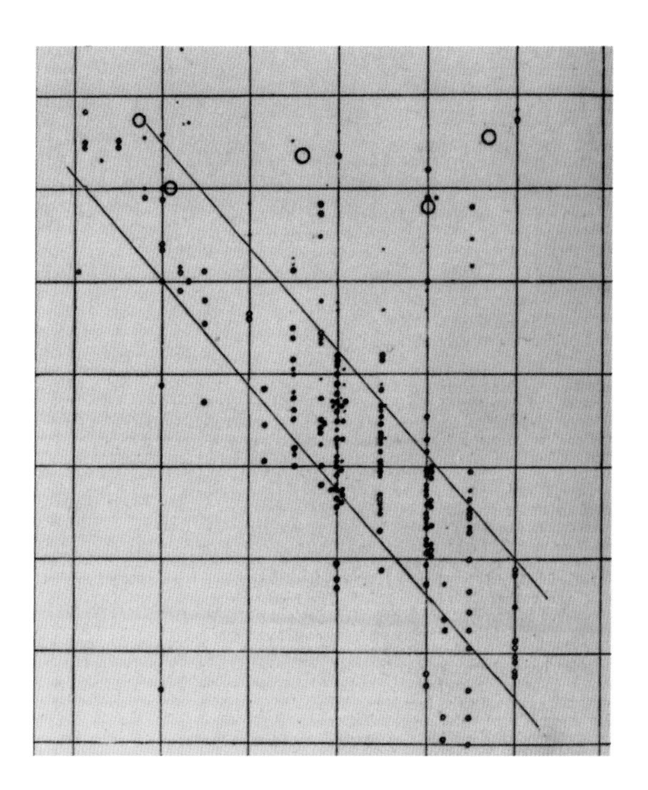

Hertzsprung's 1905 diagram showing the relationship between stars' luminosities and spectral types. Absolute magnitudes are plotted vertically and spectral types (revealing temperatures) are plotted horizontally.

that this was not the case. There were rules of physics to be discovered that governed how stars could be. Their properties were not haphazard.

If we look at the HR diagram it is clear that it is neither uniformly nor randomly populated with stars. Conventionally, the bottom axis runs from high temperatures on the left to low temperatures on the right, as the colours change from blue (O) to red (M). The vast majority of stars fall along the diagonal, or main-sequence band. Our Sun sits there, with a surface temperature of about 6000 degrees Kelvin, and reveals itself to be a typical hydrogen-burning star: a so-called 'main-sequence star'. Most stars are found on the main-sequence diagonal because this is where they spend most of their lifetimes as they evolve and change, consuming their fuel by nuclear burning of hydrogen into helium in a nice steady way.

About 10 per cent of stars are giants and supergiants and do not lie on the main sequence of the HR diagram. They must have very large diameters and surface areas from which to radiate their energy in order to explain how they can be very bright even though they are comparatively cool. Down in the other corner of the diagram is a small population of 'dwarf' stars which must have very small sizes – many similar to the size of the Earth, yet a million times heavier – in order to reconcile their faintness with their high temperatures. Undoubtedly, the HR diagram is incomplete because we know from other pieces of evidence that there is likely to be a lot of dark material in the universe that does not show up on photographic plates. Small, cool, and faint dead stars would sit in the bottom right-hand corner of the diagram and may be extremely numerous. Our inability to find them illustrates one of the ubiquitous biases that afflicts astronomy. In any survey of a wide range of objects there will be some sorts of object that will be harder to see than others, either because they are fainter or outside the reach of our detectors in some other way. Bright objects will always tend to be over-represented relative to faint ones in an astronomical catalogue. This type of bias always challenges astronomers to think very deeply about how their results might be influenced by the methods used to obtain them. But, as they do so, the HR diagram plays a starring role.

M16: The Eagle Nebula, showing dense pillars of molecular hydrogen gas and dust within which stars are forming.

THE ARTFUL UNIVERSE

NEBULAE

This tapestry ancient was hung up for you
Before time tried to reckon with space.
Cordella Lackey[40]

THE bread-and-butter images of glossy astro-magazines and picture books are neither stars nor galaxies, but nebulae: exploded stars that radiate energy out into their surroundings at high speed. The results are spectacular. The radiation interacts with clouds of gas and dust to produce a rainbow of colours and patterns that suggest all manner of mysterious cosmic events. Dark clouds of intervening dust add to the glamour by creating sharp, dark boundaries that allow our imaginations to see what they want to see – an inkblot test of significance on a cosmic scale. Just look at the names that these nebulae have attracted: the Tarantula, the Horsehead, the Egg, the North American, the Necklace, the Triffid, the Dumbbell, the Cat's Eye, the Pac-Man, the Apple Core, the Flame, the Heart and Soul, the Butterfly, the Eagle, the Crab – no imaginary stone is left unturned.

Not all the members of this astronomical light-show produce their spectacular images for the same physical reasons – some emit light, some absorb it, some merely reflect it – but their appearances have defined the non-specialist's image of the Universe. The invention of colour photography made these objects more spectacular than they could ever have been for their discoverers in the past. The modern era of the World Wide Web has put them all at our fingertips with high-resolution graphics that were once available only to an exclusive few. Things have come a long way since the late eighteenth century, when early observers such as Charles Messier used the term 'nebula' for any fuzzy patch of light on the sky, regardless of its source. At first, this led to a mixture of stars, galaxies, and even comets being labelled as 'nebulae'. Even in the late 1920s, Edwin Hubble would refer to galaxies as 'nebulae', but today the word is reserved for clouds of luminous dust and gas around remnants of stars.[41]

Nebulae create regions of 'plasma' in interstellar space. They consist of high-temperature regions containing free electrons, which have been stripped from neutral atoms, and the positively charged ions that constitute the left-over parts of the atoms.[42] Plasmas exist in the Universe under a very wide range of temperatures and densities and they behave in distinctive ways, quite different to most gases, liquids, and solids – you possibly have a plasma-screen TV.

These modern astronomical images have a fascinating subtext, which has been noticed by Elizabeth Kessler, an art historian at the University of Chicago. Looking at the Hubble Space Telescope images of nebulae through the eyes of an art historian rather than those of an astronomer, Kessler sees echoes of the great nineteenth-century Romantic canvases of the old American West by artists such as Albert Bierstadt and Thomas Moran. They expressed the grandeur of a landscape that inspired the pioneering spirit of the first settlers and explorers of these challenging new frontiers. These representations of places like the Grand Canyon and Monument Valley created a Romantic tradition of landscape art that hung on important psychological 'hooks' within the human psyche. Artists accompanied the expeditions into the West to capture the natural wonder of the new frontier and convince the folks back home of the importance and greatness of the adventure; an honourable tradition continued by war artists and photographers today.

How can this be, you ask? Surely astronomical photographs are astronomical photographs? Not quite. The raw data that is gathered by the telescope's cameras is rather one-dimensional. It is digitised information about wavelength and intensity. Often those wavelengths lie outside the range of sensitivity of the human eye. The final pictures that we see involve a choice about how to set the colour scales and create an overall 'look' for the image. If you were floating in space you would not 'see' what the Hubble photographs show in the sense that you would see what my passport photograph shows if you met me (well almost). Various aesthetic choices have to be made, just as they were by the landscape artists of old.

Thomas Moran's *Cliffs of the Upper Colorado River, Wyoming Territory* (1882).

Technically, the Hubble observers take three different filtered versions of the raw image data, in different colour bands, removing flaws or unwanted distortions and then adding the colour chosen for the photographic representation before reducing everything to a nice four-square image. This needs skill and aesthetic judgement. As we have already seen, the appearances of nebulae are extremely evocative and they are sensitive to embellishment and interpretation.

One resonant example is the Eagle nebula, a famous Hubble image that has a double claim to fame. The great pillars of gas and dust that stretch up like stalagmites in space are places where new stars are forming out of gas and dust – the cosmic incubators of life. And then the image itself is beautiful and unexpectedly structured. Here, Kessler recalls[43] Thomas Moran's painting *Cliffs of the Upper Colorado River, Wyoming Territory*, painted in 1882. The representation of the Eagle nebula could be oriented any way 'up' or 'down'. The way it has been created and the colours used are reminiscent of these great Western landscapes which draw the eyes of the viewer to the luminous and majestic peaks. The great pillars of gas are like a Monument Valley of the astronomical landscape; the twinkling, over-exposed, foreground star a fortuitous substitute for the Sun.

Indeed, one can venture further than Kessler along this line. We appreciate one particular type of landscape image that dominates our galleries of Western art. There is a particular form of composition which we find so appealing that it informs the creation of our ornamental gardens and parks. It enshrines a deep-laid sensitivity and desire for environments that are safe. Millions of years ago, when our ancestors were beginning the evolutionary journey that would lead to us, any liking for environments that were safer or more life-enhancing than others would tend to survive more frequently than the opposite tendencies. We see these effects in our liking for landscapes that allow us to see without being seen. These 'prospect and refuge' environments allow the observer to see widely from a vantage point of safety and security.[44] Most landscape art that we find appealing contains this motif. Indeed, the imagery extends beyond representative art as well. *The Little House on the Prairie*, the Inglenook, the tree house, the resonance of a 'Rock of Ages cleft for me' are all examples of prospect and refuge. This is the signature of the African savannah environment within which our earliest ancestors evolved and survived for millions of years: wide, open plains with scattered small clumps of trees – just like our parks – which allow us to see without being seen.

The dense dark forest, with its winding paths and dangerous corners where who knows what dangers lurk, is the complete antithesis of this image, like a 1960s tower block with its unappealing walkways and dark staircases. These are not inviting environments. Prospect-and-refuge environments are those which you feel drawn to enter. This test is one you can apply to all sorts of modern architecture. It will often reveal why you do and don't like different structures.

Moran's painting is classic prospect-and-refuge aesthetic. There is a wide, open plain and a mountain range that will permit a clear view. In the right foreground is a little refuge of trees and – another life-enhancing image – there is water, too. The Hubble images of nebulae are unable to use this imagery fully because they are, ultimately, representing an alien environment. They have open prospects and plumes that re-create monumental natural mountain peaks, but refuges are almost absent. Their place is taken only by crevices that look like the sort of refuge a mountaineer would seek out on a high climb if caught out by the weather or fading daylight. The soft lighting and multi-tone shadowing is evocative of the comfortable and unthreatening fire in the hearth rather than the inferno of a stellar conflagration.

Hubble images are spectacular and they are created by skilful techniques so as to be scientifically accurate and yet look beautiful. What these insights about their representation remind us of is that 'looking beautiful' has a meaning in the human context that is not always subjective and idiosyncratic. There are deep-laid human aesthetic preferences that have their origins in the environments in which our ancestors' evolutionary survival and multiplication largely took place. They still inform our preferences today. It is no accident that so much art shows flowers and fruits, other people, or appealing safe environments. We should not be surprised if scientific images reflect a little of this evolutionary history as well.

A mosaic image of the Crab Nebula assembled from twenty-six exposures taken by the Hubble Space Telescope (1999–2000). The blue colours in the filaments are created by oxygen atoms, the greens by sulphur ions, and the reds by oxygen ions. The orange filaments are the ragged remnants of the star and are mostly made of hydrogen. The blue colour near the centre is created by electrons moving close to the speed of light in a magnetic field.

OMEN 1054
THE CRAB NEBULA

Is this part of Bush's secret intergalactic war?
littlegreenfootballs[45]

OF all the nebulae in the sky, there is one that has a special status. It used to be said that every astronomy lecture had to include a picture of the Crab Nebula and all astronomers could be divided into two simple classes: those who studied the Crab Nebula and those who did not. Nowadays, this is a considerable exaggeration but, nonetheless, the Crab is certainly the most photographed astronomical object. Its spectacular appearance makes it a prime candidate for book covers, posters, and astronomy magazines.

The Crab is a supernova remnant, the cloud of gas and debris produced by a star that exploded 6300 light years away within our own Milky Way galaxy. It was first discovered in 1731 by John Bevis, an amateur astronomer from Wiltshire in England, who included it in the plans for his own star atlas, *Uranographia Britannica*. Unfortunately, the printers went bankrupt during the production process and it never brought him the fame he deserved. Indeed, he was not the luckiest of astronomers in many respects, and died from injuries sustained when falling off his telescope in 1771.

Charles Messier rediscovered the Crab on 28 August 1758, while searching for Halley's Comet. It was impressive enough to inspire Messier to begin compiling his famous catalogue and it now bears the label 'M1', to signify its first place in his observing project. Subsequently, Messier learnt from Bevis of his earlier discovery and credited it to him.

We now know that the supernova that formed the Crab was observed by Chinese astronomers at the Imperial Court on 4 July 1054. It was an unmistakable astronomical event: the exploding star brightened to about four times the brightness of Venus, and remained visible in broad daylight for twenty-three days, and for a further 630 days to the naked eye at night. It would have been interpreted as a favourable portent – a celestial endorsement of the Emperor's wisdom and stellar brilliance. It seems likely that several Central and North American Indian civilisations also recorded it, but Europeans seemed to have lost their interest in the sky during this period and it passes unnoticed under the noses of their leading scholars and political advisors.

It was first called the 'Crab Nebula' following its observation and representation in a drawing made by the Earl of Rosse around 1844, while he was observing with the 36-inch reflecting telescope at Birr Castle. We have already seen the fruits of Rosse's pen in his drawing of the Whirlpool galaxy, and here we again see his meticulous work from the pre-photographic era. The nebula's crab-like appearance is obvious, although this impressionistic picture led Rosse to make all sorts of premature deductions about the filaments issuing out from the sides of the nebula. He thought it showed the stars were all individually identifiable there.

Over many years it became clear that the Crab was changing. We know that it was formed by the stellar explosion that we first saw in 1054. What we are now seeing is the material that was ejected by the exploding star, blasting out into space, interacting with the cool gas and dust in its path. It is expanding at about 1800 kilometres per second. Part of the radiation it emits has a special form known to physicists as synchrotron radiation. It is emitted when electrically charged particles, notably fast-moving electrons, find themselves in a strong magnetic field. They accelerate, following a spiral around the line of their forward path, and emit radio waves. In 1948, it was first discovered that the Crab was a source of strong radio waves. Subsequently, it was also found to be a profuse emitter of x-rays and optical radiation. Its total energy output across all wavebands of radiation is more than 100,000 times greater than that of the Sun.

The Crab's next claim to fame came on 9 November 1968, when one of the stars near the centre of the nebula was discovered by the Arecibo radio telescope in Puerto Rico to be a pulsating source of radio waves, a 'pulsar', flashing

Lord Rosse's drawing of the Crab Nebula, first published in 1844. The Crab Nebula got its name from its appearance in this drawing, with filaments extending out from the southern extremity (*top*). It was created by the Irish astronomer William Parsons (1800–1867), the Third Earl of Rosse, using the 36-inch (91.4 cm) reflector at Birr Castle. The Crab Nebula was first observed on 4 July 1054 when a series of gas shells were cast off during the explosive death of a massive star. The shells, ejected at extremely high speeds, collided with the interstellar medium and became compressed and heated, emitting light. The core of the star that exploded as a supernova became a pulsar, a rapidly spinning neutron star that emits regular pulses of radiation.

to us, like a rotating lighthouse beam, every 33.085 ms precisely. Later, pulses of optical light were found to be emitted with the same period of variation as the radio pulses.

These pulsed emissions are one of the great wonders of astronomy. Huge numbers have been found since they were first discovered serendipitously by Jocelyn Bell, in January 1968, while she was working in Cambridge with Anthony Hewish. At first, for a short period, it was suspected that they might even be signals produced by intelligent sources ('Little Green Men'), but very quickly astronomers homed in on the one possible candidate that was small enough and rigid enough to explain the variations.[46] When stars that are between about 1.4 and 3 times the mass of our Sun exhaust their fuel and die, the residue of their explosive demise is a super-dense, rapidly spinning remnant in which all the atoms have been crushed into one another by their own gravitational attractions, the protons annihilating the electrons, to leave a rigid, solid body of neutrons crammed together as closely as possible. These 'neutron stars' may have a mass that is about two or three times that of our Sun, but in size they may be less than a few kilometres across. This fantastic compression in size from the scale of the original star means that any rotation they started out with gets vastly amplified (just like the skater pulling her arms in so as to increase her rate of spin).[47] The Crab is not an especially rapidly spinning pulsar, but its neutron star rotates thirty times every second. The fastest pulsars spin more than ten times faster still. Neutron stars are certainly dramatic: imagine a region the size of a big city but a million times more massive than the Earth spinning around 400 times per second!

The pictures of the Crab reveal a network of filaments which tell us the chemistry of the exploding material. A mixture of hot and cold regions containing atoms such as oxygen and sulphur produce different responses owing to the complexity of the explosion. But the most remarkable feature of the filamentary chemistry is to remind us that all the elements in the Universe heavier than hydrogen and helium gases are produced and dispersed throughout space by the nuclear conflagration of the supernova process. All the carbon and oxygen and nitrogen in our bodies has emerged from supernova explosions such as the Crab. This picture reveals how the Universe gets to contain chemical elements big enough to sustain complexity and life – how it passes from being dead to being the cradle of life, how astronomy creates astronomers.

M31: the Andromeda galaxy.

A FOR ANDROMEDA

THE GALAXY NEXT DOOR

Galaxy for sale: has great potential, has a strong history. Avoid scams and fraud by dealing locally.

US advertisement[48]

IMAGINE that you could go far out into space, far beyond the Earth and the solar system, far beyond our Milky Way galaxy, and then turn and look back. What would our galaxy of a hundred billion stars look like? We have a pretty good idea, because our galaxy is one of a pair. Its partner, the Andromeda galaxy, is 2.9 million light years away, about 1000 light years across, and about twice the mass of the Milky Way, but in most respects it is a twin. They rotate in opposite directions and are moving towards one another – every second they get 100 km closer. One day, 3 billion years from now, our worlds will collide. Over the next billion years they will merge and mingle, eventually settling down to be a giant elliptical galaxy quite different from the well-ordered spirals they are today.

In the meantime, we can admire the photographs of Andromeda that mirror the beauty of our own Milky Way. In the photograph reproduced here, the speckle of stars all over the picture are stars from the foreground of the Milky Way and the two bright concentrations of light on either side of Andromeda are two of its small satellite galaxies.[49] The colours of Andromeda are instructive as well as impressive. The yellowy-red light at the middle comes from the older yellow and red stars that populate those central regions. If we move towards the outer rings, which we view askew, we see the bluish light of young blue stars that were born in the spiral arms. These stars take nearly 100 million years to complete an orbit of the galaxy. This division of stars into two distinct populations was first noticed by Walter Baade, who took advantage of the wartime blackout conditions in Los Angeles to observe Andromeda with the Mount Wilson Telescope and resolve the two different populations of stars.

If this were the Milky Way then our solar system would be located on the inside edge of one of the outermost spiral arms. If there are astronomers in Andromeda, it is most likely that they will have a rather similar location. Stars form preferentially in the 'traffic jams' of material that trace out the spiral patterns. And so, if you live around a star you will most likely live near the spiral arms too.

Andromeda is the only galaxy that can be seen from Earth with the naked eye. All else in the night sky is comets, stars and planets. It is known to have been observed and identified by Persian astronomers as early as AD 905. The earliest hand-drawn picture is to be found in Abd-al-Rahman Al Sufi's *Book of Fixed Stars* of AD 964, in which he calls it the 'Little Cloud'.[50] Its close proximity to us means that Andromeda is the most studied and most photographed galaxy. It has acted as an exemplar in all sorts of studies of star motions, distances, and colours. Fortunately, like our Milky Way, it is in most respects a rather typical spiral galaxy, and more than 70 per cent of all galaxies are spirals.

Andromeda takes its name from a character in a classical legend. Andromeda was the daughter of Cassiopeia and Cepheus. Sadly, her mother was proud and thought herself more beautiful even than the daughters of Nereus, one of the gods of the sea. As a punishment for her pride Cassiopeia felt the wrath of the sea god Poseidon, who had her daughter Andromeda chained to a rock and left as a sacrifice for the monsters of the deep. But she was saved by Perseus, who asked that her parents give Andromeda to be his wife, which they did. And they all lived happily ever after.

Charles Messier first observed and catalogued Andromeda as the 'nebula' M31 on 3 August 1764, and described it in his notebook as 'the beautiful nebula of the belt of Andromeda, shaped like a spindle'. The sketch shown here

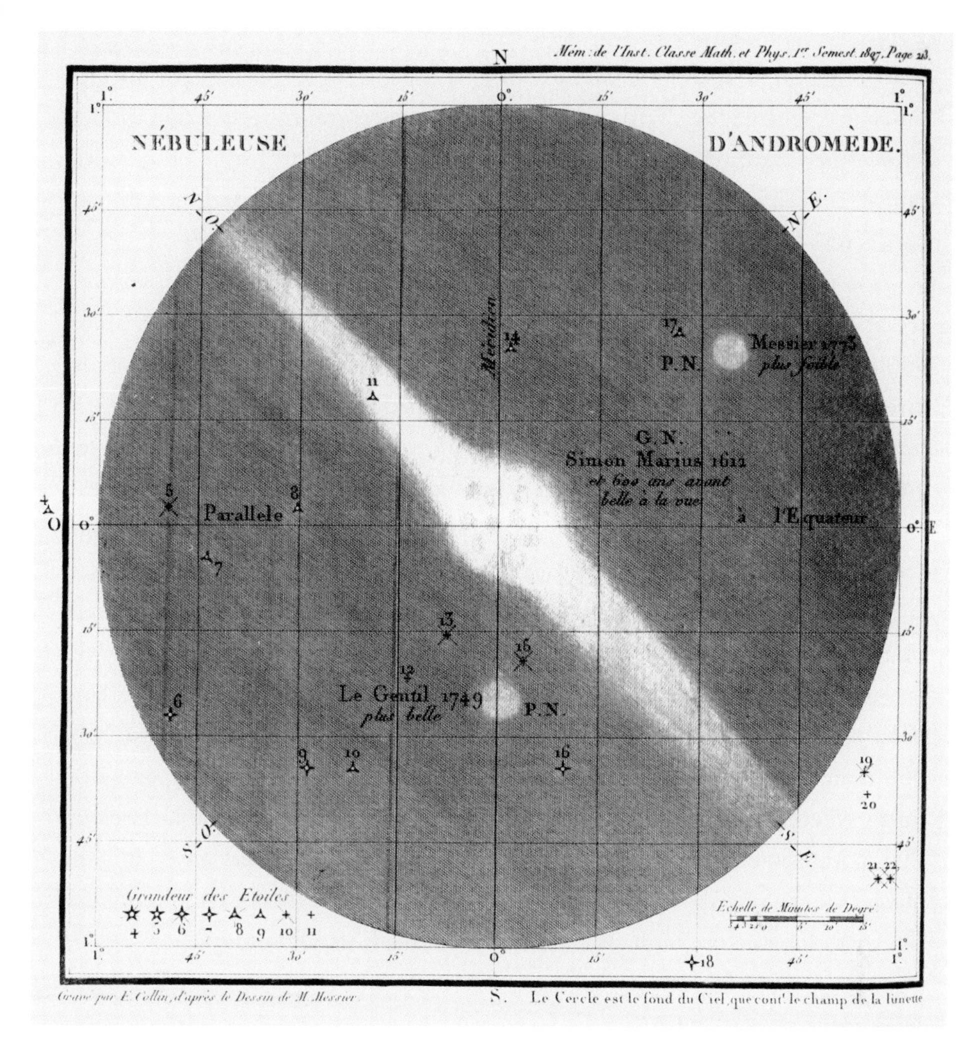

Historical artwork by Charles Messier (1730–1817), showing Andromeda's two companion galaxies M32 (*bottom*) and M110 (*top*). The image has some descriptions of the three galaxies (then called nebulae), including their discoverers, thus documenting Messier's discovery of M110 in 1773. The discovery of Andromeda itself he attributed to Simon Marius, who gave a detailed description of it in 1612 but never claimed its discovery. In fact, Al Sufi, a Persian astronomer, described and depicted it as early as AD 964. M32 was discovered in 1749 by Le Gentil.

was published by him in 1807 and also shows the galaxies M32 (a small 'dwarf' elliptical galaxy) and M110 (an elliptical galaxy).[51]

The Hubble Space Telescope discovered that Andromeda has a double nucleus. Nearby, two concentrations of millions of stars lurk close to the geometrical centre of its spiral shape. This could be the result of Andromeda 'eating' another one of its small satellite galaxies long ago, causing it to fall into the centre, where it began to orbit the pre-existing central concentration formed from other captured galaxies. Alternatively, there might really be a single central concentration that is made to look like two because a dark line of dust cuts across it.

On the darkest nights, far from the interference of artificial lights, you can see at most about 2000 stars with the naked eye. None are more than about 4000 light years away. But in November, in the north, if you look almost directly overhead you can identify the four stars that form a square that we call 'the Great Square of Pegasus' because, with greater imaginations than ours, the ancients saw how other nearby stars could add four legs and a neck and head to the square to form a winged horse. If you look below the Square to the north you will see an 'M' shape traced out by five bright stars that form the constellation of Cassiopeia. Draw a line from the brightest of these stars to the bright star of the square. Go about two thirds of the way up the line and to the right you will see a faint cloud. This is Andromeda. The visible stars around it are no more than 1500–4000 light years away, but Andromeda shines with the light of a hundred billion stars. We know quite a lot about what was happening on Earth when their starlight began its journey towards us from the little cloud that is Andromeda. But that little cloud is 2.9 million light years away. It is the most distant thing that can be seen with the naked eye. It sent its first light to the Earth long before humans existed.[52] And this is its photograph.

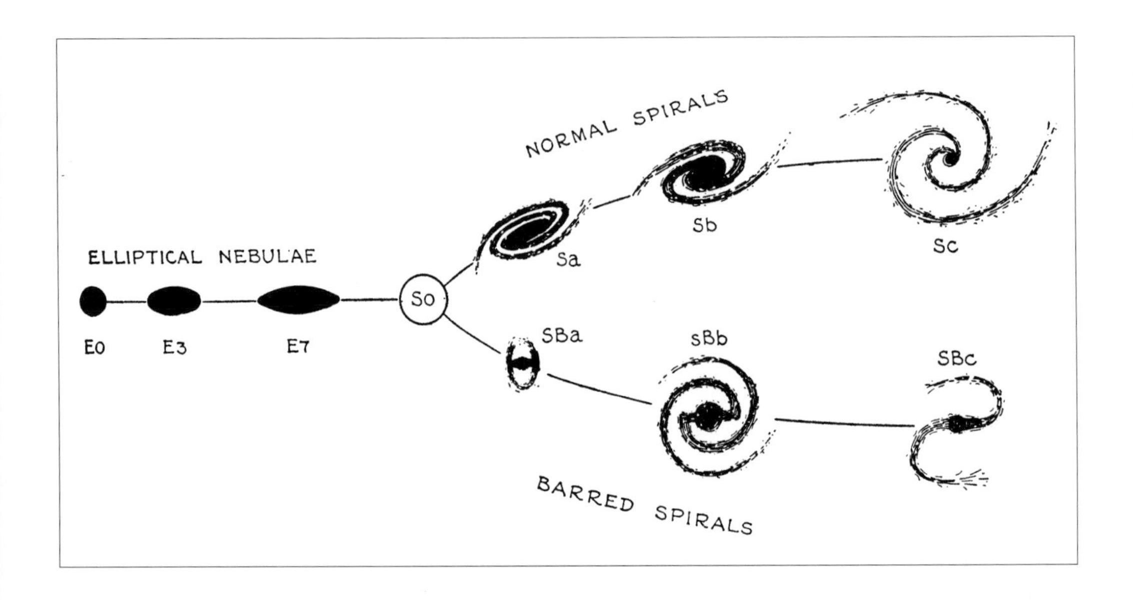

The Hubble Sequence (also called the Hubble 'tuning fork'). Published in 1936 by the US astronomer Edwin Hubble (1889–1953), this diagram shows the classification of galaxies (here called nebulae) into different types. This system is still widely used today. Elliptical galaxies, on the left, have a number to show increasing ellipticity. Shown here are the E0, E3 and E7 types. Spiral galaxies are at the right, and are separated into normal spirals (*upper right*, Sa, Sb and Sc) and barred spirals (*lower right*, SBa, SBb and SBc). Barred spirals have a 'bar' structure in their nucleus, compared to the spherical nucleus of a normal spiral. Their spiral arms begin at the ends of the bar. The subscripts a, b and c mark increasingly open spiral winding. A transition type of 'lenticular' galaxy, denoted So, resemble disk-like spiral galaxies without spiral arms.

PERFECT PITCH

HUBBLE'S TUNING-FORK DIAGRAM

To suppress the pictures is to suppress a powerful source of suggestion. Pictorial representation is essential for discovery and rapid understanding.

John L. Synge[53]

EDWIN HUBBLE'S programme of observing distant galaxies with the Mount Wilson reflecting telescope in California during the 1920s amassed hundreds of high-quality images of galaxies. Earlier observers in Europe had attempted to make sense of the plethora of shapes and brightness profiles displayed by images of galaxies ever since the Earl of Rosse's first swirling hand-drawing of the Whirlpool galaxy and Charles Messier's great *Catalogue of Nebulae and Star Clusters*. Unfortunately, Hubble's predecessors seemed to be discouraged by the seeming complexity of the zoo of galaxies. If you looked closely enough each one seemed at first to be an individual rather than an obvious member of a small number of distinctive varieties. Hubble managed to pick on a simplified picture of the variations in the family of galaxies that was straightforward enough to be useful as well as memorable enough to serve as a rule-of-thumb classification scheme.

At first, in a long 1926 paper,[54] he divided galaxies into three broad classes according to the appearance of their photographic image on the sky. They were 'ellipticals' (E), 'spirals', and 'irregulars' (Irr) – and the last category was created to catch anything that didn't fall into the other two. The spirals were subdivided into two subfamilies – the normal spirals (S), whose spiral arms emerged from the central core of the galaxy, and the barred spirals (SB), whose spiral arms began at the two ends of a barlike feature running through the centre of the galaxy. Each of these subdivisions was given a linear development into so-called 'early' to 'late' spiral types as the spiral winding became looser and the arms more numerous and less well defined: Sa, Sb, Sc and SBa, SBb, SBc. Likewise, there was a development in the shapes of the elliptical shapes, from a circular profile (E0) to the flattest shapes observed (E7).[55]

By 1936, Hubble was convinced that a further type of 'lenticular' galaxy existed. It was like an elliptical galaxy in having a smooth distribution of light, no resolution into bright stars, and no spiral arms, but provided a transition in his mind's eye between the ellipticals and the spirals. He called them 'armless spirals' but later they would be called 'lenticular' galaxies and labelled like a new zero-type of spiral, S0. Later, Hubble and other observers were able to find a good number of these.

Hubble's position as the leading observer of the day, backed up by his access to all the telescope time that he wanted on the world's best instrument at Mount Wilson, helped establish his simple and logical classification as the standard. But the real impetus for its adoption was provided by his memorable and highly suggestive picture of what became known as the 'Hubble Classification of Galaxies'.

The Hubble 'tuning-fork diagram' shows the sequence of galaxy types from left to right. It appeared in the extremely influential book that Hubble wrote in 1936 for a wide audience, extending from the informed layperson to his fellow astronomers. *The Realm of the Nebula*[56] became a classic of science and it cemented the tuning-fork diagram in the minds of astronomers everywhere. It appears still in every lecture course about galaxies delivered anywhere in the world. It even seems to have had a far-reaching effect upon literary figures such as Virginia Woolf, who found aesthetic motivation from Hubble's writings and discoveries.[57]

Despite its longevity, the information content of the diagram has undergone a number of transformations. It is a classic example of how the representation of information in a suggestive way can lead to a theoretical picture in readers' minds to justify it. The cladistic representations of the evolutionary process are in some ways analogous. A good picture can mislead as well as lead.

Hubble's tuning fork suggested to many, including Hubble, that there was

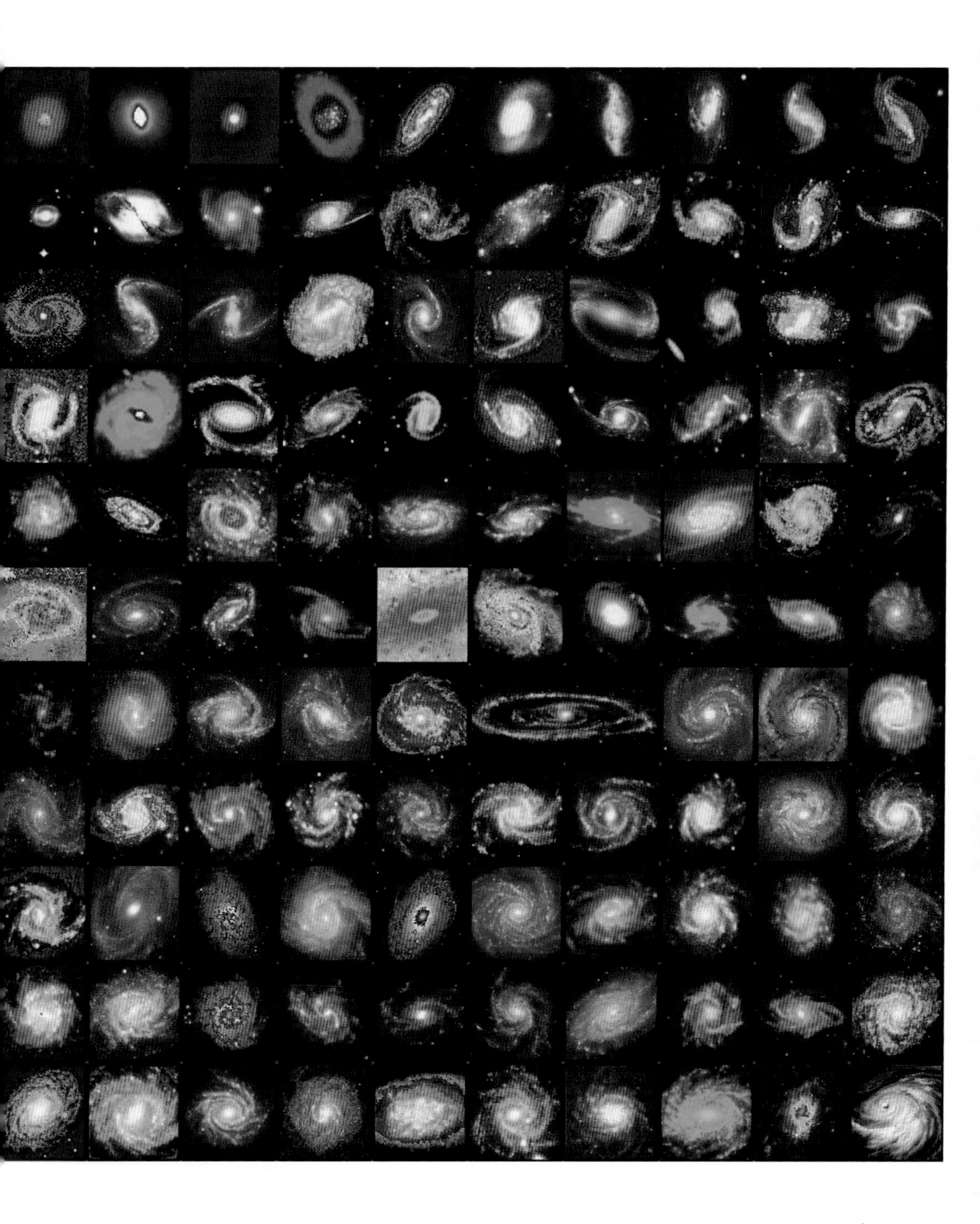

A gallery of modern images of galaxies of all Hubble types taken in different wavebands.

a smooth and continuous evolution of galaxies from ellipticals into spirals and barred spirals. Elliptical galaxies would have various degrees of flattening according to how fast they were rotating, and the spirals would evolve in time from the a to b to c types as they aged and their spirality became more ragged. Alas, none of these entirely reasonable expectations turned out to be true. The Hubble tuning fork was not due to increasing speed of rotation as one moved from left to right through the galaxy types. Quite unexpectedly, it was only realised finally in the early 1970s that the flattening of elliptical galaxies is not due to rotation at all. They rotate very slowly, far too slowly for the rotation to explain their flattening in the way that the rotation of the Earth explains its slightly oblate shape, as Isaac Newton first explained more than 300 years ago. Nor do the galaxies on the left of the tuning fork contain all the oldest stars and those on the right the youngest, as would be expected if galaxies evolved in time from left to right. Things are rather more complicated, and reflect different motions that existed in the dust and gas from which the galaxies formed. Astronomers also began to appreciate that elliptical galaxies are likely to be triaxially ellipsoidal in shape, so the classification of their apparent shapes as seen on the sky might be losing a lot of information about their true three-dimensional shapes. If you look at an egg-shaped galaxy end-on, its profile will appear circular, even though its intrinsic shape is far from spherical. Last, but not least, we have learned that what we see in the photographic images of most galaxies is but a drop in the ocean compared to the material of which they actually consist. The gravitational field of a galaxy reveals itself through the speeds at which the luminous stars are moving within it. Strangely, the gravity fields of all spiral galaxies indicate that they are being generated by the presence of far more mass than can be seen shining in stars. Galaxies are about ten times bigger than their photos make them appear. They are made of dark stuff; the stars are just for decoration.

Yet, despite all these developments in our knowledge of the shapes of galaxies and the processes that fashioned them from primordial gas and dust, Hubble's tuning fork retains its appeal. Search the internet for 'galaxies' or 'Hubble' and you will find countless lecture courses and educational websites telling you about the different species of galaxy that inhabit our universe. Almost without exception they will use Hubble's diagram to bring order to this magnificent menagerie.

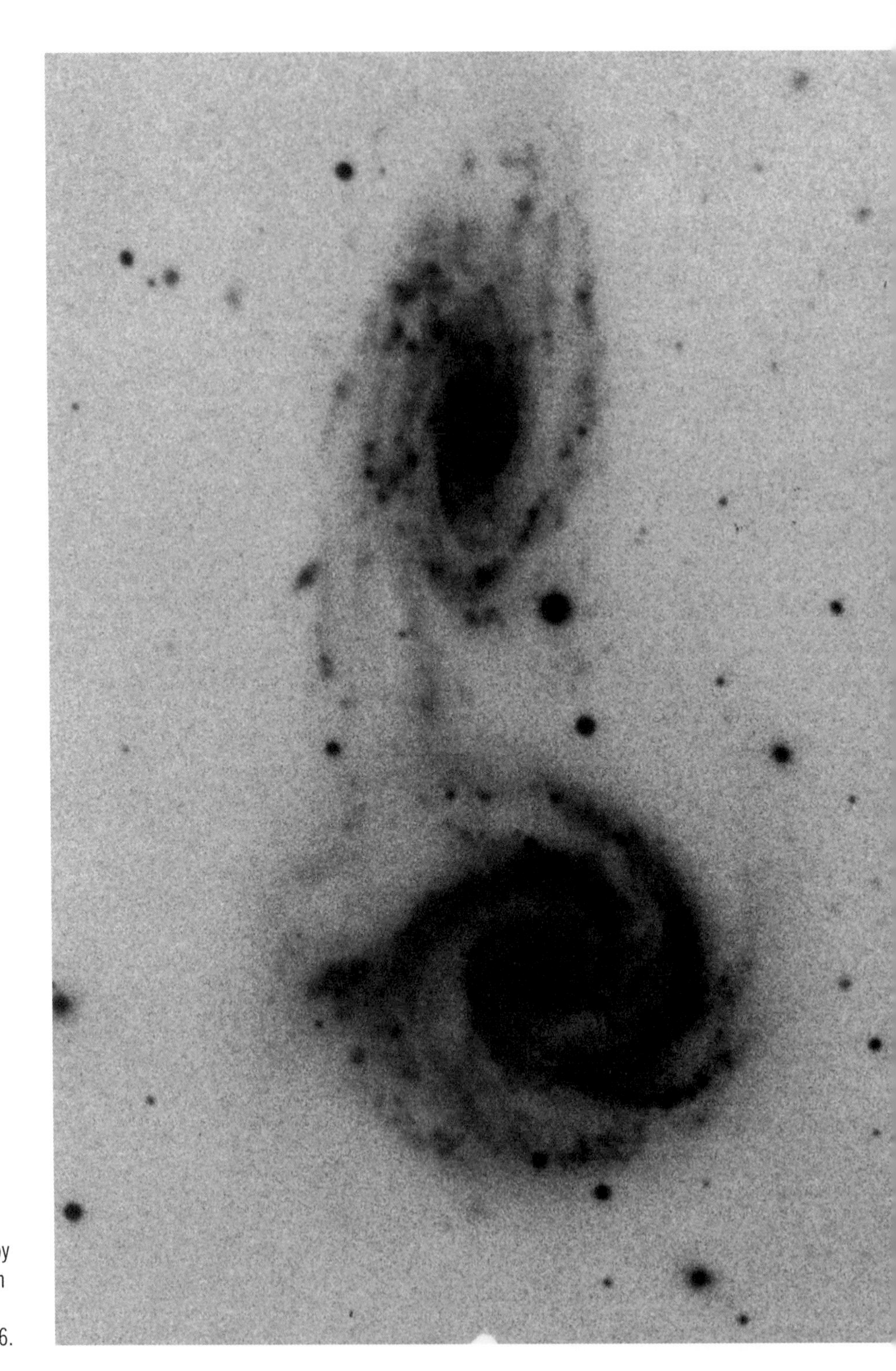

The Peculiar Galaxy Arp 271 by Halton Arp, taken with a 200-inch telescope in 1966.

SOME STRANGENESS IN THE PROPORTION

PECULIAR GALAXIES

Singularity is almost invariably a clue. The more featureless and commonplace a crime is, the more difficult it is to bring it home.

Arthur Conan Doyle[58]

THE unusual is invariably a clue to what is special about the usual. Our understanding of the normal human brain has progressed most rapidly by the study of defects and injuries. So it has been with galaxies. Edwin Hubble created an appealingly simple way of classifying galaxies by inspecting their overall shapes on photographic images. A division into elliptical and spiral galaxies, with a further recognition of an in-between class of lenticular galaxies and a division of spirals into normal and barred types is sufficient to catalogue all but about 5 per cent of known galaxies. Those few maverick galaxies that don't quite fit are by no means swept under the carpet and ignored as misfits that threaten the simplicity of the Universe. They are among the most interesting galaxies of all and these 'peculiar' galaxies, with their misshapen and irregular forms, have provided the stimulus to discover some dramatic things about the past history of galaxies.

In 1966, the American astronomer Halton ('Chip') Arp published his famous *Atlas of Peculiar Galaxies*.[59] It contained photographs of 338 fields containing 591 peculiar galaxies taken with the Hale Telescope; his original object number 271 is shown opposite. Arp selected the galaxies in his catalogue from earlier works by Zwicky[60] and Vorontsov-Velyaminov.[61] It became an

M82: a ‘Starburst’ galaxy. A near miss by its neighbour has compressed material and triggered a burst of star formation that sends plumes of hot gas thousands of light years into space.

influential and controversial compilation in many ways. First, with regard to the shape of galaxies, it acted to counter the prevailing bias towards simplicity that had grown up in the minds of astronomers who were not galaxy specialists. The most symmetrical and picturesque spiral galaxies tend to be reproduced most often, because of their aesthetic qualities, and so it is easy to forget that even spiral galaxies are not exclusively symmetrical and simple. But Arp had another agenda too: he was one of a small minority of astronomers who were not persuaded that the redshifting of the light from galaxies was a Doppler-like effect due solely to their recession velocity away from us. He thought that galaxy redshifts might have a much more exotic explanation due to 'new physics' and was also generally sceptical of the whole Big Bang theory of the expanding universe, although he didn't offer any alternative in which all the supposed anomalies became natural occurrences. As a result, Arp was interested in objects that didn't easily fit into the accepted wisdom.

Such peculiar objects had always turned up from time to time in observing programmes and some were not really mysterious – just collisions or close encounters between ordinary galaxies – but Arp actively sought them out and collected them together into a rogue's gallery of misfit galaxies. He was especially interested in situations where two galaxies appeared to be linked to one another by lines of stars but the redshifts of the two galaxies were very different. If they were truly close by, as the bridge of stars implied to him, then the redshift could not be a reliable indicator of their distance away, as conventional wisdom would have it. The contentious point was, of course, that these galaxies might not be close in space at all. They might be at very different distances – just as their different redshifts imply – and the bridge is not really linking them at all. It is just something in between them that is made to look like a physical link when its image is projected on the sky. If we had a real three-dimensional map we could see whether the two objects were truly close in space and linked by a real structure. As observational techniques have grown more accurate it has been possible to unpick these images and show that none of these neighbouring galaxies on the sky with very different redshifts are joined by physical bridges in three-dimensional space.

The catalogue of peculiar galaxies highlighted the entire problem of deriving general conclusions from special cases that have been picked out solely because they *are* special. But unless the story they are telling supports a consistent alternative scenario they are of little help. Statistically, the number of unusual objects is not unexpected, and if they were used to support a new non-Doppler explanation for the observed redshifts then that theory would have a huge problem explaining the 95 per cent of galaxies where the Doppler-like explanation provides a beautifully consistent picture of their distances (for a full explanation of 'redshifting' and 'the Doppler effect' see pages 73–78).

The most interesting type of object that can be found in Arp's catalogue is a wrecked galaxy that has undergone a collision or close fly-by with another galaxy. In the case of direct hits you can easily get a wrecked galaxy or, in the case of a direct hit right through the centre, a symmetrical loop of material emanating from the centre rather like what happens when you drop a stone in a pond (the 'cartwheel' galaxy is a spectacular example). Gradually, astronomers became persuaded of the importance of direct and close encounters between galaxies in the history of the Universe. In an expanding universe things would have been much closer together in the past when galaxies first formed and so encounters would have been more common than in the recent past. They wondered at first if the explanation for Hubble's tuning fork might lie here. What if all galaxies had started as spirals and those that collided and merged early enough to settle back down became ellipticals? Unfortunately, galactic life cannot be that simple. It would mean that the stars in spirals were all older than those in ellipticals and that is not true.

As modern computer power has grown, it has become possible to simulate the close approach and collision of galaxies containing large numbers of stars and associated gas and dust. The results are spectacular and among the largest computer simulations that are performed by scientists in any field of study today. They allow us to understand many of the strange hybrids that we see in the galaxy population. We have also begun to understand more clearly the whole process by which many galaxies are formed. They are built up by the accumulation and merger of many smaller clouds of material and so the whole process of interaction and merger is an integral part of the formation process. A large fraction of the galaxies that we see are in pairs and undergo significant gravitational interaction with their close companions. In the future, our ability to scrutinise these close encounters will continue to improve in resolution and scope while the power of our largest computers will make the simulation of galaxy formation a subject in its own right. The study of peculiar galaxies and

Arp's atlas of the anomalies and eccentricities on view in the Universe have played an important role in focusing astronomers' attention on the processes that disrupt normal galaxies. Once again, peculiarity has provided the key insights needed to understand normality. Only pictures could have made that possible.

Pictures of the Large Magellanic Cloud containing the supernova SN1987A before (*left*) and after (*right*) the explosion observed in 1987.

WHEN WORLDS COLLIDE

SUPERNOVA 1987A

Familiarity is the enemy. It slowly turns everything to wallpaper.
Gary Hamel

ON 23 February 1987, Ian Shelton, a young Canadian astronomer from the University of Toronto, had the shock of his life. He saw what no human being had been privileged to see for 383 years. He was working routinely at the Las Campanas Observatory in the high Andean mountains of Chile, making observations of a dense star field called the Tarantula Nebula, which lies within a small satellite galaxy close to the Milky Way called the Large Magellanic Cloud,[62] about 167,000 light years away – virtually our next-door neighbour by astronomical standards. He looked at the new exposure he had taken of a part of sky that was very familiar to him – he had seen it hundreds of times before. But this time what he saw was so fantastic that he did what astronomers of the technological age don't much do any more – he rushed outside to look up at the sky. And what he saw needed no telescope to record: a nearby supernova, the first to be visible with the naked eye since the time of Kepler in 1604.[63] An ageing massive star had exploded 167,000 years ago; now its brilliant light was just arriving, dominating that of all the stars in the night sky. From now onwards it would be known simply as SN1987A, the first supernova (hence 'A') to be observed in the Universe in the year 1987. In most years there are hundreds but they can only be seen by using powerful light detectors to enhance the light-gathering powers of big telescopes.[64]

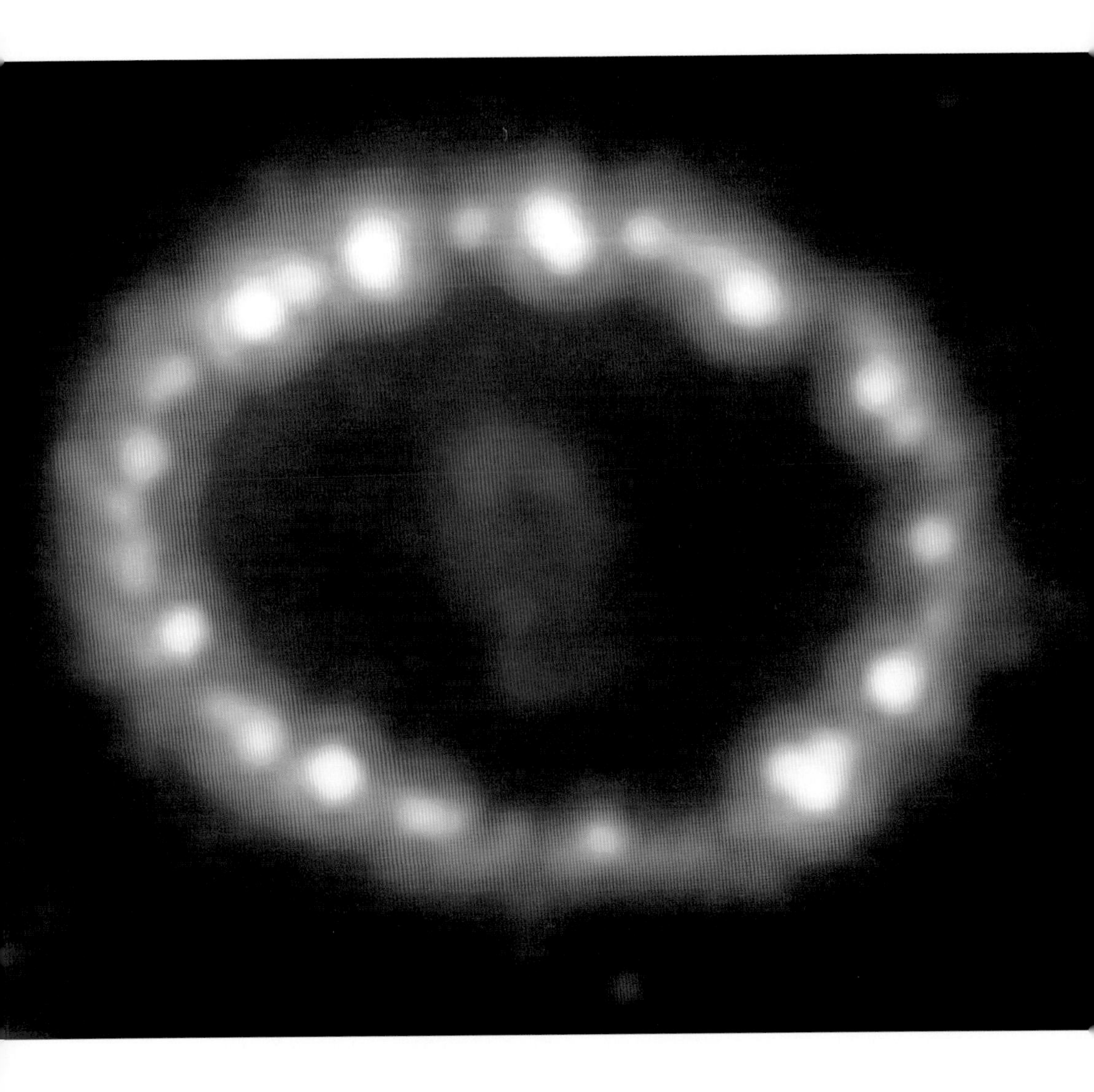

A ring of material blown out from the dying star 20,000 years before the original explosion of SN1987A, observed by the Hubble Space Telescope in November 2003. The ring is about one light year in diameter and the bright spots are caused by blast waves hitting the outlying material around the dead star.

Every telescope in the southern hemisphere was soon trained on the supernova and they were joined by satellite-borne observatories such as the International Ultraviolet Explorer (IUE), which was able to pinpoint the site of the exploding star as one previously occupied by the star Sanduleak-69 202, a blue supergiant in the Hertzsprung–Russell diagram about twenty times more massive than our Sun. It was believed that this star first swelled up in size, puffing out some of its outer material, which led to its subsequent contraction and dramatic heating. In little more than a second, its central core imploded and shoals of high-energy neutrinos suddenly heated it to more than 10 billion degrees. The star was torn apart by a cataclysmic shock wave and the neutrinos burst out into space to signal that the star was about to die. Remarkably, some of these neutrinos were detected on Earth. Nineteen of them passed through the Earth and were captured as they moved upwards towards the surface by deep underground detectors, in Ohio and Japan, primarily designed to see neutrinos coming downwards to the Earth from the Sun.

Supernova SN1987A brightened to exceed the energy output of a hundred billion Suns for a few weeks and then steadily dimmed in the characteristic way that is followed by many of the supernovae we have observed far away in the Universe. Yet it remained bright enough to be seen with the naked eye for nearly a year after its initial discovery. Its peak brightness was close to 3rd magnitude, whereas the faintest object in the night sky visible to the naked eye would be of about 6th magnitude.

Ever since 1987, the remnant of the explosion of SN1987A has been expanding. It was observed in fine detail in 1994 by the Hubble Space Telescope, which saw rings of material, about 1.3 light years in diameter, that had been thrown off before the main explosion.

Astronomers would love to see another nearby supernova. It provides a wonderful opportunity to test out our theories of the detailed sequence of events that triggers these stellar eruptions, as well as some of our beliefs about the properties of light and elementary particles. However, while a nearby supernova is a treat, one that is too close will signal the extinction of life on Earth. It may be that nearby supernovae millions or billions of years ago played an important role in changing the course of evolution on our planet. Perhaps they act as a cosmic brake on the evolution of intelligence in the Universe at large, periodically irradiating promising forms of life in their early stages. To survive long term you need a little luck – and a good location. You don't want to take too many pictures like this one.

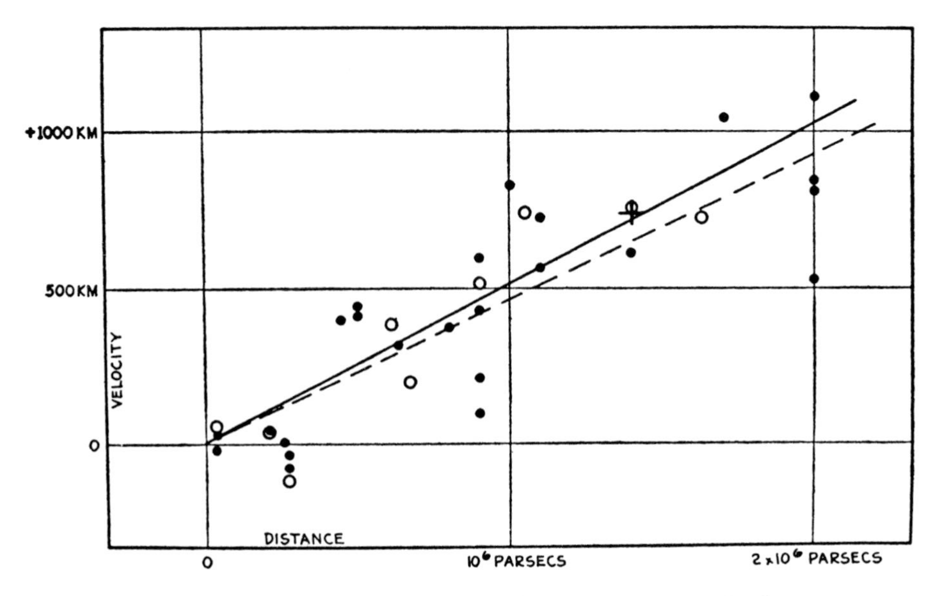

The Formulation of the Velocity-Distance Relation.

The original Hubble Law (1929) showing the velocities (in kilometres per second) of distant sources of starlight against their distances (in parsecs) away from us.[65] The solid line and black discs arise from using individual stars. The dashed line and circles arise if stars are combined into groups.

RUNAWAY UNIVERSES

HUBBLE'S LAW

Nature is a dull affair, soundless, scentless, colourless; merely the hurrying of material, endlessly, meaninglessly.

Alfred North Whitehead[66]

IN the 1920s, after it slowly became clear that Einstein's new theory of gravity predicted that our Universe should be expanding, with distant clusters of galaxies receding from each other at ever-increasing speeds, there was a slow and slightly confused accumulation of observational data, which culminated in the acceptance of the expanding-universe picture. Most, but not all, of the credit for gathering and interpreting that evidence generally goes to the American astronomer Edwin Hubble, in whose honour NASA's Hubble Space Telescope is named. Hubble used the world's most powerful telescope to gather light from stars in distant galaxies. The incoming light displays a sort of atomic barcode, or 'spectrum', that reveals the composition of the atoms that emitted it, or absorbed parts of it en route towards us. Hubble carried out the painstaking job, night after night, of gathering enough light from each distant galaxy to identify its spectrum unambiguously. What he recognised about each of the spectra was quite remarkable. The barcode looked just like it did when light from the same atoms was measured here on Earth, except that the whole of the pattern was shifted systematically by the same factor in the direction of the 'red' end of the spectrum of visible light. This indicated that the light from the distant galaxy was being stretched in wavelength, each colour by the same amount.

Hubble measured many of these 'redshifts', as they became known, and compared them with the distance away of the emitting galaxy in each case. Unlike the redshifts, these distances are not so easy to determine accurately. Hubble just assumed that he was looking at objects that were roughly the same, so that their intrinsic brightnesses were the same (like all 100 Watt light bulbs would be), so that a comparison of their *apparent* brightnesses would give their relative distances.[67]

Hubble took one further important step. He interpreted the redshifts in the light from the galaxies as arising from a Doppler effect. This is familiar to us in many situations where waves are emitted from a moving source. When a source of sound waves passes by, we hear the pitch rise as the source approaches (because the waves arrive more frequently than they were emitted) and fall as it recedes (because then the waves arrive less frequently than they were emitted). Recall the characteristic rise and fall in pitch of the 'eeee-ow' sound of the motor bike that roars past our bedroom window in the middle of the night. The pitch rises ('eeee') as it approaches, and falls ('ow') as it moves away. (Similarly in old films when the 'baddie' falls to his death from the cliff top you can discriminate between the poor special effect with the same pitch of scream all the way down and the good one where the pitch drops as the villain falls rapidly away from us). The same happens with light waves. If a source of light approaches us, then we see the light to be bluer than when it was emitted. But if the source is receding, we see its colour shift towards the red. Hubble interpreted the 'redshifting' of his spectra as arising because he was receiving light from galaxies that were moving away from us. This enabled him to make the greatest scientific discovery of the twentieth century: the expansion of the Universe.

By measuring the size of the shift in wavelength, Hubble could work out the speed of recession of the galaxy being observed. The bigger the redshift, the quicker the galaxy was moving away. When Edwin Hubble plotted a graph of the speed of the expansion against the distance of the emitting galaxy, he created one of the most famous diagrams in science and found that the speed of recession is directly proportional to the distance of the galaxy. Hubble submitted his paper containing the landmark graph, which was to become known as Hubble's Law, on 17 January 1929. It displayed the velocities and distances of twenty-two 'extragalactic nebulae',[68] and the graph of their

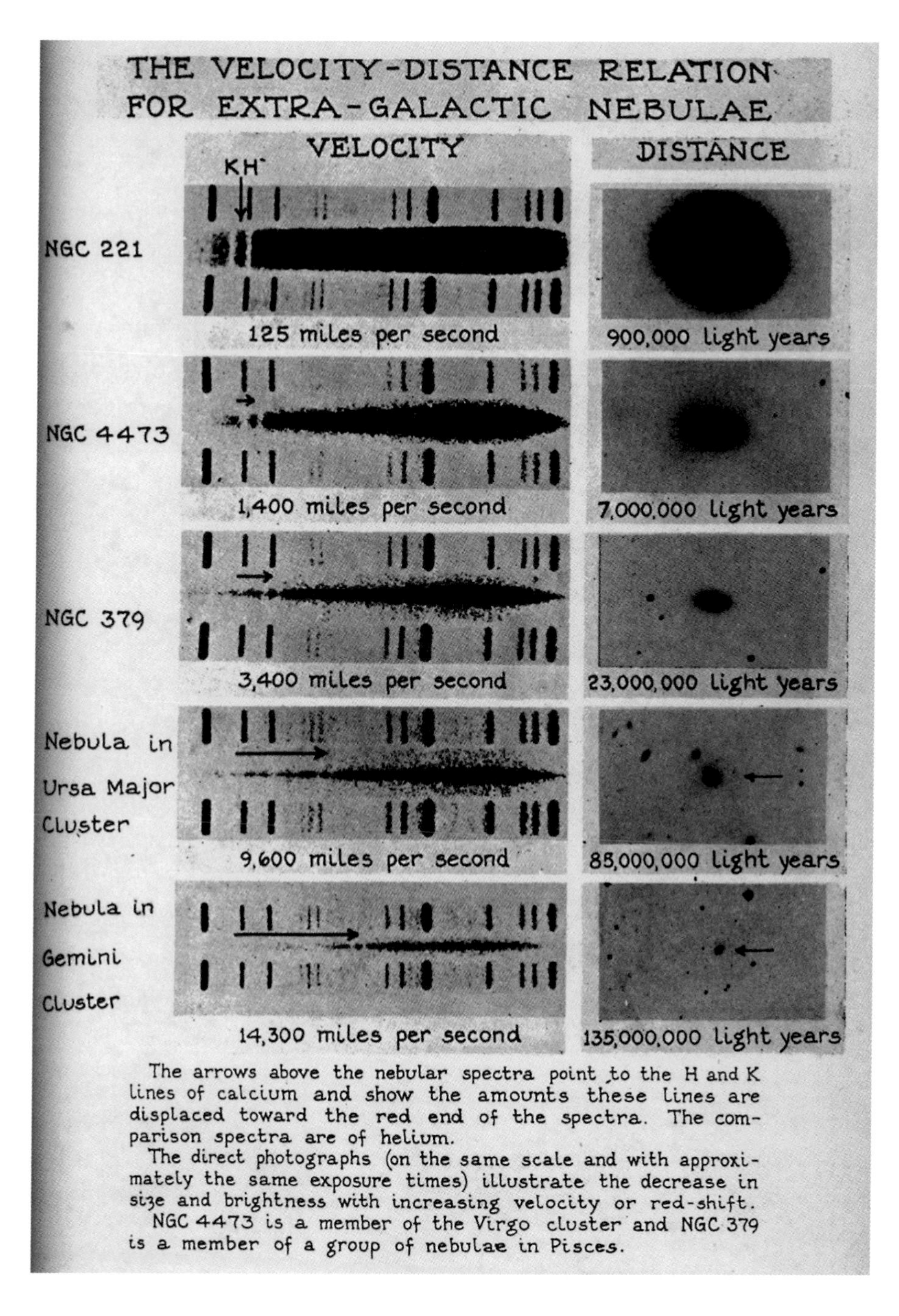

THE VELOCITY-DISTANCE RELATION FOR EXTRA-GALACTIC NEBULAE
VELOCITY
DISTANCE
K H
NGC 221
125 miles per second
900,000 light years
NGC 4473
1,400 miles per second
7,000,000 light years
NGC 379
3,400 miles per second
23,000,000 light years
Nebula in Ursa Major Cluster
9,600 miles per second
85,000,000 light years
Nebula in Gemini Cluster
14,300 miles per second
135,000,000 light years
The arrows above the nebular spectra point to the H and K lines of calcium and show the amounts these lines are displaced toward the red end of the spectra. The comparison spectra are of helium.
The direct photographs (on the same scale and with approximately the same exposure times) illustrate the decrease in size and brightness with increasing velocity or red-shift.
NGC 4473 is a member of the Virgo cluster and NGC 379 is a member of a group of nebulae in Pisces.

The redshifted light spectra and distances used by Hubble in the 1930s. The velocities were deduced from the Doppler shift in the H and K absorption lines that arise from the presence of ionised calcium (*marked by arrows*).

velocities against their distances displays a pretty good straight-line proportionality.[69] The slope of the line – which became known as 'Hubble's Constant' – was 465 ± 50 km per sec per Megaparsec (Mpc).

Interestingly, Hubble did not work alone, although he tried to give that impression for posterity. Many of the new redshift observations were made by Hubble's assistant, a remarkable man called Milton Humason. Humason left high school in his early teens and then worked as a mule driver on the Sierra Madre trail and as a ranch hand, before taking a position as janitor at the Mount Wilson Observatory. He took an interest in the astronomical work and was soon helping out, revealing himself to be a skilled observer, working with Hubble and publishing his own work as well.[70] It was only extraordinary bad luck – a flaw on the photographic plate hiding the object – that prevented him discovering the planet Pluto eleven years before the same planet was discovered by Clyde Tombaugh in 1930 at the Lowell Observatory. By 1931 Hubble and Humason had extended the velocity–distance diagram out to beyond 30 Megaparsecs.

Ever since Hubble made this pioneering discovery, astronomers have sought to see fainter and fainter galaxies that lie farther and farther away. They have been helped by the invention of exquisitely sensitive electronic detectors that detect more than half the light that hits them – whereas the traditional photographic film that Hubble and later astronomers used captured only 1 per cent of the incoming light. Gradually, Hubble's Law has been confirmed to a high degree of precision, although the value of Hubble's Constant has been found to be enormously different. In 1953, a radical revision of the determination of the distance to all the galaxies, thanks largely to the work of Humason and Alan Sandage, made a change to the scale along the bottom axis.[71] The law of direct proportionality between speed and distance was unaffected, only the slope of the graph – Hubble's Constant – changed. This was just the beginning of half a century of searching for the true value of the Hubble Constant. The difficulties of determining the distances to faraway objects confused astronomers for a long time. Through the period from 1965 to 1995 there was a vigorous debate between different observers – one group arguing that the data supported a value close to 100 km per sec per Mpc, while the other group argued for a result nearer 50 by using a different way of determining distances. Different methods of working out the distances tended to bias the answer you got for the Hubble Constant in different ways. One of the prime motivations for the construction of the Hubble Space Telescope was to solve this problem by enabling us to see intrinsically similar light sources in distant galaxies at hitherto unreachable distances.[72] These sources are the so-called Cepheid variable stars, whose rate of variation is determined by their brightness.[73] Today, the best determined value for the Hubble Constant is 70±3 km per sec per Mpc, very different to Hubble's first value, all because of the better determination of the distances to faraway galaxies. So, in some sense, neither the 50-school nor the 100-school was right in the end. The generally accepted value obtained by Wendy Freedman and her team using the best Space Telescope data is shown below. The adjacent part of the chart tracks the value of the Hubble Constant that the data converges towards out to increasing distances of observation.

Then, in 1997, something dramatic happened. Ironically, it was aided and abetted by the power of the Hubble Space Telescope, used in conjunction with powerful ground-based telescopes. Astronomers were able to see almost to the edge of the visible Universe and monitor the flaring and fading of exploding supernovae. To their surprise, the pattern of flaring and fading of all these

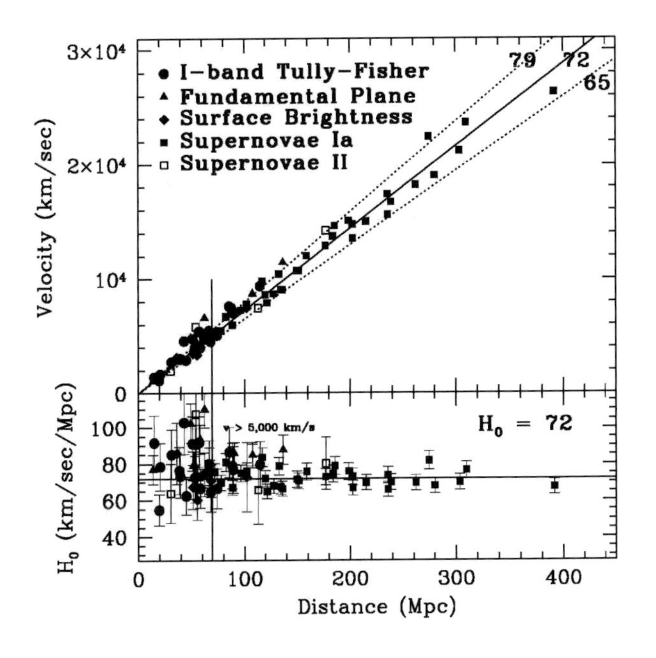

Modern determination of Hubble's Law by Wendy Freedman and collaborators (*top of panel*). As the data is examined out to increasing distances the slope of Hubble's Law approaches a best estimate of 72 km per sec per Mpc (*bottom of panel*).[74] Hubble's original observations only went out to 2Mpc.

objects mirrored very closely the pattern found in a wide class of nearby supernovae. This told them that they were most likely looking at faraway objects that were intrinsically the same as those nearby. It was like finding a population of 100-Watt light bulbs at the edge of the Universe. The redshifting of their light was easily measured to high accuracy as usual, but their distances could be determined as well. The result was as spectacular as it was unexpected. When Hubble's law was followed out to these record distances, it changed its form when the expansion of the Universe reached 75 per cent of its present extent. At that scale the expansion ceased decelerating and began *accelerating*.

This discovery has dominated cosmology ever since and has been confirmed by two separate observational teams. It has meant that Hubble's Law, a picture that looked done and dusted for a while, has re-emerged to take centre stage. Albeit with a 'twist' in its shape at large distances.

The reason for the change of gear of the Universe remains a big mystery. Cosmologists can describe it with great accuracy but they have no idea why it occurred when it did and whether the expansion is going to continue accelerating like this for ever. Physicists believe that the acceleration arises because of the 'vacuum energy' of the Universe – the lowest residual energy that the Universe can possess. They had hoped that this would turn out to be zero, but quantum uncertainty ensures that it has a positive value. The mystery is why it should take on a value that leads to it dominating the expansion of the Universe so late in cosmic history. So far, no one has any idea. In some theories of the physics that determines how the Universe behaves near the beginning of its expansion, the vacuum energy level falls out at random and there will be no further explanation for it other than that it happens to be small enough to allow galaxies and stars to form, and life to exist. If it were ten times bigger then the Universe would have begun accelerating earlier and galaxies and stars would have been unable to beat the fast expansion and condense out to form structures bound by gravitational forces. A universe with no stars and galaxies, and so no elements heavier than helium, would be a simple place – too simple for anyone to know.

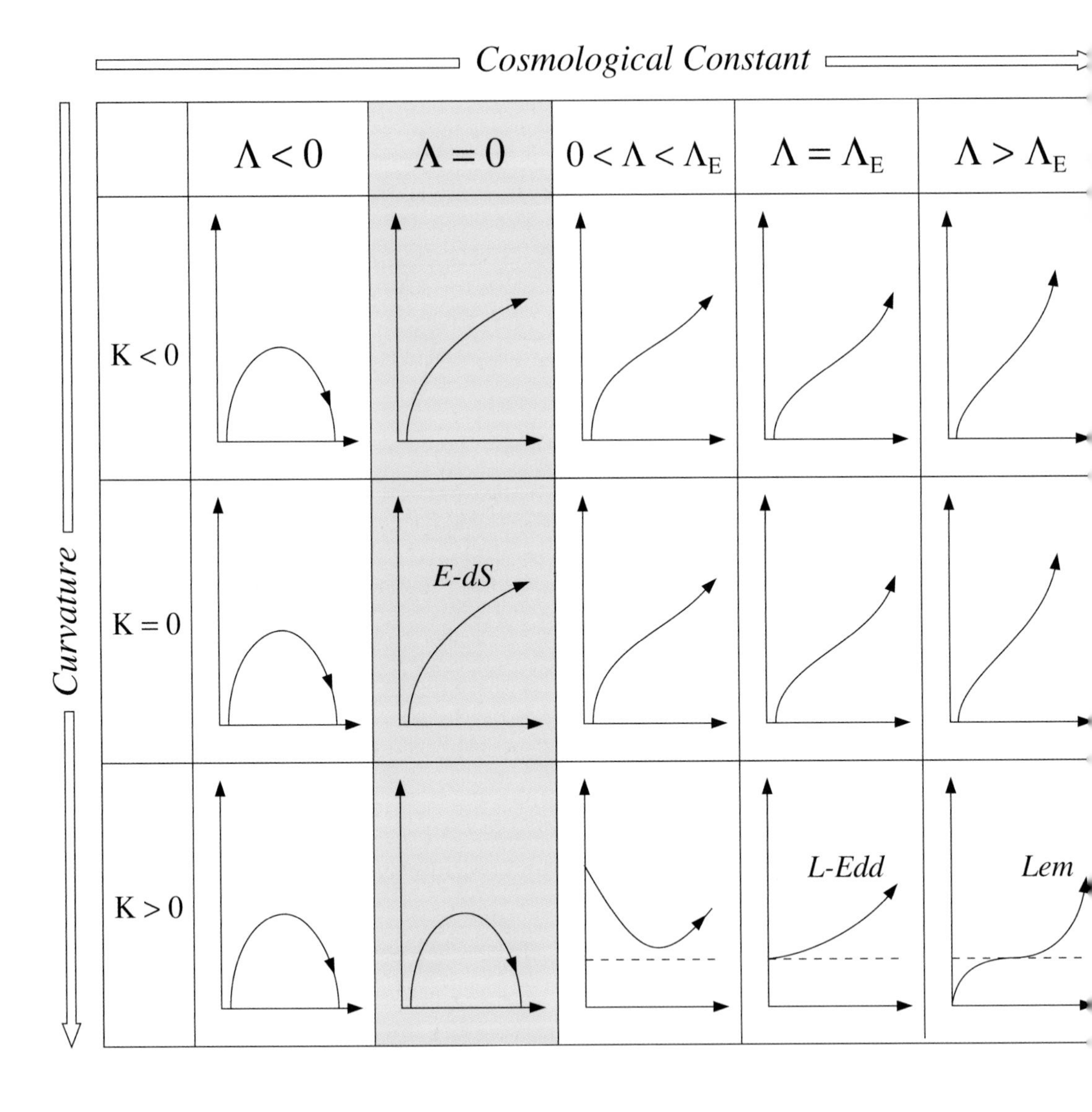

The different varieties of expanding universes according to their curvatures of space (K negative, zero, or positive) and the possible ranges of the cosmological constant (Λ negative, zero, or with any of three types of positive value determined by the special value Λ_E). Our universe appears to reside in the $\Lambda > \Lambda_E$ category, wit K = 0 or K < 0.

TOMORROW'S WORLDS
FRIEDMANN'S UNIVERSES

The irresistible, mind-boggling fantasy comes to just about everyone, sooner or later: What if everything we knew, our whole universe, was just a speck of dust on someone's shoulder?

Andrew Chaikin[75]

THE discovery that the Universe is expanding was a dramatic prediction of Einstein's 1915 theory of gravity – general relativity – that enlarged and superseded Newton's great theory of 1687. But there are many ways that a universe can expand and not all universes that expand today will continue to expand for ever. Einstein's theory described many possible universes. Some expanded at different rates in different directions, others could rotate, some contracted, some oscillated endlessly, while others just kept expanding at the same rate for ever. The early observations made by Hubble, and his collaborators, soon gave support for a gratifying simplification. As Einstein had guessed, it is a very good approximation to reality if we assume that the Universe expands today at the same rate in every direction ('isotropic expansion') and has the same density of material everywhere ('homogeneity').[76] In effect, the assumptions are based on symmetry: no special directions and no special places in the Universe, in the spirit of Copernicus, but they make the mathematical solution of Einstein's equations much easier. With these simplifications what remains is a catalogue of possibilities that can be viewed as a collection of expanding balls of material (some finite and some infinite in extent), whose histories can be described by the change in their scale with time. This scale is just the distance between two reference points in the universe: if it increases with time we say the universe is expanding, if it decreases we say it is contracting, and if it stays constant we say the universe is static.

The first person to distinguish the three simplest possible universes clearly was an unusual St Petersburg meteorologist and mathematician named Alexander Friedmann. Alexander's parents were very different to their son: his father was a dancer with the Mariinsky ballet and his mother was a pianist. But perhaps this explains why he was such an adventurous character. Unfortunately, Friedmann died at an early age as a result of his heroic (reckless?) practical studies of atmospheric physics. For a while he held the world ballooning altitude record and would deliberately take instruments to very high altitudes, to make meteorological and medical studies, allowing himself to fall into unconsciousness until the balloon descended back down to lower altitudes again. He died of typhoid in August 1925 just a month after ascending to the record altitude of 7400 m.

In 1922 Friedmann managed to solve Einstein's formidable equations under the simplifying assumptions of isotropy and homogeneity. His solutions provided cosmologists with a key picture that is ubiquitous in our modern descriptions of the history of the Universe. Friedmann universes, as these solutions are called, are the mathematical models that provide a simple way to coordinate all our astronomical observations of the expanding universe today. They were the first descriptions of entire universes that expanded from a past beginning in time, either continuing to expand for ever, or eventually contracting back to a big crunch at a finite time in the future. The three types of expanding isotropic universe have always been represented by this distinctive picture of the evolution of their size in scale with time.

In universes where all the material in the universe has the property of being gravitationally attractive,[77] universes begin to expand from a singular state – a 'beginning' – that appears to have infinite density and compression. If the initial energy of expansion exceeds that of the gravitational deceleration of the expansion exerted by its gravitational pull on all the matter in the universe, then it will continue to expand for ever. This is sometimes called an 'open' universe to reflect its infinite extent. However, if the energy of gravitational attraction is the greater, then the expansion will eventually be halted and reversed into contraction, back towards a state of unlimited density and compression. This is sometimes called a 'closed' universe to reflect its finite size.[78]

The physical distinction between open and closed universes is similar to that for the launching of a rocket from the Earth's surface. There is a critical launch speed that must be achieved if the rocket is to break free of the Earth's gravity and escape into space. If the launch speed is lower than this 'escape velocity' then the rocket will return to Earth, like a tennis ball thrown through the air. Open universes are launched with more than their escape velocity, closed universes with less, and in between there is a special 'critical' universe that has exactly the escape speed needed to propel it to infinite size after an infinite amount of time. Remarkably, our Universe lies tantalisingly close to this critical divide. As you can see from the figure below, the critical trajectory has the prop-erty that the open and closed universes veer away from it as time goes on. It is what mathematicians call an 'unstable' situation. If the starting speed deviates very slightly from the universe's critical escape speed then the expansion will move ever farther away as time passes. In order to remain very close after 14 billion years of expansion (as is the case for our Universe) it would appear that the expansion had to begin extraordinarily close to the divide.

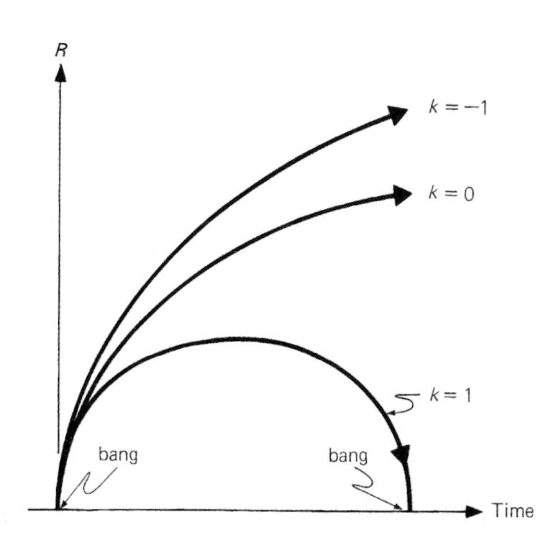

All the simple Friedmann universes with zero cosmological constant begin with big bangs.

This picture encapsulates and provokes some of the big questions of cosmology. Did the Universe have a beginning as Friedmann's universes suggest? Will the Universe continue expanding for ever? Or will it come to a dramatic end at some future time?

In 1934, the American cosmologist Richard Tolman speculated that the closed universes might have more than one cycle, each one different, as time goes on.[79] They could bounce back to create a never-ending series of cycles akin to those imagined in Hindu creation myths, in which a new universe arises Phoenix-like from the ashes of the old. Tolman injected a further piece of physics into this scenario, though, asking what would be the impact of the second law of thermodynamics on the cycles of an oscillating universe. The second law enshrines the fact of life that things go from bad to worse in the sense that the disorder of a closed system tends to increase. This is not really mysterious. There are so many more ways for order to degenerate into disorder than there are for disorder to turn into order that we tend to see the former happen in practice, even though the latter situation is allowed by the laws of motion and gravity. This one-way ticket for energy to devolve from ordered forms into disordered radiation means that each cycle of the oscillating universe differs[80] from its predecessor and they grow in maximum size and total lifetime for ever, getting closer and closer to the critical state of expansion all the time.

Einstein originally proposed in 1915 that a new form of stress could exist in the Universe which could act like a form of gravitational repulsion. His intention had been to avoid predicting that the Universe was expanding at a time when such an idea would have been completely new and unwarranted by local astronomical observations. By ensuring that the new repulsive stress exactly balanced the attractive force of gravity, any expansion or contraction of the universe could be prevented. A *static* universe resulted. Soon, however, Einstein's stress was found to be unsuccessful in stopping the expansion of the universe. Although it allowed for an exact counter-balance to the force of gravitational attraction exerted by all the matter in the universe, this situation was unstable, like a pencil balanced on its point. The slightest disturbance or non-uniformity in the static universe launched it into a state of overall expansion or contraction. Friedmann's solutions also showed this directly because Einstein's static solution was evidently not the most general solution of his equations. For a while Einstein believed that Friedmann had made a mathematical error and his non-static universes were not really solutions of the equations, but eventually Einstein recognised his mistake and recommended that Friedmann's work be published.

We can ask what will happen to Tolman's oscillating universes if Einstein's stress is included. It turns out that the oscillations must always end and the universe will end up in one of the ever-expanding, accelerating universes that appear in the gallery of possibilities pictured below.[81]

The first person to understand fully all the cosmological possibilities that were permitted when Einstein's stress was present was the Belgian cosmologist and Catholic priest Abbé Georges Lemaître. He was the first to extend the picture of open and closed universes into a further family of possibilities which allowed for ever-expanding universes that accelerated in the future.[82] Lemaître's first sketch of these possibilities, taken from his notebook of 1927, is shown below.

Lemaître was the first scientist to conceive of the physical picture of a universe beginning in a hot dense state ('the primeval atom') and expanding into its present quiescent form. In that sense he is the father of the Big Bang concept, even though the term was invented by Fred Hoyle in 1950. Lemaître's view of cosmology was eloquently put:

> The evolution of the world can be compared to a display of fireworks that has just ended; some few red wisps, ashes and smoke. Standing on a cooled cinder, we see the slow fading of the suns, and we try to recall the vanishing brilliance of the origin of the worlds.

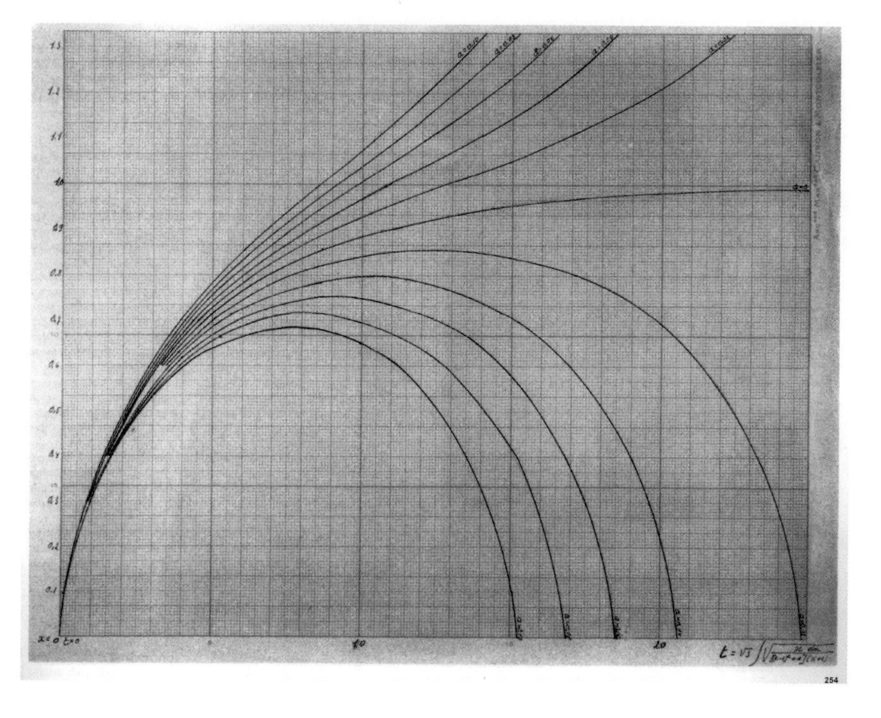

Georges Lemaître's first sketch in 1927 of the range of possible expanding universes, from his notebook.[83]

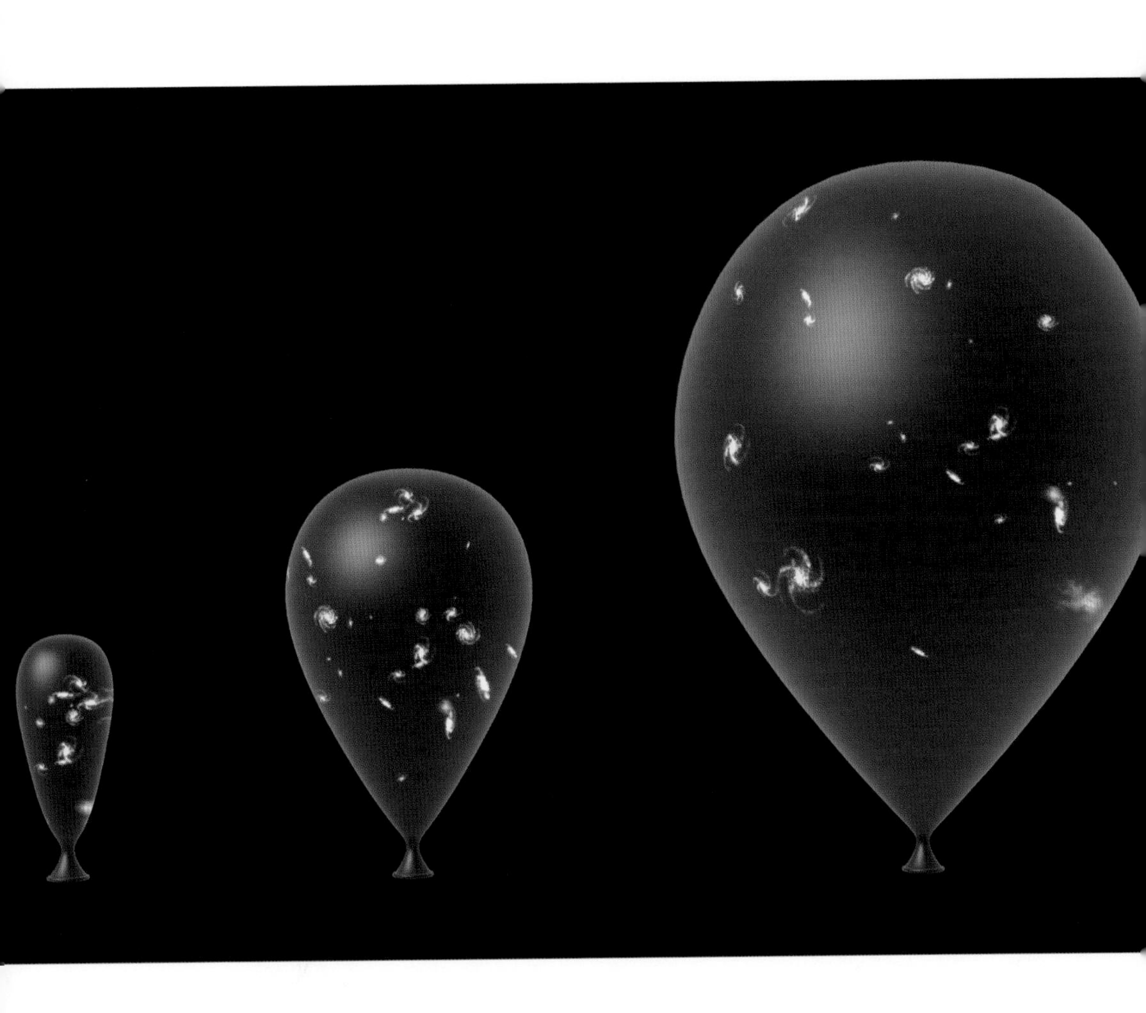

Eddington's balloon. A two-dimensional finite universe is represented on the surface of a balloon. It expands as the balloon inflates. Different points on the balloon recede from each other. Each point sees itself as the centre of the expansion, which does not lie on the surface of the balloon but in a third, 'other', dimension.

BLOW UP

THE BALLOON UNIVERSE

> But the theory of the expanding universe is in some respects so preposterous that we naturally hesitate to commit ourselves to it. For it contains elements apparently so incredible that I feel almost an indignation that anyone should believe in it – except myself.
>
> Arthur Eddington[84]

THE idea of an expanding universe is a challenging concept to grasp. All the expansions that we are familiar with are explosions *in* space and so they have a centre and an edge. As time goes on, the blast-wave front ploughs on through space, moving away from the point of detonation at its centre. Despite its pyrotechnic title, the 'Big Bang' of an expanding universe is not like that. It has neither centre nor edge. How can this be?

The case of an infinite universe is the easiest to imagine. The best guide is Maurits Escher, who in 1952 made a striking perspective drawing of an endless lattice of girders. Imagine that you are located at one of the intersections. As you look around things look the same on the average in every direction. Moreover, they would look like that from any other intersection point that you viewed from. An infinite universe has no centre. And, of course, because it is infinite it has no edge either. In fact, the promotion of an infinite universe with no need for a centre was what led to Giordano Bruno being burnt at the stake for heresy in 1600. The ancient philosophy of Aristotle, adopted for 1500 years

Maurits Escher's *Cubic Space Division*, 1952.

by the Church, maintained that the Universe of matter was finite, like a spherical ball, and so it had a centre. It needed that centre so that we could be located there in a special location. Bruno argued that this was all wrong. An infinite universe needed no centre.

But what if the Universe is finite? Here our imagination is stretched. Until Einstein came along it would always have been supposed that a finite universe must have an edge. But that intuition was based upon a picture of space that assumed it to be flat. To visualise things more easily let us think about a universe with only two dimensions of space. If it is flat, like this page, then it seems that it must have an edge if it is finite. But if it is curved everything changes. The surface of a ball is finite – you need only a finite amount of paint to colour it – but if you move around on this two-dimensional spherical surface you will never hit an edge.

This is the picture that Eddington introduced in 1933 in his book *The Expanding Universe* to help us visualise an expanding universe:

> We can picture the stars and galaxies as embedded in the surface of a rubber balloon which is being steadily inflated; so that, apart from individual motions and the effects of their ordinary gravitational attraction for one another, celestial objects are becoming farther and farther apart simply by inflation.

Imagine that the surface of a balloon is marked with crosses to depict galaxies. Inflate the balloon and watch what happens. All the crosses move away from all the other crosses. If you were located on any one of the crosses you would see all the others expanding away from you as if you were at the centre of the Universe. Actually, the centre of the expansion does not lie on the surface of the expanding balloon at all. And although the expanding surface is finite, nothing moving on it will ever reach an edge. A three-dimensional finite expanding universe is similar. It behaves like the three-dimensional surface of an expanding four-dimensional ball.

One of the first attempts to explain Hubble's discovery of the expanding Universe to the general public in the USA was Donald Menzel's strikingly illustrated article from the December 1932 issue of *Popular Science Monthly*.[85] He explains that the recession of the nebulae seen by Hubble is due to the

Blast of Giant Atom

By
Donald H. Menzel
Harvard Observatory

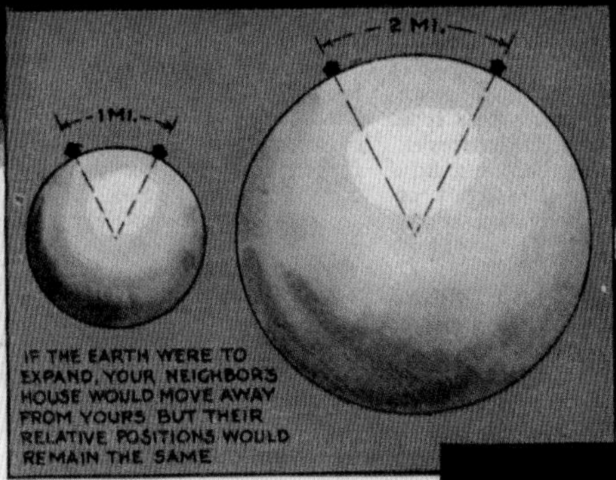

OUT of a single, bursting atom came all the suns and planets of our universe!

That is the sensational theory advanced by the famous Abbe G. Lemaître, Belgian mathematician. It has aroused the interest of astronomers throughout the world because, startling as the hypothesis is, it explains many observed and puzzling facts.

According to Lemaître's theory, all the matter in the universe was once packed within a single, gigantic atom, which, until ten thousand millions years ago, lay dormant. Then, like a sky-rocket touched off on the Fourth of July after having remained quietly for months on a store shelf, the atom burst, its far-flung fragments forming the stars of which our universe is built.

The manner in which certain kinds of atoms explode can be seen easily in a simple experiment. If you take a radium watch into a dark room and look at the dial through a magnifying glass, you see what appears to be a brilliant display of miscroscopic fireworks. While you are looking at the showering sparks, remember that each flash comes from an exploding atom. In each spark, you see a small-scale reproduction of the new theory of the birth of our universe.

On the average, every radium atom lies dormant for about 1,730 years, after which time it explodes and shoots out particles in much the same way as the parent atom gave birth to the stars.

The new theory provides an explanation for one of the most extraordinary scientific facts ever discovered. Our telescopes show us that there are, out in space, millions of disk-shaped star-clusters known as extra-galactic nebulae. It is generally believed that our Milky Way is such an object and that our sun is but one of billions of stars that go to form it. One of the larger members of the class, the spiral nebula in Canes Venatici, is so far away that light from it takes almost a million years to reach us. Furthermore, observations indicate that every second it moves still farther away from our solar system by some 170 miles.

For every large, bright nebula there are thousands of small, faint, and presumably much more distant ones. Surveys out to one hundred million light years are in progress. The extraordinary feature referred to above is not, however, the magnitude of the figures, but the discovery that the more distant the nebula the more rapid is its motion *in a direction away from us!* The present record-holder is a tiny nebula whose cosmic speedometer registers in excess of twelve thousand miles a second!

Why, astronomers have asked, are the

1 MI.
2 MI.
IF THE EARTH WERE TO EXPAND, YOUR NEIGHBOR'S HOUSE WOULD MOVE AWAY FROM YOURS BUT THEIR RELATIVE POSITIONS WOULD REMAIN THE SAME

YOU CAN SEE AN ATOM BOMBARDMENT IF YOU LOOK AT THE NUMERALS ON A RADIUM DIAL WATCH UNDER A MAGNIFYING GLASS IN THE DARK expansion of space and asks what happened before the apparent beginning of the Universe, which he dubs 'that prehistoric Fourth of July, when space came into existence'.

Arthur Eddington was one of the greatest popular science writers, and the foremost astrophysicist of his day, responsible for many great discoveries about the structure of stars and the motions of stars in galaxies. He also led the famous 1919 expedition by a group of astronomers to Principe, in Africa, where they first verified Einstein's great prediction that light would be bent by the Sun by an amount twice as great as Newton's theory of gravity predicts (see pages 169–172). His writings abound with wonderful analogies and apposite quotations (but not many pictures). However, it is his word-picture of the expanding balloon that has been the most used, the most valuable, and the most appreciated by generations of cosmologists.

This remarkable popular article by Donald Menzel about the expanding Universe appeared in 1932, only three years after Hubble's discovery.

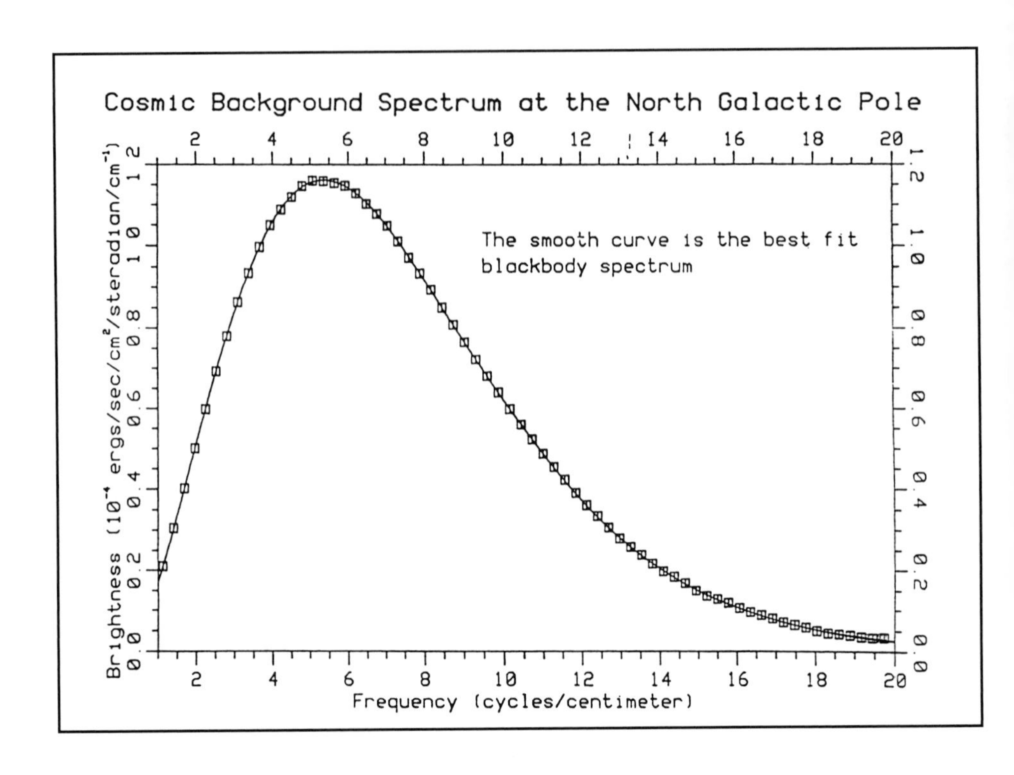

The black-body heat spectrum of the cosmic background radiation measured from space by NASA's COBE satellite and announced on 13 January 1990.

DEEP HEAT
THE COBE SPECTRUM

I awoke on Friday and because the universe is expanding it took me longer than usual to find my robe.

Woody Allen[86]

ONE of the greatest discoveries of the twentieth century was made inadvertently in 1965 by two American radio engineers working at the Bell Telephone Laboratories in Holmdel, New Jersey. Arno Penzias and Robert Wilson were rejigging a receiver originally designed to track the Echo communications satellite so as to use it for radio astronomy. As they tuned the receiver they found that they picked up a persistent low-level source of microwave radio noise with the same intensity all around the sky. After meticulous checking of all the possible local sources of this signal Penzias and Wilson started talking to local cosmologists at nearby Princeton University who, by coincidence it seemed, were planning to launch a search for a remnant radiation field from the early stages of the Universe. They believed they had made an important new prediction about what could be seen as a relic of the hot beginnings of a Big Bang universe. However, nobody in Princeton had been reading the research literature carefully.[87] Back in 1948, Ralph Alpher and Robert Herman,[88] two research students of George Gamow, had begun to publish a series of articles in which they developed our understanding of what should have gone on during the first few minutes of an expanding Big Bang universe. They had already predicted that there should be a low level of radiation 'fall-out' still remaining in the universe, cooled to very low temperature by the billions of years of expansion that the universe had undergone. They expected the radiation would have a present-day temperature of about 5 degrees Kelvin (that is, minus 268 degrees Celsius). What Penzias and Wilson had found was a microwave signal which, if it was heat radiation with a black body spectrum characteristic of pure heat radiation, had a temperature of 3.5±1.0 degrees Kelvin. It soon became clear that they had found the Alpher–Herman radiation echo from the Big Bang. Two papers were published, back to back: a low-key announcement of the measurement by Penzias and Wilson under the unassuming title *A Measurement of Excess Antenna Temperature at 4080 Mc/s*, and a theoretical paper by Robert Dicke, James Peebles, Peter Roll, and David Wilkinson that explained the place of the Penzias and Wilson measurement in the context of the expanding universe.[89] It became known as the 'cosmic microwave background radiation', or the 'CMB' for short, because it lies in the microwave band of the radiation spectrum. The impact of the discovery was very great. It confirmed the general picture of an expanding Universe that was once much hotter and denser than it is today, and the detailed structure of this radiation provided a cosmological Rosetta Stone from which cosmologists could decode information about the past history of the Universe. Penzias and Wilson received the Nobel Prize for physics in 1978.

For many years after the discovery of CMB, observers repeated the observations of Penzias and Wilson in slightly different ways, sensitive to different frequencies of microwaves. They were seeking to confirm one last great prediction – that this signal really had the signature of pure heat radiation.[90] If the CMB was truly a signal from the beginnings of the Universe then it should have a variation of intensity versus wavelength (what we call its 'spectrum') of a particular 'black body' shape that was first calculated by the great German physicist Max Planck in 1900.[91] The key feature of Planck's spectrum is that the intensity increases to a maximum and then decreases as the wavelength of the radiation changes further.

Unfortunately, when the early observations of the CMB were made from the Earth's surface they could only probe the part of the wavelength spectrum longwards of the maximum at about 0.3 cm. The Earth's atmosphere intervened to change the incoming radiation at shorter wavelengths because it was absorbed and re-emitted at a different wavelength by some of the molecules, such as water, that it contained. Gradually, observers started to fly instruments high into the atmosphere, using balloons, to mitigate these effects. But these observations are very delicate: you need to compare the temperature of the background radiation close to 3 degrees above absolute zero with a reference source of liquid helium near the same temperature, while all around the instruments the Earth is reflecting and re-radiating heat from the Sun at a temperature hundreds of degrees greater.

As these balloon-borne and ground-based observations built up, they suggested that there might be a significant distortion of the Planck form of the spectrum near the peak. This was very controversial. Some saw it as evidence for lots of violent activity in the history of the Universe associated with the formation of stars and galaxies. Others, such as Fred Hoyle and his collaborators, even clutched at this straw as evidence that the whole Big Bang picture might be wrong after all and the CMB was entirely made by stars in a messy way rather than in the past inferno of a Big Bang, as everyone else maintained. Others distrusted the ground-based observations and thought that the distortions found by Paul Richards and his graduate student, Dave Woody,[92] although real, were simply due to the atmospheric interventions compounded by the limited amount of the sky that any single ground-based or balloon-borne experiment can see.

It was against this background of uncertainty, speculation, and controversy that NASA's long-awaited COBE satellite mission to study the CMB was launched into orbit by Delta rocket into a heliosynchronous orbit[93] on 18 November 1989, fifteen years after the project's conception. Its mission was to observe the CMB around the whole sky and map the spectrum far into the infra-red region shortwards of the intensity maximum to establish whether the claimed distortions were present before the radiation entered the atmosphere and to settle the issue of whether the radiation had a spectrum with Planck's special signature. If so, by locating any peak of intensity precisely, the temperature of the radiation could also be determined with unprecedented accuracy.

The results from COBE were keenly awaited and were announced publicly by John Mather on 13 January 1990 after being guarded carefully by the observing team (although that process does not seem to have been without internal friction and controversy).[94] Mather announced that the COBE satellite's FIRAS (an acronym for Far Infra-Red Absolute Spectrophotometer) detector had produced a complete spectrum of the CMB radiation. He then displayed the picture, which produced spontaneous applause from the assembled astronomers. It was the most perfect Planck heat spectrum ever observed in Nature. No distortions; no deviations of any sort from the signature of pure heat radiation at a temperature of 2.725 degrees Kelvin. The data points match the Planck heat curve so precisely that it is impossible to distinguish the two in the picture. The errors on the data points are about a hundred times smaller than the thickness of the curve. This was the final proof that in its youth our Universe was a hot dense inferno. Perhaps sadly, all those interesting distortions to the spectrum that had been seen from the ground by Woody and Richards, and others, were not telling us in the end about complicated explosive events in cosmological history, simply about the complexities of atmospheric chemistry. The Universe was for once plainer and simpler than we had imagined.

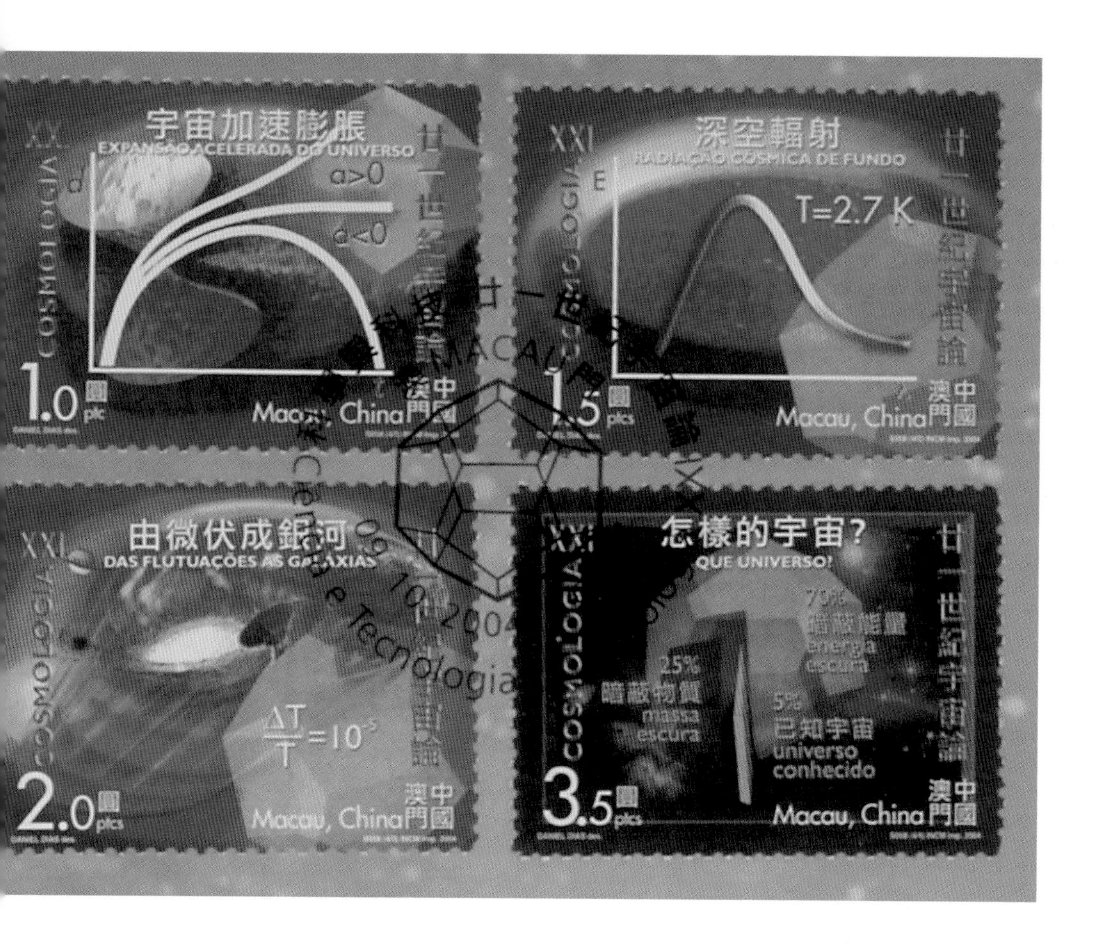

COBE stamps, issued in 2004 in Macau, China, show the three possible expansion histories of the Universe (*top left*); the spectrum of the background radiation (*top right*); the uniformity of the background radiation temperature around the sky with small fluctuations measured by COBE (*bottom left*); and the COBE satellite (*bottom right*).[95]

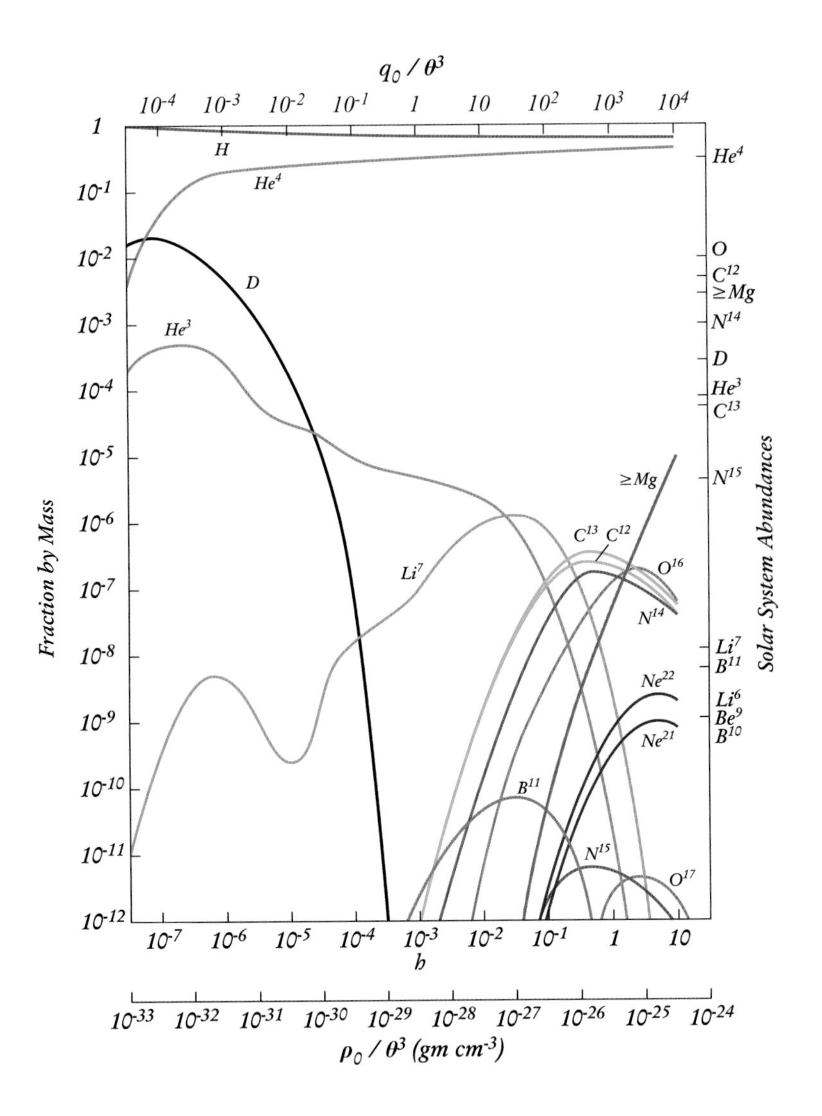

The abundances, as a fraction of the total mass of all elements, of the elements produced in the first few minutes of the Universe's history according to the present density of nuclear matter in the Universe shown horizontally. The abundance of helium-4 is almost constant over a very wide range of possible densities. The abundance of deuterium (D) is very sensitive to the density and provided cosmologists with a new method of determining the density of nuclear matter in the Universe. This picture was first computed by Robert Wagoner, Fred Hoyle and William Fowler in 1967. The abundance levels observed in the solar system are shown for comparison down the right-hand vertical axis.

WHEN A PROTON MEETS A NEUTRON

BIG BANG NUCLEOSYNTHESIS

Things are more like they are today than they have ever been before.

Dwight Eisenhower

ONE of the defining features of an expanding universe appears to be that its past must have been very different to its present.[96] Things were hotter, denser, and more concentrated then than they are now. After the discovery, by Penzias and Wilson in 1965, of the radiation left over from the hot beginnings of the Universe, there was renewed interest in reconstructing the history of the Universe back to very early times, a task begun fifteen years earlier by Ralph Alpher, Robert Herman, and George Gamow. One particular epoch of cosmic history was a focus of attention. After the expansion had been going on for between one second and about three minutes the temperature and density of the whole Universe would still have been too great for any atoms or molecules to exist; nor could there have been any stars or galaxies. The whole Universe would have been like a great nuclear reactor with spontaneous nuclear reactions filling the whole of space until, after a few minutes, the expansion had lowered the temperature and density enough for the reactions to shut down. But after all the nuclear action was over, the remnants would remain in the expanding space like fossils, frozen in time until we arrive to collect them nearly 14 billion years later.

The exciting thing about this scenario is that we can check whether our reconstruction of this very early interval of cosmic history is correct by comparing our predictions about the fossil remnants with the abundances of the same elements that we find in our Galaxy and elsewhere today.

At first, making predictions of the products of the first nuclear reactions seemed to be impossible. Surely the results would depend upon the initial ingredients and they would depend upon the completely unknowable state in which the Universe began (if indeed it ever did 'begin'). All nuclei are made up of different numbers of protons and neutrons. What were the relative proportions of these two types of particle when the nuclear reactions began? Does the outcome just depend on the unknowable make-up at the beginning? Everything hinges on the answers to these questions.

This dilemma held up our study of the early Universe for a long time until, in 1950, the Japanese astrophysicist Chushiro Hayashi[97] noticed a simple but crucial feature of the Universe. When the Universe is less than one second old, and the temperature of the radiation in the Universe exceeds 10 billion degrees, there is a complete equilibrium between the neutrons and protons. The weak force of Nature responsible for radioactivity maintains a complete population balance between the neutrons and protons: when the temperature is much greater than 15 billion degrees they will always be present in equal numbers. If protons were to become more numerous then this would simply stimulate the creation of more neutrons to remove the imbalance, and vice versa. As the temperature drops towards 10 billion degrees the radioactive exchanges struggle to keep pace with the expansion of the Universe that is pulling the protons and neutrons apart. The population balance swings slightly towards the protons because they are slightly lighter particles than neutrons and it requires a little less energy to make a proton from a neutron than a neutron from a proton. Even so, the radioactive exchanges continue and everything depends only on the temperature of the Universe until the temperature falls to 10 billion degrees. Then something important occurs. The radioactive exchanges can keep pace with the expansion no longer. The neutrons and protons cease to change into each other and their abundance is settled, with roughly one neutron remaining for every six protons, everywhere in the Universe. After that, it falls only by a small amount because there is a very slight radioactive decay of neutrons (their half-life is about ten minutes) until the temperature falls to a billion degrees when the Universe is two minutes old. Then the nuclear fireworks begin. Very rapid nuclear reactions transform protons and neutrons into deuterium nuclei, and then combine them all to produce the two isotopes of helium and lithium. Some much heavier elements, such as boron, beryllium and carbon, are made but their abundances are tiny. The final outcome is that about 77 per cent of the mass of the Universe remains as hydrogen, about 23 per cent is burnt to helium-4, while tiny traces of deuterium and helium-3 (one thousandth of 1 per cent) and lithium (one hundred millionth of 1 per cent) evade being burnt to helium. Helium-4 nuclei are very tightly bound by nuclear forces and difficult to disrupt, and so almost all the nuclear products end up in these nuclei, with just a few traces of the heavier ones escaping. Remarkably, these are the abundances of these lightest elements that we find all over the Universe when we look today.

There were many early attempts to make detailed predictions of how much helium would be made in this way in the Big Bang, notably by Ralph Alpher, James Follin, and Robert Herman in 1953,[98] by Fred Hoyle and Roger Tayler in 1964,[99] and by James Peebles in 1966.[100] But the most detailed studies, including all the nuclear reactions, computing all the nuclear abundances, and extracting almost all the interesting conclusions for the structure of the Universe were made by Fred Hoyle, William Fowler and their research student, Bob Wagoner, in 1967.[101] They produced two enormously influential and much-reproduced pictures that showed how the abundances of the lightest nuclei built up during the first few minutes of expansion and how the final abundances depended upon the density of matter we observed in the Universe today.

These pictures showed cosmologists how a measurement of the abundance of deuterium in the Universe could tell us the density of matter in the Universe because it is the matter density that determines the speed of the nuclear reactions that burn the deuterium into helium-4 nuclei. In a high-density universe the destruction goes faster and little deuterium remains, but if the density of the sort of matter that takes part in nuclear reactions is low then much more deuterium will survive. Deuterium is the cosmic densitometer.

In 1973 atomic deuterium was discovered for the first time in interstellar space by the *Copernicus* satellite.[102] Its abundance was two thousandths of 1 per cent and showed that the universal abundance of atomic matter was ten times too low to slow the Universe enough in the future to make it collapse back in upon itself (a density exceeding 2×10^{-29} gm per cc was needed for that).

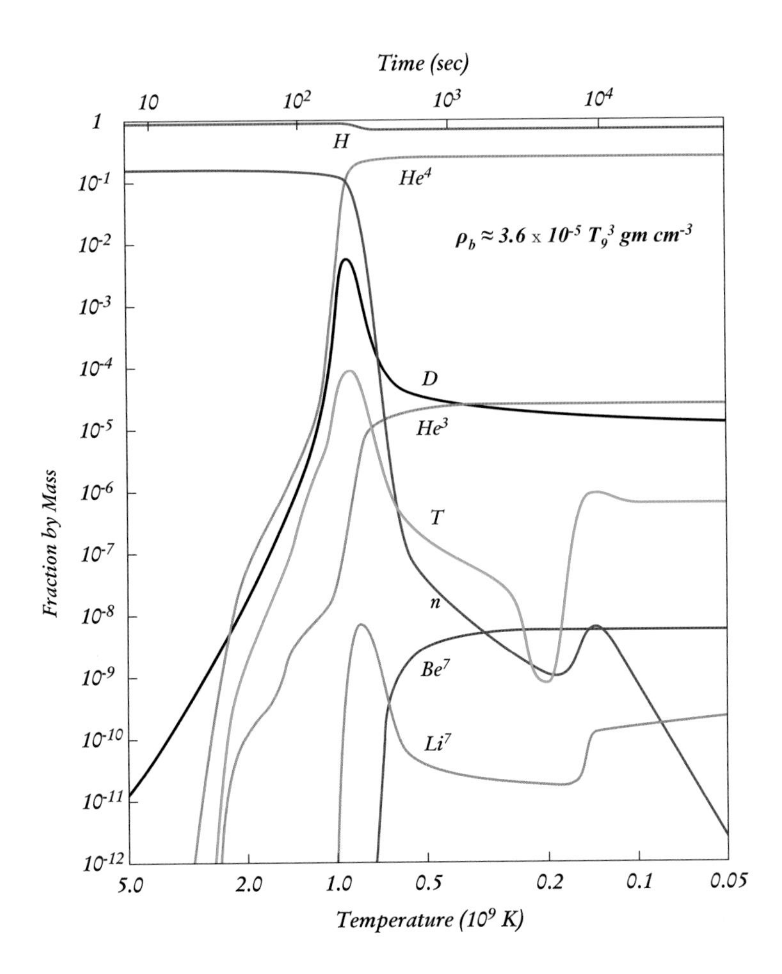

The build-up of the lightest nuclei with time increasing and temperature falling in an expanding universe. After a rapid change due to nuclear reactions, the abundances of helium, deuterium and lithium stay constant after 1000 seconds because the density and temperature are then too low for the nuclear reactions to operate.

In the decades that have followed we have gradually found evidence for considerably more matter in the Universe than the deuterium densitometer allows – about ten times more. It reveals itself by the strength of its gravity, which dictates how fast visible stars and galaxies are moving, and by its effect on the paths of light rays through space. By putting all this evidence together we have been driven towards a pleasing self-consistency – but with a strange twist. The density of ordinary nuclear matter made out of protons and neutrons (like you and me) that the deuterium abundance says should exist in the Universe is about the same as what we find in all the ordinary stars and planets and debris that we can detect. But the strengths of the gravitational forces being exerted on moving stars and galaxies in the Universe show that there must be ten times as much matter again hiding in some invisible form that cannot have taken part in nuclear reactions in the early universe, or we would have too little deuterium today. This 'dark matter', as it is known, is suspected to be made up of new neutrino-like elementary particles. If they do provide the dark matter in the Universe, then we can predict what their masses will be (a prime candidate would have a mass about thirty times that of a hydrogen atom) and go out and search for them with sensitive detectors.

Neutrinos feel only the forces of gravity and radioactivity so do not take part in nuclear reactions or contribute to the destruction of primordial deuterium. They interact very weakly, and huge numbers of these dark matter neutrinos are hitting your head at 250 km per sec at this moment. Less than one per day will be seen to disrupt an atom, so weak are their effects. Today, many experiments are searching deep below the Earth's surface, where the confusing effects of cosmic rays can be shielded out by kilometres of solid rock and other, man-made, barriers, for these occasional collisions with ordinary atoms. Steadily, the size and sensitivities of these experiments are approaching the level at which they must see the neutrino-like particles if they really do constitute the dark matter. We hope that some day soon they will find the dark, ghostly forms of matter that the dramatic pictures of Wagoner, Fowler, and Hoyle led us to expect.

Johannes Kepler's 'M picture', from his *Epitome of Copernican Astronomy* (1618–21). The M, for *mundus* (world), marks the location of the Earth in the starry universe.

BLACKOUT

THE DARK NIGHT SKY

Sirius, the brightest star in the heavens . . .
my grandfather would say we're part of
something incredibly wonderful – more marvelous
than we imagine. My grandfather would say we
ought to go out and look at it once in a while
so we don't lose our place in it.

Robert Fulghum

THERE is no greater shared human experience than the sight of the night sky that envelops us. Ever since our ancestors had eyes to see, they witnessed the daily rising and setting of the Sun bounding their horizons. Their lives were structured by the dangers and limitations of darkness, safeguarded by fire, and extended by fire's artificial light. As we saw in our opening chapter, human imagination has always been stimulated by the patterns in the dark night sky. Changing cycles of the Moon and the Sun gave structure to time, and the unchanging positions of stars gave confidence in the reliability of reality.

The night sky is sprinkled with stars and planets, and dominated by the waxing or waning each month of the brightness of the Moon shining in the reflected light of the Sun. Astronomers are fascinated and challenged by these twinkling, much-photographed distant worlds, but in between them lies a greater and deeper mystery – the darkness of the night sky.

Why is the sky dark at night? It is a question that most people would answer rather confidently, almost embarrassed that the questioner could pose such a seemingly stupid question. 'Have you not seen, have you not heard,' they would ask, 'that at the end of the day, as the football managers say, the Sun goes down?'

But the darkness of the night sky has little to do with the Sun. The first person to be puzzled by the matter was Johannes Kepler, who, upon receiving a copy of Galileo's new book *The Starry Message* in 1610, wrote immediately to him[103] to provide what he believed was a telling argument against our Universe being infinite. For if the Universe was filled with stars, stretching away endlessly, then:

> the whole celestial vault would be as luminous as the sun . . . this world of ours does not belong to an undifferentiated swarm of countless others.

What Kepler imagines is that looking out into an infinite starry universe is like looking into a great forest. Your line of sight ends everywhere on the trunk of a tree and, as you pan across the landscape before you, you see only a phalanx of trees. If this is what happens when you look into a small forest, should it not also occur when you look out into the night sky? Everywhere your line of sight should end on the surface of a star. And so the whole of the night sky should look like the surface of a star. Day and night, night and day, the whole sky should shine like the surface of a star.

Kepler was a strong proponent of the view that the Universe was finite. In his work of 1605, entitled *The New Star*, he states his deep revulsion for the very idea of a limitless sea of stars because it has no centre:[104]

> This very cognition carries with it I don't know what secret, hidden horror; indeed one finds oneself wandering in this immensity, to which are denied limits and centre and therefore all determinate places.[105]

The 'M picture' (shown on page 104), from his book *Epitome of Copernican Astronomy*, shows part of an endless pattern of stars. Everything looks the same to Kepler, and our Sun would be nothing more than an undistinguished typical star like the one labelled with an 'M'. On page 109 is Kepler's accompanying picture showing the solar system surrounded by the sea of distant stars.

This was a deeply puzzling observation that became the more troublesome with the idea, which appeared in Newton's model of the world in the late seventeenth century, of an infinite universe. It was Newton's friend Edmund Halley – of Halley's Comet fame – who was the first to address the problem in two short articles, published in 1720.[106] Halley tried to resolve the paradox by arguing that the far-distant stars were so faint in appearance that they do not contribute significant light to the sky. Alas, this argument does not work. The distant stars all contribute just as much light as the nearby ones because their greater number exactly cancels out the effect of their apparent brightness diminishing with distance from us.

Over the next hundred years, all sorts of ideas were tried to solve the problem. Perhaps the starlight was absorbed by dark material in between the stars? Or perhaps the light gets 'tired' and loses energy en route to us? None of these suggestions worked.[107]

Today, we know why. The key to understanding the darkness of the night sky is our discovery that the Universe is expanding. As a consequence, the Universe appears big and old, dark and cold, and superficially seems rather hostile to life as we know it. Yet, these properties are vital. The huge age of the Universe is important for our own existence. We are each made of complicated atoms of carbon, nitrogen, and oxygen, along with a raft of others, such as phosphorus and iron, and artificial forms of life rely on others still, like silicon. The nuclei of these atoms do not come ready-made with the Universe. They are made by a long, slow sequence of nuclear reactions in the stars. It takes almost 10 billion years for hydrogen to be transformed by this stellar alchemy, first to helium, then to beryllium, and then on to carbon and oxygen, and beyond. The dying stars explode in supernovae and spread their life-giving debris around the Universe, from whence it finds its way into grains of dust, planets, and, ultimately, into people. The nucleus of every carbon atom in your body has been through a star. You are made from the dust of the stars.

So you begin to understand why it is no surprise that the Universe is so big and so old. It takes at least 10 billion years to make the building blocks of living complexity in the stars, and because the Universe is expanding it must be at least 10 billion light years in size. We could not exist in a universe that was significantly smaller. An economy-sized universe, just the size of our Milky Way galaxy, with its 100 billion stars and accompanying planetary systems, seems room enough. But it would be little more than a month old – barely enough time to pay off your credit card bill, let alone evolve complexity and life.

Any universe that is a home for life must be big and old; so it must also be dark and cold, because an expanding universe gets cooler and cooler, and energies fall as space expands. The inferno of the past Big Bang must, after billions of years, be replaced by the dark night sky we see around us containing a faint glimmer of microwaves, just three degrees above absolute zero of cold: an echo of its hot beginnings. The huge amount of expansion and cooling required to make the Universe habitable means that there is too little energy around today to illuminate the night sky. Even if all the mass in the Universe were suddenly transformed into light by Einstein's $E = mc^2$ formula, all that would happen is that the temperature of space would rise by about ten degrees. There is just too little light energy in a large life-supporting universe to illuminate the night sky. But more than 13 billion years ago, when the Universe was 1000 times more compressed than today, its temperature would have exceeded 3000 degrees, and the whole sky would have been bright like the Sun. Alas, in such an environment there could be no atoms, no planets, no stars, and no astronomers to tell of it.

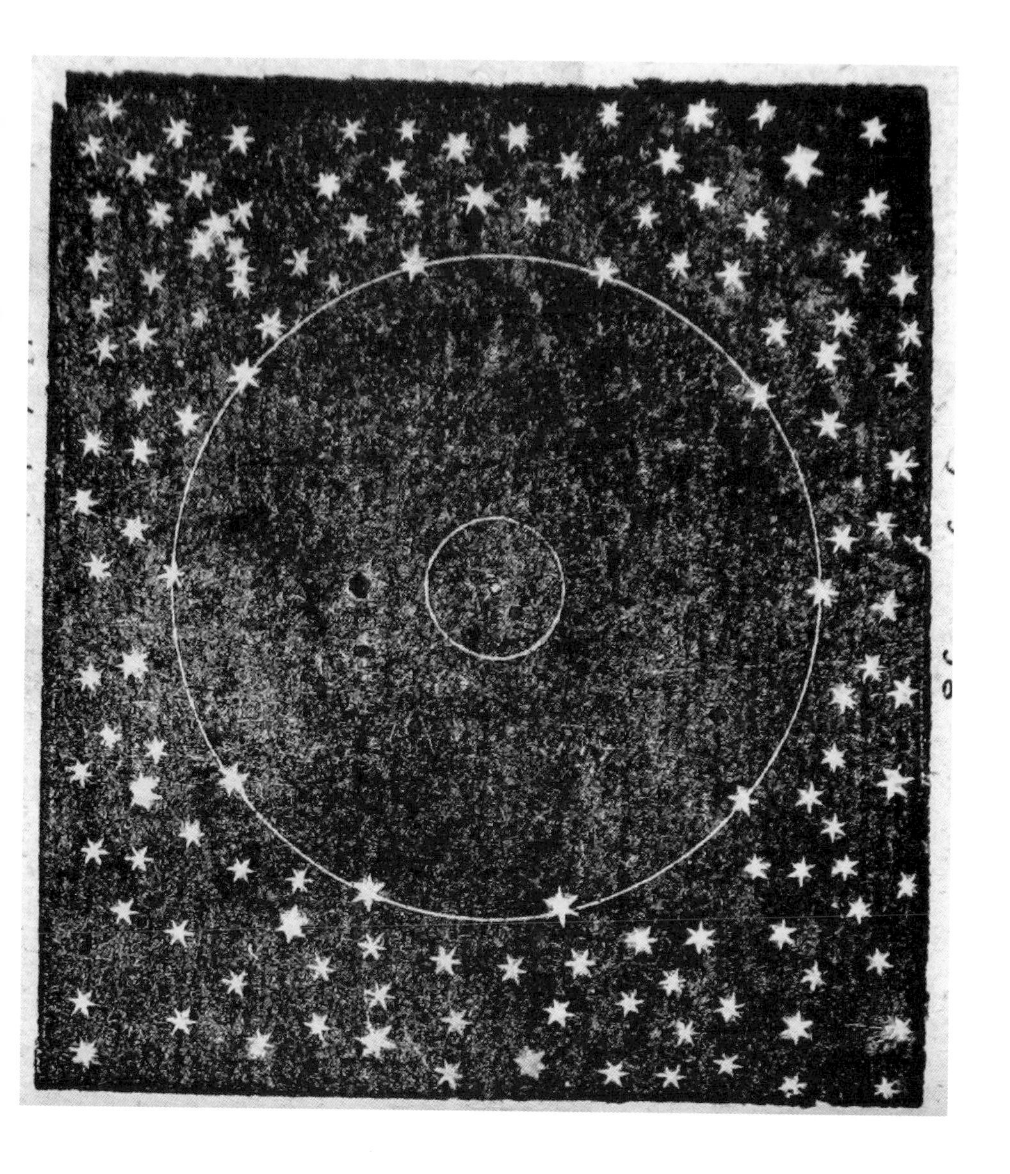

Kepler's figure from his *Epitome of Copernican Astronomy* showing the solar system surrounded by a Universe of other stars in a dark night sky.

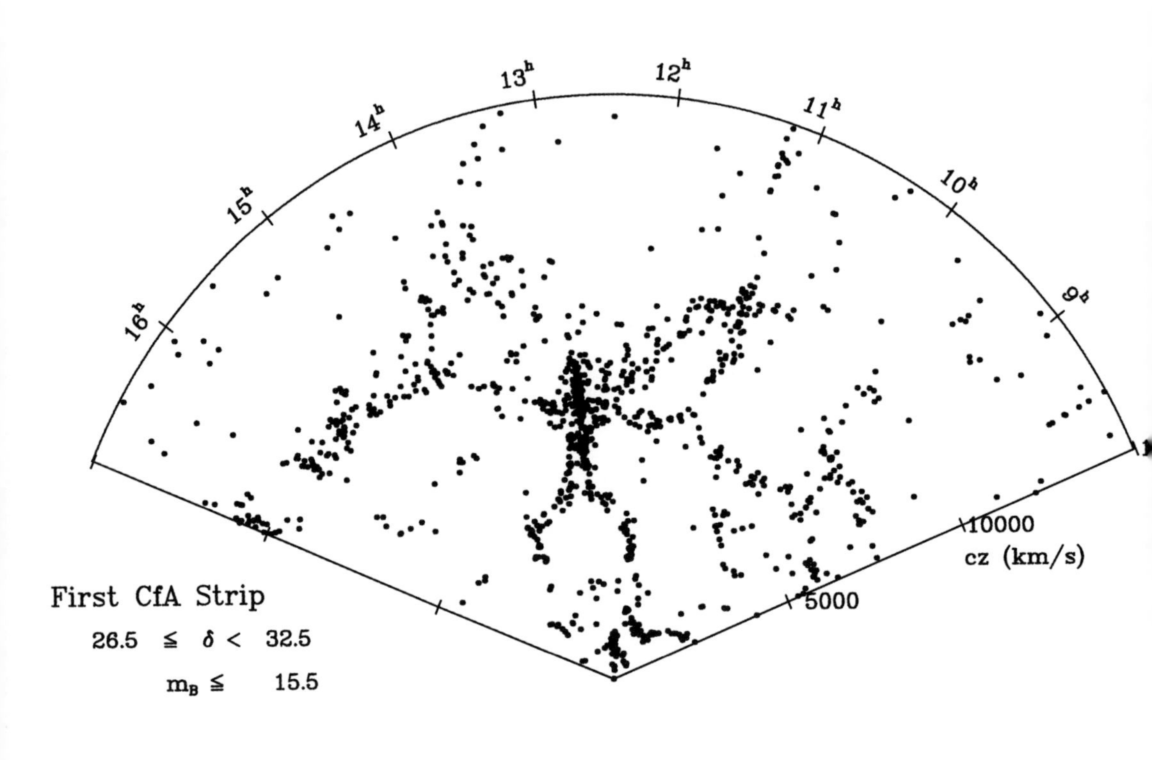

The Center for Astrophysics (CfA) survey of 1985, showing the clustering of galaxies in a fan-shaped volume of the Universe about us. Distance is marked by velocities using Hubble's Law to convert them to distances. The striking 'Stick Man' feature lies in the centre and suggested for the first time that luminous galaxies were distributed in a cobweb network of sheets and filaments.

THE TENTACLES OF TIME

THE CFA REDSHIFT SURVEY

In fact I consider the man who is dissatisfied with a universe containing ten thousand million million million stars rather grasping.

Arthur Eddington[108]

WHEN Edwin Hubble was making his pioneering measurements of galaxy redshifts during the 1920s, it could take most of the night to gather enough light from a distant source for its redshift to be determined. As the years passed, the number of measured redshifts grew rather slowly. Astronomers were not terribly excited by the prospect of a long programme of measuring hundreds, or thousands, of similar objects all over the sky in order to create maps and catalogues. It was much more exciting to search for new types of celestial object or the most distant quasar, peculiar colliding galaxies, or evidence for black holes and dark matter.

The slow business of measuring redshifts, and from them inferring the distances of galaxies, meant that the early great maps of galaxies kept track of positions on the sky only. There are three particularly famous compendia. In 1932, Harlow Shapley and Adelaide Ames completed a catalogue of galaxies brighter than 13th magnitude which contained about 1200 objects. In the 1960s, Fritz Zwicky and his colleagues identified over 30,000 galaxies from photographic sky-survey plates, all of them within a billion light years of the Earth. In 1967, Donald Shane and Carl Wirtanen[109] completed a catalogue of galaxies brighter than 17th magnitude. The Shane–Wirtanen catalogue contains about a million galaxies and its creation was a heroic display of dedication and patience. The plates were studied with the unaided eye (although the authors' children played an important role in searching the plates as well). It even appears that the number of the very faintest galaxies that they identified fell in a statistically predictable way as the researchers aged and their eyesight became weaker. These catalogues probed deeper and deeper into the Universe, cataloguing fainter galaxies, and mapping their positions on the sky. The Shane–Wirtanen catalogue was digitised in 1977 by Jim Peebles and Ray Soneira[110] to show the variation in galaxy density on the sky.

When Shane and Wirtanen started to present the early results of their cataloguing, a few statisticians became interested in understanding what the maps of the Universe were telling us. Most notably, Jerzy Neyman – one of the world's greatest statisticians – and his colleague Elizabeth Scott, at the University of California at Berkeley, took up the challenge and began to develop a range of sophisticated statistics to determine the degree of galaxy clustering on the sky.

The Peebles–Soneira map gave further huge stimulation to this effort because it contained information about more than a million galaxies. The sky was divided into small cells, which were coloured a different shade of grey according to the number of galaxies inside them on the digitised map. When the results were printed they revealed a shaded map that told us the pattern and density of visible galaxies over a large portion of the sky. The overall impression was rather intriguing. There seemed to be hints of filaments and chains of galaxies running through the map. Here one had to be rather careful. The human eye is very good at seeing lines where no true patterns exist. We tend to join up the dots between points and their nearest neighbours to create the impression of intrinsic patterns. Perhaps our earliest ancestors evolved in environments where there was real survival value in seeing tigers in the bushes? And to add to the dilemma of whether the patterns were intrinsic or manufactured by the eye, two of us[111] showed that tinkering slightly with the grey scale coding of the photograph – and even the dot size in the printing of the map – could produce big changes in our perception of the clustering.

Yet, whatever the right interpretation of these sky maps, the principal difficulty was their two-dimensional character. Two galaxies may be close to each other on the sky map, but be at very different distances away from us and not clustered together in real space. Statistical assumptions could be made in order to produce a likely three-dimensional map that we were seeing in projection but those assumptions are unverifiable. The real desideratum was a true three-dimensional map of galaxies in space.

In the 1970s, it first became possible to measure the redshifts of galaxies quickly and easily using automatic measuring techniques – eventually the whole process was turned over to robotic telescopes. Redshifts could be measured in a few minutes rather than hours. In 1977, Marc Davis, John Huchra, David Latham, and John Tonry at the Center for Astrophysics, Harvard University, began to create a map of a significant slice out of the volume of the northern celestial sphere by finding the distances to the galaxies in the appropriate part of Zwicky's catalogue of galaxies and their positions on the sky. The CfA Redshift Survey, as it was known, entered its second stage in 1985, with Huchra and Margaret Geller and their students beginning a huge programme of redshift measurement. The first map, which they presented with their student Valerie de Lapparent, of a slice of the Universe containing 1100 galaxies, had a dramatic effect on the world of astronomy. The pattern of galaxies in real space revealed something completely hidden in old maps of the two-dimensional sky.

We are located at the centre of the circular arc of the fan shape. As we move out to the edge of the fan we are at a distance of about 700 million light years (remember, speed of expansion away from us is proportional to distance according to Hubble's Law – this is why the 'distance' is labelled by speeds on their picture), and as we rotate around the fan we see different positions on the sky labelled by a coordinate that is the celestial version of longitude on Earth.

The appearance of this map stimulated enormous amounts of discussion and research because of its completely unexpected form. Instead of the galaxies being dotted about the Universe at random, they seemed to trace out great lines and walls around empty cells in space. In the centre was a distinctive reference point, looking like a cosmic scarecrow, that became known as the 'stick man'. Suddenly, all the debate about whether the apparent filaments in the Shane–Wirtanen maps were real or not was superseded by a recognition that there *were* patterns in the clustering of galaxies which suggested that a great cobweb pattern of galaxies existed in the Universe. All sorts of ideas were proposed to explain the patterns and, echoing the earlier response of statisticians to the orig-

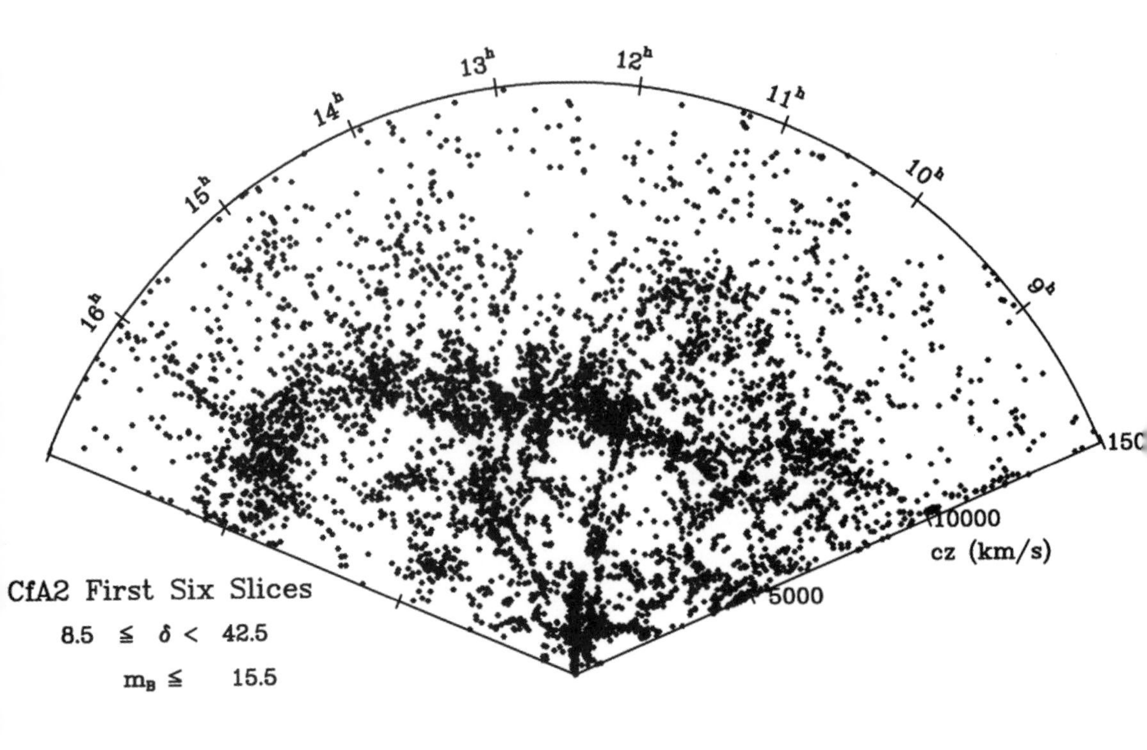

The 1989 CfA redshift survey contains many more galaxies than its predecessor and reveals a 'Great Wall' of galaxies running across the sky.

inal Shane–Wirtanen sky maps, there was huge interest amongst statistical astronomers in devising the best ways of quantifying and evaluating the significance of the apparent three-dimensional patterns.[112]

The other major effect of seeing this map was to focus attention on the great voids between galaxies. Previously, all attention had been on the pattern of galaxy clustering and the explanations for it, but now it was clear that the huge dark regions devoid of visible galaxies needed explanation as well. Were they really empty? Or were they simply filled by very faint galaxies, too dim to be seen in the survey?

In the years that followed, the CfA team measured more and more redshifts in different contiguous slices of the sky. By superimposing them, you got a more detailed mapping of a larger sky volume. By 1989, the map had grown in a fascinating way.[113] A huge band of galaxies was now running across the map and the stick man was fading out. The voids were still there though, the more impressive for the sheer number of galaxies that now surrounded them. The huge swathe of galaxies became known as the 'Great Wall' and was the largest single structure ever seen in the Universe. Its three dimensions are 600 x 250 x 30 million light years. It contains all types of galaxies and clusters of galaxies. Again, its visual impact was significant, spurring astronomers to consider all sorts of ways in which to explain such an unusual set of patterns in the sky.

This is far from the end of the quest to map the Universe. Ever-larger automatic surveys have been developed. Some look very deeply into the Universe to see faint galaxies down a very narrow 'pencil' beam, while others look at a very large expanse of sky but record the brighter galaxies which can be seen out to greater distances. The pattern of galaxy clustering is tangled and non-random. It is telling us about the physical processes that led to their formation. It may be that material tended to collapse under the pull of its own gravity into great pancake-like sheets. As these ran into each other, the effect would be like the formation of a foam of lopsided bubbles. At the corner points of contact between several bubbles, the density of material would be especially high and the brightest, richest clusters of galaxies would form. However, the whole process has been complicated by a growing recognition of the importance of the darkness between the shining galaxies. The motions of the visible galaxies reveal that they are responding to the forces of gravity exerted by a huge quantity of dark unseen material. There is about ten times more dark matter than luminous galaxies inhabiting the Universe. The maps don't show it, but the dark materials dictate the form of the maps in all sorts of crucial ways.

Astronomers now construct huge computer programmes that mimic the clustering process that draws galaxies together into clusters in the presence of different possible candidates of dark matter with different distributions. The results are a sort of identity parade of possible universes to be compared with the actual one shown in the sky maps. They not only tell us what the universal landscape looks like, but also set the agenda for the types of predictions that theoretical astronomers will try to make in the future, and the types of artifical universe that their computers will be programmed to create.

Original Hubble Deep Field of 15 January 1996.

THE FINAL FRONTIER

THE HUBBLE DEEP FIELD

The universe may be like Los Angeles.
It's one-third substance and
two-thirds energy.

Robert Kirshner[114]

THE Hubble Space Telescope (HST) has revolutionised public perceptions of astronomy and the Universe. Since it first went into orbit on 25 April 1990, it has taken over 700,000 photographs of the Universe. Its spectacular copyright-free images are recognisable all over the world and are much used by the press, on CD covers, and even for postage stamps. Time and again they have brought astronomy on to the front pages of the world's press outlets.

The most famous of these images was in some ways the most unexpected. In 1995, Robert Williams, the Director of the Space Telescope Science Institute in Baltimore, accepted a recommendation that he use the special allocation of personal observing time that the Director is given to use as he thinks fit on a single project. The idea was to aim the telescope at a single dark spot in the northern sky, where there were no bright objects nearby to dominate the image, and gather light for ten consecutive days between 18 and 28 December 1995. During this time, the HST would make 150 consecutive orbits[115] of the Earth and take 342 separate exposure frames, of which 276 were processed to create the picture of what became known as the 'Hubble Deep Field'. The result was the deepest photograph ever taken of the Universe.

The results were as spectacular as they were unexpected. Several thousand galaxies were found that had never been seen before and the whole image was covered in galaxies of different shapes, colours, and brightnesses, revealing them at vastly different distances and at different stages in their evolutionary history. Most of the galaxies seen are between 2.5 and 10.5 billion light years away. Many of the galaxies look peculiar and are likely to have suffered close encounters or collisions with other galaxies.

Yet, despite the plethora of galaxies on view, the Deep Field is a tiny snapshot of the Universe, like looking at it through a keyhole. It covers a portion of the sky about thirty times smaller in diameter than the full moon. The part of the entire Universe that can be seen by us, in the sense that light has had time to reach us since the Universe began expanding 13.7 billion years ago, contains about 100 billion galaxies, each containing about the same number of stars. So despite the wealth of celestial activity captured in the Hubble Deep Field, it is but a drop in the ocean of stars and galaxies. Even larger surveys, like the Sloan Digital Sky Survey (SDSS), have revealed large-scale structures across the Universe that are vastly greater that the region scanned by the Hubble Deep Field.

On 14 December 1998 the Hubble Telescope released another Deep Field image,[116] this time of the southern sky. It was created over ten days in the same way and provides a complementary deep photograph of the Universe in the opposite direction, near the south celestial pole, revealing thousands of new galaxies, but confirming the general pattern of diversity found in the original Deep Field image.[117] In the future we will undoubtedly improve upon this classic Hubble image, but I suspect we will never again be so surprised by a deep optical image of the Universe in the large.

It gives us the most comprehensive glimpse of a small part of the Universe around us, simultaneously revealing galaxies near and far, at different stages of their evolution, with different colours and shapes, a fabulous cosmic vista that has come to symbolise the spectacular capability of the Space Telescope. Its pictures have done far more than simply inform astronomical researchers: they have created a new public perception of the Universe.

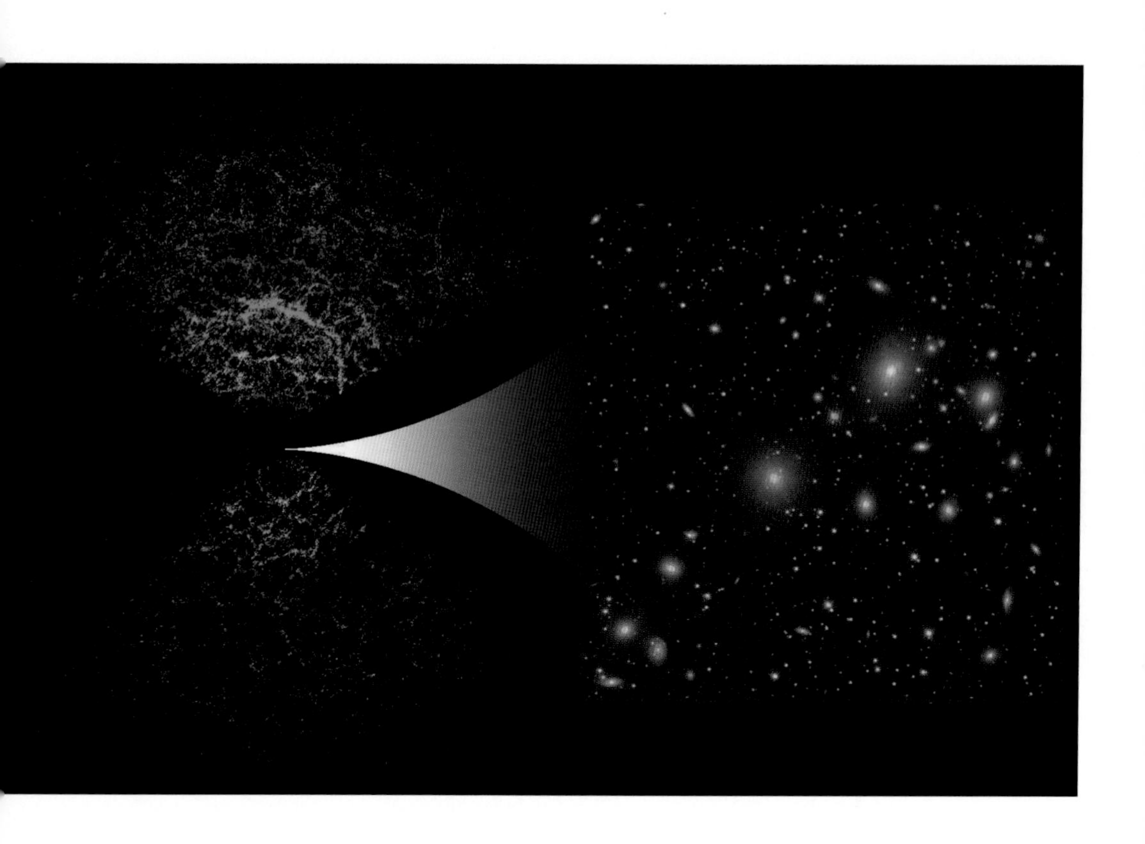

The Hubble Deep Field shown as a tiny part of the current Sloan Digital Sky Survey (SDSS) of the Universe. The Sloan Survey has mapped more than one quarter of the sky, creating a three-dimensional map that contains more than 200 million stars, galaxies and quasars. It is the biggest astronomical survey ever undertaken.

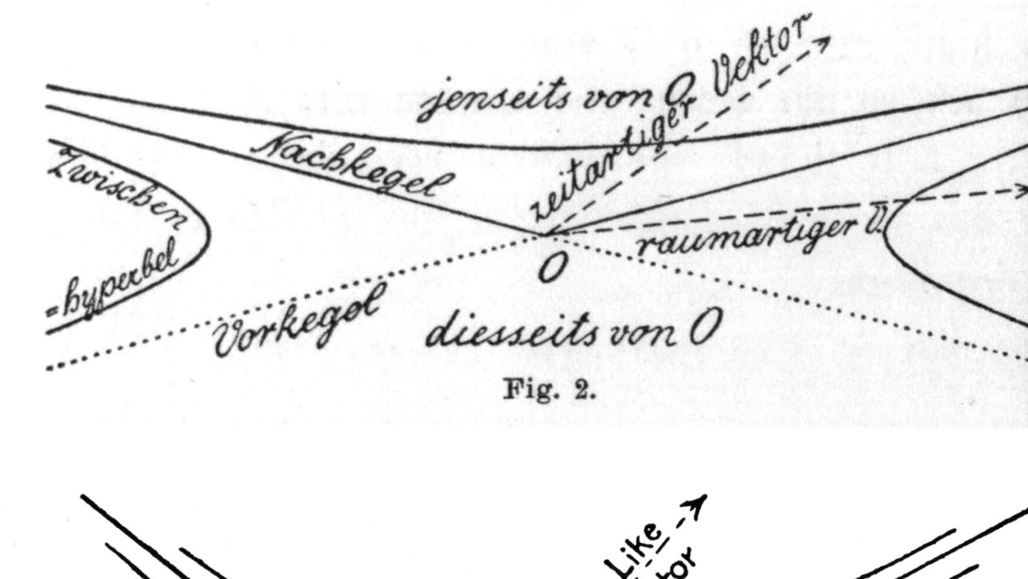

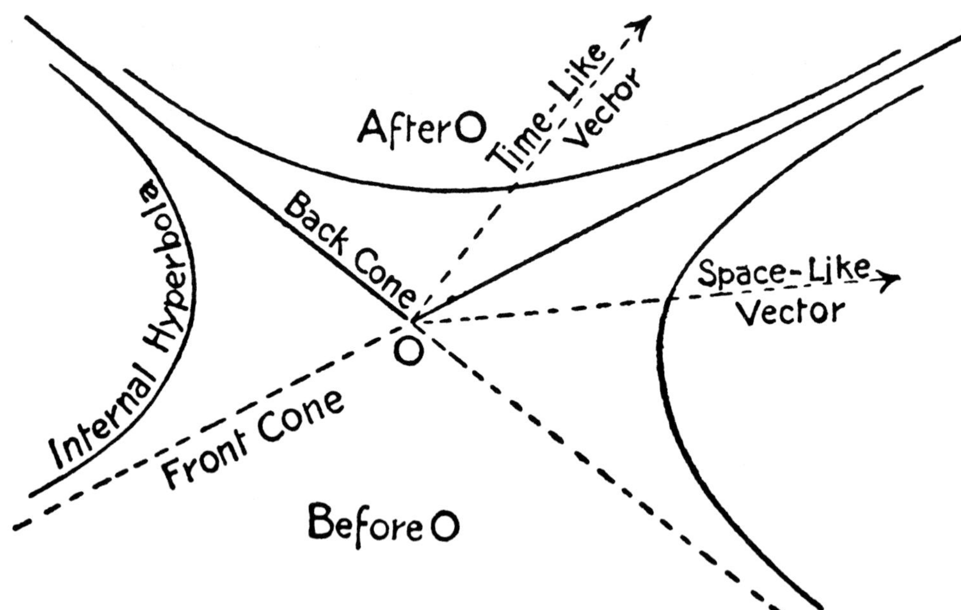

Hermann Minkowski's first space-time diagram from his 1908 book *Raum und Zeit* (*Space and Time*), with the translated version below. Time and Space are in the directions indicated by the dashed arrows. The observer is located at O in space and time.

THE TIMES THEY ARE A-CHANGIN'

SPACE–TIME DIAGRAMS

I can embrace the whole all at once like a painting or a statue. In my imagination, I do not hear the work as it unfolds, as it must play out, but I hold it all as one block so to speak . . . and when I succeed in thus 'super-hearing' the entire assembly, it is the best moment.

Wolfgang Amadeus Mozart

IN the old Newtonian way of looking at the world, space was an arena in which events were placed, and time was an unbendable arrow of progress distinguishing the present from the future and the past. Space was a cosmic stage on which happenings played out, and which nothing could shake. Time was linear and straightforward. Nothing could alter its rate of flow. No events that occurred in the Universe could affect the nature of space and the flow of time. Time and space were not linked; they were separate and different. You could move in any direction in space but only forwards (we generally supposed) in time.[118]

In 1905, Einstein completed work begun by Hendrik Lorentz and Henri Poincaré to establish the special theory of relativity. Its central premise was that the speed of light in a vacuum is a universal constant of Nature. It will be measured to have the same value by all observers *no matter how they and the light source are moving*. So, if someone shines a torch at you from an approaching rocket you will measure the light to be coming towards you at the same speed that you would if the rocket was travelling away from you when the light was emitted.[119] Very strange, but true.

Once there is an invariant quantity in Nature such as the speed of light with the units of distance per unit time then there must be an intimate connection between space and time. Einstein forged these links and showed also that the speed of light in vacuum should be a cosmic speed limit – no signal can travel faster. In 1907, Hermann Minkowski,[120] who had been one of Einstein's teachers at Zurich Polytechnic in 1896, made a famous claim that:

> From this hour on, space as such and time as such shall recede to the shadows and only a kind of union of the two shall retain significance.

In his written account of these developments he explains a little more:

> The views of space and time which I wish to lay before you have sprung from the soil of experimental physics, and therein lies their strength. They are radical. Henceforth space by itself, and time by itself, are doomed to fade away into mere shadows, and only a kind of union of the two will preserve an independent reality.[121]

The new limits on communication created by the speed of light gave rise to a profound series of space-time diagrams that delineate the structure of the universe of space and time that Minkowski melded into a single block of space-time. In order to make such a picture manageable we usually suppress one or two of the three dimensions of space and plot them against one dimension of time on a graph, or 'Minkowski diagram'. In our pictures here, we are showing one dimension of space against one of time in Minkowski's original (page 120), and then a perspective drawing showing three dimensions of space projected onto two, both with time going vertically in the picture at right angles to the directions of space (opposite).

If we sit at the point marked 'Observer' then the world of space and time is divided into four parts after we have drawn two cones. These cones are the paths of light rays. They come in to the Observer from the past in many possible directions and they go out from the Observer in many directions into the future. These two cones are called our *past* and *future light cones*, respectively. The

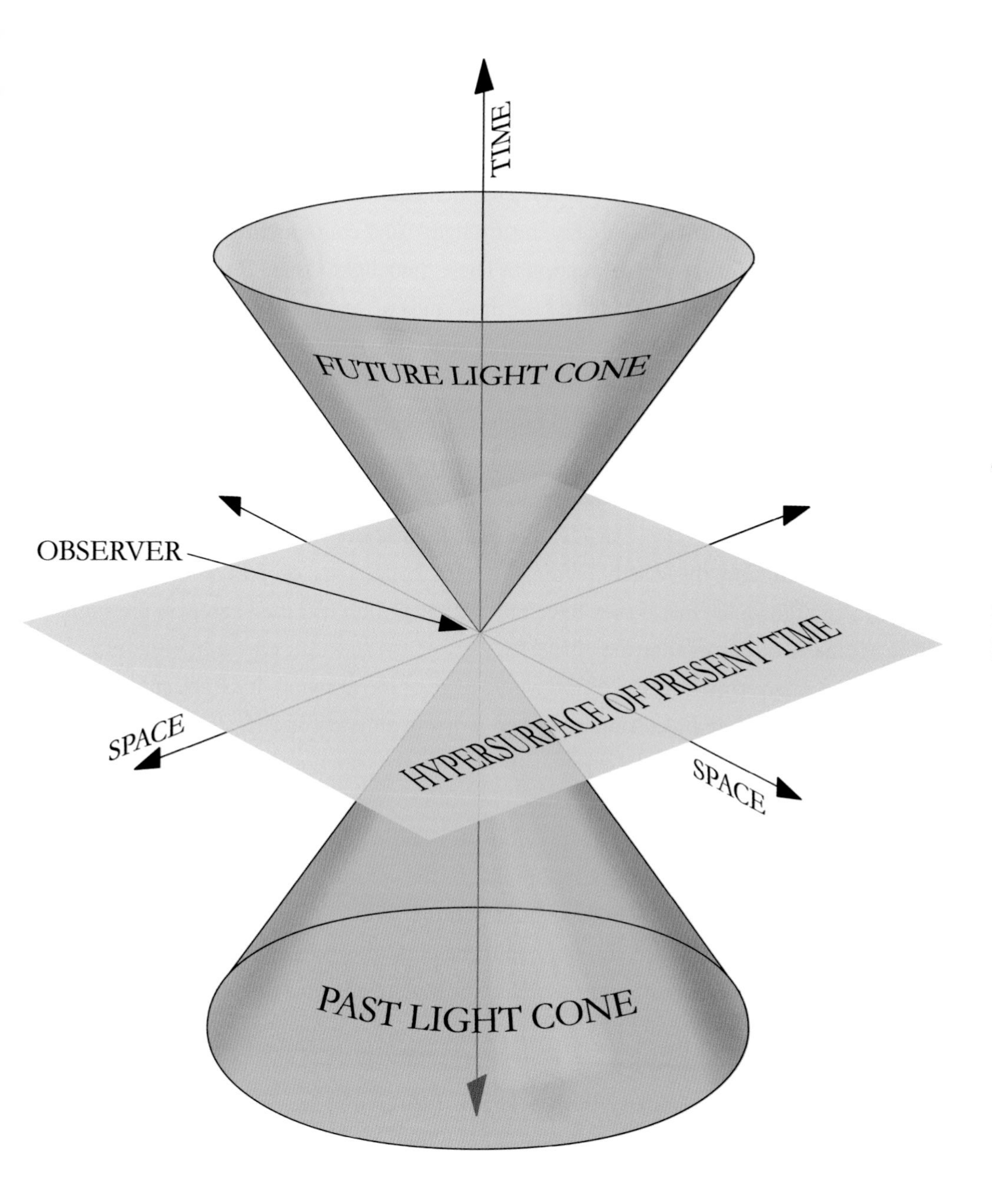

Space-time diagram for an Observer showing the present, and the conical regions to the future and to the past to which, and from which, signals travelling at the speed of light, or slower, can go or be received. These two regions are called the Future and Past Light Cones, respectively.

unbounded regions of space and time that lie outside the two light cones are inaccessible to us. We could only 'see' them, receive signals from them, or influence them if instantaneous signalling occurred at infinite speed.[122]

This picture reveals our relationship to the Universe. The past light cone contains all those events that have happened and which I can no longer change or influence. It is passive. The future light cone contains the region that I can actively influence by my actions. These two light cones do not border one another as people had believed for thousands of years. For you, they touch only at the point marking your location in space-time.

This simple picture hides many strange paradoxes. It shows that simultaneity is in the eye of the beholder. If I see two events simultaneously then someone moving relative to me will see them happen at different times. If two people move along different paths through space-time which intersect, then at that meeting we can consider the past light cones of each of them. They are different and they each define different collections of past events as being simultaneous. Only if they have no relative motion will they agree on the simultaneity of events that they observe.

Space-time consists of the whole page of the drawing. It can be split up into space and time by cutting it up into strips in all sorts of different ways. We can draw lots of horizontal parallel slices and label them as lines of constant time. Or, we could slice up the picture in a different way by drawing a collection of parallel lines inclined to the previous set. This is the essence of Einstein's theory of relativity. Each slicing corresponds to the splitting of space and time seen by a different observer whose relative motion with respect to other observers is what distinguishes the different splittings that they see. As the relative speed of the observers increases so the inclination increases and the effects of relativity become more pronounced.

This relativity in the splitting of space and time suggests that space and time are each secondary things. The real thing is space-time – the whole undivided picture. And that suggests something very unusual indeed. It makes us think that the true reality is the whole block[123] of space-time and every part of it, even that part that lies in our future, already exists.[124] The future – our future – is laid out before us like a fixed road map. It is not created by us as we go along our way. We do not really appear to have the sort of free will that we feel we have.[125] But maybe there are constraints on what space-time diagrams can be drawn. If you try to move into the future and enter a part of the Universe that has already 'happened' then it would be rather like undergoing time-travel into the past.

The study of space-time has also become greatly influenced by another version of Minkowski's space-time diagram, introduced by the British mathematical physicists Brandon Carter and Roger Penrose. They are known variously as Carter–Penrose diagrams, Carter diagrams, Penrose diagrams, or conformal diagrams. They play a central role in research into the structure of black holes, universes, and complicated space-times. They allow the consequences of cause and effect and the motions of light rays and heavy particles to be seen very easily from a picture rather than calculated mathematically. They are classic examples of how physicists use pictures to understand the answers to questions which are too complicated to work out in exact mathematical detail without major human or computer-aided effort. For simplicity, reduce the number of dimensions of space in the diagram to one (this is like assuming that the other two dimensions of space just behave in the same way) along with one of time. The Minkowski diagram looks a little simpler and we can highlight some of the distant regions that lie infinitely far away.

There are five different infinities. Future lightlike infinity is where all light rays go to the infinite future and past lightlike infinity is where they have all come from. Future timelike infinity is where all particles which move at speeds less than that of light go to and past timelike infinity is where they have come from. Spacelike infinity is inaccessible to light or particles moving slower than the speed of light. The Carter–Penrose trick is to apply a mathematical distortion to this diagram that has a special property: it leaves the paths of light rays (straight lines at 45 degrees) unchanged, but it brings all the five infinities in to finite distances so the picture is a convenient finite-sized diamond. The timelike and spacelike infinities just become the four points at the corners of the diamond while the past and future lightlike infinities are formed by the four sides of the diamond, which lie at 45 degrees to the horizontal. Nothing exists outside the boundary of the diamond. All light rays move along 45-degree lines and all particles with mass move along less-tilted lines that end up converging on future timelike infinity. The picture shown on the following page is the first Penrose diagram drawn by Roger Penrose.[126] This is the simplest Carter–Penrose diagram but others are much more complicated once black holes or other objects are introduced into the space-time. Yet, whatever the situation, they have become the lingua franca of discussion of space-time structure and causality in physics and astronomy over the past forty years.

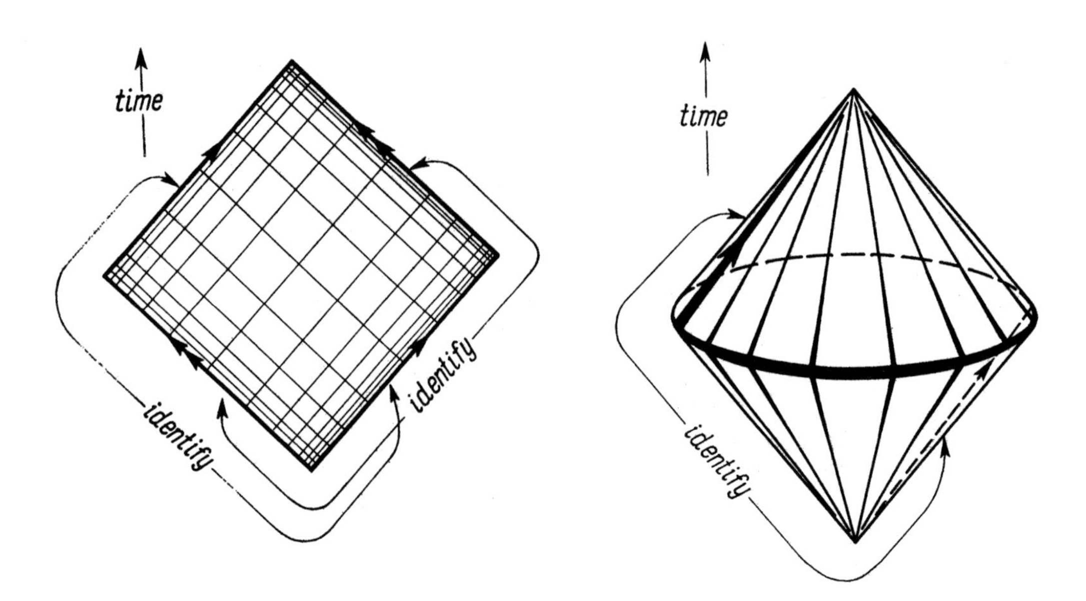

First conformal space-time diagram drawn by Roger Penrose for (*left*) a space with one dimension of space running horizontally where the time axis is vertical and light rays follow the diagonal paths at 45 degrees to the vertical. Infinity has been brought in to make the diagonal sides of the diamond. The same construction is also shown (*right*) for a universe with two dimensions of space shown in perspective.

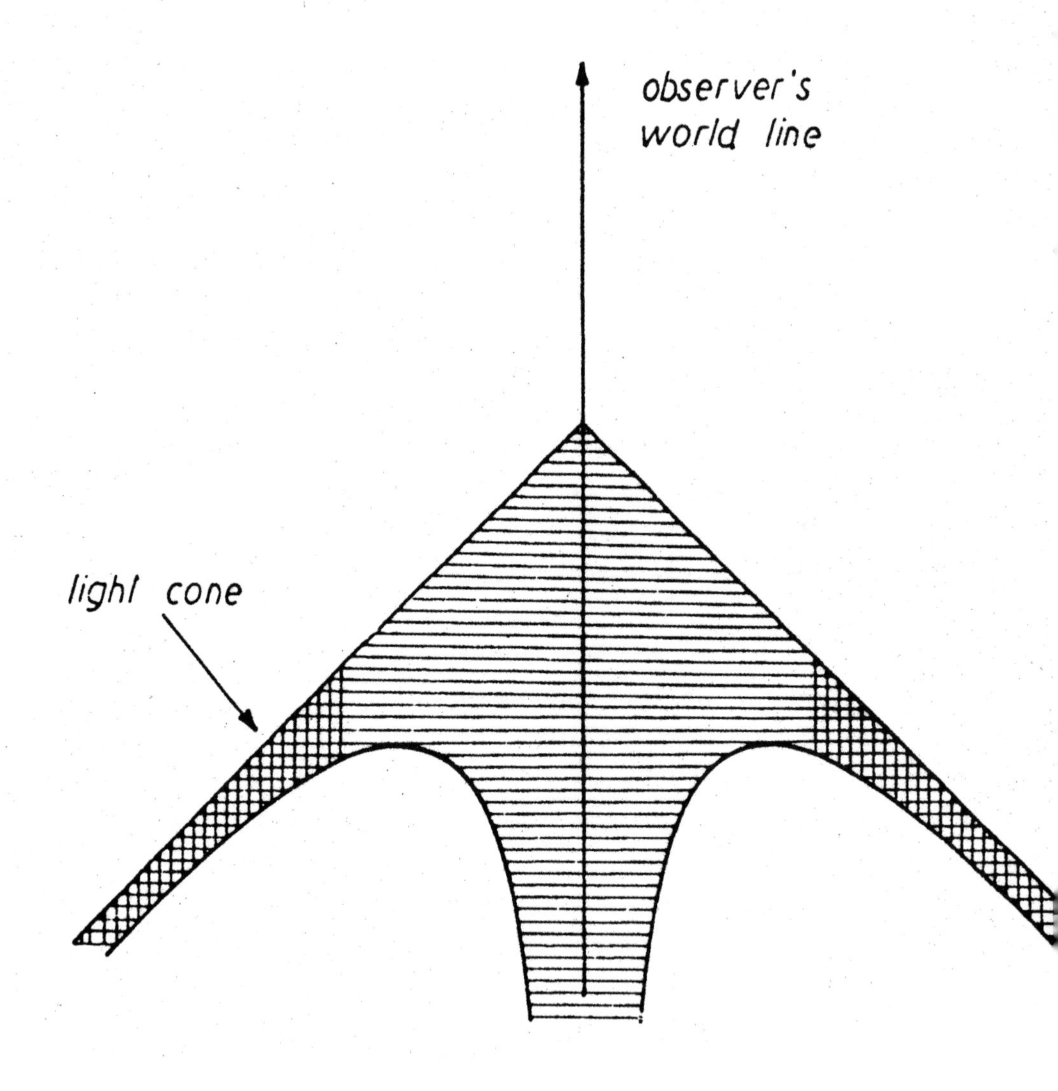

Fred Hoyle's 1960 picture of the past light cone of an observer today showing (shaded) the regions just inside its boundary which are accessible to us by geology and astronomy, as well as the light cone boundary along which travels all the light and radiation that astronomers observe.[127]

ALL OUR YESTERDAYS
OUR PAST LIGHT CONE

I think we agree, the past is over.
George Bush[128]

THE creation of space-time diagrams with which to map the possible histories and futures of light rays and particles entered a new phase when astronomers started to think about the effects of gravity. If we look out into the Universe we gather information in two ways. We collect photons of visible light, or radiation of other wavelengths such as x-rays, using our telescopes and receivers. And we also detect massive objects such as meteorites, comets, and cosmic rays[129] which move at a speed less than that of light because of their masses. If we look at this incoming information on the space-time diagram we see that astronomy helps us determine the structure of the surface of the light cone stretching back into the past. This tracks the histories of the light rays and radiation we detect today back to their point of origin, or until something obscures our view. The fast-moving cosmic rays, terrestrial fossils, and other debris from space allow us to look back about 4.6 billion years into the region just inside the edge of our past light cone. This gives us direct access to the shaded region in the picture that Fred Hoyle was the first to draw in 1960. Thirteen years later, George Ellis elaborated the picture in much greater detail, showing the different epochs in cosmic history and the observationally accessible regions of space and time.

If we follow our sources of incoming radiation backwards in time we run into some barriers. The first occurs when the Universe was about a thousand times more compact than today and only 372,000 years old. Its temperature was over 3000 degrees Kelvin and far too high for any atoms or molecules to exist. It was a sea of photons, electrons, and light nuclei interacting together in a vast ionised plasma, with huge numbers of ghostly neutrinos and gravitons passing through unperturbed except for the pull of gravity. If our telescopes could look back to that distance which corresponds to when the temperature of the radiation was 3000 degrees, they would be able to see no further. It would be like looking into frosted glass. The photons would be scattered around by the free electrons, whereas at later times and lower temperatures the electrons will be caught up in atoms and the photons will pass freely by. It is like looking at the Sun. What we see as a fairly sharp edge of the Sun is just the surface where the scattering of photons produced by nuclear reactions deep within the Sun ends. The density has fallen enough for them to fly unhindered towards us. When we 'see' the Sun we see this surface where light scattering last occurred – the Sun's 'photosphere'. It stops us seeing into the interior of the Sun by using ordinary light. The same processes stop us seeing directly the photons of light from the early stages of the Universe.

But photons of light are not the only things we can detect from the Sun or from the early stages of the Universe. For many years astronomers have been able to detect neutrinos from deep within the heart of the Sun. Neutrinos interact by the weak force of radioactivity and do not respond to the electromagnetic force because they possess no electric charge themselves. They stream out from the centre of the Sun and can be captured occasionally when they hit our specially designed detectors deep below the Earth's surface after other, more strongly interacting, particles have been absorbed by the mass of the Earth's crust they have failed to penetrate.

We might one day be able to look directly into the early Universe using neutrinos. Unfortunately, although it is most likely that the bulk of the matter in the Universe consists of as yet undetected types of neutrino that are much heavier than those we see in detectors on Earth, we are a long way from building a neutrino telescope. The energies of the neutrinos from the early Universe are too low to produce any significant interaction with our detectors. But if we could see them we would be able to reconstruct our past light cone all the way back to the time when the Universe becomes opaque to neutrinos as well as photons. This occurs when the Universe is just one second old and 10 billion times hotter and more concentrated in size than it is today.

If we want to look directly into the first second of time then we have to be

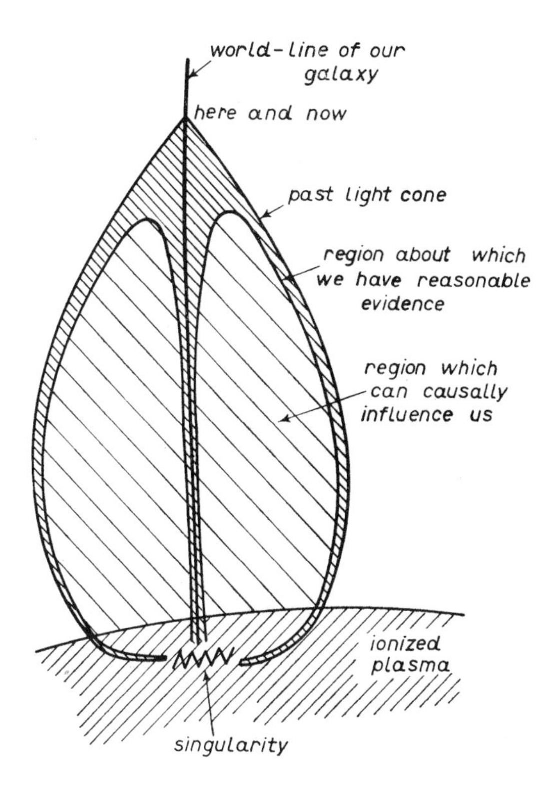

A more detailed picture of our past light cone drawn by George Ellis showing the effects of gravity which cause it to focus into the past towards an initial Big Bang singularity (unless other forces intervene). We cannot see directly into the ionised plasma era with light-gathering telescopes.

more rarefied still in our choice of probes. We can 'see' before one second of cosmic history using gravitons, packets of gravitational-wave energy that behave partly like particles and partly like waves, and that passed unhindered through the early universe after they were produced just 10^{-43} seconds after the apparent beginning, when the Universe was 10^{32} times hotter and more compressed than today.

As we ponder these barriers to our attempts to reconstruct our past we begin to appreciate how little of space and time is accessible to us. The astronomical Universe that we can study by the methods of observational science is only a small part of the whole. In order to make assertions about what the universe is like in those inaccessible regions we have to take some steps of faith and make assumptions that cannot be tested. We might assume that, today, the

Universe elsewhere is on the average pretty much the same as it is locally, where we can see it. For a long time this assumption was characterised as a modern edition of Copernicus' stricture that we should not place ourselves in a special position in the Universe. Alas, this may not prove to be a correct extension of Copernicus' picture of our Sun-centred solar system to the Universe as a whole. In recent years we have come to appreciate that the 'eternal' versions of the inflationary-universe cosmology lead us to expect that, in the large, the Universe will be very complicated in structure and very different from region to region. We do not expect the Universe to be similar from place to place if we go far enough beyond the current reach of our telescopes. Moreover, we have also begun to understand that the region of the Universe we inhabit needs to possess many unusual features in order to meet the requirements for life (perhaps just ourselves, but maybe others too) to have evolved within it.

The pear-shaped past region of space and time that can influence us witnesses to another remarkable feature of our Universe. The way in which the incoming light rays deviate from straight lines, turn around and start converging together as they are traced back into the past, is a result of gravity. Einstein taught us that gravity acts on everything – even light – and it is the gravitational attraction of all the material inside our past light cone that causes it to converge. This property proved to be a crucial ingredient in the famous deductions by Roger Penrose and Stephen Hawking[130] that showed what properties a universe would have to possess in order to have had a beginning to time. If gravity is attractive and there is enough material in the Universe to create this pear-shaped convergence of our past light cone (and there is), then history cannot be extended indefinitely into the past. When these mathematical theorems were first proved it was generally believed that all forms of matter in the Universe would experience gravitational attraction and so they provided grounds for believing that the Universe did have a beginning in time.[131] Subsequently, new forms of matter were predicted to exist which need not exert gravitational attraction, and the discovery that the Universe's expansion is accelerating shows that such a form of gravitationally repulsive matter is the dominant constituent of the Universe today. This means that we can no longer use these mathematical theorems to deduce that there was a beginning to time. This doesn't mean that time didn't have a beginning, only that it didn't have to, and we can't prove that it did.

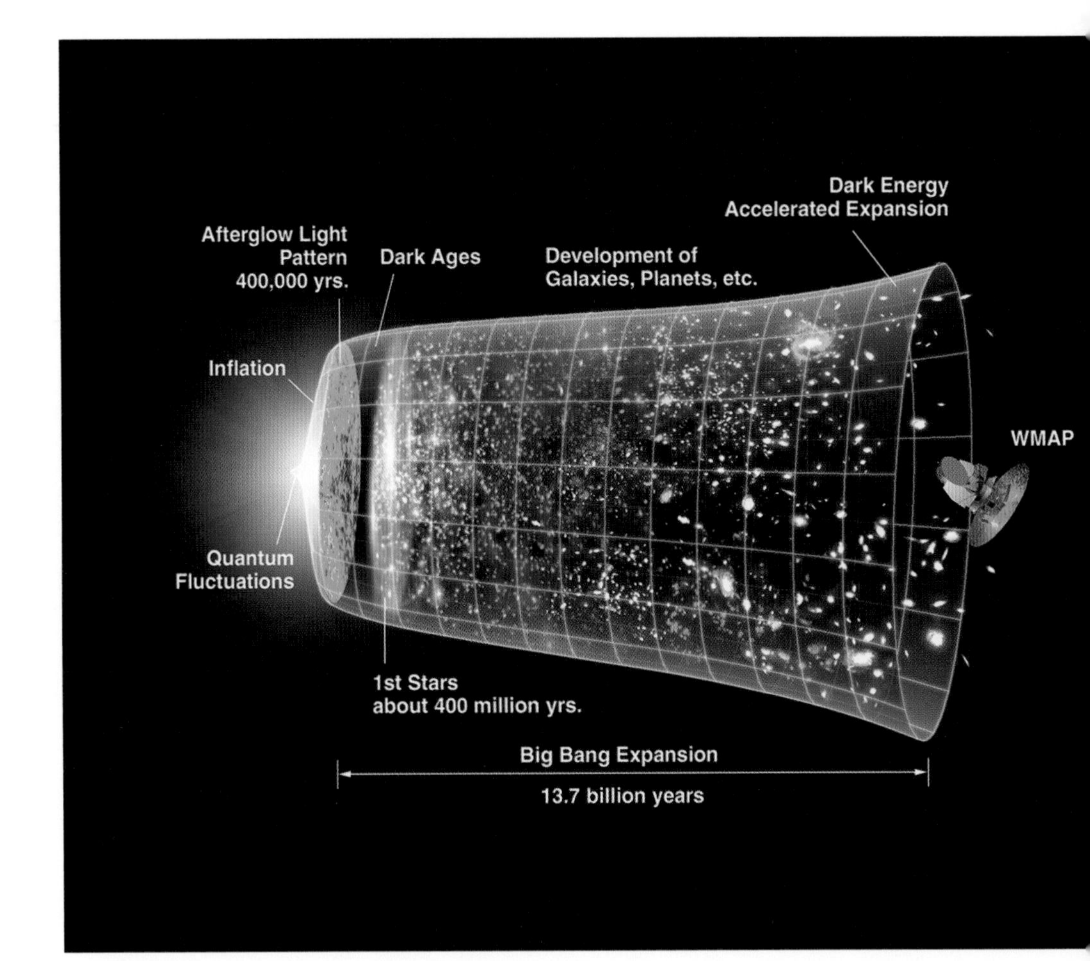

A modern picture of the history of the Universe from an apparent beginning, through a period of 'inflationary' expansion into periods dominated by radiation plasma and then by forms of dark matter within which the galaxies and stars formed. Today we live in the most recent era, which began about 4.5 billion years ago, in which the Universe's expansion is accelerating because of the presence of a mysterious dark form of energy that makes up 70 per cent of the Universe.

GRAVITY'S RAINBOW

THE SPECTRUM OF THE INFLATIONARY UNIVERSE

If you can look into the seeds of time,
And say which grain will grow and
which will not
William Shakespeare[132]

IN 1981, a new theory of the Universe hit the headlines. It has formed the central paradigm around which our understanding of the large-scale structure of the Universe is organised ever since. It was dubbed the 'inflationary Universe' by its creator in tune with the economic problems of the times, but it was an appropriate name in many other ways. It created an explosion of interest amongst cosmologists, generating huge numbers of scientific papers, conferences, and popular expositions. It changed the profiles of many astronomy departments around the world and made new calls on the budgets of funding agencies and universities. All this because of one simple idea, originally introduced by Alan Guth, a particle physicist then working at Stanford University.

The idea was simple. The accepted picture of the expanding Universe is modified in one respect. There is a very brief interlude in the very early history of the expansion when it accelerates. In the usual picture the expansion is always decelerating, dragged back by the gravitational attraction of all the material in the Universe. Acceleration requires a new form of matter to exist that exerts a repulsive force on other matter of the same sort.[133] Fortunately, new theories of elementary-particle physics emerging at that time suggested that forms of matter of this repulsive type may be very common in ultra-high-energy environments such as those found in very early stages of the Universe.

If this type of anti-gravitating matter exists, then it soon comes to dominate over conventional sorts of matter and radiation, and creates a surge in the universal expansion. If, as expected, it is short-lived, and quickly decays away into radiation and other particles as the temperature falls, then the expansion will resume its usual decelerated expansion after any short bout of inflation is over.

The consequences of an inflationary interlude of expansion are immense. The Universe grows much bigger, and does so more quickly than it otherwise would have; the expansion is driven very close to the tantalising divide that separates ever-expanding universes from those that eventually collapse back upon themselves towards a Big Crunch; and the Universe we see becomes very smooth and approximately similar in every direction in space. These were all mysterious unexplained features of the visible Universe before the idea of inflation emerged on the scene. The short period of inflation provides a natural explanation for them all.

The most interesting consequence of inflation is that the period of accelerated expansion enables the whole of the visible part of the Universe today (more than 14 billion light years across) to have expanded from a small primordial fluctuation of mass and energy; a fluctuation small enough to be kept smooth by light rays moving from one side to another.[134] This coordinated primordial region will have predictable statistical variations which will end up being inflated in scale to become the large-scale variations in temperature and density that we observe in the Universe today.

Thus, inflation provided an explanation for the existence of galaxies, and predicted that a very special signature should be found in the variation in the temperature of the cosmic background radiation in the Universe if inflation really happened. Cosmologists have been looking keenly with an array of instruments for this 'smoking gun' from inflation and the evidence is building impressively to confirm it. The COBE and WMAP satellites flown by NASA have searched for the distinctive pattern of variations expected in the temperature of the incoming cosmic background radiation as we compare its intensity between different directions of the sky. Satellites do better than Earth-based instruments because they do not have to look through the changing atmosphere

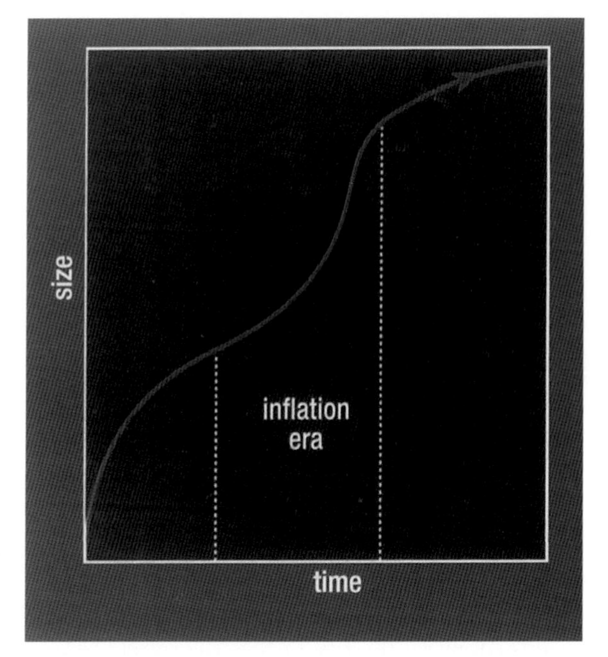

Inflation means that the expansion of the Universe accelerates for a short period in its very early history. This means that the Universe has expanded by a larger amount at any later time than it would have done in the absence of inflation.

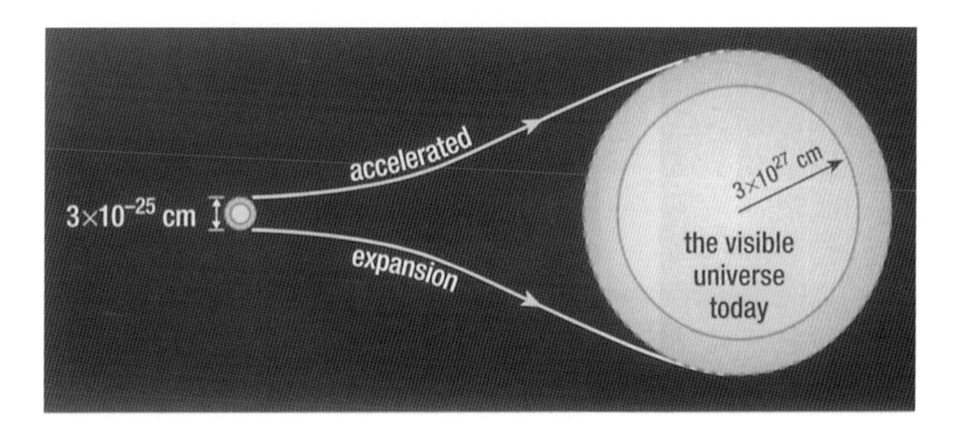

Our visible Universe today is the part of the entire Universe from which light has had time to travel since the expansion began 13.7 billion years ago. The accelerated expansion provided by inflation enables the whole of our visible Universe today to be the expanded image of a primordial region that is small enough to be smoothed by radiation transfer from hot regions to cooler ones.[135]

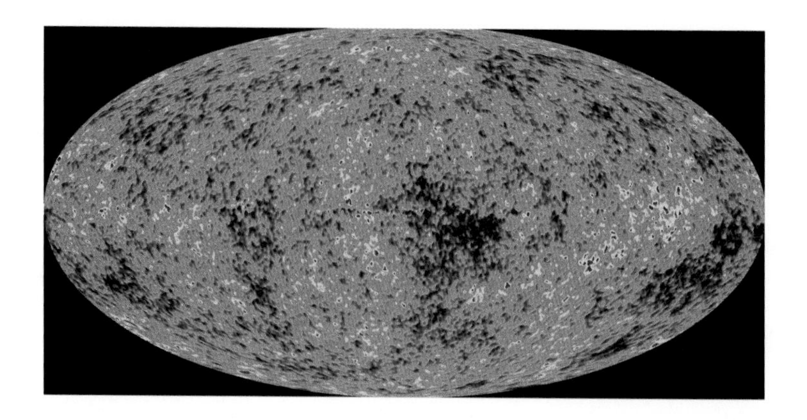

A microwave temperature map of the sky created by the WMAP team. The red regions are 200 micro Kelvins hotter than the average; the darkest blue regions are 200 micro Kelvins cooler than the average temperature across the sky.

and they can scan the entire sky, building up a huge number of temperature comparisons between different directions on the sky so that the purely chance variations that arise in any data set are rendered comparatively insignificant.

These observations and the theoretical predictions they are testing have been represented on an important picture which plots the strength of the temperature differences against the angular separation on the sky over which they are measured. The predictions of the standard inflationary universe model on small angular scales are shown with the solid line. There are distinctive features – oscillations, for example, that decay away like the ringing of a bell. As we go to the right we are probing smaller and smaller scales and eventually all the fluctuations will be ironed out by physical processes that transfer energy from hotter regions to cooler ones. If we extended the solid line to the left then it would be horizontal and in very good agreement with the observations taken by the old COBE satellite, which were confined to separations on the sky exceeding 10 degrees. The data points which accurately trace out the solid curve come primarily from the WMAP satellite mission.[136] We notice that there is a very close agreement with the predicted pattern over the first few bumps but then the observational uncertainties become large as we reach the limits of the instrument sensitivity. There is a peculiar 'dip' in the signal as it straightens out on large angular scales on the left which has been a subject of considerable argument and interpretation by astronomers. It may be statistically less significant than appears or might be caused by the suppression of very large fluctuations close to the size of the Universe because they have to 'fit' into the space.

In 2008 a further satellite mission, called Planck, will be flown by the European Space Agency to improve the accuracy of the observations on small angles even further. In the meantime astronomers observing from the ground are taking advantage of the continual march of technological progress in electronics to build ever more sensitive detectors in the race to construct the full and detailed radiation signal from the first moments of the Universe's history. Did inflation really happen? This could be the picture that decides and we will see a lot more of it in coming years as the observers fill out the gaps in exquisite detail. It is the music of inflation. It allows us to see right back to when the Universe was little more than 10^{-35} of a second old. Massively reproduced in the science news media and in technical and public talks about the state of play in modern cosmology, one day it will be seen as the first evidence that we had about the earliest conceivable moments of our Universe. It will be the cosmological counterpart of that new-born baby photo.

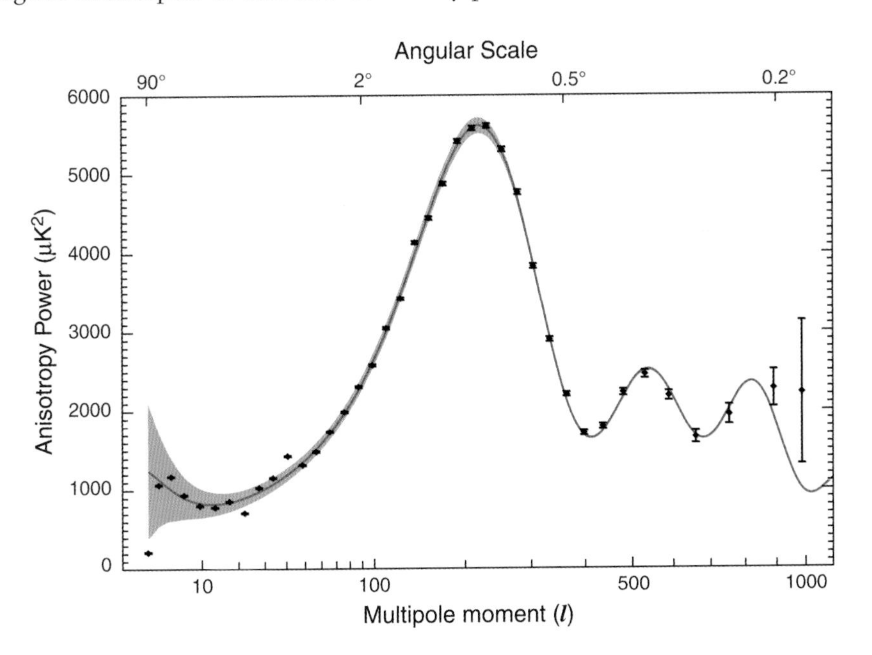

The magnitude of the average temperature differences in the background radiation measured at points separated by different angular separations on the sky (the angular size of the full Moon is 0.5 degrees). The standard inflationary model predicts the red curve. The data points are from the WMAP measurements.

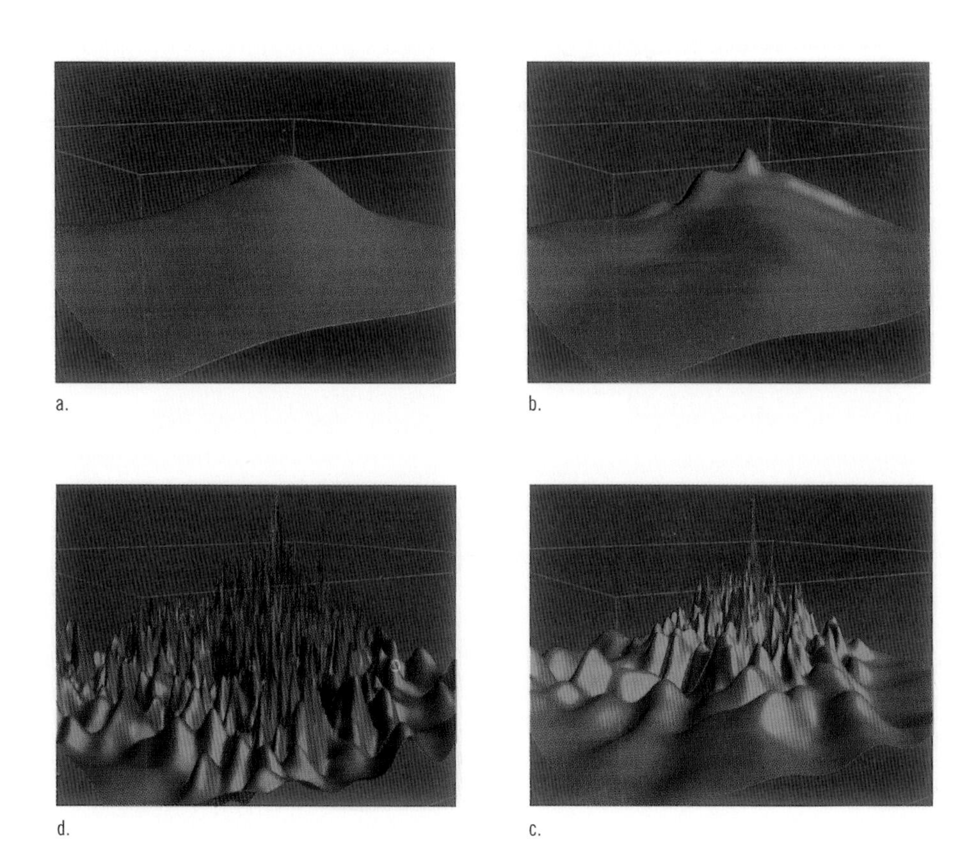

The development of self-reproducing inflation is shown with the passage of time (*clockwise from top left*). The formation of 'hills' in space shows the onset of inflation. Hills then form on hills, like stalagmites, showing the runaway self-reproduction of inflationary universes.

THE WORLD IS NOT ENOUGH

ETERNAL INFLATION

The baby figure of the giant mass
Of things to come at large.
William Shakespeare[137]

THE inflationary picture of the early history of the Universe gave rise to two unusual elaborations that were entirely unsuspected. The inflationary theory proposes that the Universe underwent a brief surge of accelerated expansion very early in its history. This enables the whole of the portion of the Universe that is visible to us today to be the expanded image of a region that was once small enough to be kept uniform and coordinated by light rays and other simple smoothing processes that arise naturally in the early stages of the Universe. At first, our attention was focused almost completely on how successfully this simple proposal explained the gross properties of our visible Universe: its overall uniformity peppered with the small irregularities destined to become galaxies, its special expansion rate, and the extreme similarity in the expansion rate from one direction to another. All had previously been unexplained coincidences. Now they are seen to be possible consequences of a single simple hypothesis.

However, this simple picture soon revealed some unexpected complications. It is all very well to consider how our visible portion of the Universe evolved to become the smoothly expanded image of one tiny primordial region, but what about all the neighbouring regions? Each of them would undergo a slightly different amount of inflation and create large smooth regions of their own, but with different properties to our own. We can't see them because light has not had time to reach us from them as yet,[138] but one day trillions of years in the future our descendants might see that far, and what they will find is a Universe whose geography is enormously complicated and irregular over very large scales, even though it looks very smooth and relatively simple over the scales that are visible to us today. Inflation leads us to expect that the local part of the universe that we can see is not typical of the whole (and possibly infinite) Universe; its geography is more complicated than we thought.

As if this wasn't unnerving enough, it was then recognised by Alex Vilenkin and Andrei Linde that the inflationary universe has another awkward feature: it is self-reproducing. Once inflation takes hold in one small region of the Universe, and causes it to accelerate its expansion, it creates the conditions needed for further inflation to occur in subregions of this inflating region. The result is a self-reproducing process with inflating regions spawning further inflating regions, which in turn produce others, and so on, it would appear, *ad infinitum*. And if *ad infinitum* to the future, then why not also to the past?

Eternal self-reproducing inflation means that while our little inflating 'bubble' universe may have had a 'beginning' when its expansion commenced, the whole 'multiverse' of all these bubble universes need have had no beginning and will have no end. We inhabit one of those (perhaps rare?) bubbles in which expansion persisted for long enough to allow stars, planets, and life to evolve. History is a more complicated subject than we thought.

This picture of 'eternal inflation' introduces a new complexity into our understanding of the history of the Universe. It is intriguing because, as with the recognition of the geographical complexity of an inflationary Universe, we find ourselves faced with the likelihood that we live in a Universe of huge diversity and historical complexity, most of which is totally inaccessible to us. We inhabit a single, simple patch of an elaborate cosmic quilt.

Andrei Linde, one of the architects of this conception of the eternal inflationary Universe, created a vivid picture of the self-reproduction process in action. The landscape of curved space begins to develop low hills as different amounts of inflation occur in different places. Then spikes form on the hills, and spikes on the spikes, and so on, as inflation self-reproduces in a fractal display of complexity growth. We live in one of the rare pinnacles where inflation has been completed and the expansion calmed down. Yet this appears to be an atypical state to be in. Most of the space in an infinite multiverse should still be undergoing inflation. This colourful sequence of inflationary stalagmites, which he characterises as the 'Kandinsky universe' (although they are more reminiscent of the work of the American conceptual artist Sol Lewitt, below), is perhaps the nearest we have to looking at one of God's home movies.

Sol Lewitt's *Splotch 15.*

John Wheeler at the blackboard.

GRAVITATIONAL ANONYMITY

BLACK HOLES HAVE NO HAIR

These people are under continual disquietudes, never enjoying a minute's peace of mind; . . . Their apprehensions arise from several changes they dread in the celestial bodies . . . that the sun, daily spending its rays without any nutriment to supply them, will at last be wholly consumed and annihilated; which must be attended with the destruction of this earth, and all of the planets.

Jonathan Swift[139]

THE American physicist John Archibald Wheeler is famous for inventing memorable terminology and striking pictures to illustrate key properties of gravity. He is the inventor of the term 'black hole' and the co-author of one of the most ambitiously designed and beautifully illustrated texts to be found in any area of science.[140] His lectures were picturesque, illustrated by beautifully inscribed coloured blackboards that required meticulous preparation ahead of time. Memorable images that appeared first on the Wheeler blackboards then found their way into his articles and books. There are many that we could have chosen but we have picked a few that will be instantly recognised by astrophysicists everywhere.

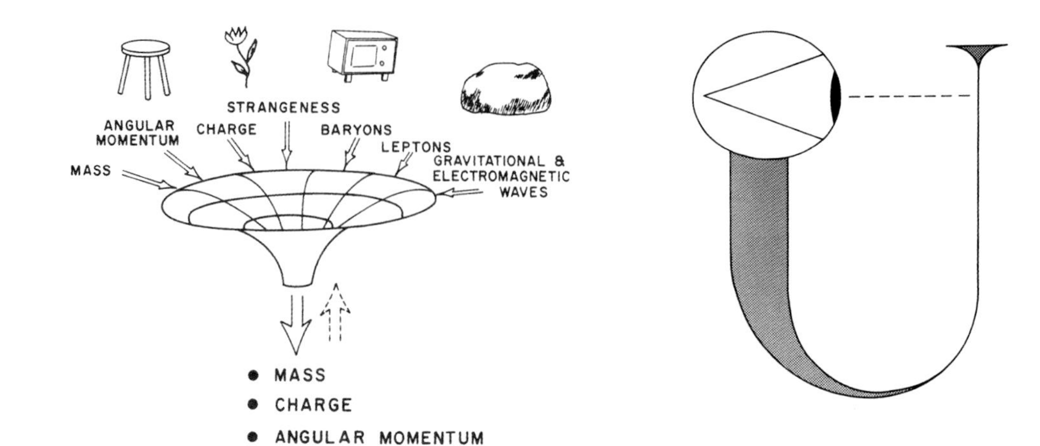
STRANGENESS
ANGULAR MOMENTUM
CHARGE
BARYONS
LEPTONS
MASS
GRAVITATIONAL & ELECTROMAGNETIC WAVES
MASS
CHARGE
ANGULAR MOMENTUM

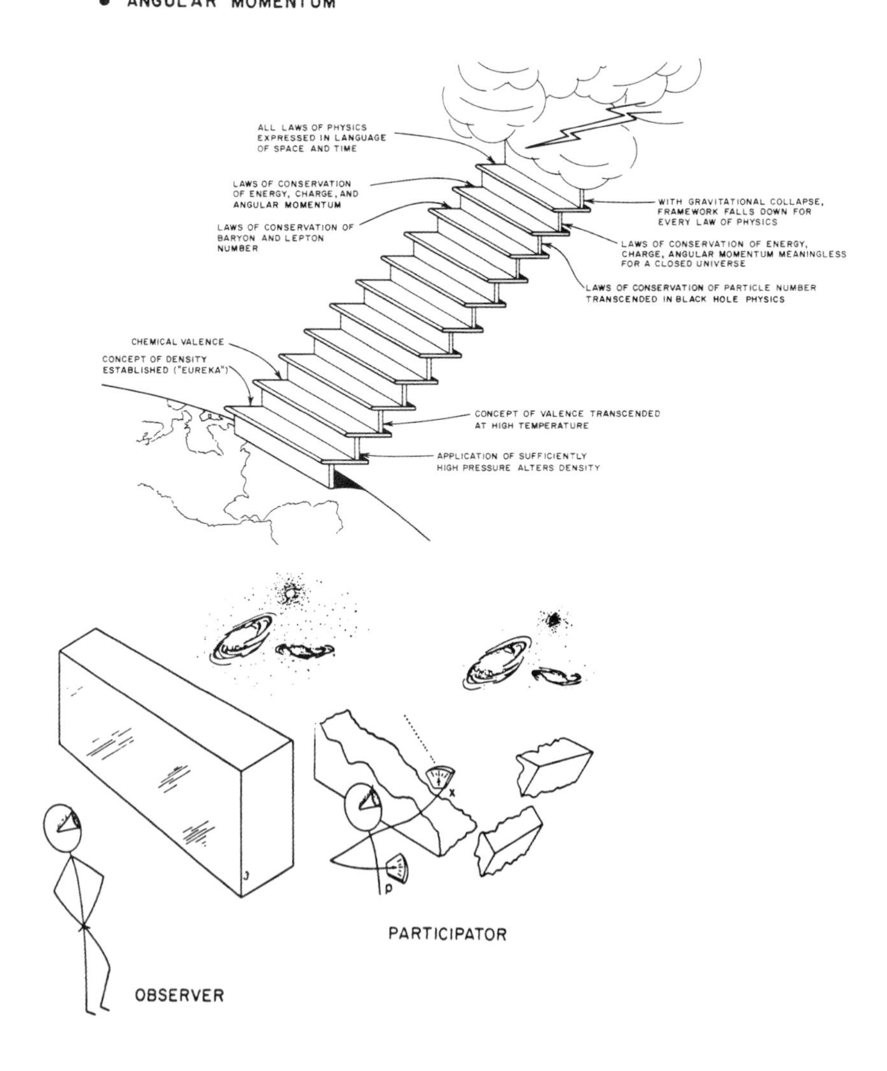
ALL LAWS OF PHYSICS EXPRESSED IN LANGUAGE OF SPACE AND TIME
LAWS OF CONSERVATION OF ENERGY, CHARGE, AND ANGULAR MOMENTUM
LAWS OF CONSERVATION OF BARYON AND LEPTON NUMBER
WITH GRAVITATIONAL COLLAPSE, FRAMEWORK FALLS DOWN FOR EVERY LAW OF PHYSICS
LAWS OF CONSERVATION OF ENERGY, CHARGE, ANGULAR MOMENTUM MEANINGLESS FOR A CLOSED UNIVERSE
LAWS OF CONSERVATION OF PARTICLE NUMBER TRANSCENDED IN BLACK HOLE PHYSICS
CHEMICAL VALENCE
CONCEPT OF DENSITY ESTABLISHED ("EUREKA")
CONCEPT OF VALENCE TRANSCENDED AT HIGH TEMPERATURE
APPLICATION OF SUFFICIENTLY HIGH PRESSURE ALTERS DENSITY
PARTICIPATOR
OBSERVER

Black holes are now part of modern culture as well as modern astronomy. They are in the movies and high-school physics classes. The perennial image of the black hole is of a great trap in the sky. One will form whenever a large enough mass of matter enters a small enough region.[141] When a critical concentration is reached then a surface of no return forms around it. We call it the *event horizon* of the black hole. If you cross it on the way in, then no bells will sound, or other local happenings arise, to signal the event. Only if you try to reverse your path to escape will you discover that something odd has happened. You will always fail to get back out through the event horizon. You are trapped in a region of space and time. As time passes, you will fall inexorably in towards the centre of the black hole, where the tidal forces will eventually tear you and your spaceship to pieces. At first, you may notice nothing unusual at all. The very large black holes, lurking in the centres of many galaxies have masses which are billions of times greater than our Sun, and densities that are less than that of air. We could all be passing through the event horizon of one of these big black holes at the moment and we would feel nothing unusual at all. Everything might look just as it does here and now. By contrast, for outsiders far away from the black hole's horizon, the situation is very strange. As our spaceship falls towards the black hole's horizon they can still receive signals from us and watch us with their telescopes. However, the light they receive will look redder and redder as it uses its energy to climb out of a stronger and stronger gravity field in order to travel back from us to them. Eventually, a last photon of light and a last radio signal would be received from the spaceship. It has passed through the event horizon and into the black hole. After this happens, life becomes much simpler for the astronomers watching from far away. There are now only three things that can be known about what is inside the black hole: its total mass, its total electric charge, and its overall rotation (specifically, its angular momentum). The material on the inside may

A montage of some of John Archibald Wheeler's striking images of physics: (*clockwise from top left*) the disappearance of accessible information behind the black-hole horizon; the self-observing universe as a quantum paradox; the ascending staircase of our deepening understanding of the laws of Nature; and the role of the 'observer' who is also an observed 'participant' in quantum reality.

have countless other properties – it may be made of matter or antimatter, coloured black or white, radioactive or inert, round or square, gold or silver – but the outside astronomers can no longer know any of these things once the horizon is crossed. It is important to recognise that the things inside the black hole don't cease to have these other properties – if we were falling alongside them we could still determine them – it is just that they are not part of the information that is accessible to those outside the horizon.

From the viewpoint of the astronomer on the outside, black holes are the simplest objects in the universe. Once you know their mass, their electric charge, and their spin, you know everything there is to know about them. By contrast, a typical star has countless properties and it is impossible to know all but a small number of them.

Wheeler's picture captures this dissolution of information as objects with all sorts of information-rich properties fall into the black hole leaving only information about the hole's mass, charge, and angular momentum to get out. These three quantities appear in the great conservation laws of physics. If the Universe lost memory of these quantities when a black hole formed, then it would be possible to violate the conservation of mass and energy, or of electric charge, or of angular momentum in any part of the Universe that included black holes.

One of the great unsolved problems of modern physics, which goes by the name of the 'information paradox', is to determine what happens to the information content of material that falls into a black hole. Does it disappear from the Universe down a singularity at the centre of the black hole; does it get back out of the black hole by means of the quantum evaporation process that Stephen Hawking discovered in 1974; or does something more exotic happen?

Wheeler encapsulated this property of black holes with typically memorable terminology: 'black holes have no hair'. The theorem that black holes can have only these three properties is usually described as the 'no hair' theorem. Black holes with the same mass, charge, and angular momentum have no other individually distinguishing characteristics ('hair'). If two of them have the same values of these three quantities they are indistinguishable. In the 1980s, in Sproul Plaza at the University of California at Berkeley, there used to be an enterprising street trader who sold real estate on the Moon (you were supplied with the title deeds for your plot) and black holes. His black holes were the simplest sort: they had neither spin nor charge, only mass. They all seemed to have the same mass and I used to ask him how we could tell which one was ours.

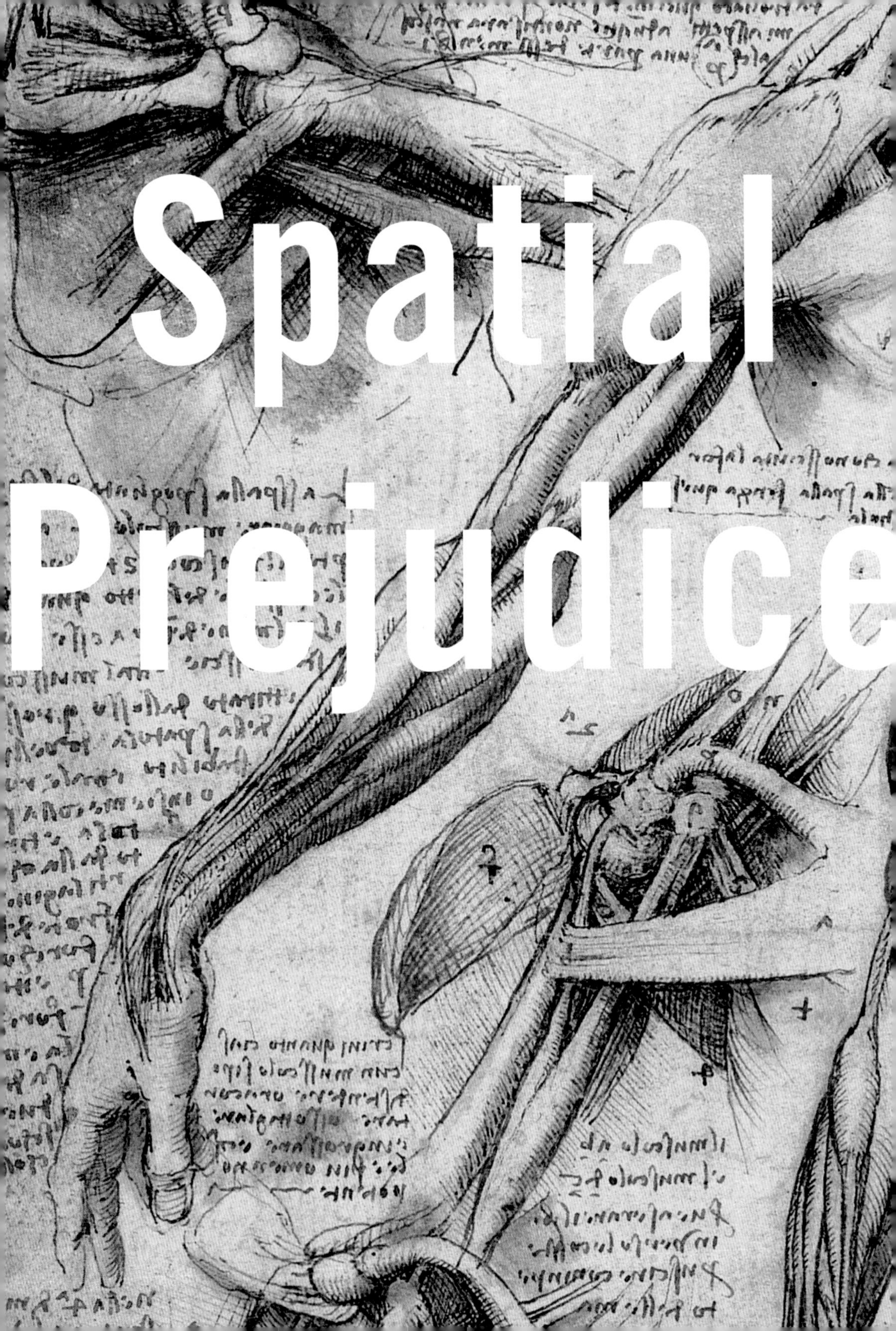

Spatial Prejudice

If you want to make an apple pie from scratch, you must first create the Universe.

Carl Sagan

Leonardo da Vinci's anatomical drawings annotated with 'mirror' writing, *c.*1510.

WHEN we wanted to broadcast our Earthly presence in the Universe, we used pictures as common currency to introduce ourselves to intelligences of other sorts on distant worlds. The images we sent were encapsulations of our bodily form, our knowledge, our spirit, and our location in space and time. But what of those who might be 'out there'? Earlier, convinced that there were canals on Mars, astronomers helped to create a fictional image of extraterrestrials as Martians, and the flying saucer would appear in the press as the transportation device of choice for advanced extraterrestrials, who journey hundreds of light years across the Galaxy to land in a potato field in Idaho, before departing to whence they came.

In this Part, we look at the Earth in the large; we see it for the first time from space in the Apollo pictures, and contemplate the impetus that these pictures of the Earth, a blue jewel beside the arid Moon, gave to the environmental movement. We look at a range of canonical images of the Earth that defined new sciences. The first geological maps, the first weather charts, contour maps, three-dimensional maps of flora and fauna, and the signatures of appealing terrestrial landscapes. We chart a period of history in which the European map-makers were the supreme craftsmen of their cartographical art.

Maps reveal a desire to know where we stand as well as to determine where we might go. The distortions needed to convert any maps of the Earth's almost spherical surface to one on plane paper are inevitable. The political implications of the distortions that were conventionally used are interesting, along with the responses of those who have created new projections to supersede the distortions of their predecessors.

Maps of all that lies below, above, and upon the Earth have remained an impressive influence on how we see the world. When we first saw the Earth from space, we not only found inspiration from Apollo's images, but we recognised the threat that we presented to our planet's continued existence as a life-supporting environment. The ozone hole showed for the first time what the care and maintenance of a small planet required of us.

We will look also at some of the enduring images of life on Earth, from creations of the modern conception of dinosaurs, the earliest footprints of our ancestors, to the evolution of illustrations of the human body; from works of art to the first textbook diagrams – and the ubiquitous beguiling symmetrical presence of the human face.

The first microscope revealed unsuspected wonders in the microworld. Robert Hooke's *Micrographia* displayed the microcosmos of small-scale intricacy to the world at large. Hooke looked at a few snowflakes through the eyepiece of his microscope, but it was not until the twentieth century that the full complexity of snowflakes was appreciated. Their tiny differences in the face of a unifying ground-plan of symmetry reveal much about the meaning of emergent complexity. Such microscopic complexity becomes more challenging still to capture when it is in motion. We will look at the splash made by the first images taken by high-speed photography.

Here, we also see the transition from the Renaissance world of fabulous artful illustrations to the era of the 'diagram' and the 'figure' – accurate but lifelessly bland. The early illustrations in books of botany, anatomy, and cartography are works of art, filled with life and individuality, pregnant with meaning about the brave new philosophical world that they were revealing to us – a world that we could now explore and copy for ourselves. No inherited wisdom from ancient Greece was required as an entry ticket. Some of these pictures were created to take advantage of new printing technologies, which enabled scientific illustrations to be duplicated in quantity for the first time. Previously, they were all laboriously hand-copied. As a result, only a handful of copies possessed the original detailed drawings. Other copies were unillustrated or contained poor approximations to the originals. Suddenly printing meant that scientific images could become part of truly public knowledge: a giant leap for humankind.

Earthrise, taken by Apollo 8 astronaut William A. Anders on 24 December 1968.

ONE SMALL STEP FOR MAN

THE EARTH FROM THE MOON

Yes, the surface is fine and powdery. I can kick it up loosely with my toe. It does adhere in fine layers, like powdered charcoal, to the sole and sides of my boots. I only go in a small fraction of an inch, maybe an eighth of an inch, but I can see the footprints of my boots and the treads in the fine, sandy particles.

Neil Armstrong's second words after first setting foot on the Moon

THE Apollo programme to land a human being on the surface of the Moon was energised by the speech made by President John F. Kennedy to the United States Congress on 25 May 1961. After ten preparatory missions, its goal was successfully reached on 20 July 1969 at 4:17:40 Eastern Standard Time, after a four-day journey from Earth. Neil Armstrong and Buzz Aldrin left the Lunar Module and Armstrong first set foot on the surface of the Moon at the Sea of Tranquillity landing site, taking 'one small step for man, one giant leap for mankind'.[1] They left the lunar surface the following day to rejoin Michael Collins in the orbiting *Apollo 11* and begin their long journey home. The trio splashed down safely in the Pacific Ocean on 24 July to complete the longest human journey ever made, watched throughout by the largest television audience ever achieved.

The motivations for NASA's programme were primarily political and technical, but the accomplishments and consequences were far wider and less foreseeable. It completed the most complicated technical project ever created on Earth, primarily staffed by recent graduates in their twenties, and managed by the remarkable James E. Webb.[2] It was a huge organisational achievement. Technically, it accelerated the development of computer technology in a dramatic way. The control and reliability demands were immense and set the US microelectronics industry on a trajectory that would culminate in a long period of world domination. The Soviet programme, as a consequence of its total secrecy, had no spin-off for the Soviet Union's industrial and business growth. However, these American technical and management successes – along with the development of better non-stick saucepans – are not the reasons for including the Apollo programme in a discussion about influential images. For most of us, the Apollo programme means pictures – pictures *of* the Moon and pictures *from* the Moon. And they are some of the most spectacular photographs ever taken.

The images of the sky from the lunar surface are striking. There is no lunar atmosphere and so there is no scattering of light by its molecules, no diffraction, and there are no sky colours.[3] The sky always looks black except when you look at particular sources of distant light. And there is no sound either – this environment is completely silent.

The nearest and most impressive source of light was the Earth, glowing in the reflected light of the Sun, spinning on its axis 238,857 miles (384,403 km) away. The two classic *Apollo 11* pictures of the Earth show two thirds of the crescent Earth and Earthrise over Smyth's Sea. Earthrise had been photographed by *Apollo 8* six months earlier which had been the first picture of the Earth from space.

These crescents in the sky are familiar to us because we are used to seeing the phases of the Moon from the Earth. To see the roles reversed was dramatic. But the most enduring legacy of these pictures was nothing to do with the Moon at all. Rather, it was the beginnings of a deeper concern for the Earth's environment. Such concerns had been first awakened by the publication of Rachel Carson's book *Silent Spring*, in 1962, which raised a storm of protest about the use of agricultural pesticides and ultimately led to the formation of several national environmental protection agencies and new legislation to protect birds and animals from the risks of chemical pollution.[4] It brought these issues out into the public arena for the first time. The Apollo images resonated with this new spirit of environmental concern. The *Apollo 8* mission had arrived in orbit around the Moon on Christmas Eve, 1968, and as part of the preparation for *Apollo 11* sent back pictures of Earth from space by using a portable TV camera. For the first time, in spectacular vivid isolation, we could all see what the Earth looked like from space. Whereas the Moon looked arid and dead, the Earth was colourful and alive: a beautiful glass marble in the sky. The beauty and changing complexity of its environment created the impression of a jewel in space: something different, something unique, something irreplaceable, something of ultimate concern.

Apollo's pictures of the Earth from space were copyright-free and were reproduced all over the world in books and on posters. Their imagery and subliminal messages about the nature of the Earth reached out directly to the public in Western democracies. They reinforced and resonated with the environmental concerns being expressed by a growing band of authors, such as Ernst Schumacher,[5] Barbara Ward, and René Dubos[6] – concerns that reached their zenith with the arrival of the epoch-making images from *Apollo 11*.[7]

MAPPING THE MONEY

THE EARTH AT NIGHT

What if everything is an illusion and nothing exists? In that case, I definitely overpaid for my carpet.

Woody Allen

LOOK at a picture of the Earth at night taken from space[8] and you will find it illuminating. The most noticeable places are those with the greatest wattage of artificial lights. The huge cities of the developed world – London, New York, Los Angeles, Paris, Frankfurt, Tokyo, Beijing – are brightly illuminated, but great swathes of Asia and Africa are dark. The image is a powerful reminder that light does not necessarily trace mass – in this case, the mass of humanity. The dark regions of Africa, Asia, South America, and India are actually the places where the bulk of the human population resides. But the light traces the money. The illuminated regions pick out the regions of greatest affluence, conspicuous consumption, and technological advancement. They create a picture of two Earths rather than one. If we look at a different sort of map, one of internet traffic, we are seeing a chart of information highways. Information processing and transmission pick out the same pathways again and again; just as the most ancient trade routes forged routes between the great ancient centres of commerce, so the airways link the commercial, political, and academic hubs around the globe. Again, the internet hubs and bright spots are in places where the population density is often small. The mass of humanity is poor and almost invisible to the internet.

The Earth at night. The bright regions show the appearance of different parts of the Earth's surface during the hours of darkness.

The Universe is a little like that too. As we have steadily mapped out the

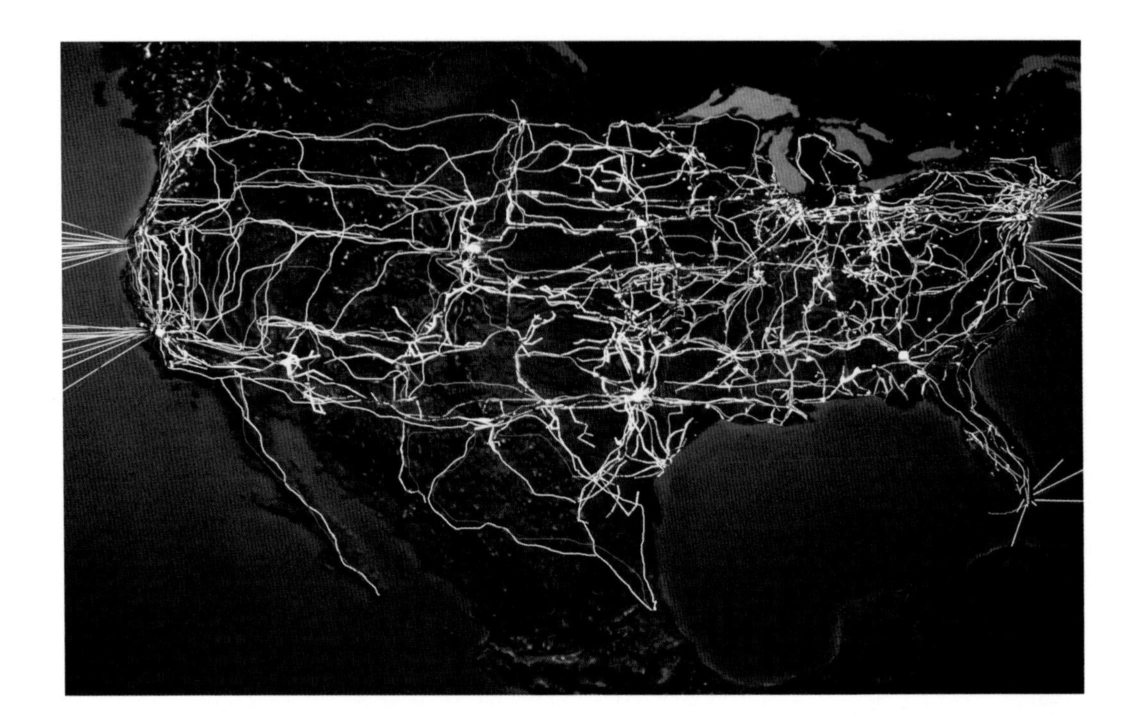

distribution of matter and energy in our visible Universe, we have discovered that the spectacular patterns of luminous stars and galaxies we admire so much are not a true reflection of cosmic demographics. All forms of matter and energy both create and respond to gravity, and so we can map the distribution of invisible dark matter by observing its gravitational effects on the things that we can see. The results of this exploration are rather striking. The regions that shine in the dark, the stars and galaxies that populate our photographs of the Universe, are but a tiny trace element. They pick out the places where the density of matter became very high – high enough to initiate nuclear reactions, so stars could form and shine brightly in the dark. But these are not the places where most of the matter and energy in the Universe resides. They are merely the most visible. It is like looking down on the ocean from a low-flying aircraft. The most noticeable features are the white crests of the waves, but these are not the places where the bulk of the water is: they are merely the places where air bubbles have accumulated in the surface foam and so are the most visible from afar.

Internet traffic map of the USA.

About 72 per cent of our Universe resides in a mysterious dark form of energy that is smoothly distributed everywhere. It reveals its presence by accelerating the overall expansion of the Universe.[9] Of the remaining 28 per cent, only about 4.5 per cent consists of material in stars and planets – some of which are luminous, while others are burnt out and faint. The rest is also dark, but different in character, most probably in the form of ghostly elementary particles[10] that rarely collide or interact with ordinary atomic material.

This gradual discovery that most of our Universe consists of dark materials of unknown provenance has steadily emerged over the last fifteen years, aided by the Hubble Space Telescope's ability to see virtually to the edge of the visible universe.[11] It is a sobering cosmic revelation. For it reveals that what astronomers have been studying and philosophers have been interpreting for thousands of years is merely a tiny piece of icing on the cosmic cake. The real substance of the Universe has only recently shown itself. And it is a form of energy that is nothing like the atoms and molecules of which we are made. The stars are inspiring but they are exceptional. They form only in special places and at special times in the history of the Universe. Yet, they have come to dominate our interpretation of the Universe throughout history.

We live in one of those special places where stars can form. It must be so, because life is only possible in regions where stars form because the nuclear reactions within them spawn the biochemical building blocks of life, and provide the stable climatic conditions that planets seem to require if they are to become the abodes of living complexity in all its diverse forms. The deceptive appearance of the Earth at night is an object lesson for us not to jump too quickly to conclusions. Things are not always as they might at first appear.

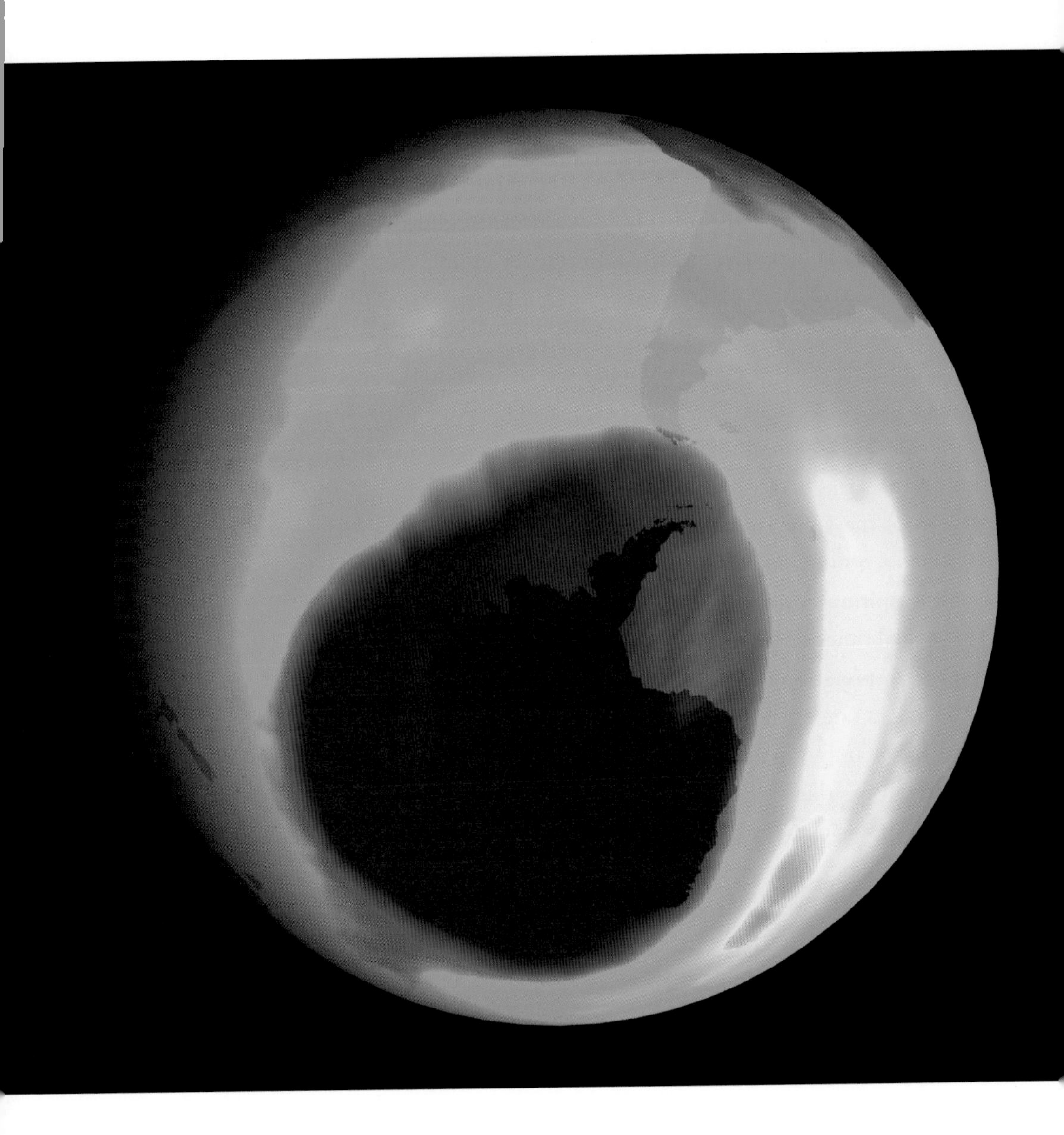

The ozone depletion hole over the South Pole, monitored from space in September 2000.

THE CARE AND MAINTENANCE OF A SMALL PLANET

THE OZONE HOLE

They were doomed. We all are: technological civilization represents a species' desperate attempt to build a bubble to keep hostile environments at bay. Sentient species also never cooperate with one another over the long term, because the environments they need in order to live are incompatible. Destruction or devolution are the only choices.

Karl Schroeder[12]

OZONE is a molecule that is built from three atoms of oxygen.[13] It forms a tiny trace element in the Earth's atmosphere but its presence there is vital for our survival. Although we can't see it, we can occasionally smell it. After an electric storm there is often a slightly sulphurous smell in the air, which is caused by the lightning discharge breaking one of the many oxygen O_2 molecules into two free O atoms, which then combine with another oxygen O_2 molecule to make two triatomic ozone molecules of O_3. This chemical process creates most of the ozone in the atmosphere, although strong ultra-violet light from the Sun, rather than lightning, is the general cause of the break-up and recombination process. The biologically harmful ultra-violet light from the Sun is absorbed by ozone molecules and prevented from reaching the Earth's surface with a sustained intensity that would destroy the fragile molecules of life.[14] Ironically, most light-skinned humans are happy to bathe in the sunlight and allow the remaining transmitted ultra-violet light to provide them with a suntan. For millions of years we have lived in an environment where the average ultra-violet radiation levels were safe, but once industrialisation began to create new forms of chemical waste the fragility of our environmental balance became alarmingly evident.

In 1974, two Californian chemists, Mario Molina and Sherwood Rowland, predicted that some of the chemicals that were being widely used in refrigeration plants and aerosol spray cans were depleting the natural abundance of ozone far more quickly than it could be replenished in the atmosphere.[15] These widely used chemicals are chlorofluorocarbons – or CFCs as they became known – and are much lighter than air, so they rise into the upper atmosphere, where they threaten the survival of the ozone shield in the stratosphere around the Earth. Molina and Sherwood urged governments and industries using CFCs to take urgent remedial action. They took little notice. And when they did they said there was no positive evidence that this ozone depletion was happening at all.

Joseph Farman, a scientist working for the British Antarctic Survey at Halley Bay (latitude 75.5 degrees south), with Brian Gardiner and Jonathan Shanklin, had been monitoring the Earth's atmosphere over the South Pole since 1957. Aware of this controversy, he started to notice a depletion in the atmospheric ozone levels around the Pole in 1977. He continued to measure an increasing depletion over several years, but was reluctant to make any public claim because NASA's total ozone mapping spectrometer (TOMS) on the *Nimbus 7* satellite, along with all its high-powered computing systems back at base, were evidently not seeing this as they were making no claims about ozone levels falling. Farman wondered if he was measuring something wrongly. Finally, in 1985, and still with no American announcement, the effects became so huge that Farman, Gardiner, and Shanklin could hold back no longer and their discovery was soon published at length in the journal *Nature*.[16] Despite their simpler observational methods, they claimed there was a huge depletion. Astonishingly, it turned out that the computers analysing data from *Nimbus 7* had been programmed to ignore ozone levels as low as those Farman was finding because they were treated as forms of instrumental error and discounted as physically impossible. Embarrassingly, a re-analysis by *Nimbus 7* showed that Farman's observations were correct: ozone levels in the stratosphere were 35 per cent lower than in the 1960s. There was a large depletion of ozone over the South Pole, manifested by a huge 'hole' in the ozone shield that subsequently became known as the 'ozone hole'.

The ozone hole appeared first over the South Pole because the ozone-depleting chemical reactions work fastest in cold conditions. Although a hole has subsequently been found over the North Pole, the South Pole is colder because it is a landmass rather than only frozen water like the Arctic. The hole lasts for several months of the year and persists in its largest extent for about two months in the southern spring, moving off to the tip of South America and having strong ultra-violet effects on plants and small animals emerging from their winter hibernation. Its size was enormous. It typically reached a maximum area of about 30 million square kilometres, roughly 6 per cent of the Earth's surface cover.[17] By the end of 2005, its extent had been significantly reduced.

Even after Farman's discovery, industries were still reluctant to act, but chemists warned governments of the dramatic rate of depletion that observations were showing year on year. Finally, significant international bans[18] on CFCs and related chemicals were negotiated by the United Nations in 1996, one year after Molina and Rowland received the Nobel Prize for chemistry.[19] It will take the rest of the twenty-first century to restore stratospheric ozone levels to the values they had in the 1950s.[20]

The image on page 162 displays the remarkable extent of the ozone hole and also shows the power and importance of satellite images of the Earth's surface from nearby space. For the first time in human history we have the capability to evaluate the impact of global effects around the Earth's surface in real time.

The first eclipse daguerreotype,
taken by Berkowski in July 1851.

DARKNESS AT NOON

ECLIPSE

And it shall come to pass in that day, saith the Lord God, that I will cause the sun to go down at noon, and I will darken the earth in the clear day.

The Book of Amos[21]

SURROUNDED by the nocturnal glare from city lights, we see little of the stars. For the ancients, things were very different. The spectacle of hundreds of shining stars would have been the most impressive natural sight they ever saw. And they saw it every night. The regularities in the heavens offered important practical benefits and provided subtle psychological influences about the reliability and regularity of Nature. In such a climate of predictability, how much more impressive are the exceptions, the disasters,[22] and the cataclysmic. These are the things that make the record books.

Ancient eclipses were famed for their influence upon human affairs. The word 'eclipse' derives from the Greek *ekleipsis*, meaning 'an omission' or 'an abandonment', and in several cultures, such as the Chinese, 'to eclipse' is 'to eat' or 'to devour', and a dragon is usually the diner. We can find many eclipses which had dramatic effects that changed human history. The total eclipse that occurred on 28 May 585 BC was so unexpected that it ended the five-year-old war between the Lydians and the Medes. Their chronicles tell us that in the midst of battle 'the day was turned into night'; their fighting stopped and a peace treaty was hurriedly signed. Many centuries later, Christopher Columbus exploited his astronomical knowledge of an imminent eclipse of the Moon by the Earth (in which the Earth's shadow covers the surface of the Moon) to enlist the help of the Jamaicans after his damaged ships were stranded on their island in 1503. At first, he traded trinkets with them in return for food; eventually, they refused to give him any more supplies, and his men were facing starvation. His response was to arrange a conference with the islanders on the night of 29 February 1504 – at the time when he knew an eclipse of the Moon would begin. Columbus announced that his God was displeased by their lack of assistance, and He was going to remove the Moon as a sign of his deep displeasure. As the Earth's shadow began to fall across the Moon's face, the terrified islanders quickly agreed to provide him with anything he wanted, so long as he restored the Moon. Columbus informed them that he would need to withdraw and persuade his God to return the Moon. Retiring with his hour-glass for the appropriate period, he returned, in the nick of time, to assure them of God's pardon for their bad attitude and His reinstatement of the Moon. Soon afterwards, the eclipse ended. Columbus had no further problems in Jamaica and he and his men were eventually rescued and returned triumphantly to Spain.

Complete eclipses are remarkable events which occur because of an accident of Nature. The true diameter of the Sun is about 400 times greater than that of the Moon, but their relative distances from the Earth are such as to make their apparent sizes on the sky almost the same.[23] As a result, the passage of the Moon in front of the Sun can cover the face of the Sun almost exactly, to produce a total eclipse of the Sun's light. By way of contrast, if we were to look from the surface of any of the other planets in the solar system, their moons would generally appear much larger in their skies than the disc of the Sun. The only other place in the solar system where a complete eclipse could be seen is from Prometheus, an irregularly shaped moon of Saturn. But, viewed from the surface of Saturn, the complete eclipse would be very brief in duration and extend over a tiny area of the sky, because Saturn is so far from the Sun and Prometheus is very small.

The distance from the Earth to the Moon is very slowly increasing, by a few centimetres every year, and in 500 million years' time complete eclipses of the Sun will no longer be seen from Earth.

The first photograph of an eclipse of the Sun was made by Berkowski on 28 July 1851 at Königsberg and was used repeatedly to illustrate this dramatic phenomenon over the next fifty years.[24] The image shows active solar flares from the solar surface as well.

Indirectly, eclipses have played an important role in human evolution. The appearance of total eclipses is largely a by-product of the large size of the Moon.[25] If the Moon did not exist, or was much smaller, there would be dire consequences for life on Earth. There would be no tides, but we might be able to live with that and might even have been able to evolve all the same. However, the direction of the Earth's rotation axis (approximately the line through the north and south magnetic poles) would be destabilised by the gravitational perturbations from the other planets coming into resonance with the wobble in the direction of the Earth's rotation axis (it completes one complete revolution every 25,000 years as the Earth rotates like a precessing top). Normally, the rotation axis is inclined at about 23.3 degrees to the vertical from the plane in which the Earth and other planets orbit the Sun. If the Moon did not exist, then this inclination angle would change significantly, in a chaotically unpredictable fashion, over millions of years. There would be massive seasonal changes and a very challenging environment for biological evolution to adapt to.

This chaotic sequence of events has occurred on Mars, whose two tiny moons, which are merely captured asteroids, are too small to play any stabilising role. However, our Moon is so large that its gravitational field dominates the destabilising effects of the external perturbations from the other planets, and the direction of the Earth's rotation axis barely changes over long periods of time.[26]

The last and more unusual consequence of the astronomical coincidences that enable us to see total eclipses is the intellectual benefits that we have gained from doing so. Besides allowing us to view activity on the surface of the Sun in its corona, it has enabled us to make advances in fundamental physics which would otherwise have been much slower.

Einstein's general theory of relativity extended Newton's great theory of 1687 to deal with strong gravitational fields and motions of light in a self-consistent way. One of its great predictions was that the paths followed by rays of distant starlight should be 'bent' by the gravitational attraction of the Sun as they pass it by en route to our telescopes. The bending angle was calculated by Einstein and was quite different (twice as big) from what might be expected if Newton's theory was true. However, the measurements needed to test this crucial prediction are only possible during a complete eclipse of the Sun, because otherwise the brightness of the Sun overwhelms everything we want to see.

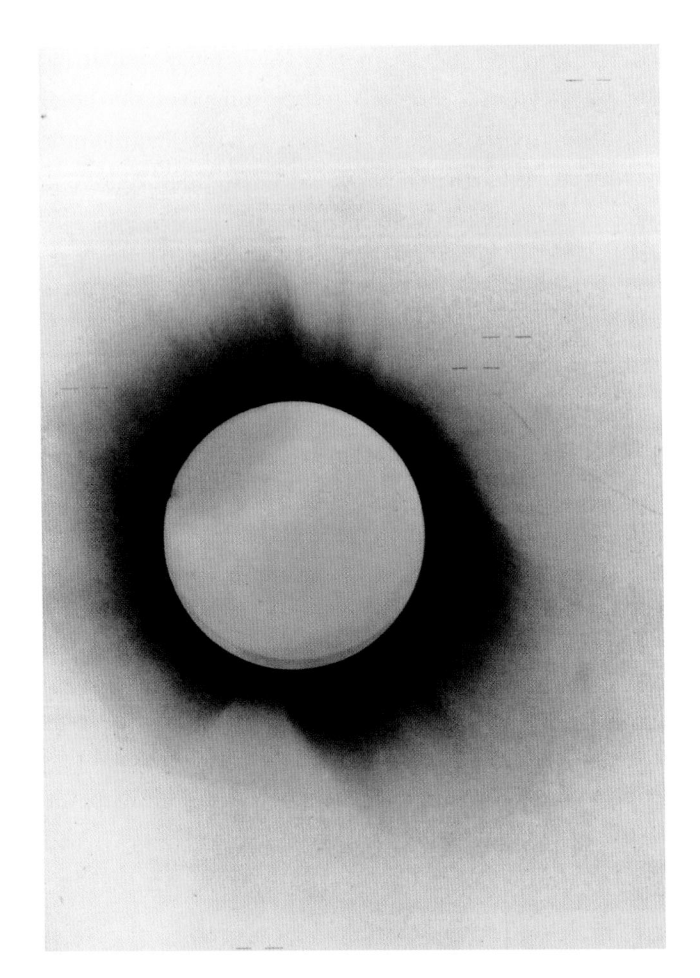

One of the negatives taken with the 4-inch lens at Sobral, Brazil in 1919. The horizontal lines mark the images of the stars whose apparent positions were measured to determine any shift due to the gravitational effect of the intervening Sun.[28]

A photograph was taken of a bright star field at night when the Sun was not in the sky and then repeated during a complete eclipse when the Sun was in front of the same star field. These stars would only be visible to us with the Sun in front of them during a complete eclipse. If light does not travel in perfect straight lines, but is bent by gravity as Einstein predicted, then we should be able to see some stars very close to the Sun's position on the sky that would otherwise be blocked by the face of the Sun on the sky. In effect, we would be able to see some stars behind the Sun because the bending of light allows us to see behind its face.

Einstein predicted that the small bending angle should be just 1.75 seconds of arc (the angular site of the Sun on the sky is about 32 minutes of arc – about 1097 times bigger than the predicted deflection). After the First World War had ended, the British astronomer Arthur Eddington[27] led a famous expedition to the island of Principe, in the Gulf of Guinea off the West African coast, and Andrew Crommelin led another, to Sobral in Brazil, to make the crucial observations. On 29 May 1919, an unusually long period of totality (six minutes) was observed, during which the Sun was right in front of the bright Hyades cluster of stars. The results obtained by the Eddington and Crommelin teams, despite the rather poor weather, were revolutionary and the negative of one of Crommelin's Sobral plates can be seen opposite (sadly, all Eddington's Principe plates are lost). The horizontal lines mark the stars whose positions were measured in order to compare their positions when the Sun was in front of the star field to when it was absent. Eddington and Crommelin's expeditions confirmed the distinctive bending of light predicted by Einstein's theory, and their dramatic announcement at a meeting of the Royal Astronomical Society in London, on 15 November 1919, that his predictions had been confirmed, propelled Einstein into the limelight in a way that would turn him into the biggest scientific celebrity of all time.

The original caption for the graphical explanation of the experiment in the *Illustrated London News* of 22 November 1919 read as follows:

> The results obtained by the British expeditions to observe the total eclipse of the sun last May verified Professor Einstein's theory that light is subject to gravitation. Writing in our issue of November 15 [1919], Dr A.C. Crommelin, one of the British observers, said: 'The eclipse was specially favourable for the purpose, there being no fewer than twelve fairly bright stars near the limb of the sun. The process of observation consisted in taking photographs of these stars during totality, and comparing them with other plates of the same region taken when the sun was not in the neighbourhood. Then if the starlight is bent by the sun's attraction, the stars on the eclipse plates would seem to be pushed outward compared with those on the other plates . . . The second Sobral camera and the one used at Principe agree in supporting Einstein's theory . . . It is of profound philosophical

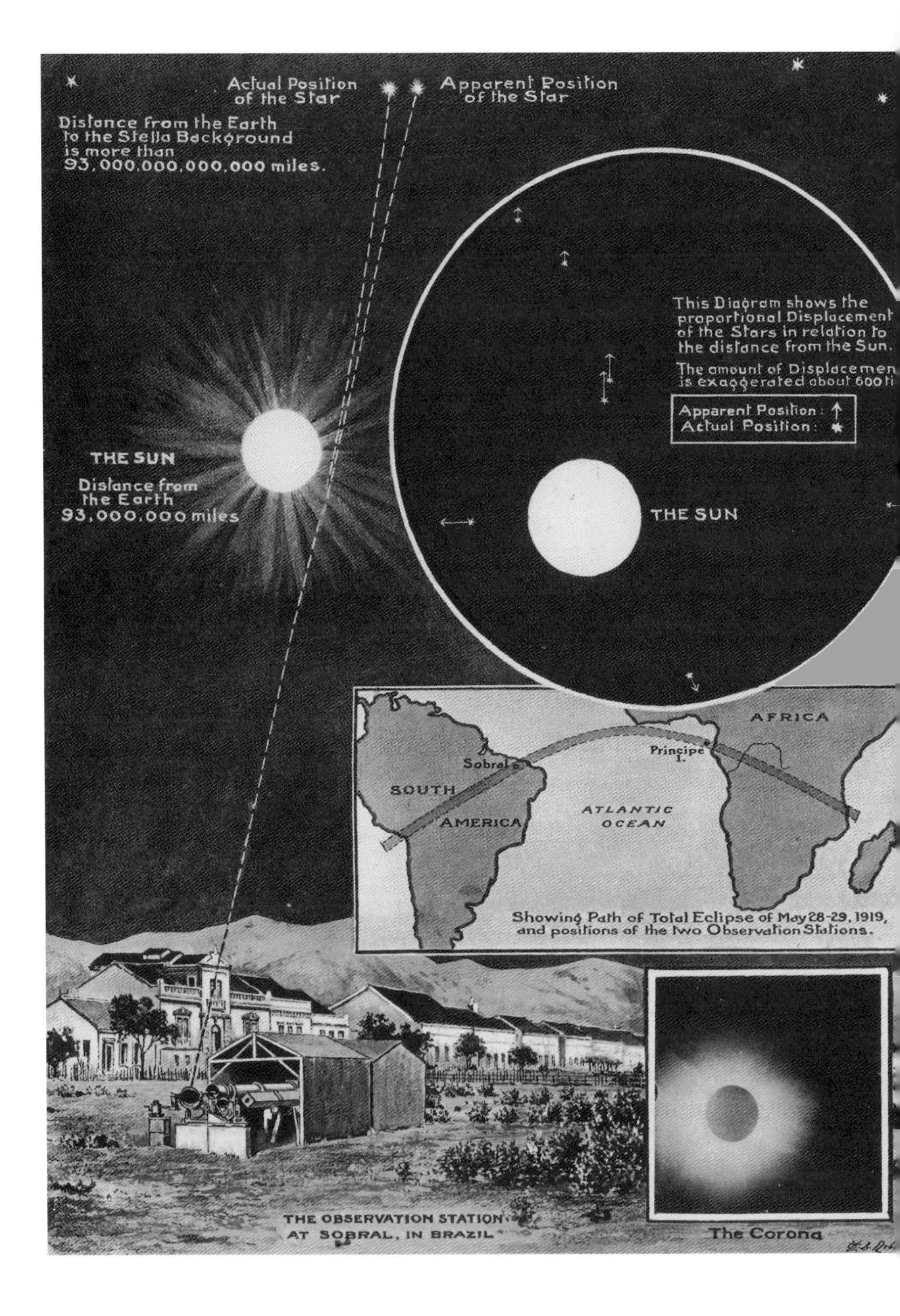
Actual Position of the Star
Apparent Position of the Star
Distance from the Earth to the Stella Background is more than 93,000,000,000,000 miles.
THE SUN
Distance from the Earth 93,000,000 miles
This Diagram shows the proportional Displacement of the Stars in relation to the distance from the Sun.
Apparent Position:
Actual Position:
THE SUN
AFRICA
Principe I.
Sobral
SOUTH AMERICA
ATLANTIC OCEAN
Showing Path of Total Eclipse of May 28-29, 1919, and positions of the two Observation Stations.
THE OBSERVATION STATION AT SOBRAL, IN BRAZIL
The Corona

interest. Straight lines in Einstein's space cannot exist; they are parts of gigantic curves.

Without a total eclipse we would not have been able to test Einstein's theory of gravity in this way. Without a planet like Mercury orbiting very close to the Sun we would not have been able to carry out the other crucial test of his theory – its predicted wobble in the orbit of planets.[29] One lesson we learn from these accidents of our astronomical situation is that it should not be presumed that extraterrestrials, no matter how intellectually capable, will inevitably discover all the laws of physics that we have. Many of our discoveries have arisen because of helpful geographical coincidences. A cloud-covered sky scuppers simple astronomy; an absence of magnetic ores and a planetary magnetic field stymies the understanding of magnetism and electricity; the absence of radioactive elements near the surface means that the study of radioactivity and nuclear disintegration is hindered. The paths to scientific development on a given planet will be significantly entrained by the type of environmental challenges that need to be overcome in order to survive, and by the type of sky that is visible overhead. In our own case, the occasional visibility of complete eclipses of the Sun is a wonderful natural light-show, a reminder of the Moon's vital role in our existence, and a profound aid to our understanding of the Universe's deepest laws.

The original depiction of the observations and theory that appeared in the *Illustrated London News* in 1919. It refers specifically to the Crommelin expedition to Sobral and shows how Einstein's prediction of light-bending by the gravitational pull of the Sun will lead to displacements in the apparent positions of the stars when the Sun lies along the line of sight to the star field being observed.

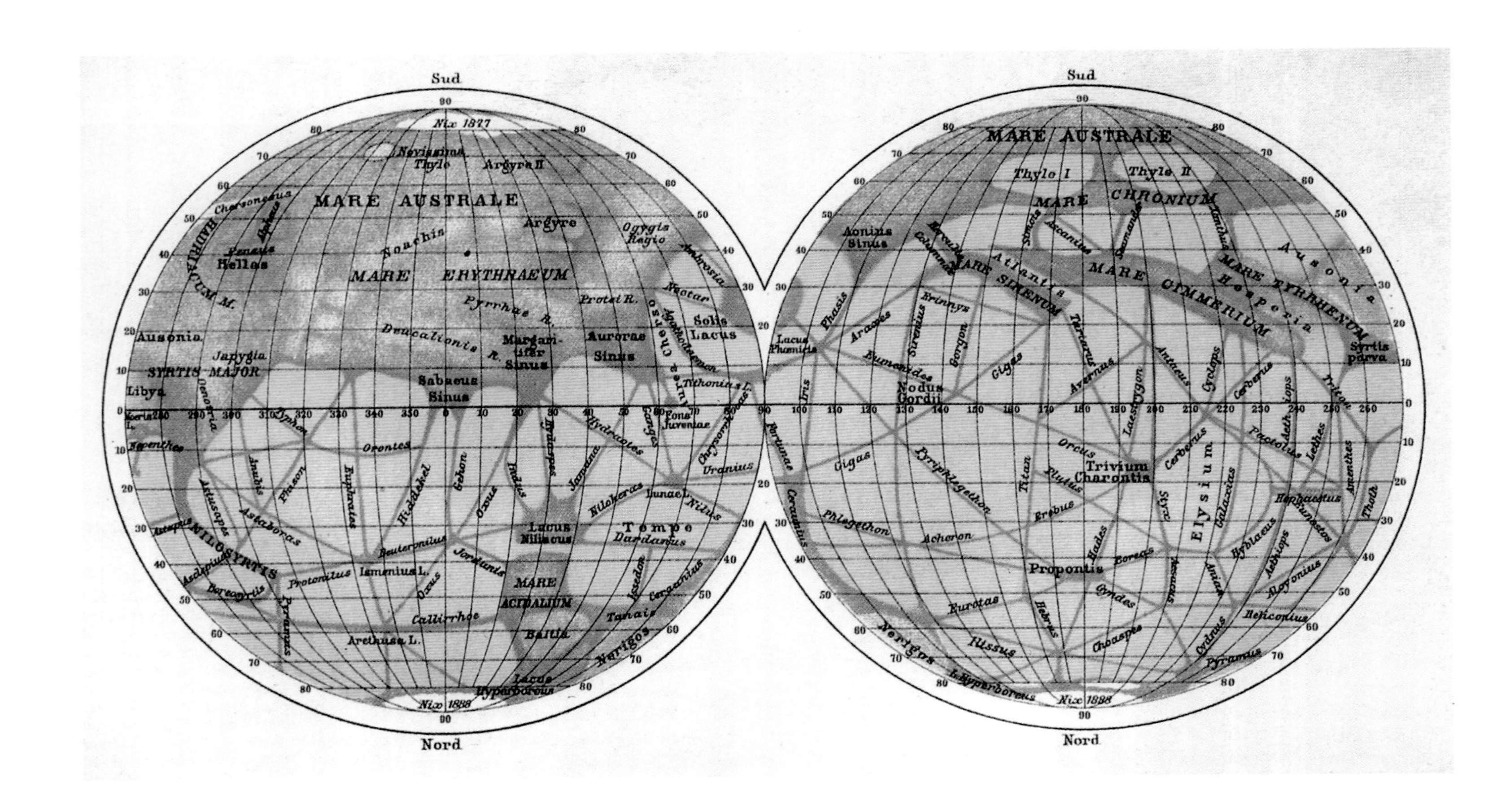
Sud
Nix 1877
Novissima Thyle
Argyre II
MARE AUSTRALE
Chersonesus
HADRIACUM M.
Hellas
Noachis
Argyre
Ogygis Regio
Ambrosia
MARE ERYTHRAEUM
Pyrrhae R.
Protei R.
Nectar
Solis Lacus
Ausonia
Japygia
SYRTIS MAJOR
Deucalionis R.
Margaritifer Sinus
Aurorae Sinus
Libya
Sabaeus Sinus
Tithonius L.
Fons Juventae
Nepenthes
Oenotria
Typhon
Orontes
Ganges
Hydraotes
Hydaspes
Anubis
Phison
Euphrates
Hiddekel
Gehon
Oxus
Indus
Jamuna
Uranius
Nilokeras
Lunae L.
Nilus
Astusapes
Astaboras
NILOSYRTIS
Deuteronilus
Lacus Niliacus
Tempe
Dardanus
Jordanis
Protonilus
Ismenius L.
MARE ACIDALIUM
Issedon
Ceraunius
Boreosyrtis
Pyramus
Callirrhoe
Tanais
Arethusa L.
Baltia
Nerigos
Lacus Hyperboreus
Nix 1888
Nord
Sud
MARE AUSTRALE
Thyle I
Thyle II
MARE CHRONIUM
Aonius Sinus
Simois
Ascanius
Xanthus
Ausonia
Atlantis
MARE SIRENUM
MARE CIMMERIUM
Hesperia
MARE TYRRHENUM
Phasis
Araxes
Sirenius
Erinnys
Tartarus
Lacus Phoenicis
Gorgon
Gigas
Eumenides
Nodus Gordii
Avernus
Antaeus
Cyclops
Cerberus
Syrtis parva
Triton
Iris
Fortunae
Gigas
Pyriphlegethon
Titan
Orcus
Laestrygon
Trivium Charontis
Cerberus
Elysium
Pactolus
Lethes
Amenthes
Ceraunius
Phlegethon
Acheron
Erebus
Styx
Galaxias
Hephaestus
Eunostos
Thoth
Hades
Propontis
Boreas
Hyblaeus
Aethiops
Eurotas
Hebrus
Gyndes
Aesacus
Alcyonius
Heliconius
Choaspes
Cydnus
Nerigos
Ilissus
L. Hyperboreus
Pyramus
Nix 1888
Nord

WAR OF THE WORLDS
MARTIAN CANALS

Yet across the gulf of space, minds that are to our minds as ours are to those of the beasts that perish, intellects vast and cool and unsympathetic, regarded this earth with envious eyes, and slowly and surely drew their plans against us.

H. G. Wells[30]

MARS is the fourth planet from the Sun and the seventh-largest in the solar system. It is named in honour of Ares, the mythical Greek god of war, for reasons that are not entirely clear, but perhaps because of its blood red colour. In our cultures, Mars is synonymous with the extraterrestrial. The word Martian is common. The month of March derives from Mars, and whole sections of the confectionery business are devoted to selling Mars chocolate bars – because 'A Mars a day helps you work, rest and play' as their old advertising slogan claimed. Curiously, great planets such as Saturn, Jupiter, or Neptune have a much lower terrestrial profile. Only the romantic claims of Venus come close. Do their inhabitants just employ the wrong marketing company or could there be something about Mars that makes it so much more fascinating for the average Earthling? How did it become the by-word for alien worlds?

Giovanni Schiaparelli's drawings of the surface of Mars (1877), which suggested the presence of *canali* to other readers.

Mars is easily visible with the naked eye in the night sky. Its brightness varies a lot, as does its distance from the Earth. Every twenty-six months Mars is at its closest to the Earth and we can send a space probe there with a minimum of fuel. This is why in 2004 we saw spacecraft from Europe and America queuing up to land on, or orbit, the red planet. It was one of those times of closest approach. Although it is rather smaller than the Earth, Mars has about the same amount of surface land. It has two tiny moons, Phobos and Deimos, that look like misshapen potatoes: Phobos is a mere 22 km and Deimos a trifling 12 km in diameter. Both are simply asteroids that got too close to Mars and found themselves captured by its gravity.

Our fascination with Mars has been fed by the captivating patterns visible on its surface. In the autumn of 1877, when Mars was also close to Earth, the great Italian planetary astronomer Giovanni Schiaparelli (the uncle of the famous fashion designer Elsa Schiaparelli), at the Brera Observatory in Milan, thought he saw channels (*canali*) on the surface of Mars.[31] When his reports were translated into English, *canali* became 'canals', suggesting that they had been artificially constructed by local Martian residents for purposes of irrigation or transport. The dark and light areas on the planet surface he named after terrestrial seas, capes, and peninsulas, coining exotic and euphonious names like the *Herculis Columnae* (Columns of Hercules), *Mare Tranquillitatis* (Sea of Tranquillity), *Aurorae Sinus* (Bay of the Dawn), and *Solis Lacus* (Lake of the Sun). By these leaps of imagination, Schiaparelli had re-created Mars in the image of ancient Earth, pregnant with myth and meaning. It was never the same again.

Intrigued by Schiaparelli's drawings and detailed observational reports, the leading American astronomer Percival Lowell added weight to the growing misconception. In 1894, he claimed that the intricate grid of surface markings was the result of work by intelligent beings who were inhabiting the planet even now. He inferred the presence of clouds and a temperate climate and identified vast engineering projects on the planet's surface. Lowell developed his views in three books: *Mars* (1895), *Mars and Its Canals* (1906), and *Mars as the Abode of Life* (1908), and by then Mars had become the most fascinating place in the solar system.

This speculative groundwork laid the foundations for the work of great science fiction writers such as H. G. Wells and Olaf Stapleton, and a host of their successors who continue, imaginations unabated, today. So enthusiastically did the American public seem to take up the idea of intelligent Martians that, on Sunday, 30 October, the day before Halloween 1938, the young Orson

Welles was able to create panic among millions of Americans who tuned into a portion of his radio adaptation of Wells's *War of the Worlds*. Listeners quickly became convinced that they were hearing reports of a *real* Martian invasion of America, that a huge flaming object had landed in New Jersey. Welles had incorporated realistic news broadcasts read by actors describing the Martians as they emerged from their spaceships. Commentator Carl Phillips interrupts his interview with Professor Richard Pierson at Princeton Observatory to give his eyewitness report:

> They look like tentacles to me. There, I can see the thing's body. It's large as a bear and it glistens like wet leather. But that face. It . . . it's indescribable. I can hardly force myself to keep looking at it. The eyes are black and gleam like a serpent. The mouth is V-shaped with saliva dripping from its rimless lips that seem to quiver and pulsate . . . The thing is rising up. The crowd falls back. They've seen enough. This is the most extraordinary experience. I can't find words. I'm pulling this microphone with me as I talk. I'll have to stop the description until I've taken a new position. Hold on, will you please, I'll be back in a minute.[32]

Eventually, the real newsreaders had to appeal for calm and explain the mass panic had been created by a mere story.

Today, it is we who are 'invading' Mars. Detailed observations long ago revealed that Lowell's canals were just tricks of the human eye, which has evolved to detect patterns wherever it sees them, joining up neighbouring points to make straight lines whenever it can. Yet some meandering channels are real. In the past few months we have had evidence from the *Mars Express* space probe that there is frozen water at the South Pole of Mars and flowing water probably once eroded great channels in its surface. Perhaps, deep below the surface, the pressure of pack ice will be great enough to sustain liquid water even now. NASA's *Spirit* rover took dramatic high-resolution pictures that give a real feeling for the nature of the red planet's surface.

For astronomers, Mars teaches us about the wonderful properties of Earth. Mars has no plate tectonics: its terrain is simple. Also, unlike the Earth, Mars has no magnetic field. This deficiency left the Martian atmosphere at the mercy of the 'solar wind' of fast-moving electrically charged particles that are blown towards it from the Sun. Gradually they blew away the Martian atmosphere, leaving almost nothing behind. Earth's atmosphere would have suffered the same fate had it not been for our magnetic field. It deflects the incoming wind of solar particles around our atmosphere, and gravity is strong enough to hang on to it.

Mars had a far more extreme climatic history than the Earth. The reason is again remarkable, as we have just seen in the last chapter. Both the Earth and Mars rotate with an axis of rotation inclined at about 23–24 degrees from the vertical to the plane of their orbits around the Sun. But without the benefit of the stabilising effect of a large moon, and unable to hang on to its atmosphere in the face of the solar wind, Mars has been subjected to a chaotically variable climatic history, as is witnessed by the huge variations in ice and temperature around its surface. Without the Moon, complex life on Earth would perhaps only have been able to exist, like that on Mars, in the minds of other beings and on the pages of their science fiction books.

Our future exploration of the solar system will focus with ever greater emphasis upon the surface of Mars. And, as it does so, the aura of Mars will be embellished with new images of a world that was once living but which died, and which may have played a part in seeding life on Earth and could play one last part in fashioning worlds that can know about themselves.

True-colour image of Martian rocks in the Gusev Crater. Three types of rocks are seen here: thin, jagged rocks buried under the sand; light grey, rounded rocks; and dark, pock-marked rocks. The dark rocks are thought to be volcanic in origin and rounded rocks often owe their smoothness to water. This image was obtained by the Mars Exploration *Spirit* rover. The colours of this image were created by combining camera filters to approximate what a human would see. The false-colour equivalent image is R360/248. *Spirit* landed in the Gusev Crater on Mars in January 2004, and obtained this image on 12 April 2006, the 809th Martian day of the rover's mission.[33]

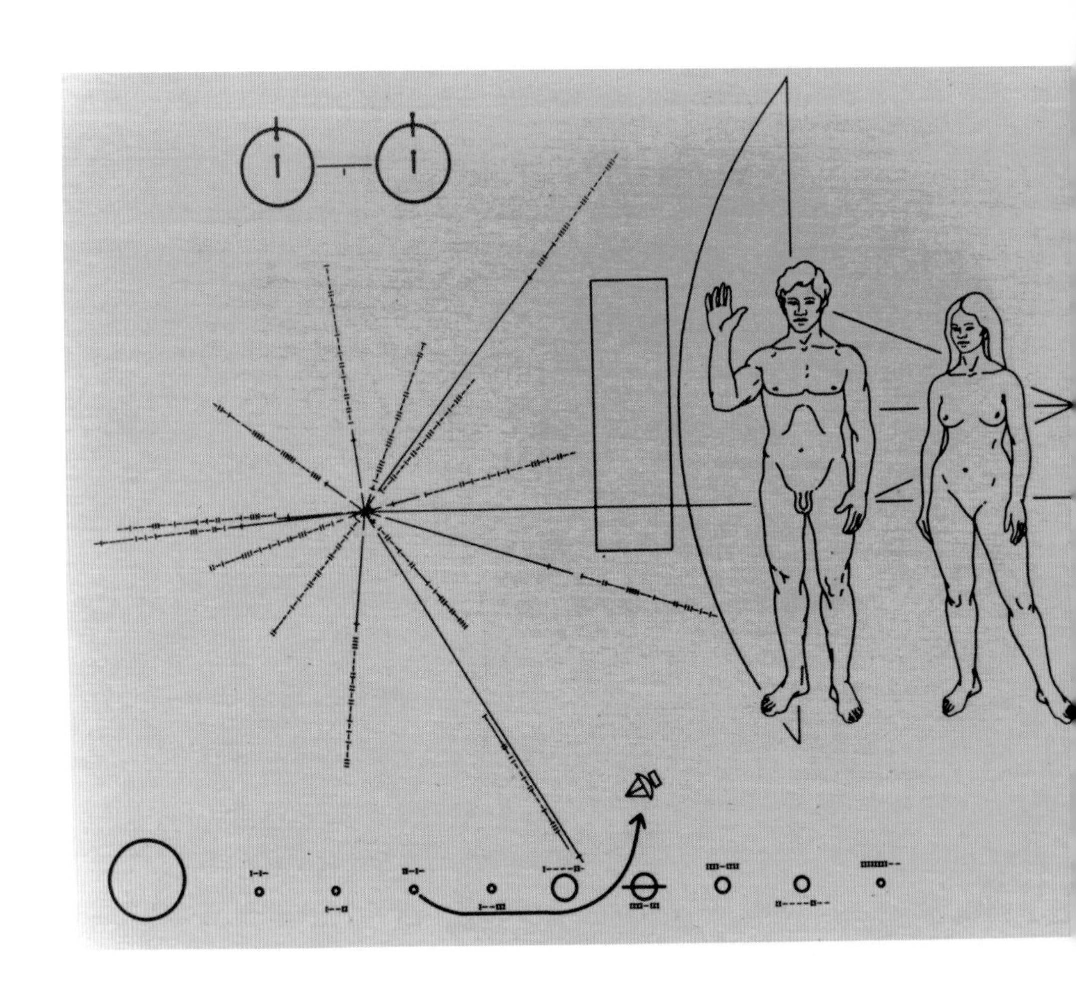

The *Pioneer* Plaque, designed by Frank Drake and Carl Sagan and drawn by Linda Salzman Sagan in 1972.

HUMANITY'S GOLDEN DISCS

THE *PIONEER* PLAQUE AND THE *VOYAGER* RECORD

The best classified ad I ever read was one that said 'Unused tombstone for sale. Would suit family called Nesbitt.'

David Frost[34]

THE *Pioneer 10* and *11* planetary missions launched[35] by NASA in 1972 and 1973 have a special place in human history. *Pioneer 10* passed Jupiter on 3 December 1973, *Pioneer 11* followed on 2 December 1974 and headed on towards a rendezvous with Saturn on 1 September 1979. The last contact with *Pioneer 10* was received on 27 April 2002, and the last with *Pioneer 11* on 1 October 1995 when, science mission accomplished, its antenna was left pointing away from Earth. In the 1980s the Pioneer probes had become the first human constructs to leave the solar system. Since then, both probes have been heading towards the nearby stars. *Pioneer 10* is set towards the star Aldebaran in the constellation of Taurus and will arrive in 2 million years' time. *Pioneer 11*, meanwhile, is heading towards the constellation of Aquila, close to Sagittarius.[36] It will arrive at the closest stars in the constellation after another 4 million years of interstellar travel. But both must survive the debris that litters the way – the asteroids, rocks, and dust that threaten to destroy them – if they are to complete their journeys to the stars. The *Pioneer* space probes are also special because they are carrying messages into space from the human race. Both have identical gold-anodised aluminium plaques attached to their sides, identifying their date of launch and planet of origin, in the hope that one day they might be read by recipients on other worlds.[37]

The idea of a message being attached to the *Pioneer 10* spacecraft was first proposed by the space-science writer Eric Burgess. Following his suggestion, a plaque was created from scratch in less than three weeks by Carl Sagan and Frank Drake, two human pioneers in the search for extraterrestrial signals of intelligence, and drawn by Linda Salzman Sagan. Its meaning is designed to be 'obvious' to a scientifically literate extraterrestrial. The man and the woman pose next to an outline of the spacecraft so as to give a sense of scale. The man's hand is raised in greeting showing our opposed thumb and some of our limb joints (apparently, more explicit representation of the female genitalia was rejected by NASA).[38] Running horizontally along the base of the plaque are scale pictures of the Sun and the solar system planets. The Earth is marked as the point of origin of a trajectory followed by the spacecraft past Jupiter and Saturn (whose rings are marked as an added identification clue). The relative distances of the planets from the Sun are shown alongside the pictures in a binary representation which employs units of a basic distance equal to one tenth of Mercury's orbital distance from the Sun.

The position of the Sun is also shown relative to the positions of fourteen pulsars (identified by their distinctive relative pulsing periods in binary units along the radial lines pointing to their sky positions) and the centre of our Galaxy.[39] The cosmic time and position of the launch of the spacecraft can be deduced from this information as the length of the lines radiating out to the pulsars scale in proportion to their distances from the Sun. A mark at the end of each line indicates the distance of the pulsar above the plane of our Milky Way galaxy. Only three of these pulsars are needed for a triangulation on the Earth's position but lots of 'spares' were built into the picture in case the receivers only knew a few of them, or because some of them might no longer exist in the same form when the plaque is read. In addition to the fourteen radial lines to the pulsars there is one further horizontal line that passes behind the human figures. This marks the distance and direction to the centre of our Milky Way galaxy relative to the distances to the pulsars.

The quantitative key is provided by a binary signature of the number 8 and the element hydrogen, whose characteristic 'hyperfine' transition is represented in the barbell figure at the top left. This transition between two different energy states of the hydrogen atom occurs when the single electron orbiting the proton in its nucleus changes its spin direction from pointing 'up'(parallel) to pointing 'down' (antiparallel). The 'up' state has the slightly higher energy, so the transition results in the emission of radiation to take away the excess energy. The emitted radiation has a wavelength of approximately 21 cm[40] – a fundamental quantity in astronomy that would be known to any serious observers of the Universe able to practise radio astronomy.

This 21cm wavelength is the defining unit of length for numerical quantities on the plaque.[41] The binary representation of 8 between the two horizontal bars marking the height of the woman defines the height of the woman as 8 units of 21 cm, or 168 cm. The actual size of the spacecraft relative to her picture provides a cross-check.

The follow-up to this anthropocentric piece of interstellar advertising was far more ambitious. The *Voyager 1* and *Voyager 2* probes were launched on 5 September and 20 August 1977, respectively. The plan this time was to prepare a much more substantial 'message in a bottle' to accompany these grander missions to the outer solar system planets. It aimed to convey far more information than the hastily prepared *Pioneer* plaque. Instead of merely locating us in space and time, it sought to encapsulate humanity in sound and pictures with an encyclopedic record about our world and our cultures. The information was cut into a twelve-inch gold-plated copper gramophone record that looked a little like the reward for selling a million terrestrial pop records, and enclosed in an inscribed protective cover.

What should go on humanity's Golden Record? Naturally, a committee was needed to decide, and NASA convened one under the chairmanship of Carl Sagan.[42] They assembled a medley of 115 analogue-encoded images, natural sounds from the Earth, greetings and messages in fifty-five human languages from ancient Akkadian and Sumerian to modern Chinese and Punjabi, words of friendship from President Jimmy Carter and the Secretary General of the United Nations, and a ninety-minute selection of music from many cultures. At the more practical level, the disk was accompanied by a cartridge and needle, together with playing instructions in symbolic form.

The 115 images cover a range of subjects. First, there are astronomical

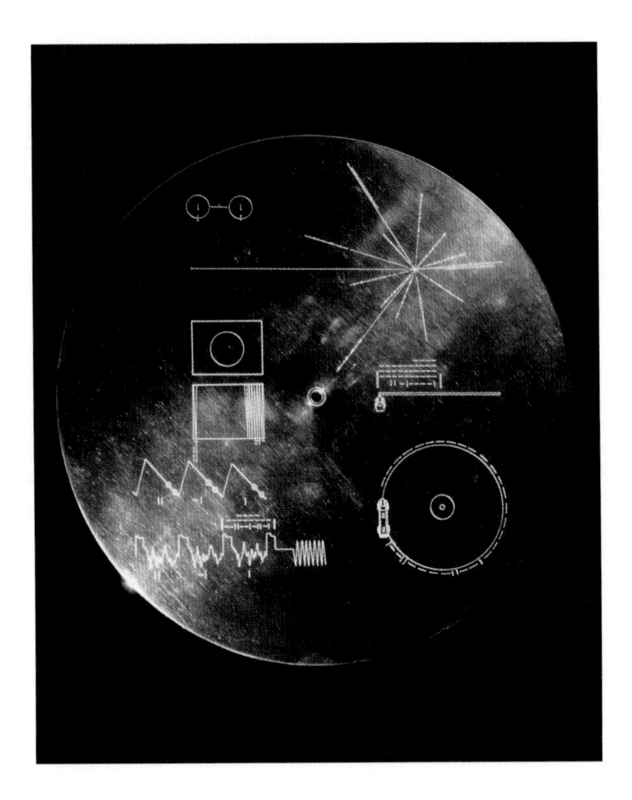

images that identify our Sun and the neighbouring planets, along with pictures of the Earth from space. Then there are lists of our units of physics and mathematical symbols, followed by the chemical symbols and the structure of our DNA. A guide to our anatomy and physiology follows, showing how human birth takes place, followed by pictures of families with mothers, fathers, and children of many ages. Next, the extraterrestrial viewer is taken on a photographic tour of planet Earth to see natural terrains and seascapes of all sorts, together with a diverse menagerie of birds, fish, insects, and animals. The next section covers many examples of different ways of human life, from Bushmen to modern city dwellers, active sportsmen and women, social activities, farming, shopping, running, cycling, rush-hour traffic and busy manual workers, food and drink, houses and human constructions, some of the architectural wonders of the ancient and modern worlds, transport systems, telescopes, artists, orchestras, and schools. All human life is here: a moving encapsulation of what it meant to live on planet Earth at that time.

Humanity's Golden Record cover, which was carried by the *Voyager* spacecraft in 1977.

The sound recording attempted to pack human acoustic experience into its

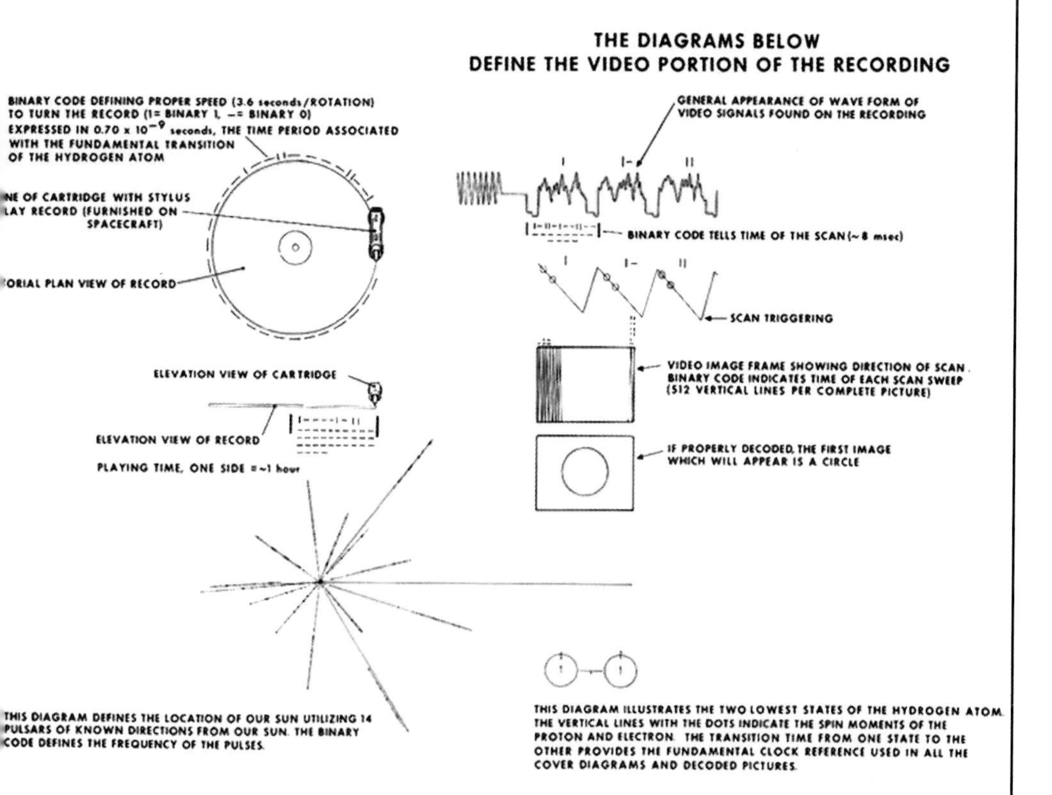
EXPLANATION OF RECORDING COVER DIAGRAM
THE DIAGRAMS BELOW DEFINE THE VIDEO PORTION OF THE RECORDING
BINARY CODE DEFINING PROPER SPEED (3.6 seconds/ROTATION) TO TURN THE RECORD (1= BINARY 1, —= BINARY 0) EXPRESSED IN 0.70 x 10^{-9} seconds, THE TIME PERIOD ASSOCIATED WITH THE FUNDAMENTAL TRANSITION OF THE HYDROGEN ATOM
NE OF CARTRIDGE WITH STYLUS LAY RECORD (FURNISHED ON SPACECRAFT)
ORIAL PLAN VIEW OF RECORD
ELEVATION VIEW OF CARTRIDGE
ELEVATION VIEW OF RECORD
PLAYING TIME, ONE SIDE ≅~1 hour
THIS DIAGRAM DEFINES THE LOCATION OF OUR SUN UTILIZING 14 PULSARS OF KNOWN DIRECTIONS FROM OUR SUN. THE BINARY CODE DEFINES THE FREQUENCY OF THE PULSES.
GENERAL APPEARANCE OF WAVE FORM OF VIDEO SIGNALS FOUND ON THE RECORDING
BINARY CODE TELLS TIME OF THE SCAN (~8 msec)
SCAN TRIGGERING
VIDEO IMAGE FRAME SHOWING DIRECTION OF SCAN. BINARY CODE INDICATES TIME OF EACH SCAN SWEEP (512 VERTICAL LINES PER COMPLETE PICTURE)
IF PROPERLY DECODED, THE FIRST IMAGE WHICH WILL APPEAR IS A CIRCLE
THIS DIAGRAM ILLUSTRATES THE TWO LOWEST STATES OF THE HYDROGEN ATOM. THE VERTICAL LINES WITH THE DOTS INDICATE THE SPIN MOMENTS OF THE PROTON AND ELECTRON. THE TRANSITION TIME FROM ONE STATE TO THE OTHER PROVIDES THE FUNDAMENTAL CLOCK REFERENCE USED IN ALL THE COVER DIAGRAMS AND DECODED PICTURES.

am explaining the images and symbols engraved on the Golden Records carried by *Voyagers* 1 and 2. It was led as a greeting to any intelligent life that may find them in the future and gives audio and visual information Earth and its people. At the upper left are details of how to play the record to listen to the sounds, with nation on viewing the images at the right. The circle in the rectangle is the first image, so aliens will know interpretation is correct. The lines at the lower left show Earth's position relative to 14 pulsars. The two circles lower right represent the two spin states of the hydrogen atom. The *Voyagers* were launched in 1977, and will nearby stars in 40,000 years.

allotted recording time. Ethnic music of all sorts was included. There is African and Javan music, Bach, Mozart's *Magic Flute*, Peruvian pipes, Indian Ragas, Solomon Island panpipes, Beethoven and Chuck Berry, choral music, Louis Armstrong and the blues – but neither the Beatles nor any other pop music made an appearance.[43]

The Golden Record was placed inside an aluminium cover that contained images reminiscent of Sagan and Drake's earlier designs for the *Pioneer* plaque. In the top-left corner is an illustration of the record and the accompanying stylus, correctly located to begin playing at the beginning of the recording. The time for the record to rotate once, 3.6 seconds, is also recorded in special time units of 0.7 billionths of a second – the atomic transition time of the hydrogen atom that corresponds to the 21 cm wavelength. Underneath, we find a side-view of the playing record with the binary symbols ('1' denotes binary 1 and '–' denotes binary zero) equal to the playing time of one whole side of the record (about sixty minutes). In the top right-hand corner of the cover are some drawings that indicate how the video pictures are to be constructed from the sound-wave signals on the disk by converting them to 512 vertical lines, just like conventional TV signals. There is a picture showing the general shape of the wave form for the video recording on the disk with some binary code to show the 8 milliseconds time extract and diagrams to show how the images are constructed from the scans. Underneath, there is a copy of the first picture on the record (a circle) so that the decoding can be checked and the correct scaling of horizontal and vertical sizing can be fixed. In the lower left-hand corner of the cover are the pulsar maps from the old *Pioneer* plaque and the picture of the hydrogen spin-flip transitions that define the units of frequency and time referred to above. Finally, there is an overall electroplated 2-square centimetre area of the cover made out of pure uranium-238. This radioactive element slowly decays with a half-life of 4.51 billion years, and by examining the relative abundance of uranium-238 to its decay products in this area of its surface it is possible to calculate the amount of time that has passed since it was first placed on the spacecraft – a nice cross-check against the deduction of the launch time that is possible by a more elaborate analysis of the pulsar map.

These images, simple but rich in information, are our first attempt to encapsulate much of what we (or, at least, what Carl Sagan, Frank Drake, and their colleagues) would like extraterrestrials to know about us. There is much that is missing but everything that was included is worthy of its place. It is this spectrum of images alone that may one day transmit our likenesses to worlds that we will never find, minds that we will never know, faces that we will never see. It is a sobering thought. President Jimmy Carter's message put it thus:

> This is a present from a small, distant world, a token of our sounds, our science, our images, our music, our thoughts and our feelings. We are attempting to survive our time so we may live into yours.

For a long time it was believed that these two messages were the first detailed attempts by humans to compose a message that would encapsulate the essence of our existence here on Earth and communicate it in a generic way to an unknown form of intelligence. But we now know that there was a forerunner at work nearly a century before Drake and Sagan.

The first person to create a special symbolic message to send out to extraterrestrials was a little-known nineteenth-century Finnish army officer and mathematician named Edvard Englebert Neovius who taught astronomy, map reading and navigation to cadets at the naval school in Hamina, on the southern coast of Finland, about 350 km east of Helsinki. He had studied at the school as a student before spending time at the Engineering Academy in St Petersburg, and then returned to teach at Hamina in 1845.

Neovius was greatly interested in the question of intelligence in the Universe and the consequences of believing in a universal rationality. His primary inspiration came from reading a book by the famous Danish physicist Hans Christian Oersted entitled *Anden i naturen* (*The Soul in Nature*), published in Swedish in 1869.[44] In its pages Oersted argued for a convergence of thinking and knowledge to be expected in all intelligent beings in the Universe (something implicitly assumed by Sagan and Drake). Although their senses might differ, according to their environmental histories, they would all deduce in their own way the essential laws of dynamics and physics in order to understand and exploit the ways in which Nature worked. As a result, the development of life would follow similar patterns in many different places in the Universe. This all has a rather modern sound to it, and Oersted mentions Darwin and Wallace's theory of evolution by natural selection as an argument for believing in the universality of senses such as sight. Oersted's arguments actually had some theological underpinning. He believed that the source of universal rationality was God, and the similarity to be expected in the laws of Nature, life, and rational activity, in different parts of the Universe, was one of the manifestations to be expected of this singular Divine Reason.

Neovius accepted Oersted's arguments and set about appealing directly to the rational beings that he confidently expected to be watching and listening for our signals. Neovius' message to extraterrestrials is rather fascinating in its construction, and was published in a short booklet, in Swedish, in 1875 under the uplifting title *Vår tids största uppgift* (*The Greatest Mission of Our Time*), before being translated into French and Russian; a year later, a slightly extended second edition appeared in Swedish.[45] But despite attracting this early attention it seems to have had little lasting impact – this is not entirely surprising: Hamina was hardly a centre of intellectual influence and Neovius seems to have done nothing to promote his ideas beyond its boundaries.

Neovius' prime candidates as sites for extraterrestrial life were the planets Mars and Venus. Of these, he picked Mars as the better because of the appearance of continents and ice caps on its surface, and the daily and seasonal variations that were not unlike those of Earth. Appealing to some popular contemporary theories of the origin of planetary systems, which argued that planets should form first far from the Sun, he believed that any solar-system civilisation on Mars would therefore be older and more advanced than one on Venus – and even than us. Indeed, he takes it for granted that there will be intelligent creatures on Mars and that they will share many of our senses and patterns of thought. While still believing in a theistic direction of all things, he is also aware of Darwinian evolution and appeals to it as further support for his assumptions about various types of evolutionary convergence.

The centrepiece of Neovius' short book is the creation of a message that can be read by intelligent recipients on other worlds. He has in mind sending signals by pulsing bright lights and urges that we should be searching for incoming signals from Mars as well as transmitting them. He spends some time estimating luminosities that would be achievable using many sources (900,000 light elements!) directed by mirrors from a high-altitude site such as Quito or the Sierra Nevada mountains. He even works out a budget of around 100 million French francs for the whole project to get it up and running. By the time of the second edition, the broadcasting system is simplified and the budget is down to 30 million francs – under the assumption that we merely need to assume that the Martians have the ability to detect a signal of brightness equal to that of a 16th-magnitude star. A further huge reduction could be brought about by assuming the Martians had a telescope like the Earl of Rosse's Leviathan at Birr Castle in Ireland, in which case only 970 light elements would be needed in order for the signal to be seen on Mars – a thousand-fold decrease on the original technical and financial requirements.

Neovius plans a message that will be sent by a sequence of light pulses, analogous to the Morse code, which had been invented by the German telegrapher Friedrich Clemens Gerke in 1848, and was probably among the topics he taught to the naval cadets at the school in Hamina. His aim was to begin by defining a system of counting and then move on to relations of logic and geometry, before describing our location amongst the system of planets. The light pulses are all uniform except for one that increases and another that decreases in intensity. His 'alphabet' consists of twenty-two basic characters:

1. A short flash that he writes as zero or o.
2. Seven characters marked by short flashes. The number of flashes gives the number 1, 2, 3, 4, 5, 6, or 7. This is unexpected. It means that Neovius is going to work with a base-8 arithmetic rather than with our usual base-10, or 'decimal', system. He actually thinks a binary system would be better still but starts with a base-8 system so as to have a good number of characters available for storing information.
3. Eight characters denoted by longer pulses (one to eight) and represented on paper by a, b, c, d, e, f, g, h.
4. Three characters denoted by much longer pulses (one, two, or three of them) and represented on paper by the capital letters A, B, and C.
5. Two characters with the same (but longer) pulse length. One, denoted by *Tillt*, increases in intensity while the other, denoted by *Aft*, decreases.
6. One very long pulse, denoted by **𝔄.**, which will be used to mean 'and so on'.

1, 2, 3, 4, 5, 6, 7, 𝔄. —— 1 A 1, 2 A 2, 3 A 3, 4 A 4, 𝔄; 1 Tillt 2, 2 Aft 1, 1 Tillt 3, 3 Aft 1, 2 Tillt 3, 3 Aft 2, 𝔄. ——

2 a 3 A 5, 5 b 3 A 2, 2 c 3 A 6, 6 d 3 A 2. —— 7 a 1 A 10 A 2 c 2 c 2; 10 c 10 A 100; 10 c 100 A 1000, 𝔄; 2 c 10 A 20, 3 c 10 A 30, 𝔄; 2 c 100 A 200, 𝔄; 10 a 1 A 11, 10 a 2 A 12, 𝔄; 100 a 1 A 101, 100 a 10 A 110, 100 a 10 a 1 A 111, 𝔄. —— 17 c 26 a 3 A 515, 17 c 7 a 1 A 152, 515 a 26 A 543, 152 a 7 A 161. ——

e Aft 3, e Tillt 26 d 7, e Aft 515 d 152, e Tillt 543 d 161 𝔄; e A f d g; g A 2 c h, f A 2 c e c h. —— e c h c h A aa. —— 4 c e c h c h c h d 3 A ab. 4 c e c h c h A ac. ——

ad, ae, af, ag, ah, 1ba, 2ba, 3ba, 𝔄 224ba, bb, bc, bd, be; ad A 40000000 c ah, ad A 4634300 c ag, ad A 1700 c bb; ad ab A 4634300 c ag ab, ad ac A 26620 c ag ac, ad h A 154 c ag h. – ah bf A 1235 c ah bg, ag bf A 555 c ag bg a ag bg d 4. ——

1 A 1, 1 oA 2, 2 oA 1, ag A ag, ag oA ad. —— ag B ab, ag oB aa, ad B ab, ad oB aa, e oB f, e B bh. —— 1 B bh, 2 B bh, 3 B bh, 𝔄, 1 d 2 B bh, 1 d 3 B bh, 2 d 3 B bh, 𝔄. – 1 B ca bh, 2 B ca bh, 3 B ca bh, 𝔄; 1 d 2 oB ca bh, 1 d 2 B cb bh; 2 d 3 B cb bh, 3 d 2 B cb bh, 𝔄. —— e oB ca bh, e oB cb bh, e B cc bh. —— ca bh oB cc bh, cb bh oB cc bh, ca bh B cd bh, cb bh B cd bh. —— e ceA f d g, e o ceA 543 d 161, e cfA 543 d 161. ag cfA af. ag bf cfA 555 ag bg a ag bg d 4. ——

ad C ad, ag C ag, 𝔄; cc bh C ocd bh; cd bh C occ bh. C cfC A; —— 1 b 1 A o, 2 b 2 A o, 𝔄; bh b cg bh A o . bh d cg bh A 1. —— ch C ocg, bh b ch bh oA o; bh d ch bh oA 1. —— da C o; da ca bh Tillt 1. —— o d o A db bh; db cb bh B cd bh. dc C odb oda; dc cb bh Tillt 1. ——

2 Tillt 3, 3 Tillt e dd 2 Tillt e. —— de bh Aft e, df cg bh Aft 3, dg e Aft 3. de bh oB cc bh, df cg bh B cd bh. —— db bh B dh cd bh, dh cc bh. db cb bh dh Aft 1, db Tillt 1, ea 2 d 3 Tillt 1, 3 d 2 Aft 1. ——

eb cfC ea; dc ca bh Tillt e, ea 2; dc ca bh Tillt e, eb 1, 2, 3. —— ec cfC a, ed cfC oec, ea 1, 2 ec 3 Tillt e, ed 4 Aft e. —— ee cfC ec, ea ah Tillt ag, af ee Tillt ag, ed bb Aft ag. —— ef cfC cg, ea bh, ef Tillt e, ee Tillt 4; cd bh, ef Tillt 1, B cb bh, ed cd bh, ef Aft 1, B dh ca bh, dh cb bh. ——

f B cg, h ee B eg. h B eh eg; f B fa eg. aa B eh fb, ac B fa fb. ab B fc. —— fc fd 3 fe, fb fd 2 fe, eg fd 1 fe, ff fd o fe. —— fg B ff; db aa fd 1 fg; db ab fd ee 1 fg. —— ff, ef fh, ga eg; eg, ef fh, ga fb; fb, ef fh, ga fc. ag fh gb ad gc 1 ag bf, ag fh gb gd gc 1 ag bg; gd B g. bf B ge, bg B ee ge. bg A gf a gg, bf A gh a ha. gf B hb ge, gg B hc ge. ad B hb fc, ag B hc fc. gh B hd ge, ha B he ge. ad B hd fc, ag B he fc. ag B hf, db hf B hc ec he fc. ad oB hf. ad B hg, db hg B hb ec hd fc. —— 1hb cfC hb; 1hb fhb hh ad aaa ag gc ge, ef A 1 ag bg d 257. ——

Neovius' message, published in 1875.[46]

The first edition of Neovius' *The Greatest Mission of Our Time* (1875), which contained his message to extraterrestrials.

Using these characters you can make 'words' such as aa3 or cdC by leaving longer gaps between words than between the pulses which make up the characters within the words. Lastly, punctuation marks . , ; and —— are used for intervals between strings. Using these simple rules Neovius composes his message to the Martians.

It would be disappointing to the reader to be given the whole translation and so we will just indicate some early features of the message which give the flavour of his thinking. The first paragraph is purely descriptive. It introduces the signs being used – remember that the mode of transmission, like the number of pulses, explains some of the symbols more explicitly. The next list of symbols such as 1A1, 2A2, 3A3, . . ., ? are designed to show that A means 'equals' = and ? means *et cetera*.

The next examples are designed to show that *Tillt* means less than, <, while *Aft* means greater than, >, a meaning that is also reinforced by the falling and rising intensity of the two pulses used to represent them.

In the next paragraph, we are introduced to the four basic arithmetical operations of +, -, ×, and ÷ in the form of the characters a, b, c, and d. They are such as 2a3A5, 5b3A2, 2c3A6, 6d3A2, meaning 2+3=5, 5-3=2, 2×3=6 and 6÷3=2 (remember that for products and quotients bigger than 7 the arithmetic is base 8). Near the end of this paragraph we see that Neovius introduces some new characters: e, f, g, h. Using the inequalities, < and >, he indicates that e means the number *pi* (= 3.14 . . .), and then defines f, g and h as the circumference, diameter and radius of a circle. Next, we find some formulae such as 'echch A aa' meaning 'multiplied by radius and by radius again equals aa', and so 'aa' must therefore mean the area of a circle. See if you can interpret the two formulae, '4cechchchd3 A ab' and '4cechch A ac', that follow.

In paragraph four, Neovius uses these circles and spheres to describe aspects of the solar system. The new signs bf and bg are introduced to denote the numerical relations between Martian and terrestrial timescales for days and years: 'Earth bf = 655 × Earth bg' and 'Mars bf = 669 × Mars bg'.

The message goes on to introduce logical relationships, a symbol, bh, meaning 'number' and another, cg, meaning 'same as', illustrated by examples such as 'bh d cg bh 𝕬. 1'. The symbol for zero 'o' appears in rules like '1b1 𝕬. o, 2 b 2 𝕬. o, 𝕬.'.

These clues to Neovius' thinking should enable you to go on to read all of his message. Originally he tested it out (without clues) on his brother-in-law, the mathematician Lorenz Lindelöf, who decoded it in full on the same day that he received it.

Alas, there are no Martians, and so Neovius' focus upon the specifics of our solar system was inappropriate. But it is interesting to consider the similarities of style and structure in Neovius' pioneering message and the message that was carried a century later on the *Pioneer* Plaque. The SETI project has long been interested in receiving messages and there have been other plans to send out pulses into space for others to see. Today, we recognise that radio waves are more pervasive than optical light for signalling and, like Oersted and Neovius, we expect all intelligent beings to know about them. Neovius thus displays the supreme optimism of the modern searchers for extraterrestrial intelligence. Although his contribution is virtually unknown outside Finland, it is remarkable in its imagination, simplicity, and logic. He would have been a worthy first contact for any Martians who were listening.

Comic cover by Alex Schomburg, 1955.

ET, PHONE HOME
FLYING SAUCERS

. . . flying like a saucer would

Kenneth Allen[47]

NOTHING captures the essence of other worlds and extraterrestrials so universally and so unambiguously as the picture of a 'flying saucer'. Propulsionally dubious it may be, but the flying saucer is the hallmark of an entire genre of science fiction and fantastic speculation about life on other worlds and the possibilities of advanced space travel.

The idea of travelling to other worlds was explored by several authors in the nineteenth century, notably by the French astronomer Camille Flammarion, whose 1877 *Les Terres du ciel (Worlds in the Sky)* tells of imaginary journeys to other worlds in the solar system, and his later speculations about *The Inhabitants of Other Worlds* were extremely popular with a public fascinated by outer space and the discoveries of the telescopic age.

Earlier there had been a tradition of fantastic stories set in imaginary worlds, such as Swift's *Gulliver's Travels* of 1727 or Voltaire's *Micromégas* in 1752, that were really parables with an intent to comment on our world and its social absurdities. We would no more regard them as science fiction than we would confuse George Orwell's *Animal Farm* with an edition of the *Farmers Weekly*.

Science fiction in the modern style emerged along with new technologies and the possibilities for far-ranging terrestrial travel. Ballooning gave the idea that if you wanted to fly to the Moon it was just a matter of aiming higher. This technophilia became entwined with a spirit of progress that looked to the future by means of a detailed form of realism that extrapolated current technology in a convincing fashion. The most famous early exponent of the genre was Jules Verne, whose 1865 novel *De la Terre à la Lune (From the Earth to the Moon)* saw his heroes launched into space from the barrel of a 300-metre cannon (shades of the infamous Iraqi Supergun Project Babylon). But while there were imaginative propulsion systems and rockets, there were as yet no flying saucers.

It was three o'clock in the afternoon of 24 June 1947 and Kenneth Arnold was flying his private plane past Mount Rainier towards his home in Boise, Idaho, when the flying saucer was born. Arnold reported seeing nine bright, shining metallic objects, all roughly circular in form, except for one that was crescent-shaped, speeding southwards through the sky. He watched them manoeuvring through the mountain ranges for several minutes until they disappeared towards Oregon. Arnold stopped at the next airstrip to refuel and gave a detailed account of the appearance and movements of the objects he had seen. He described them moving like 'a saucer would if you skipped it across the water'. By 26 June, his story had been picked up by the press and wire services all over America and was national headline news. His innocent description of what he had seen as resembling 'flying saucers' stuck. The UFO cult was born and no amount of explanation by the air force that he had almost certainly seen a group of disc-shaped clouds above the mountains could put the genie back in the bottle again. Ever since that day, extraterrestrials have been expected to drive flying saucers. Science fiction comics explored the whole raft of pictorial possibilities during the 1950s through the golden age of illustrated science fiction story-telling. They were pervasive in a host of languages, appearing all over Europe in translations as the *fliegende Untertassen*, *soucoupes volantes*, or *dischi volanti*.

The UFO fantasy lives on (as a check on the web will dramatically confirm), but our interest here is simply to highlight the remarkable power and pervasiveness of the image of the flying saucer. It met a deep need that many people clearly felt to see and believe in life beyond the Earth. Such desires were once filled by the uplifting works of art to be found in the cathedrals of Europe evoking the other worlds of heaven and hell. Science fiction began a process by which the ancient religious symbols of other spiritual worlds were replaced by the signs of the technology and science of supposedly concrete worlds. At first, they were dominated by machines and gadgets that pointed towards the promise of future progress. Later, in the second half of the twentieth century, they would become dominated by the abstractions of cosmology and particle physics – time travel, black holes, wormholes, and other universes. The fantastic possibilities spawned by the study of Einstein's theory of relativity, or the emerging theories of elementary particles, provided a new source of other worlds that were not solely in the realm of the imagination. In many cases, they were more fantastic than any creations of the science fiction writers. The flying saucer was the tip of an imaginative iceberg that linked together science fact and science fiction: a single simple shape from the kitchen cupboard that linked us to minds on the other side of the Universe.

MISTER BENTLEY'S FEELING FOR SNOW

SNOWFLAKES

Always winter, never Christmas
C. S. Lewis[48]

ONE of the most beautiful of natural patterns is the six-fold symmetry of the snowflake. All sorts of myths surround it. Supposedly, no two snowflakes can be the same and each arm of a snowflake is identical to its five companions. Alas, beautiful and diverse as snowflakes are, neither of these romantic claims about them is quite true. Identical snowflakes are perfectly possible, but highly unlikely, and each arm of a single snowflake differs significantly from its five companions when examined closely. We are so impressed by the symmetrical architecture of the snowflake that we are fooled into imagining that the arms are much more alike than they really are. Although flakes appear to be the same in length and separation from each other, if we look very closely, we will see that they are different both from arm to arm and from one side of an arm to the other.

The first European scientist[49] to become preoccupied with explaining the organised complexity of the snowflake was Johannes Kepler, the great astronomer who discovered the mathematical regularity of the planetary orbits in the solar system. In 1611, while Imperial Mathematician to the Holy Roman Emperor Rudolf II, he wrote a book entitled *On the Six-Cornered Snowflake* as a 'New Year's Gift' to his patron.[50] He was the first scientist to try to explain the six-fold snowflake symmetry. Unfortunately, his approach didn't solve the problem. Kepler looked everywhere in the Universe for the fundamental

Just one of Wilson Bentley's microscopic photographs of a snowflake, published – with over 5,000 others – in 1931.

symmetries possessed by the Platonic solids and equated symmetry with direct design laid down by the Creator. He wondered whether the snowflake's shape had a definite cause, whether it had been designed for some unknown future end, or whether it simply could not be otherwise. Did its beauty have some special function, or was Nature merely being enchantingly 'playful'? In the end, he had to confess that he didn't know, but he was convinced that there was an answer for his successors to seek out and understand:

> The cause of the six-sided shape of a snowflake is none other than that of the ordered shapes of plants and of numerical constants; and since in them nothing occurs without supreme reason – not, to be sure, such as discursive reasoning discovers, but such as existed from the first in the Creator's design and is presented from the origin to the day in the wonderful nature of animal faculties, I do not believe that even in a snowflake this ordered pattern exists at random.[51]

Snowflakes have inspired an extraordinary devotion among scientists and photographers. The most famous was Wilson Bentley (1865–1931), a.k.a. the 'Snowflake Man', a farmer from Jericho, Vermont, whose photographic gallery of 5381 microscopic views of snowflakes displays his lifetime quest to capture their ephemeral beauty.

Bentley was self-taught, careful, and patient. His mother, a former schoolteacher, gave him a microscope on his fifteenth birthday. Providentially, it was snowing on that day, and the first thing he placed beneath the lens of his special present was a snowflake. And so began his life-long fascination and study. He meticulously adapted an old-style bellows camera to operate with the microscope, honing his skill until he could reliably capture the snowflakes' six-sided beauty. Needless to say, in his entire life, he found no two to be the same. His famous book, *Snow Crystals*, published in the year of his death,[52] established him as a legendary figure in the academic and natural history worlds, and one of the world's leading authorities on the structure of snow,[53] who published more than sixty research papers on the subject, although only once did he journey outside the state of Vermont. His pictures became the standard source for research papers, Christmas cards and decorations, textbooks, and popular articles. No one would dream of writing about snowflakes without showing one of his beautiful photomicrographs.

We understand a lot more about the shapes of snowflakes today than we did in Bentley's day, but by no means all that we would like to know. Snowflakes form around specks of airborne dust. Their six-armed configuration is inherited from the hexagonal symmetry that water molecules take up when they join together in a latticework pattern, which is shown here in two different views. The red balls in the picture are oxygen atoms and the grey rods are hydrogen atoms, two hydrogens for each oxygen (H_2O).

Despite this universal ground plan, the crystals all grow in slightly different ways. As conditions of temperature and humidity change, the floating crystal grows larger, the six corners of the hexagon extend further into the surrounding vapour and accrete faster, giving the appearance of growing arms. Within each crystal, the ambient conditions are fairly uniform, and so each arm grows at about the same rate, 60 degrees from its neighbour, producing complicated branches and sub-branches all along its length like a sprouting Christmas tree.[54]

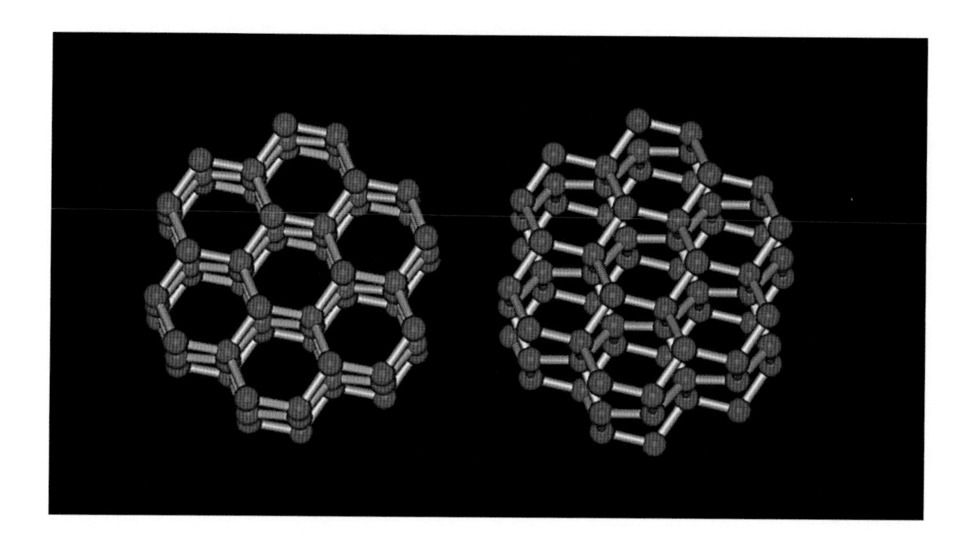

Molecular model of one of the several structures of ice, the frozen form of water. The crystalline lattice structure of ice explains why it is a hard, brittle solid. Here the six-fold hexagonal symmetry (known as Ih) explains the hexagonal symmetry of snowflakes. The oxygen atoms are shown as red spheres, the hydrogen atoms as grey rods. Two hydrogen atoms and one oxygen atom make up each water molecule. The structure at the right is the same as the one at the left, but seen from a different angle.

Although the arms look identical at first glance, the eye is being duped by the dominant six-fold symmetry. If you look closely, the detailed crenellations are different from arm to arm, and also on each side of any one arm. The longer and more complicated the history of the falling snowflake, so the more complicated its arms can become and the more differences there will be from arm to arm. Other snowflakes will not share the same thermal history as they fall through different parts of the fluctuating atmosphere, and so the patterns that each one develops will be different. As they grow bigger, they will find more ways to be different, while tiny, embryonic, snowflakes containing only a few dozen water molecules will all look virtually identical. This order in the face of diversity is what makes a snowflake so fascinating.

A typical snow crystal that acts as the seed for a flake contains about one billion billion water molecules and about one in 1000 of them will be out of the ordinary, containing a deuterated water molecule or a different isotope of oxygen, so one thousand billion of them are all likely be different from each other. These large numbers mean that you are never going to come across two the same. A billion is a very large number: if you started counting now you would not live long enough to reach a billion. For all practical purposes snowflakes are unique.

In reality, the fundamental crystal shapes that form the basis of snowflake architecture may also start out differently, depending upon the way the snow is forming and falling. All sorts of water crystals form in the atmosphere – some so small that we tend not to notice them at all – with a range of different shapes and sizes. Classic snowflakes are the most beautiful and photogenic so they tend to grab all the attention.

There are still lots of intriguing mysteries about snowflakes. How do the tips of each arm 'know' how long the others are? What is the process which keeps the growing arms in step? One theory is that vibrations occur within the crystal and become synchronised like a squad of soldiers marching in step. These coordinated oscillations can then keep distant parts of the growing flake in unison. The random nature of all these processes means that no arm gets to grab more vapour than any other and so to very high accuracy they grow at the same rates. All this sounds intuitively very reasonable, but we still lack a definitive description of the internal vibrations that do the trick, and do it so well, under rapidly changing conditions with so little variance.

Modern close-up of a snowflake, showing the 'almost' six-fold symmetry and revealing the tiny asymmetries within and between arms.

Alexander von Humboldt's geography of equatorial plants, showing the variation of flora and fauna with altitude around the Chimborazo volcano in Ecuador (1817).[55]

UP THE AIRY MOUNTAIN

BARON VON HUMBOLDT'S ECOLOGY OF PLANTS

The Baron spoke English very well, in the German dialect. Here I shall take notice that he possessed a surprising fluency of Speech, & it was amusing to hear him Speak English, French and the Spanish Languages, mixing them together in rapid Speech. He is very communicative and possesses a surprising fund of knowledge, in botany mineralogy astronomy Philosophy and Natural History: with a liberal Education, he has been collecting information from learned men of almost all quarters of the world; for he has been travelling ever since he was 11 years of age and never lived in any one place more than 6 months together, as he informed us.

Charles Willson Peale[56]

ALEXANDER VON HUMBOLDT is not a household name outside Germany, but his claims to be one of the most important scientists of the past 200 years are extremely persuasive. Born the son of an army officer in Berlin in 1769, he was well educated and socially privileged enough to be involved in a wide range of political and intellectual affairs at a time when Prussia, under Frederick the Great, was at the heart of European politics. Inspired by a meeting with Joseph Banks in London while making his maiden tour of Europe, the young Humboldt was determined to follow in Banks's footsteps by travelling the world for the benefit of science. He extended his formal education to equip himself better for the task and then began a career of exploration, discovery, cataloguing, and sheer scientific entrepreneurship that would have been remarkable had it been achieved by a dozen people. He did pioneering work in botany, geology, meteorology, animal physiology, geomagnetism, and zoology, discovered new minerals and diamond deposits, and carried it all out with a careful eye for national economic advantage and interdisciplinary scientific collaboration. Back home in Germany, he played an important role in the reform of universities, the fostering of research, and the linking of university scientists with industry. He also promoted the development of scientific conferences and the creation of learned societies. Overall, he was responsible for a massive professionalisation of science and educational practice. His wide travels to visit other scientists and learned societies around Europe helped form the first international scientific collaborations and he established the first transatlantic scientific links with the United States, where Thomas Jefferson was a great admirer of Humboldt's knowledge, curiosity, and energy.

Humboldt spent most of the period between 1805 and 1834 writing up some thirty volumes of memoirs and scientific data collected on his earlier travels, notably those in South America. He was one of the first scientists to be interested in developing big pictures of climatic trends around the world. Many others had gathered plant specimens and recorded the countries of origin, and the dates and times when they were found, but Humboldt made wide-ranging studies of temperature variations around the Earth and recognised the importance of considering other variables still, such as altitude and aspect, in these ecological maps. These were the forerunners of modern maps of average seasonal temperature variations around the Earth, which have recently become of crucial importance in the study of global warming. In 1817, one of his papers[57] reporting on the variation of the mean temperature around the Earth's surface invented the concept of the isotherm.

Humboldt created many important pictures that encapsulate vast amounts of information and allow its retrieval at a glance. He was a good artist and a meticulous draughtsman who recognised the way in which a well-constructed diagram could do the work of many pages of data and explanation. We have chosen an impressive example of this influential work. During the period 1799–1804 von Humboldt carried out extensive investigations in central and South America, accompanied by the botanist Aimé Bonpland.

This picture (page 204) shows the ecology of plants that he found in and around the extinct Chimborazo volcano in Ecuador. The slice through the mountain (which is 22,000 feet high) is used as a surface on which to mark the altitudes and distribution of the various flora he found there, which he records with both their Latin and common names on the charts alongside the artwork. He recognised that comparative studies were possible between equatorial species at high altitudes and sea-level species at temperate latitudes. The high peaks in the picture are the Cotopaxi and Chimborazo volcanoes, with the height of Vesuvius added for comparison. The cloud level and the tree line are clearly illustrated. A simple idea, elegantly conceived and painstakingly executed, that set a standard for the presentation of ecological information.

The diversity of Humboldt's interests led naturally to his introduction of comparative studies, looking at correlations between climate, diversity, flora, and topography. His creative use of visual representation provided natural scientists with vivid representations of multi-dimensional collections of data. It presaged a new way of thinking about and correlating facts, and is a fitting memorial to the great German polymath's talent for the synthesis of simplicity and complexity.

A
DELINEATION
STRATA
ENGLAND AND WALES,
SCOTLAND;
THE COLLIERIES AND MINES,
THE MARSHES AND FEN LANDS
VARIETIES OF SOIL
ILLUSTRATED
THE IRISH SEA
THE GERMAN OCEAN
CAERNARVON BAY
CARDIGAN BAY
ST. GEORGE'S CHANNEL
BRISTOL CHANNEL
THE ENGLISH CHANNEL

ROCKIN' ALL OVER THE WORLD

SMITH'S STRATA

The geological formations of the globe already noted are catalogued thus: The primary, or lower one, consists of rocks, bones of mired mules, gas-pipes, miners' tools, antique statues minus the nose, Spanish doubloons and ancestors. The secondary is largely made up of red worms and moles. The tertiary comprises railway tracks, patent pavements, grass, snakes, mouldy boots, beer bottles, tomato cans, intoxicated citizens, garbage, anarchists, snap-dogs and fools.

Ambrose Bierce[58]

SCIENCE tends to be team work. Many different people contribute their efforts, like single bricks, one by one, to build up an edifice of knowledge from simple beginnings. True, someone often manages to make a crucial contribution by discovering the solution to an important problem that is barring progress, while others may well discover how to make a better quality of brick, but the history of most sciences is a drama in which many actors play important supporting roles. However, there are notable exceptions. Our understanding of the geology of England and Wales is largely due to the painstaking efforts of a single man, with no formal education, over a period of twenty years. Mapping the Earth's surface in a three-dimensional

William Smith's original geological map of England and Wales, 1815.

way was a rather late invention. The first topographical map, with contour lines showing altitude (or depth below the sea) was drawn by Marcellin du Carla-Boniface, and included in an article written by him in 1782,[59] but no one had prepared maps of what lay under the Earth's undulating surface.

William Smith's father, a blacksmith, died in 1777, when William was only eight years old, and his young son seems to have largely brought himself up after that, attending his village school in Churchill, Oxfordshire, and immersing himself in a range of solitary activities such as collecting fossils and making maps. Upon leaving school as a teenager, he readily found employment in the business of surveying and mapping the routes for a booming new network of canals that were planned for his part of the country. Smith took the job in his stride, gaining ever-greater responsibility, and was soon overseeing the construction of canals himself. He travelled extensively around the country to learn about the different techniques being used to excavate the canal channels, and at the same time started keeping detailed notes about the rock strata revealed by the excavations and built up a fine collection of fossils in the process. From now on, his friends began to call him 'Strata Smith'.

Impressed by the systematic patterns that were emerging from his growing coverage of different parts of the country, Smith started to draw detailed maps of the geological layout in parts of south-west England around Bath – the first such maps ever drawn. Encouraged by his success, in 1805 he set himself the ambitious goal of drawing a complete geological map of England and Wales. Along the way to achieving his goal, he would develop a range of new principles for dating and ordering rock strata and fossils. From this account it would be easy to think of Smith as a geology professor gathering his data, returning to his colleagues to deliver lectures and write learned articles. But Smith had little education. He had never been near a university and writing books presented him with a considerable challenge. Instead, he published short leaflets which displayed his maps and cross-sections through the different rock strata as fine drawings. Other members of his family helped with the drawing and, although he always struggled for money, sometimes selling parts of his collection of fossils to support himself, his achievements were eventually recognised by the highest award of the Geological Society in 1831. It is fitting that if, today, you go into the entrance to the Geological Society's lecture rooms on the Royal Academy Square on Piccadilly, opposite Fortnum and Mason's, then halfway up the staircase on the wall, protected from the light by a movable curtain, is Smith's greatest achievement: the *Delineation of the Strata of England and Wales with Part of Scotland*. Completed in 1815, ten years after he first set his mind to the task, it is a huge piece of cartography, 9 ft by 6 ft, drawn on a scale of 5 miles to the inch and made from fifteen smaller, square, copperplate engravings.[60]

The painstaking work that lay behind Smith's map is extraordinary. This was an era of unmade roads, slow and uncomfortable travel on dangerous and unlit routes, and few maps to guide the stranger. Despite seemingly insuperable practical obstacles, William Smith managed to travel the length and breadth of the country and put together the big picture that formed the foundation stone on which all the geologists that followed him would build.

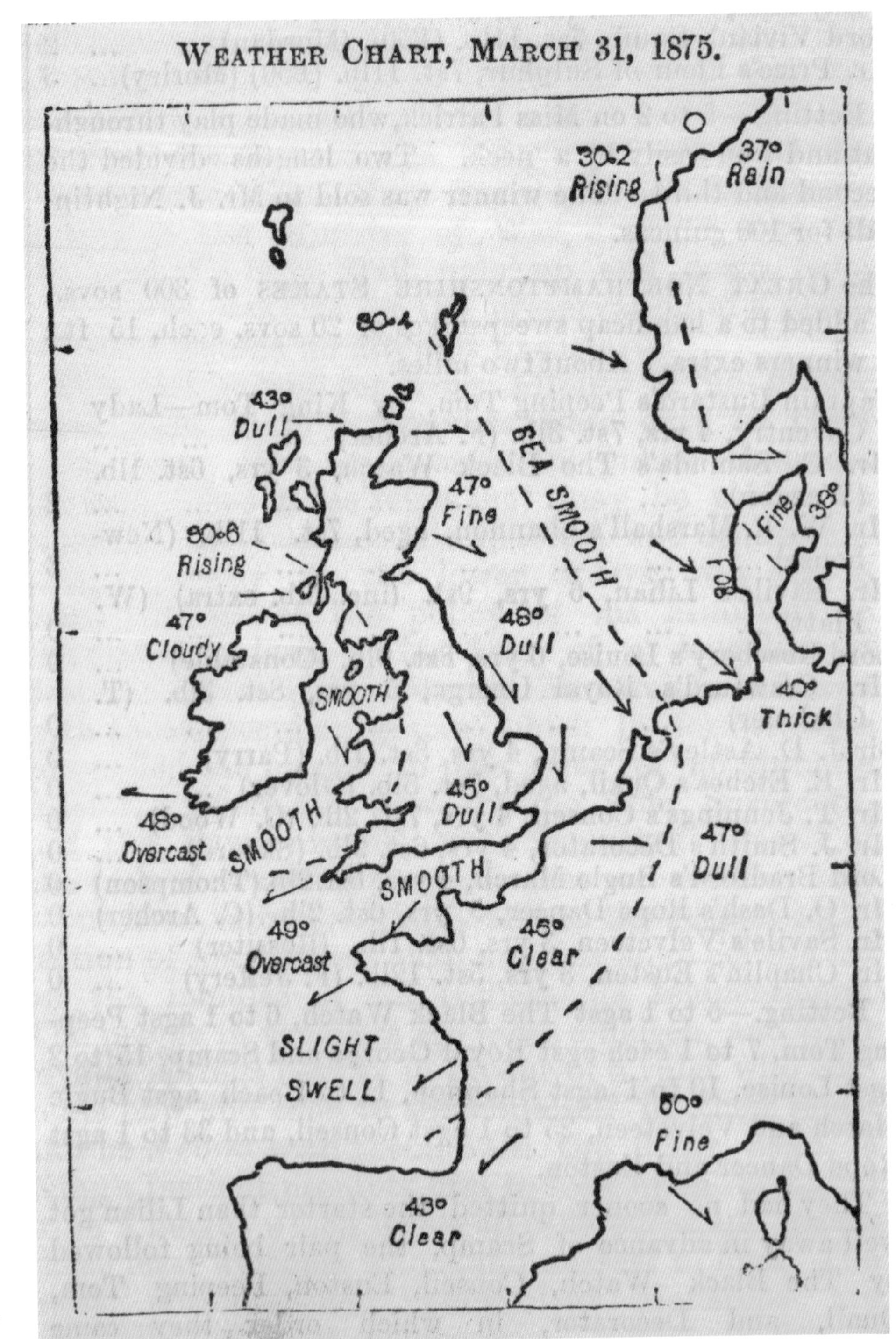

The first weather chart, prepared by Francis Galton for 31 March 1875 in *The Times*, London, on 1 April.

GONE WITH THE WIND

WEATHER MAPS

Weather, A permanent topic of conversation among persons whom it does not interest.

Ambrose Bierce[61]

THERE was a time when the most sophisticated and regular exposure of ordinary citizens to science came about when they tuned in to watch the daily weather forecast on the television or glance at the weather map in the newspaper. The maps were real scientific charts with isobars tracing contours of equal atmospheric pressure alongside the passage of cold, warm, and occluded fronts. Alas, today the weather forecast in many parts of the world is increasingly part of the entertainment industry and televised maps contain no quantitative information other than temperatures. But weather maps are still instantly recognisable, and look equally distinctive in black and white on pages of newsprint as on ancient black-and-white television programmes where weather forecasters drew the isobars and fronts with felt-tip pens. So who first drew these impressive weather maps and when?

Much of the groundbreaking work in meteorology was carried out by someone who is better known for his energetic, and often controversial, studies of statistics and heredity – and much else besides. Francis Galton was the first to recognise that fingerprints provided a way of uniquely identifying individuals and the classification of the distinctive marks and patterns that are used to distinguish them are still known as 'Galton's marks'. He was always looking for ways to correlate the abilities of parents and children by statistical study; in fact, he was the first person to use the word 'correlation' precisely in the context of statistical studies. He also played an important role in the exploration and mapping of the Nile. Galton's work on statistical interpretation and use of data was important and lasting, so much so that it might be hard to find a statistician today who thought that he did much else besides. But, from 1860 onwards, Galton did more than anyone to found the systematic science of meteorology. He identified anticyclones (high-pressure systems) and devised a system of symbols to denote directions of winds and differences in air pressure. He invented numerous instruments for automatic recording of information, such as barometric pressure, by ingenious set-ups which allowed light to shine through onto sensitive paper at different heights as the barometer level changed. In 1863, he drew his weather work together into the book *Meteorographica,* which presented the first attempts to gather weather data over substantial continental regions, represent it on charts using simple symbols and isobars, and then apply it to interpret and predict the way weather changes occur.[62]

Later, in 1875, Galton created the first of many weather maps for *The Times* newspaper in London. It appeared on 1 April and showed the weather system over north-western Europe on the previous day. It signalled the beginning of the British public's engagement with the weather forecast.

By comparison, a modern forecast for the same region thirty-one years later, from the Meteorological Office in the UK, would show barometric pressure isobars in units of millimetres of mercury.

Charts of the surface pressure are now updated every six hours and show isobars linking points with equal pressure at 4 mB intervals of pressure. Cold, warm, and occluded weather fronts are marked by triangles, semi-circles, and the two symbols together, respectively. The path of a front gives information about likely sites of rainfall, the closeness of the isobars tells us the strengths of the winds, which, in the northern hemisphere, blow in an anticlockwise direction around a low-pressure centre (depression) and clockwise around a high-pressure centre (anticyclone).

Galton's fascination with mapping and the gathering of large quantities of data for statistical analysis led him to turn meteorology into a semi-quantitative science. It would be a little longer before the study of atmospheric physics developed,[63] and a lot longer before the chaotic[64] sensitivity of weather prediction to small uncertainties in its present state was understood, but Galton's strong feeling for the power of visual information led him to create an enduring piece of visual science. The weather map does a remarkable job in combining information about changes of several quantities in time and space into a single chart. It serves the expert as well as his neighbour who merely wants to know if he should travel with his umbrella tomorrow.

A drawing of some dinosaur models made in Victorian times by Benjamin Hawkins, aided by the palaeontologist Richard Owen. When the models were finished in 1853, a grand dinner was held for the scientific great and good of the day inside the model of the Iguanodon. The models are still on display in London on an island at Sydenham, near the original Crystal Palace.

WALKING WITH DINOSAURS

THE WORDS AND THE PICTURES

Dinosaurs are Nature's Special Effects.

Robert T. Bakker[65]

NO idea in science captures children's (and adults') imaginations more powerfully than dinosaurs. Click on Amazon to see just how many children's dinosaur picture books there are – the list was so long I couldn't get to the end. Their popularity is understandable: the reconstructions of their appearance and lifestyle sit beautifully beside their fantastic names – brontosaurus, Tyrannosaurus rex, triceratops, and pterodactyl – which conjure up expectations that no mere public relations company could have created. Unfortunately, the word 'dinosaur' has become synonymous with an inability to change and adapt to new circumstances, a not entirely just legacy for a species that got along quite successfully for 170 million years before catastrophic changes in their environment ended their time, and made space for the mammals to emerge and eventually dominate the Earth. We humans, by contrast, have been on the scene for only about 2 million years.

Pictures of dinosaurs are instantly recognisable the world over, and have done more to recruit young people into palaeontology and its related sciences than any other single factor. Yet, the whole dinosaur business is surprisingly recent. Although bones had been unearthed in China, and elsewhere, by the Greeks and Romans for thousands of years, and had often been attributed to the dragons, or other fantastic creatures, starring in their myths and legends, they are ultimately a Victorian English invention.

The first fossils of giant reptiles were found and interpreted by the amateur fossil hunter Gideon Mantell and the eccentric Oxford geologist William Buckland, who gave the name megalosaurus to the first of them to be unearthed, in 1824. Mantell's description of these creatures was downbeat. He saw them as 'lowly creeping' things, closely akin to more familiar snakes and lizards. Huge sea-serpents were familiar from sea stories and many of these early fossils, such as Hydrarchos, matched these expectations quite well. They didn't really capture the imagination until they underwent a makeover at the hands of the anatomist Richard Owen (1804–1892). Owen introduced the collective name 'dinosaurs' ('terrible lizards') to distinguish them from living reptiles and this raised their profile dramatically. Even before huge bone fossils had been found, he envisaged them growing to five or six times the size of the largest elephants. Owen recognised several of their common structural features, such as the way that five of their vertebrae were fused to the pelvic girdle, and the way in which their legs supported their bodies underneath their bellies rather than spread out on jointed limbs to the side of their torsos. Probably most striking of all to the uninitiated was his creation of the image of a past age of titanic struggles between huge aggressive dinosaurs such as Tyrannosaurus rex and their smaller cousins, which superseded even our mythological images of fantastic creatures like centaurs and unicorns battling with each other to the death.[66]

Owen seems to have had a reputation for being a little unscrupulous in his scientific dealings and was certainly an adept self-publicist. He managed to wrest the focus of attention away from Mantell and Buckland by creating a spectacular new public image for his rebranded 'dinosaurs'. They made their first big impact on the public imagination at the Great Exhibition of 1851. Owen succeeded in placing an exhibit of re-created dinosaurs in the great glass Crystal Palace at Sydenham, just south of London. With the help of the sculptor Benjamin Hawkins, Owen oversaw the design and construction of life-sized statues of dinosaurs, made out of metal and concrete, in a Victorian theme park that was opened by Queen Victoria. The Queen's husband, Prince Albert, was fascinated by dinosaurs, and a great fan of Owen and his public lectures, and his personal support was a key factor in getting Owen's dinosaurs into the Crystal Palace exhibition. Their exhibit played host to gatherings of leading scientists and politicians of the time. Invitations to scientists to attend a grand opening dinner inside the reconstructed body of one of the larger igauanodon dinosaurs were sent out on simulated pterodactyl wings. Very large amounts of alcohol were consumed at this banquet, by all accounts.

The way that Owen and Hawkins portrayed the dinosaurs in concrete at Sydenham also set in stone the way they would be viewed for a very long time afterwards by the world at large. Although the old Crystal Palace was destroyed by fire in 1936, the models of Owen and Hawkins survived, and can still be seen in Crystal Palace Park today in artfully created surroundings on a grand scale.

Owen and Hawkins actually possessed very few fossil bones and their reconstructions of the appearance of the dinosaurs owed a lot to the imagination. Their representation of their lifestyles owed everything to it – some were shown on land, others swimming in a natural lake[67] – but we know now, with the help of vastly more complete fossils, that many of these dinosaurs looked very different from their imaginative reconstructions at Sydenham. Owen chose to set the dinosaur models in mortal combat with each other, and fixed in the public mind the image of battling titans, red in tooth and claw – a sort of Victorian *Jurassic Park*. Comics, novels, and movies eventually did the rest.

All the earliest dinosaur fossil finds were made in Europe, but in 1858 the first American dinosaur was found near the small town of Haddonfield in New Jersey, and was named *Hadrosaurus* in its honour. This was an exciting find because the skeleton was almost complete and it convinced scientists that some dinosaurs must have walked on two legs. This American find, and news of the Owen's great park at Sydenham, led to enthusiastic plans to incorporate a dinosaur exhibit into the plans for the new Central Park that was being designed for the city of New York. Hawkins was brought over from London to advise. The 'Great Paleozoic Museum of Central Park' project looked to be on course in 1870, and Hawkins' models were steadily being assembled, but, alas, political corruption and violence put a stop to it. The new city chief, William 'Boss' Tweed, couldn't find a way to profit from the dinosaur exhibit, and so was determined to stop it in its tracks and replace it by a project that offered greater scope for illegal kickbacks. After obstructing its progress in various ways, he simply sent in his gangsters to smash up all the models in Hawkins' workshops. Appalled by this turn of events in the wild east, Hawkins moved briefly to Princeton University but was soon on his way back to England, and the Central Park dinosaurs became extinct.

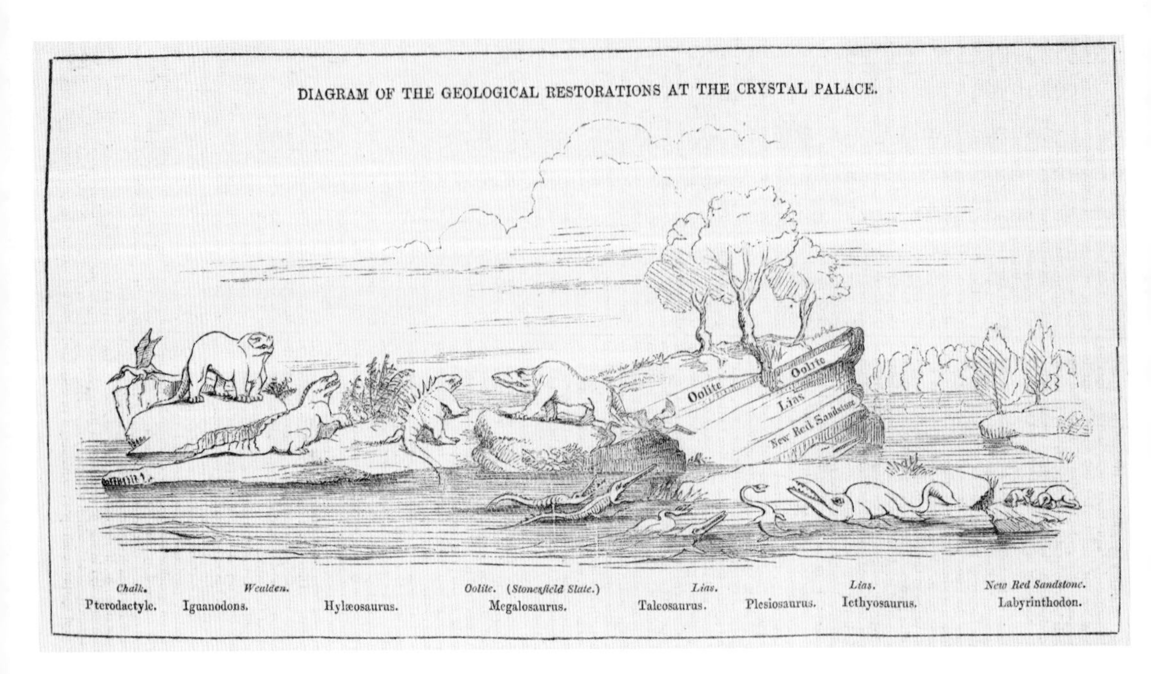

Plate by Benjamin Hawkins of the re-creation at Sydenham. From left to right, the dinosaurs are Iguanodon, another Iguanodon, Hylaeosaurus and Megalosaurus. The sequence runs chronologically from right to left, so animals found in the New Red Sandstone, at the far right, are the oldest, while animals found in the Chalk, at the far left, are the most recent. The crocodile-like animal on the shore in the centre is a Teleosaurus, a non-dinosaur Mesozoic reptile, and other Mesozoic marine reptiles, such as Plesiosaurus and Ichthyosaurus, may be seen in the water at the right.

Fortunately, the images of the dinosaurs outlived these human disasters. The first pictures and statues of these fantastic creatures shaped our picture of the prehistoric world. Once a picture is printed on our minds it is very hard to overwrite or redraw it. The dinosaurs may undergo changes of name occasionally, as palaeontologists attempt to better order our knowledge of their lives and times.[68] We still argue about whether they were warm- or cold-blooded, and there are a good deal more of them in the menagerie. But our understanding of them will always be dominated by their fantastic names and our mental pictures of what they might have been like, huge and ferocious, as they wandered the Earth hundreds of millions of years ago. And, for the most part, it is the unlikely Victorian duo of Richard Owen and Benjamin Hawkins that we have to thank for those indelible and influential pictures.

An individual footprint at the Laetoli site in Tanzania, preserved in volcanic ash for 3.6 million years.

STEPPING OUT
LAETOLI FOOTPRINTS

A feat of clay.
Anon.

IN 1976, the Yale University palaeoanthropologist Andrew Hill was working with Mary Leakey's research group on the excavation of an early-hominid archaeological site in Laetoli, Tanzania, when he unexpectedly stumbled across one of the most spectacular prehistoric discoveries ever made: a line of hominid footprints left in the mud 3.6 million years ago. Up until then, the earliest known human footprints were only tens of thousands of years old. Remarkably, Hill's Laetoli footprint trail was nearly 30 metres long and preserved in volcanic ash that had turned to a cement-like solidity as good as a plaster cast. It left us with an action replay of one of the first species of prehistoric hominids who walked upright on two legs. The ideal surface material was created by a coating of volcanic ash from the nearby Sadiman volcano that had settled on a sandy surface. When rain then fell, the sooty volcanic sediment became soft, like wet cement, and all the birds and small animals that walked on it left their small prints. But they were joined by the tracks of two hominids, one large and the other smaller, trailed possibly by a third (child?), whose tracks share some of the larger individual's footprints. A further eruption of dust from the volcano served to seal up the footprints for posterity before the rains could return and wash them away. The light rain then turned the ash into a cement, which set solid. There, they remained until they were exposed by millions of years of gentle erosion.

The Laetoli prints have led to all manner of deductions and speculations. They differ significantly from chimpanzee footprints and are not very different from those of modern humans, with an aligned big toe, heel, and an arched foot.[69] The smaller individual of the two (because they left shallower prints) is lopsided, bearing more weight on one side – perhaps carrying a baby. Also, it became immediately clear that bipedalism was not what led to tool making because tool making did not begin until a million years after the Laetoli bipeds had lived. Increased brain size was the most likely stimulus for tool making, rather than simply the freeing up of hands for other uses when walking on all fours ceased, as some had proposed.

The individual footprints also contain more interesting information about those who left them. The big toe lies close to the smaller toes, just like on our feet, and doesn't point off in a different direction, like a chimpanzee's toe. It could not have been used as a thumb. The sequence of depressions also reveal the pattern of walking, with an initial heel strike followed by a push-off by the front toes, again just like our own stride pattern. The length of the footprints along the two trails are 19 cm and 20 cm (about 'boys size 1' at the shoe store in the UK), indicating individuals who were probably about 120 cm (4 ft) and 152 cm (5 ft) tall, respectively.[70]

For the first time we had what amounted to a photograph of pre-human intelligent activity that could be readily understood in terms of our present-day movements. A summer stroll on the beach could create the same impressions on the sand; only a marvellous accident of Nature could preserve them. But will there be anyone here 3.6 million years from now?

These remarkable glimpses of prehistoric life catch our ancestors in motion, tempt us to ask questions about them that we never dreamed we could answer, and make us hope that equally sensational finds might come to light in the future.[71] Maybe Neil Armstrong's footprints on the Moon will one day fascinate someone else just as much.

The trail of footprints left by two bipedal hominids 3.6 million years ago.

The representation of a pumpkin, from *De historia stirpium (On the History of Plants)*, 1542.

THE FIRST PICTURE SHOW

THE FLOWERING OF ILLUSTRATION

As far as concerns the pictures themselves, each of which is positively delineated according to the features and likeness of the living plants, we have taken peculiar care that they should be as perfect as possible . . . We have devoted the greatest diligence to make sure that every plant is depicted with its own roots, stalks leaves, flowers, seeds, and fruits.

Leonhart Fuchs[72]

IN the sixteenth and seventeenth centuries the interplay between art and science was especially symbiotic. Many great works of art were influenced by the growth of scientific knowledge and science benefited in spectacular ways from art. The discovery of perspective by Renaissance artists allowed the accurate representation of three-dimensional figures in landscapes and the visualisation of geometries on curved surfaces. Although mathematicians would not understand that self-consistent geometries existed on curved surfaces until the nineteenth century, artists were intuitively aware of the analogues of straight lines and triangles on curved surfaces with positive and negative curvatures long before them. There were specific sciences where the use of accurate perspective was of crucial importance. Anatomical drawings were more impressive than mere words in describing the results of human dissections. The accuracy and detail that correct perspective allowed added a new dimension to the scientific understanding of the human body. Artists needed to understand the structure of the human body in great detail if they were to accurately and realistically portray the shapes of bones and undulations of muscles and sinews in different states of tension and extension. Some of the greatest artists and sculptors, such as Leonardo and Michelangelo, studied human anatomy with an eye for detail that few scientists could match and none could so perfectly re-create.

The elegance and accuracy of artistic illustration played an important role in the creation of influential texts that summarised human knowledge in forms that could be easily reproduced and widely disseminated. Vesalius' great anatomical treatise of 1543 – examined in the following chapter – was a landmark in the understanding of the human body. Just one year earlier, Leonhart Fuchs's great work of botany *De historia stirpium* (*On the History of Plants*) was published, first in Latin and then in German translation. Just as Vesalius stressed the accurate representation of the body's parts without recourse to superstition and symbolism, so Fuchs presented Albrecht Meyer's drawings of plants made from life, beautifully and accurately rendered, as copies of Nature. The text of the densely illustrated book was not especially innovative – it simply passed on received wisdom, slightly extended, from the usual ancient botanical authorities such as Dioscorides, Galen, Theophrastus, and Pliny. Remember that plants were also of prime importance for herbal medicine and Fuchs wanted his book to become the primary reference for doctors and medical students. But it was his departure from stylised and symbolic pictures of plants that was radically new. Previously, the plants had been used to decorate books of botany rather than to illustrate their texts. To press home this new emphasis, there was even a special plate in the book which showed the artist and Heinrich Füllmaurer, who transferred the hand-coloured images from paper to the woodblocks for cutting, at work on the drawing of a corn cockle, in order to

Frontispiece to *De historia stirpium*, showing the three principal artists involved in creating more than 500 beautiful illustrations from woodcuts. Albrecht Meyer drew the plants from life, Heinrich Füllmaurer then transferred them to woodblocks, and they were cut into wood by the extraordinarily skilful wood engraver Veit Rudolf Speckle.

PICTORES OPERIS,

Heinricus Füllmaurer. Albertus Meyer.

SCVLPTOR

Vitus Rodolph. Speckle.

stress the faithfulness of the transfer of information from Nature to the pages of the book via the printer's block.

The creation of such beautiful illustrations also coincided with a major technological advance. The invention of printing from movable type meant that many copies of books could now be produced, including illustrations, in large numbers in a cost-effective manner. Previously, book production was slow and expensive, and the incorporation of drawings by hand was extremely laborious. Often, only the original production and a few special copies prepared by the author would contain the illustrations. All the other copies would simply omit them. And even if they did appear in a multiplicity of copies, they were liable to become degraded in quality, gradually accreting errors and omissions as cheaper copies of the original led to copies of copies, and so on. Like rumours, the original was gradually transformed by its retelling.

Printing changed all this. Every printed copy was as perfect as the original. At first, illustrations were engraved onto metal plates and printed separately, each on a page of its own, and interleaved by hand into the text. The advent of woodcuts meant that text and illustrations could be intermingled and printed on the same page.[73] Time-consuming separate pagination was eliminated, and text could easily refer to an illustration that was adjacent to it.

These early illustrations paved the way for something that we just take for granted now. In the pre-photographic era all scientific illustrations had to be hand-drawn. The artistic skill required would be daunting for most naturalists and scientists practising today. Fortunately, in the sixteenth century, the link between art and science was strong, and the natural world provided the secular artist with subjects which were as impressive in their intricacy as the crucified human form was for European artists with a religious imperative.

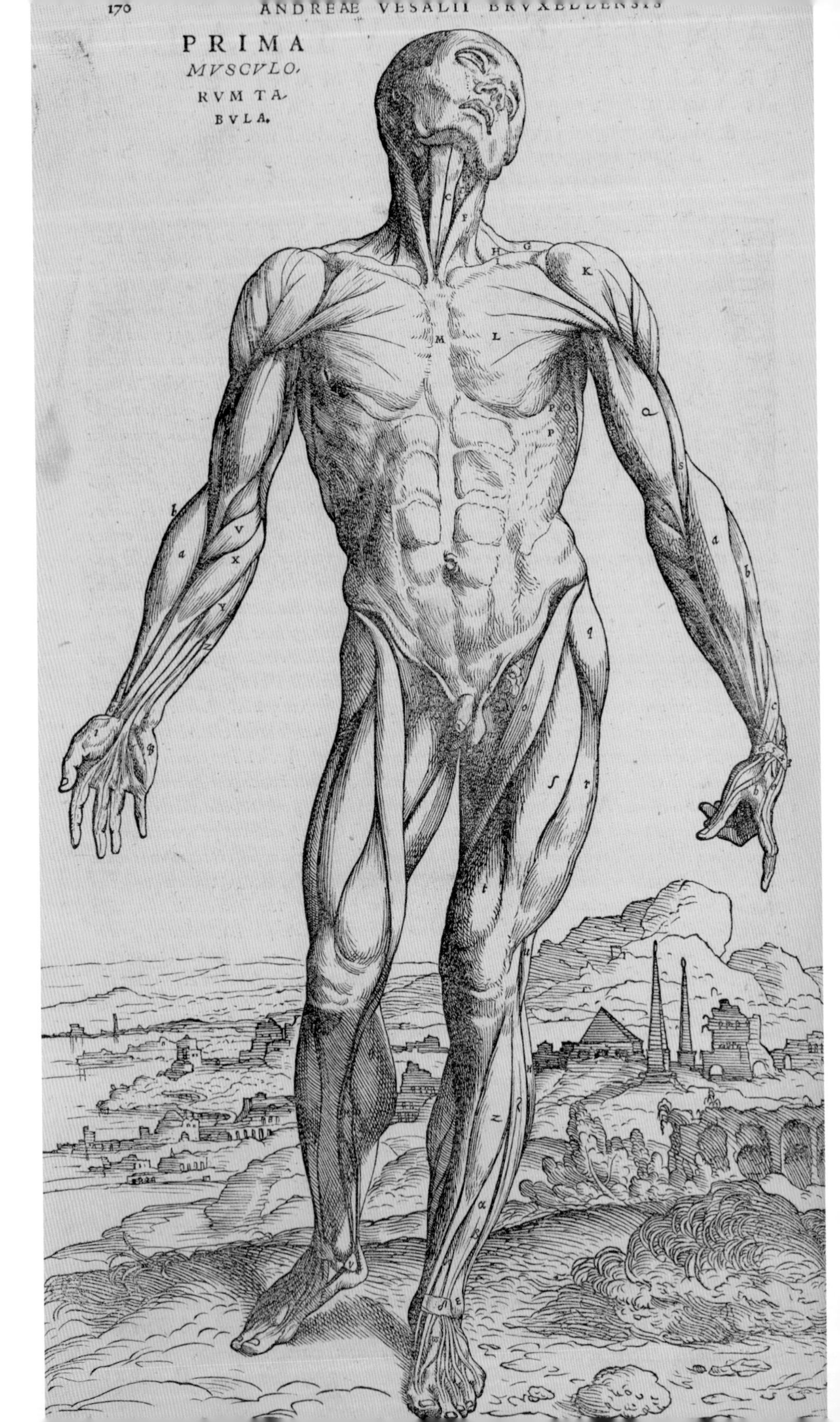
170 ANDREAE VESALII BRVXELLENSIS
PRIMA
MVSCVLO-
RVM TA-
BVLA.

SPECTACULAR BODIES

VESALIUS AND THE HUMAN FRAME

The cure for boredom is curiosity.
There is no cure for curiosity.
Ellen Parr

ANDREA VESALIUS was a Belgian anatomist who transformed our understanding of the human body through the pages of his illustrated volume *De Humani Corporis Fabrica* (*On the Fabric of the Human Body*), first published in 1543 and dedicated to the Holy Roman Emperor Charles V. Until then, human anatomy had been dominated by the received wisdom of Galen of Alexandria. There had been little imperative to check what he had to say, and Vesalius showed that a good deal of Galen's information about human anatomy had been derived from dissections of animals rather than humans. The ancients also provided only words, no pictures, in their description of our anatomy.[74] In the first plate of the human frontal muscle system, shown here, Vesalius wanted these pictures not only to educate anatomists and surgeons as to the detailed local structure of bodily organs, but to display a total view of the scheme of muscles such as 'only painters and sculptors are wont to consider'.[75] In the background to the series of plates of this type there are scenes from the Abano Terme region of Padua, which form a continuous panorama of the landscape if the plates are placed side by side.

Vesalius was the first to present a complete and detailed picture of the interior of the human body, which he illustrated by more than 200 meticulously prepared woodcuts showing bones, arteries, muscles, brain tissue, and all of the vital organs. His proud display of the tools of his dissecting trade near

Andrea Vesalius, standing man, from *De humani corporis fabrica* (*On the Fabric of the Human Body*), 1543.

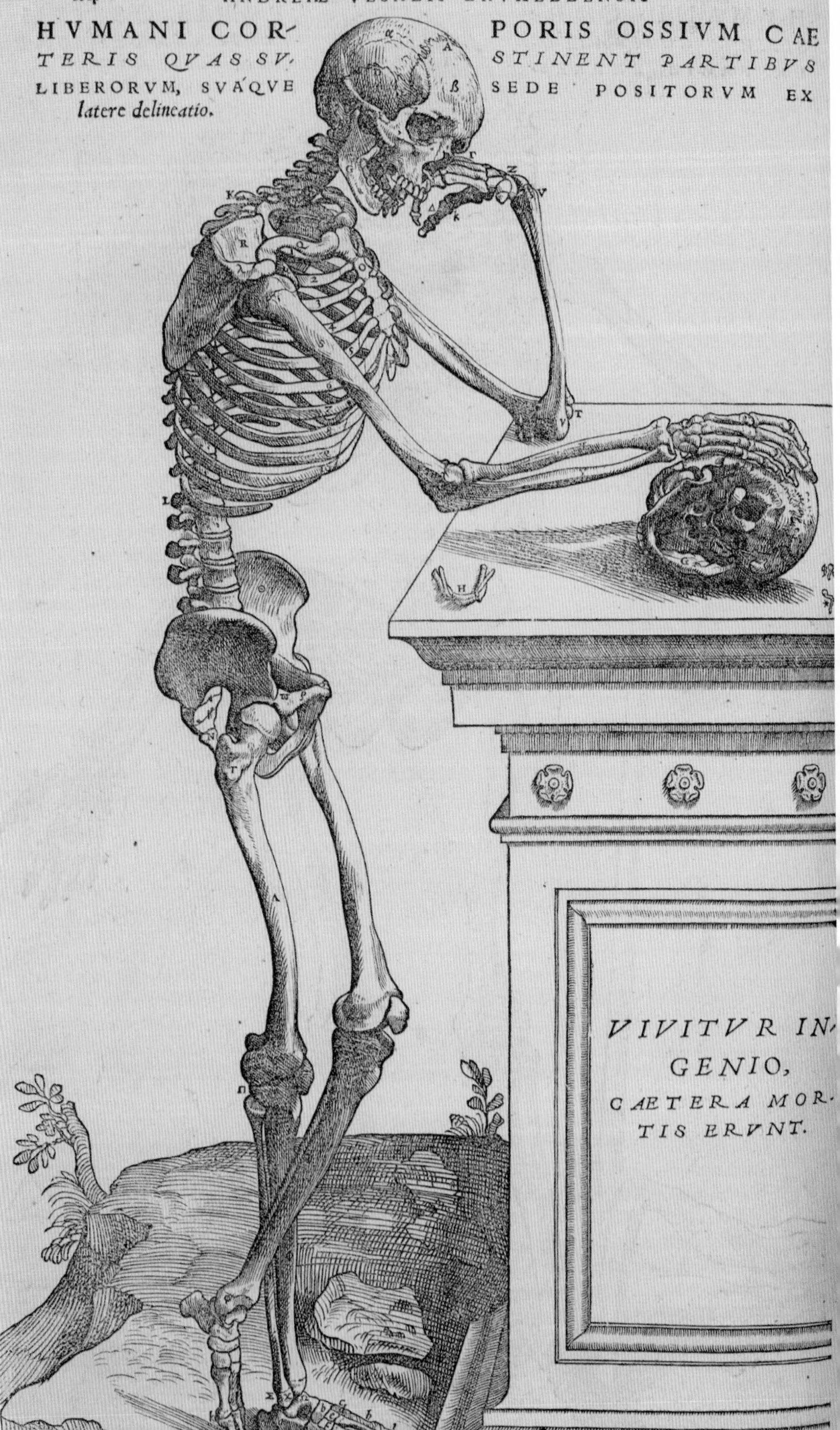

Vesalius, skeleton.

the outset was designed to show the reader that he was not merely passing on old information and hearsay. Moreover, armed with the tools shown in the picture below, readers could check what was written for themselves! Many of his dissections were open to the public, who were encouraged to come and confirm his claims, although the cadavers he used were on occasion obtained by robbing graves (a point he confesses to in the book) or accepting the victims of state executions. The dedication to the Holy Roman Emperor of what might appear a rather gruesome assembly was to show his awe and wonder at God's design of the human body in all its amazing detail.

Vesalius trained at Louvain and became a lecturer in Padua, whose great anatomical theatre – although damaged in Second World War bombing and then rebuilt – can still be seen today. He was only twenty-eight years old when his remarkable book appeared. About 500 copies were originally printed and 130 of them are known to survive today. Each contains 700 folio pages, bound according to the tastes of its owner. *De Humani Corporis Fabrica* was published in the same year as Copernicus' *De Revolutionibus*.[76] It is potentially of similar importance in the history of science. It was certainly studied and used more than Copernicus' work. Its beautifully detailed drawings exposed the

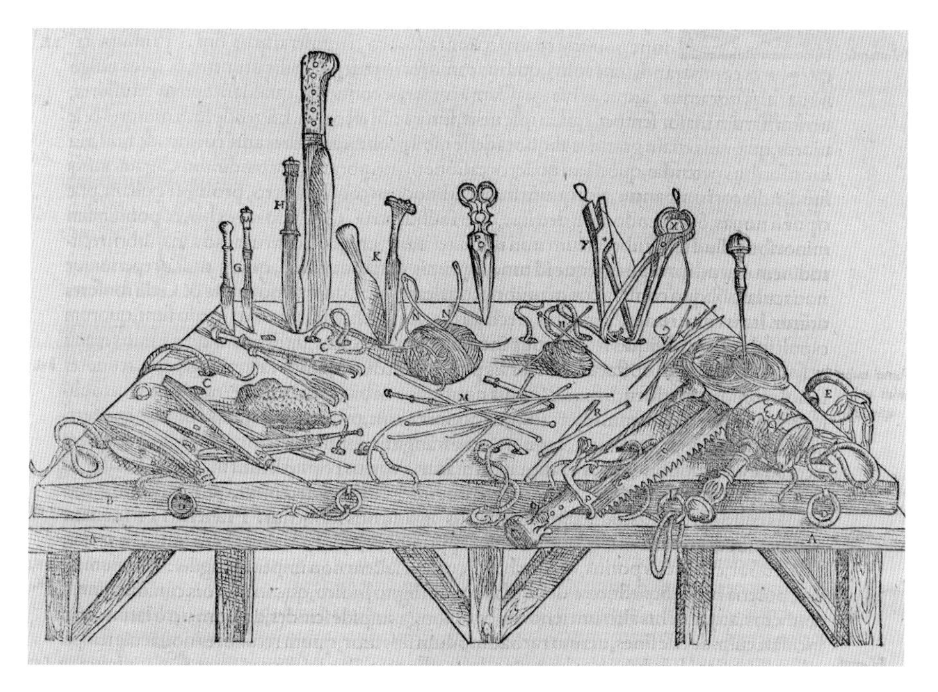

Vesalius, dissecting tools, shown to emphasise the experimental basis of his anatomical pictures.

intricacy of the human body in ways that encouraged others to emulate his detailed practical studies. He signalled the beginnings of the Renaissance in our understanding of anatomy – a *Fantastic Voyage* of the mind for the reader of its day. For students and researchers today it carries the worrying implication that to be a leading anatomist you needed to be a world-class draughtsperson as well.

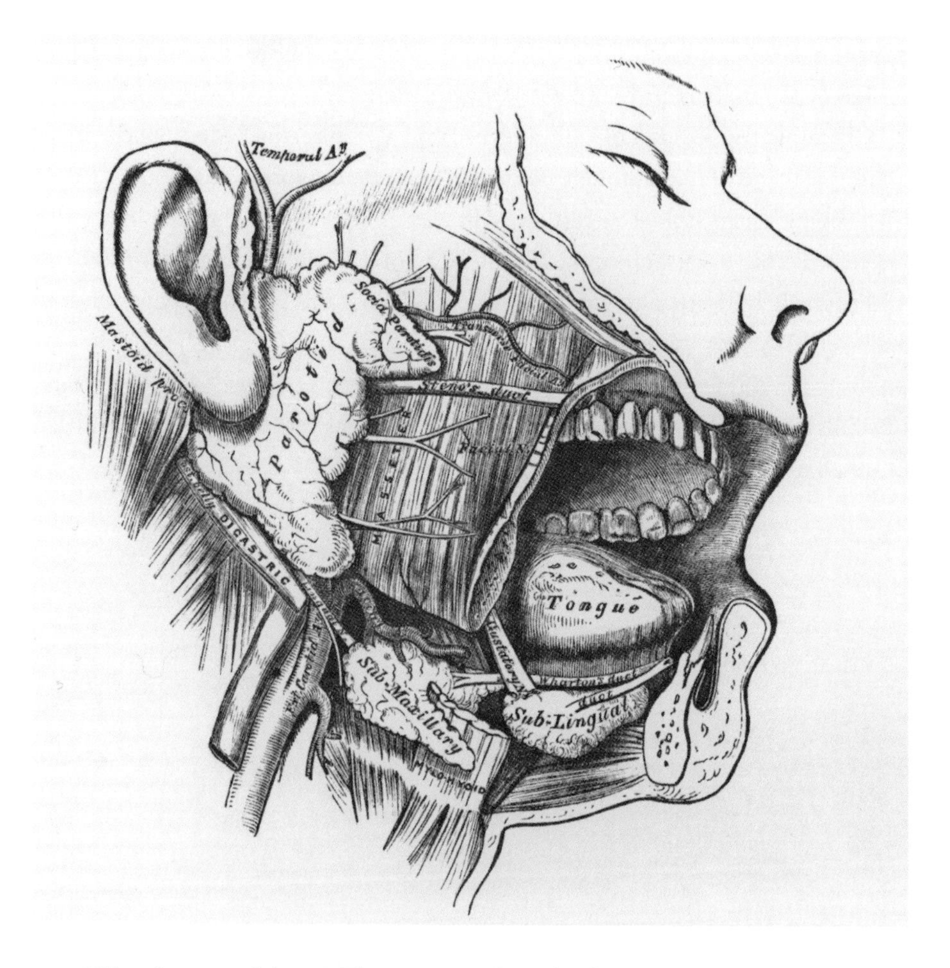

The Salivary Glands: an illustration from the 1858 edition of Gray's *Anatomy*.

What happened later? The canonical medical text today is Henry Gray's *Anatomy, Descriptive and Surgical*, first published in 1858, which has established itself as the bible of anatomy for students ever since. Known now simply as *Gray's Anatomy*, its 989 original pages have run through many editions and upgrades in presentational quality.[77] But when it first appeared, it displayed a bland technical style of drawing that signalled the end of the era of the great anatomical artists. It was a textbook, plainly bound and presented in businesslike fashion. It contained 'figures' and 'diagrams' rather than pictures.

The 363 plates, drawn by Henry Vandyke Carter, are detailed but flat, accurate but unthreatening pictures of small parts of the body, but never the whole. Where Vesalius' bodies looked like the living dead, and threatened to rise again, Gray's corpses were body parts, organs of pathological study rather than objects of wonder. *Gray's Anatomy* brought the curtain down on three centuries of dramatic anatomical art, replacing it by technically accurate but lifeless anatomy.[78]

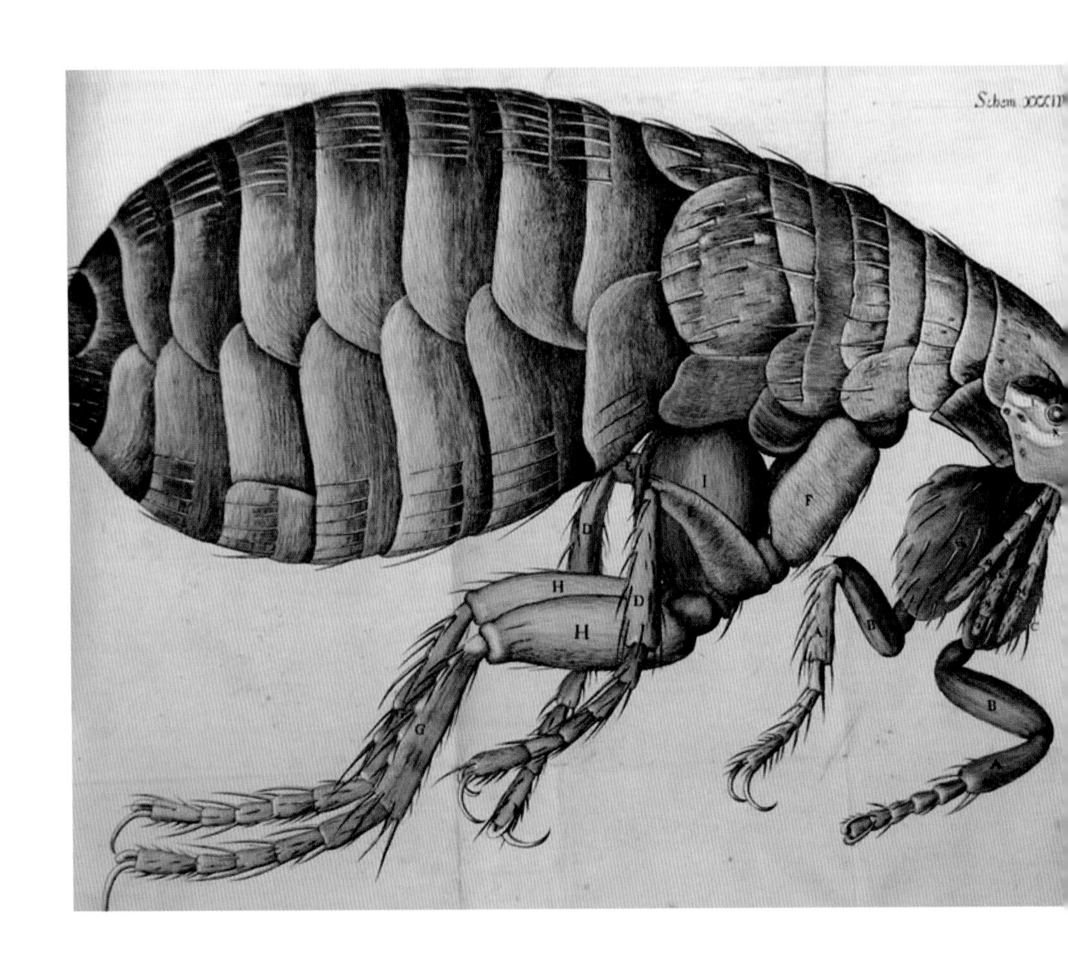

Robert Hooke's flea, from his *Micrographia* (1665).

A FLEA IN YOUR EYE

THE INGENIOUS MR HOOKE

The most ingenious book that I ever read in my life.

Samuel Pepys

MANY of the greatest discoveries in science have been brought about by the creation of an ability to see what was previously unseen. The most exciting of these inventions was the microscope because it could be used by almost anyone and what it revealed required little interpretation in order to amaze. It brought into focus a world of unsuspected complexity and beauty in the most commonplace things and changed for ever our admiration for the intricacies of Nature. It first made people gaze in wonder at the very small as well as the very large. One person deserves the credit for launching this voyage of discovery into the microworld and for illustrating it with matchless skill and precision.

Robert Hooke began his scientific career while still a student at Christ Church, Oxford, in 1653, where he met and worked for Robert Boyle.[79] He impressed Boyle greatly, and in 1662 he became the Curator of Experiments at the new Royal Society of London, which meant that he was responsible for practical demonstrations at the meetings of the Society. Hooke became the most important Fellow of the Society because of his vast range of interests and extraordinary practical skills. Over the next three years, among his many interests, he was active in developing powerful new compound microscopes, and used his own beautifully crafted instrument[80] to create a remarkable atlas of the microscopic natural world.[81] It was published in 1665 for the Royal Society under the title *Micrographia* and contained a series of sixty studies, fifty-eight microscopic and two telescopic pictures of the Moon and the stars. Each was illustrated by a meticulous drawing from Hooke's observations and accompanied by a detailed description and interpretation of what was to be seen by close examination of the drawings.[82] The majority of Hooke's microscopic studies were of living things: a louse, a fly's compound eye, sponges, herbs, a bee's sting, fish scales, snail's teeth, insects, stinging nettles, and spiders. All are striking and revelatory, but the most impressive drawing of all was Plate number 53, that 'Of a flea'. Hooke begins by telling us that 'The strength and beauty of this small creature, had it no other relation to man, would deserve a description.'

The fold-out drawing of the flea shows Hooke to be a draughtsman of the highest quality. In fact, as a young teenager he had been apprenticed to an artist but developed respiratory problems because of the oil paints. He was meticulous in his desire to represent accurately what was seen in the microscope and was well aware of the ambiguities created by light and shade and flat projection. He explains at some length how he does not begin drawing until he has developed a full three-dimensional understanding of the creature in the eyepiece. It is interesting to compare the precision of Hooke's drawing with a modern high-quality photograph.

Hooke's commentary on the wonderfully precise drawing of the flea focuses upon its self-evident beauty and strength – a masterpiece of precision engineering:

> the curious contrivance of its leggs and joints, for exerting that strength, is very plainly mainifested, such as no other creature, I have yet observ'd, has any thing like it; for the joints of it are so adapted, that he can, as 'twere, fold them short one within another, and suddenly stretch, or spring them out to their whole length, that is, of the fore-leggs, the part A lies within B, and B within C, parallel to, or side by side each other; but the parts of the other two next, lie quite contrary, that is, D without E, and E without F, but parallel also; but the parts of the hinder leggs, G, H, and I, bend one within another, like the parts of a double jointed Ruler, or like the foot, legg and thigh of a man; these six leggs he clitches up altogether, and when he leaps, springs them all out, and thereby exerts his whole strength at once.

These pictures inspired scientists to embark upon serious systematic study of the detailed structure and function of insects and other small-scale intricacies of Nature. They also created a resurgence of interest in the old natural theological arguments about 'design' in the natural world that are, curiously, still with us in certain parts of the world. The extraordinary fine tuning of the system of legs and arms revealed in pictures of tiny creatures such as the flea led many to believe this must have been designed ready-made for the functions that it now performs. This claim would ultimately be superseded by the proposals of Alfred Russel Wallace and Charles Darwin, in the mid-nineteenth century, that natural fine-tuned complexity was the result of a sequence of successive approximations steered by natural selection. Yet, the focus of attention on remarkable contrivances in living things, their intricate microscopic engineering, and the many ways in which they appear tailor-made for their environments played a crucial role in lining up a remarkable body of natural phenomena that needed further explanation. Darwin acknowledged the debt he owed to the early marshalling of that evidence, first taught to him at Cambridge as an undergraduate. Without it, there would have been no problem for the mechanism of natural selection to solve.

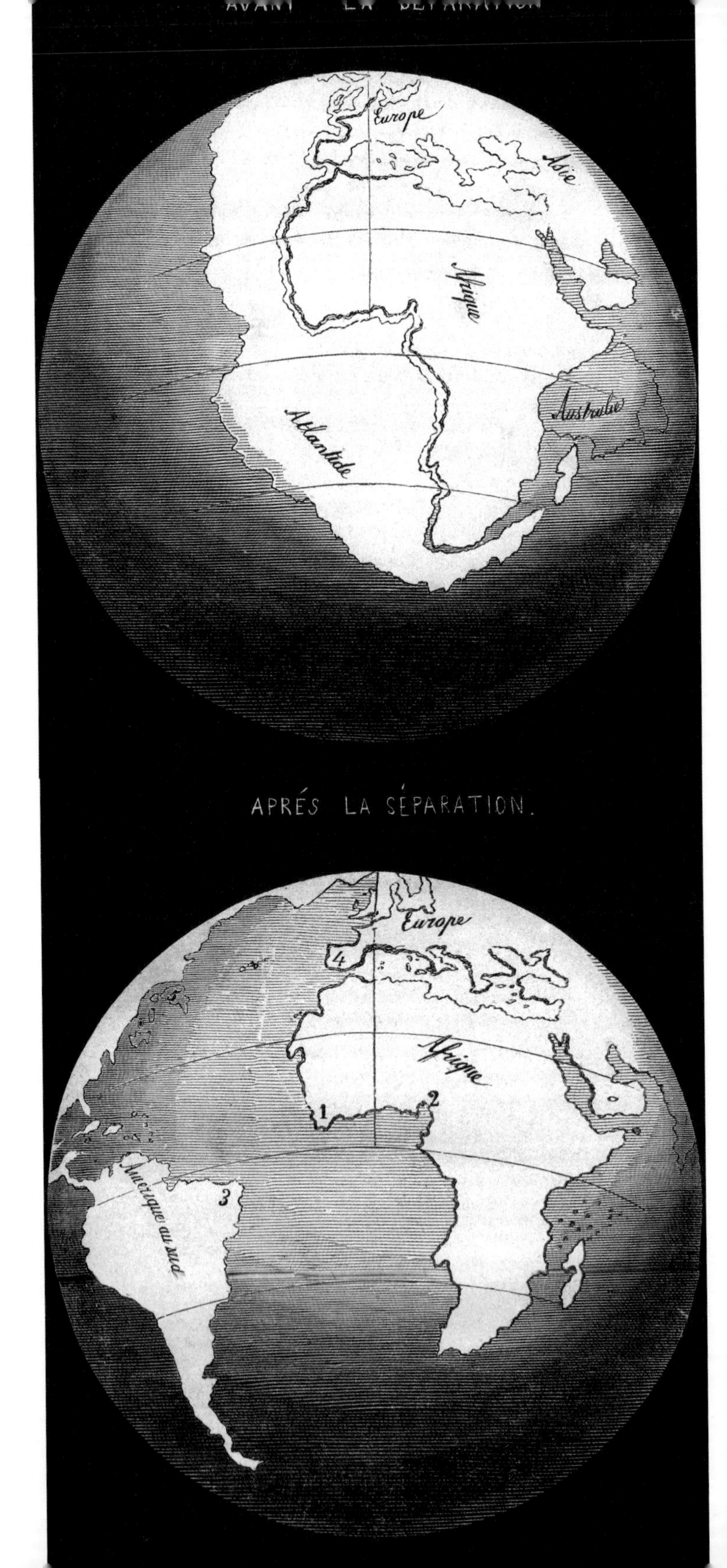

Map of the continents before their separation, as depicted in 1858 by the French geographer Antonio Snider-Pellegrini. The map shows how the coastline of the Americas fits those of Europe and Africa. Snider-Pellegrini's theory was that the continents were created by God in this form, and were then separated by a volcanic explosion on the sixth day of creation. The idea that there was once a single supercontinent was not a new one, but the first modern theory of continental drift was that of Wegener, published in 1912.

DID THE EARTH MOVE FOR YOU?

CONTINENTS IN MOTION

Indifference is the strongest force in the universe. It makes everything it touches meaningless.

Joan Vinge[83]

UNTIL the eighteenth century, most Europeans thought that the Earth's surface had been shaped primarily by catastrophic events that occurred only thousands of years in the past. They were greatly influenced by their traditional biblical beliefs that Noah's Flood immersed the Earth's surface under water at a time when human civilisation was well established in the Middle East. Even those who were not so literal in their interpretation of the Hebrew Bible were still greatly swayed by a belief that all geological change took place by means of a series of catastrophic events. By the middle of the nineteenth century, this idea had been displaced by the alternative picture of a more uniform progression of historical events in which a fairly complete understanding of the history of the Earth could be obtained by applying those laws and principles that we see at work today. Although sudden atypical events, such as earthquakes and volcanic eruptions, might occasionally occur, they were viewed as minor variations on a systematically uniform progression of physical processes rather than the dominant theme of geological evolution. Coupled with the proposal by Wallace and Darwin that living things had evolved into their present forms by a process of natural selection, and the new nebular theories of the origin of the solar system being pursued by astronomers, this created a perspective on the Universe in which time really was of the essence.

In 1596, in the last edition of his book *Thesaurus Geographicus*, the great Flemish geographer and cartographer Abraham Ortelius, a contemporary of Gerardus Mercator, became the first person to assemble large numbers of maps into a single book rather than create them on single scrolls. Later, books of this sort all became known as 'atlases'[84] after Mercator placed an engraving of Atlas, the ancient Greek Titan we saw in our opening chapter holding the world on his shoulders, on the front cover of his book of maps in 1590. In his famous *Theatrum Orbis Terrarum* (*Theatre of the World*), Ortelius suggested that the Earth's continents may have shifted in the past. Observing the plan of the Earth's geography provided by the best maps, he suggested that the Americas might have been torn away from Europe and Africa, and by considering the coastlines of these three continents he thought that 'the vestiges of the rupture might reveal themselves'.

This was the first appearance in print of an idea that suggests itself as soon as you look at a terrestrial globe. Move the continents about and they seem to fit together like the pieces of a simple jig-saw puzzle. The fit of the South American and African continents is the most compelling of all. Perhaps it is a testament to the influence of the catastrophic viewpoint that this idea attracted so little attention and we do not find it resurfacing until 1858, when the American geographer Antonio Snider-Pellegrini published a book entitled *La Création et ses mystères dévoilés* (*Creation and Its Mysteries Revealed*), in Paris. His book contained maps showing how he imagined the continents to have once been joined together before separating into their present state. He also expected there to be links between the locations of fossils laid down before the separation occurred.

Yet, the idea still did not catch the imagination of nineteenth-century geologists. The tide began to turn slowly only in 1912 when a young German meteorologist named Alfred Wegener[85] presented a full theory of 'Continental Drift', which hypothesised the existence originally of a single super-continent, Pangaea, which split apart 200 million years ago into two large landmasses, Laurasia in the north and Gondwanaland in the south. Both continued to move and divide into the continents that we see today. Wegener built his theory on the apparent fit between the shapes of different continents and the match

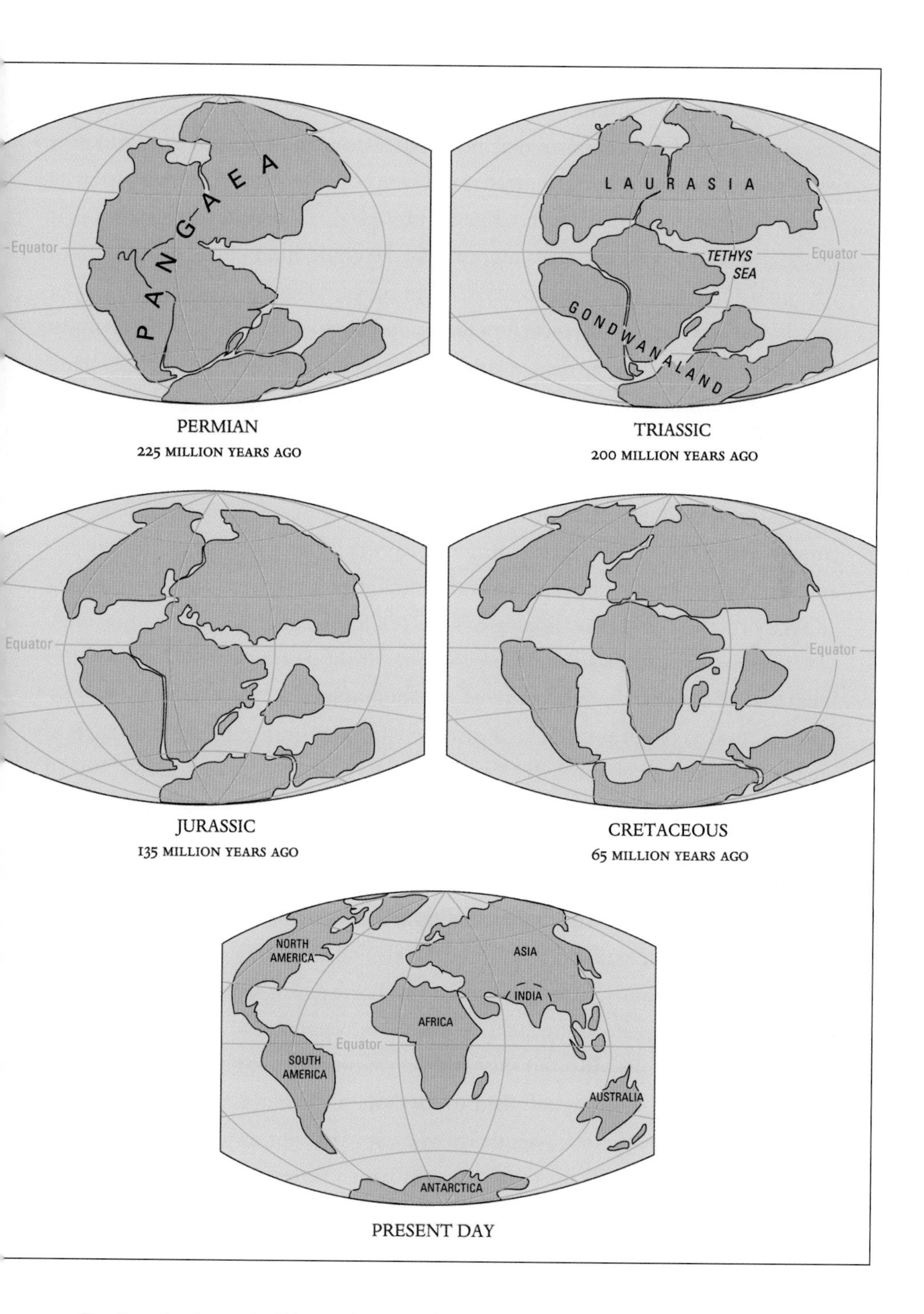

The time development of the continents and intervening oceans, US Geological Survey.

between plant and animal fossils along the coastlines of continents which, although now far distant, would have once been joined. Wegener's theory certainly gained the attention that his predecessors had failed to attract. But it was not generally enthusiastic in tenor. He was the first to challenge seriously the traditional idea that the continents are fixed in location, suggesting that they must be in constant slow motion, but he could not explain the nature of the enormous forces needed to make them move. Wegener devoted the rest of his life to developing his theory of drift, and finding the physical mechanism which created it. Sadly, he died of cold on an expedition to explore the Greenland icecap in 1930 and never saw his theory gain widespread scientific support.

Eventually, Wegener's portrayal was vindicated and the picture of the Earth's surface covered by 'tectonic plates' became widely accepted by the scientific community. The Earth's surface is built of plates of material that slide above and below each other at different speeds. The movement of the continents takes place at a speed of several centimetres a year and can be measured directly by the Global Positioning System (GPS) of satellites.

Today, the geophysics of the evolution of the Earth is a dynamic subject that can draw upon a vast array of oceanographic, astronomical, palaeontological, and geological data. It is also of vital importance to the existence of many communities who inhabit low-lying parts of the Earth, or regions close to colliding plate boundaries. Our understanding of plate tectonics may ultimately enable us to foresee in general terms the human consequences of the motion of the Earth's crust and mitigate some of its most devastating results. It was those first suggestive pictures of the Earth's nested continents that started us on the road towards understanding how the Earth's surface became like it is today, but, more importantly, it taught us that the Earth's geography is not a finished product. It is a never-ending story of steady movement. No other solar-system planet is geologically 'alive' in the same way, and the suitability of the Earth as an abode for life is strongly linked to its dynamic geological activity. There is much still to learn about the sequence of events that got us where we are today and much to be done to sift those changes which were inevitable consequences of huge natural forces from those which were mere accidents that could have gone one way or the other.[86] It is a quest for the 'big picture' about the Earth and how it got to be what it is today.

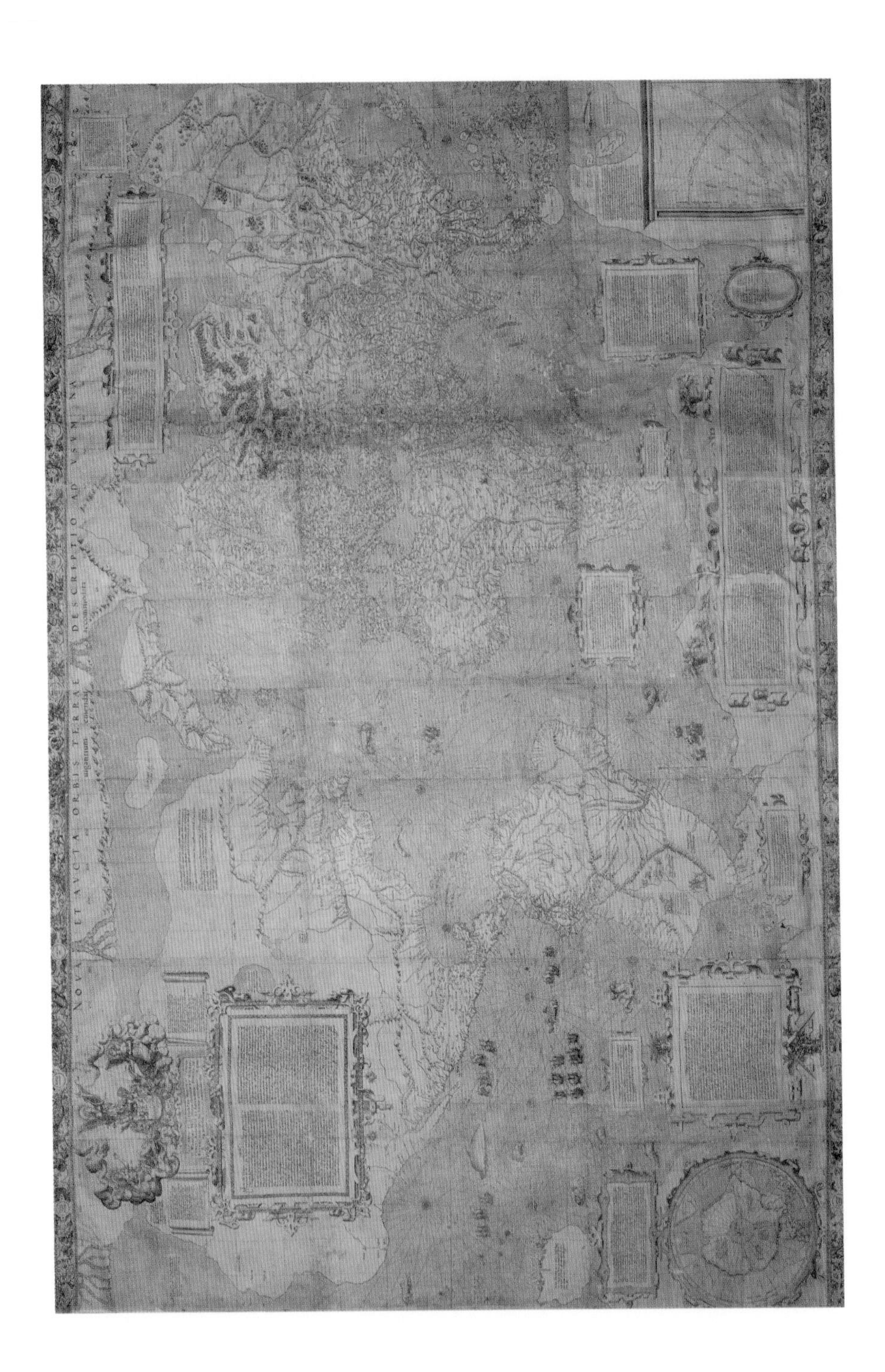

Mercator's map of the world, 1569.

SHOW ME THE WAY TO GO HOME

MERCATOR'S MAP OF THE WORLD

Maps encourage boldness. They're like cryptic love letters. They make anything seem possible.

Mark Jenkins[87]

GERARDUS MERCATOR is the most famous member of the Dutch master map-makers of the sixteenth century. His name lives on in every geography textbook in the world because in 1569 he found a way to present the spherical surface of the world on a flat page in a very special way. For most of us, Mercator's projection of the world sums up the way the world 'is'. Any representation of the almost spherical surface of the Earth on a plane sheet of rectangular paper will produce a distortion.[88] There are innumerable ways to do it, all with their own advantages and disadvantages. Mercator's projection exaggerates the sizes of regions as we move away from the equator to the poles. The parallels and meridians are all straight lines, and although the areas of regions are distorted as we move towards the poles, their shapes are not. This is because Mercator's projection preserves angles between directions on the Earth's surface – it is what mathematicians call a 'conformal' transformation.[89] These features make Mercator's projection rather unfashionable today. The distortion of areas gives a false impression of the size (and perhaps therefore the status?) of countries far from the Tropics. Its other awkward feature is that a straight line drawn on Mercator's map does not correspond to the shortest distance between two points on the Earth's spherical surface, which is actually given by the so-called 'great circle' path on the Earth's surface.[90] But a straight line on Mercator's map has a property that was arguably more important to sixteenth-century navigators: any straight line drawn on his map is a line of constant navigational bearing.[91] As a result, Mercator's map became the basis for all nautical charts. Its inability to give the shortest sailing distance between two points might seem like a serious defect, but you would only have a serious problem if you were trying to navigate near the Poles. In practice, distance is not the key factor in sailing between places.

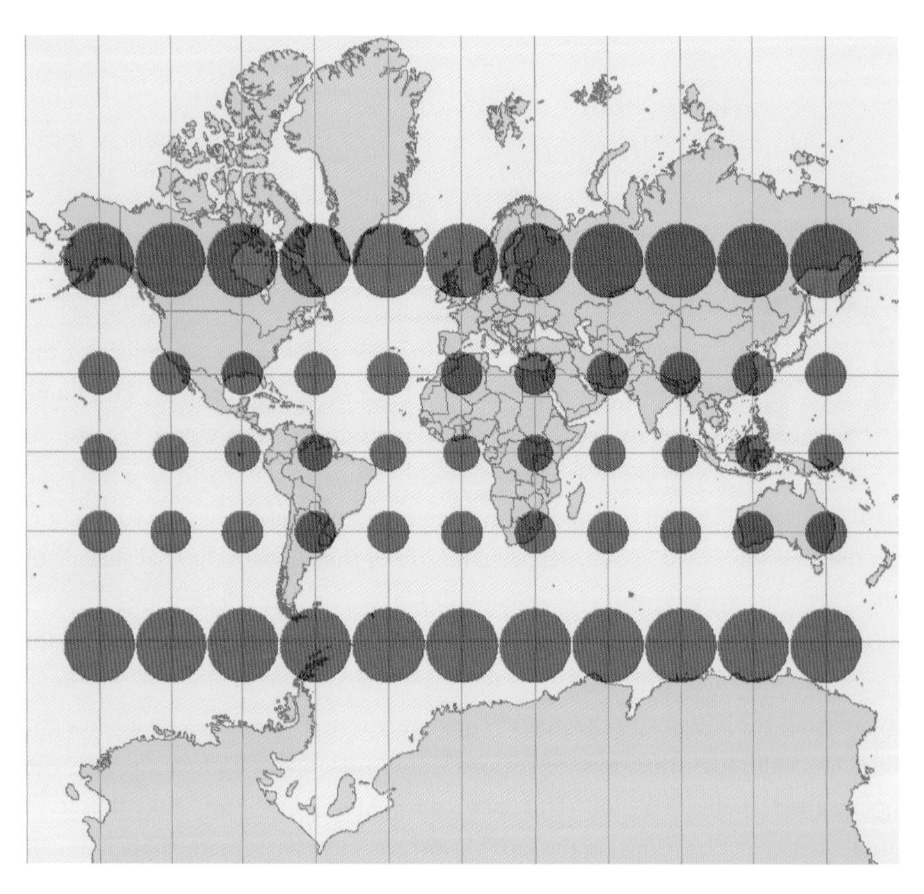

The Tissot Indicatrix applied to a modern world map in Mercator projection. It shows the area distortion of a circle as one moves away from the Equator to the Poles. Hence, countries like Greenland have their apparent sizes exaggerated. Note that the circles remain circular regardless of their sizes.

It is time. And the sailing time is determined more by winds and currents than by the shortest route that can be taken. Mercator's map is ideal for choosing and staying on a bearing so that the prevailing winds and currents can be used more advantageously. It is only people sitting at home behind desks that worry greatly about travelling the shortest distance when getting from A to B on the sea.

A simple way to gauge the variation in distortion created by a map projection was invented by Nicolas Tissot in the nineteenth century and is called Tissot's Indicatrix. It shows the distortion that is created when identical small circles are mapped from the Earth's spherical surface using Mercator's projection. The distortion of areas grows rather large as you move away from the Equator. By the time you reach the latitude of Greenland, it is more than ten times bigger than the distortion near the Equator.

The Mercator projection has influenced the way we see the world in many ways. It helped shape the world as well as describe it. It was drawn when the world was dominated by the European sea-faring powers. The overall orientation of the map therefore places Europe at the 'top' and near the centre. The projection's properties make the European and North American regions appear much larger than they truly are when viewed on a spherical globe. If a non-conformal map projection is used which preserves areas at the expense of distorting shapes then the world looks very different. The first such 'equal-area' projection was introduced by Johann Lambert in 1772 and cast into a useful form by the Scottish mathematician and map-maker James Gall, in 1855. But the most cited equal-area projection was created for overtly political reasons and presented in Germany by Arno Peters in 1973. The first English version appeared in 1984 and was adopted by UNESCO and many other international organisations for egalitarian reasons. Peters believed that his projection (now known as the Gall–Peters projection) was fairer to the under-developed regions of the world. Peters campaigned relentlessly to have his map used all over the world and struck a chord with groups wishing to see the post-colonial world afresh. Cartographers disliked both the map and the massive self-publicity that Peters generated. Arthur Robinson sourly described it as looking like 'wet, ragged long winter underwear hung out to dry on the Arctic Circle'. All this gave rise to a surprisingly vehement period of 'map wars' that set cartographer against cartographer!

A Gall–Peters equal-area projection of the world, which gives a true picture of their surface area but distorts their shapes by elongating the North–South dimensions relative to the East–West dimensions.

Following Lambert and Gall's projection, Peters also shifted the lines of latitude to the south by about 45 degrees to bring southern continents closer to the centre of the map. However, the map has its own distortion problems. Although it preserves areas, it does not preserve shapes. As a result Africa ends up with a correct representation of its area relative to Britain or Greenland, but it appears to be twice as long from north to south as it is from east to west, whereas a look at a globe shows these two dimensions are roughly the same.

One further notable attempt to turn the world upside down was made by a young Australian. Since his Melbourne schooldays, Stuart McArthur had occasionally surprised his geography teachers by responding to the task of drawing a map of the world by creating an 'upside-down' version of an equal-area projection in which Australia sat in the centre at the top surrounded by a vast expanse of Pacific Ocean. Invariably they asked him to redo his homework assignment. But when he left school he created his own 'Universal Corrective Map', whose caption proudly announces that it is, at last, 'the first step in the long overdue crusade to elevate our glorious but neglected nation from the gloomy depths of anonymity in the world struggle to its rightful position – towering over its northern neighbours, reigning splendidly at the helm of the universe'. Sales in Australia were reputedly very good.

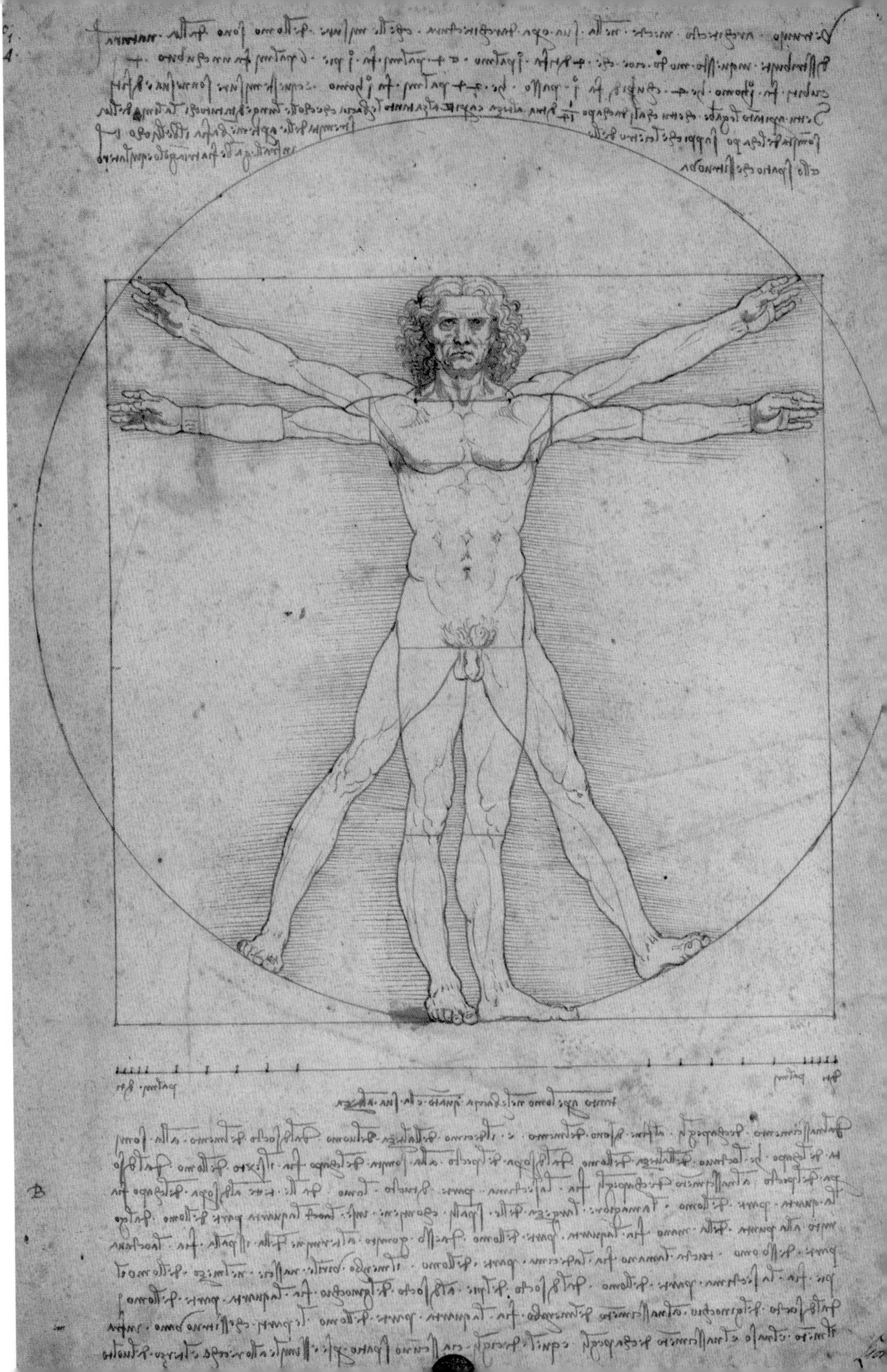

SHAPELINESS
THE SYMMETRIES OF LIFE

He had the sort of face that, once seen,
is never remembered.
Oscar Wilde

THE most memorable and the most important sights that most of us experience from birth are those of human faces. They identify us and form the basis of first impressions; they are a source of art and social significance in many cultures. But there is a tantalising mystery behind the appearances: our faces and our bodies are strikingly symmetrical. Whereas inanimate objects rarely display perfect symmetry, living things almost always possess external right–left symmetry. On the face of it, this might seem an improbable state of affairs. After all, it requires delicate engineering. By contrast, symmetry is absent in the top–down direction because bodies are adapted to deal with the variation of gravity and weight, with height, and with the need to remain stable under the influence of small perturbations that would otherwise cause them to fall over. It is rare to find back–front symmetry in animals because it is 'cheaper' to engineer the ability to turn around. Bilateral symmetry is very advantageous for movement – the imbalances caused by any bilateral asymmetry make straight-line motion tricky to engineer and the benefits of symmetry are even greater if motion has to occur off the ground, in air or water.

The classic representation of our human symmetry is displayed in Leonardo's famous drawing of the 'Vitruvian Man',[92] which has been artistically reinterpreted and reproduced on countless occasions, most recently on all

Leonardo's rendering of the Vitruvian Man (1490), with pen and ink and washover.

Italian 1-euro coins. How different it seems to Michelangelo's huge statue of David, with its exaggerated features and hands.

Deviations from a symmetrical bodily form often signal an injury or genetic impairment, and some of the worst consequences of disease arise from the loss of our delicate bodily symmetry. Many of our superficial evaluations of human beauty, or attractiveness, focus upon the symmetries of the face and body. Plastic surgeons receive large sums for restoring or enhancing it, and the cosmetics industry is based upon the desire to display and highlight it. Among lower animals, the symmetry of bodily form is an important indicator in the selection of mates and in distinguishing fellow members of your species from predators. Camouflage is often evolved so as to disguise bodily symmetry by overlaying it with asymmetrical patterns.

The human face displays a very high degree of symmetry, and the evolutionary importance of recognising creatures with faces in crowded fields of vision, between trees and boulders or foliage, has resulted in our sensitivity to lateral symmetry amid asymmetries. It is a good rough-and-ready guide to distinguishing things that are alive from things that are not. A sensitivity to this symmetry will be selected for over insensitivity because you are more likely to survive if you can identify friends, enemies and potential mates better than others.[93]

The most interesting feature of the high degree of symmetry found in human faces and our external bodies is the contrast with the squalid muddle to be found under the skin. Our bodies are not symmetrically engineered under the surface. Hearts are on the left, our brains are laid out in an asymmetrical fashion that reflects the type of cognitive activity being performed. If symmetry persisted under the skin, then vital organs would have to be duplicated to maintain it. The brain would use resources in a wasteful way if it were as symmetrical as the moving body parts that it controls. Because the brain does not move independently, but simply controls movements of other parts, it can be asymmetrically hard-wired.

Our deep-laid sensitivity to human faces is an evolutionary inheritance that aids our survival and multiplication. As with many evolutionary legacies, it has all sorts of by-products that have little to do with the environment in which we now find ourselves. We don't need to identify human beings exclusively by the symmetry of their facial appearance but we often use that information for all sorts of evaluations. The sensitivity for symmetry that we have inherited from these tendencies shows up in all sorts of other places – in the decorations that we like to use in our homes, in the types of mathematics we like to study, and the sorts of scientific theory that we like to create. All are residues of the time when we first knew we had faces.

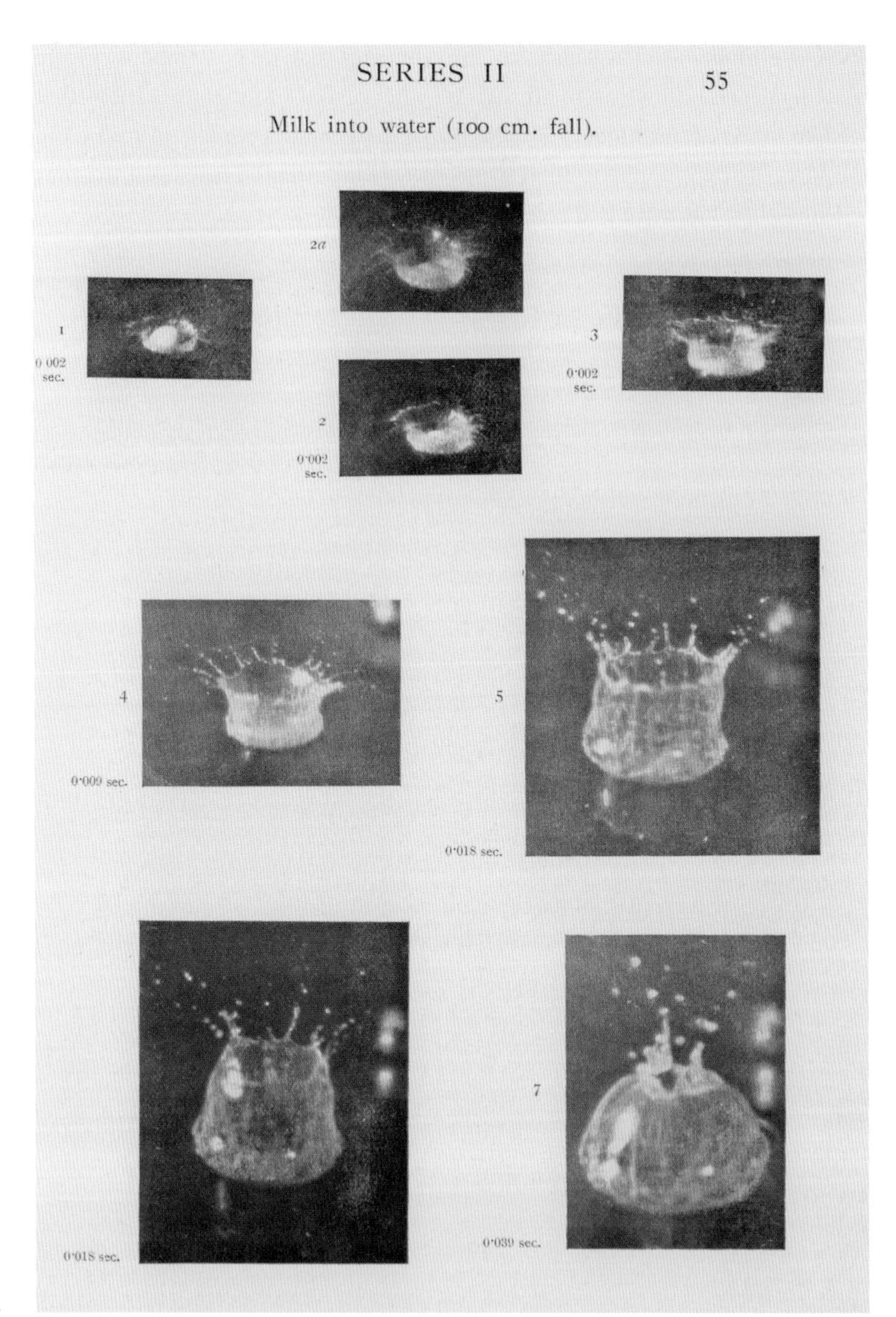

A. M. Worthington and R. S. Cole, 'Impact with a Liquid Surface, Studied by the Aid of Instantaneous Photography', 1897.

MAKING A SPLASH
HIGH-SPEED PHOTOGRAPHY

Everyone has a photographic memory.
Some just don't have any film.
Steven Wright

ONE of the pioneering figures in photographic techniques, whose work had an influential effect on the studies of fluids and human movement, was the little-known Arthur Worthington (1852–1916), who was a professor of physics at the Royal Naval College at Devonport, in the first decade of the twentieth century. As appropriate for someone holding such a position, he was interested in the motion of projectiles through water and their impacts into water. In those days before the advent of computers, the only way to study such complicated mathematical problems, which were too difficult for direct solution with pencil and paper, was to look and see what was going on in detail. Worthington did this by introducing the technique of high-speed time-lapse photography to capture events, which to the eye appear continuous, as a series of still frames. His studies of the process of formation of a splash when a body drops into water are some of the classics of early photography. Worthington was able to capture sets of splashes formed under identical conditions at intervals of one hundredth of a second. By ensuring from experience that all these splashes followed identical histories, he was able to create the individual frames of a movie.

Worthington recognised that there was a repeatable and orderly aspect to the creation of the splashes, but the whole phenomenon was too complex to describe mathematically because it involved following the path of every point in the water's eruption. Worthington was on the verge of elucidating the problem of chaos because he discovered that his splash sequences were very sensitive to the liquids and materials used. The viscosity (or 'stickiness', resisting flow) of a liquid played a sensitive role, so that the form of water splashes differed significantly from those made by milk. Similarly the roughness of projectiles hitting water produced a significant variation in the splash they produced. Worthington's pictures were first published in the scientific literature in 1897 (and then in book form in 1908)[94] but only became widely known in the scientific community after they appeared in D'Arcy Thompson's classic book *On Growth and Form*, which appeared in 1917.[95]

Alas, Worthington never really got the credit for his innovation because the next actor in the story, Harold Edgerton (1903–1990), a professor at MIT, in the 1930s, had an eye for the commercial possibilities opened up by the invention of strobe photography, capable of capturing 3000 images per second. Edgerton was an electrical engineer who had needed to develop a form of high-speed photography during his PhD thesis work to monitor changes in a motor's speed. He reported the invention of the first stroboscopic photographic device in the May 1931 issue of the journal *Electrical Engineering*. This launched him on a remarkable career of scientific and photographic firsts: the 'flash' system of wartime night photography, Jacques Cousteau's underwater cameras, the side-scan sonar which found the USS *Merrimac*, sunk in 1862, and the wreck of the *Titanic*. His iconic pictures of bullets cutting playing cards in half, the sequential movement of athletes, his book *Flash*, and his film *Quicker 'n a Wink*, which all came out of the work done in his 'Strobe Alley' research and development group at MIT, are some of the most remarkable in the history of photography[96] and he was showered with scientific and photography awards, including the National Medal of Technology from President Reagan in 1988.

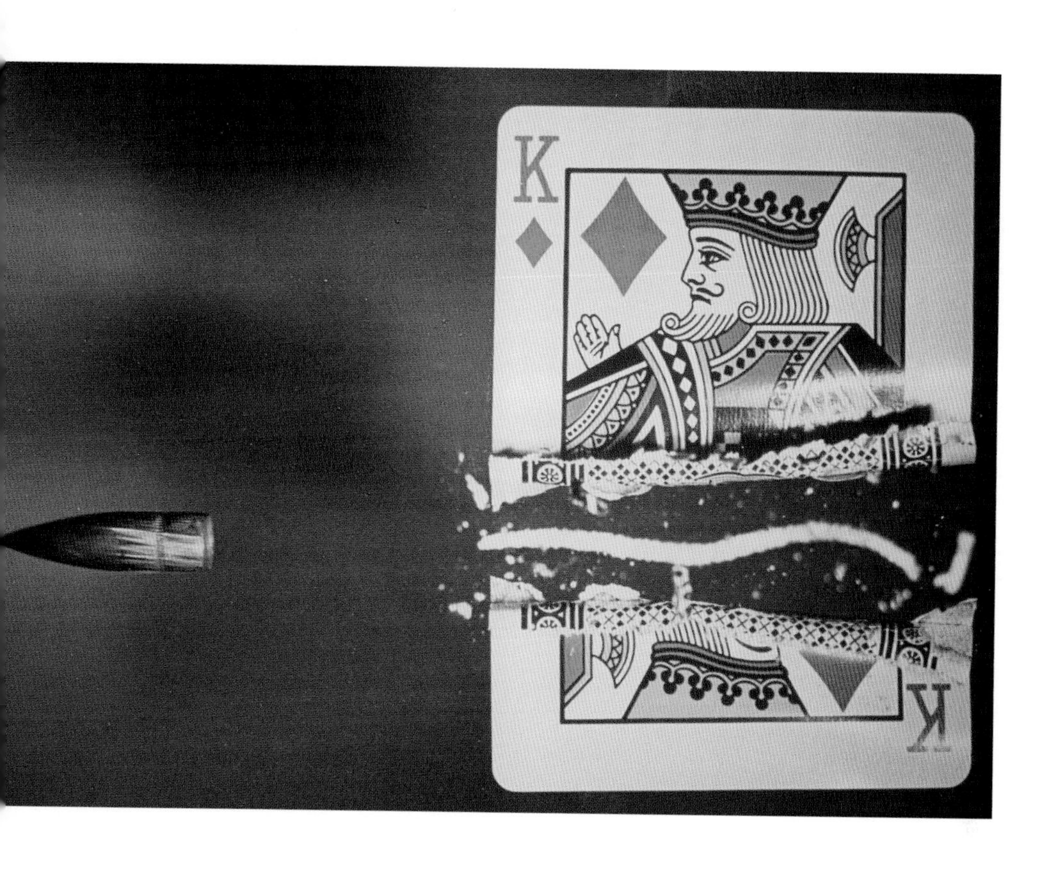

Harold Edgerton,
King of Diamonds
playing card hit by
a 30-calibre bullet
(1970).

Painting by Numbers

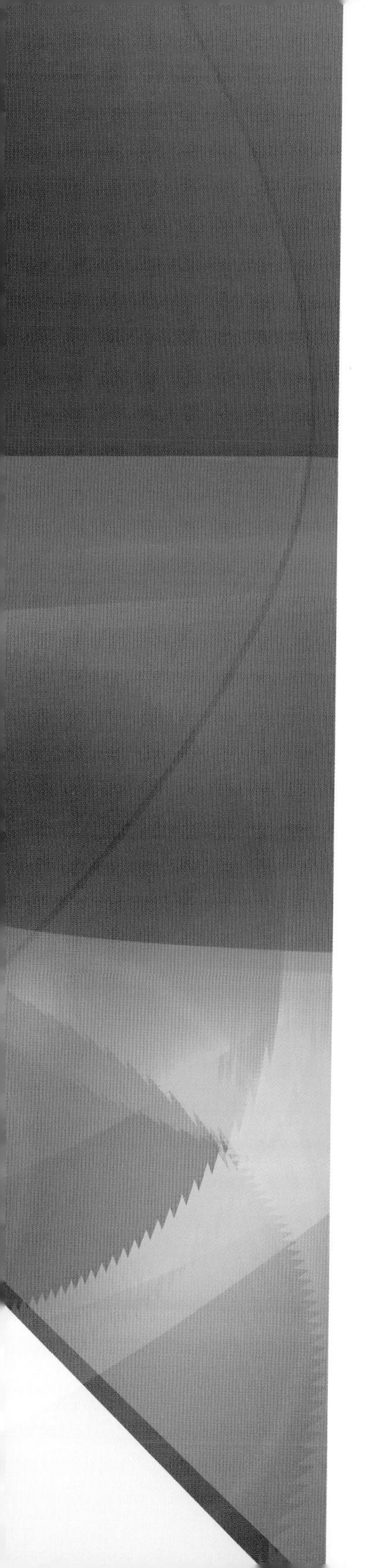

Numbers written on bills within the confines of restaurants do not follow the same mathematical laws as numbers written on any other pieces of paper in any other parts of the universe.[1]

Douglas Adams

Image by Andy Burbanks.

MATHEMATICS is the catalogue of all possible patterns. It is the first place to look for the most spectacular man-made images and natural symmetries. Mathematicians have always used pictures to illustrate their new-found truths of geometry, and as an aid to the reasoning process. Thousands of years of transmitting Euclid's *Elements* from the Greeks through the Middle Ages and the Renaissance to modern times carried with it a tradition about mathematical diagrams and their purpose. Yet, at times, there have been purists, such as D'Alembert and the French Bourbaki consortium in the twentieth century, who have shunned pictures as an unwanted crutch to logic, liable to mislead. Proof by picture was not allowed. Most mathematicians declined to put on this straitjacket and have continued to exploit pictures to the full, seeing them as an intuitive guide to logic, and a shorthand with which to provide instant demonstrations or fatal counter-examples.

We have chosen some of the most important visual examples from mathematics that often illustrate deep geometrical truths about the nature of space. Some involve only shapes, others reveal truths about the organisation or division of space, while others still reveal the panoply of structures that can exist with appealing symmetries. Some of these images reflect the ubiquity of mathematical symbols; they are far more universal today than any written alphabet. Who first devised the symbols of arithmetic, or the signs for infinity and equality, and why are they as they are? There are many other images here which, although they have a mathematical side, are much more familiar for other reasons. Dice have a long history as gaming devices but have also come to symbolise chance and randomness. The map of the London Underground train system is an icon of twentieth-century design, a symbol of London, but it was also the first topological map – a circuit board of lines and exchanges that reshaped the sociology of London. Graphs appear in every newspaper, but who drew the first one, and why? Or perhaps there are other types of symbolic representation, such as the first musical score, that could lay claim to being 'graphs' of sound versus time. Bridges and arches on great buildings are vivid examples of our great powers of engineering, using shape to defeat gravity. We appreciate their sweeping symmetries but what are these shapes and why are they ubiquitous?

We shall see how the first explorations of the world of intricacy at the beginning of the twentieth century produced intriguing mathematical pictures with counter-intuitive properties. Eventually, they would be called 'fractals' and be mined out for their riches by new technology. The advent of inexpensive fast computers enabled mathematicians, such as Benoit Mandelbrot at IBM, to explore this brave new world with unprecedented computational power. Like the impossible figures of Maurits Escher, fractals have become the wallpaper of mathematics, the aesthetic graphic face of algorithmic computation. Remarkably, all these beautiful pictures are finite attempts to picture the infinite. Structures such as the Mandelbrot set exist only in the limit of an infinite number of computations, but computers can take us a long way on the long winding road to infinity, and there is a lot to see on the way.

One of the great realisations of the past thirty years has been that very simple rules, starting without any randomness and uncertainty, can soon give rise to situations which are, for all practical purposes, completely unpredictable. Sometimes, that unpredictability is chaotic and structureless, but it can also lead to organised complexity that seems to be 'alive' in some simple sense. The best way to explore these patterns is to look and see – the eye is a more sensitive analyser than any statistic.

Finally, we take a look at some of the 'impossible' figures and striking visual illusions that mathematicians and artists have created. They exploit the projection of three-dimensional reality onto the flat page and enable us to visualise what higher-dimensional objects would look like if we saw them. Illusions feed on the way the brain responds to partial patterns and colour; they can be used to deceive, to perplex, and to entertain. Believing shouldn't just be down to seeing, but when it comes to mathematics we do need to see to believe that we really understand. It is no linguistic accident that understanding is so often accompanied by the statement that 'I see.'

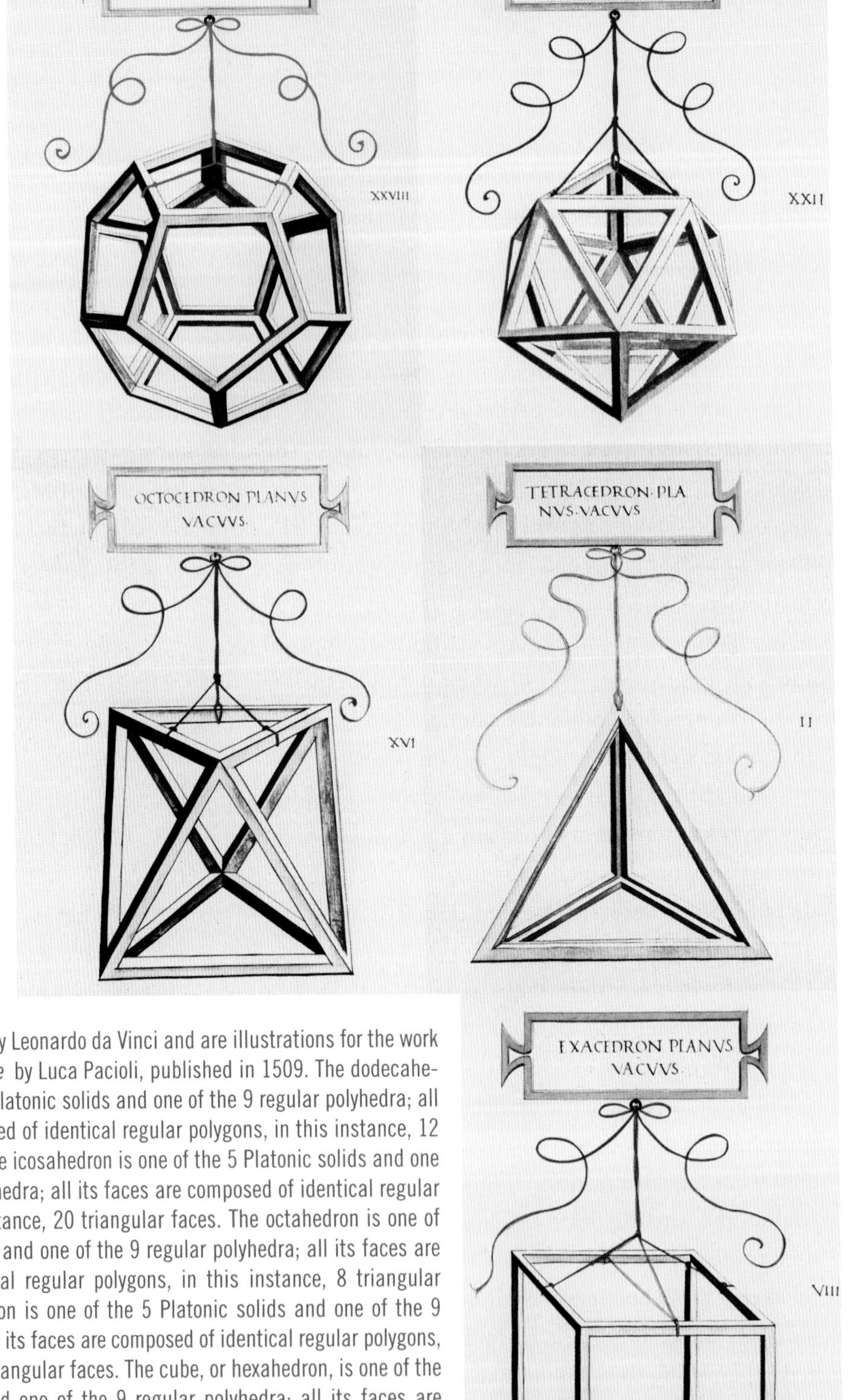

These drawings are by Leonardo da Vinci and are illustrations for the work *De Divina Proportione* by Luca Pacioli, published in 1509. The dodecahedron is one of the 5 Platonic solids and one of the 9 regular polyhedra; all its faces are composed of identical regular polygons, in this instance, 12 pentagonal faces. The icosahedron is one of the 5 Platonic solids and one of the 9 regular polyhedra; all its faces are composed of identical regular polygons, in this instance, 20 triangular faces. The octahedron is one of the 5 Platonic solids and one of the 9 regular polyhedra; all its faces are composed of identical regular polygons, in this instance, 8 triangular faces. The tetrahedron is one of the 5 Platonic solids and one of the 9 regular polyhedra; all its faces are composed of identical regular polygons, in this instance, 4 triangular faces. The cube, or hexahedron, is one of the 5 Platonic solids and one of the 9 regular polyhedra; all its faces are composed of identical regular polygons, in this instance, 6 square faces.

THE FAMOUS FIVE

PLATONIC SOLIDS

. . . one of the most beautiful and singular discoveries made in the whole history of mathematics.

Hermann Weyl[2]

POLYGONS are figures with straight sides that you can draw on a flat sheet of paper. Regular polygons have sides that are equal in length, and their interior angles are also equal. This is very restrictive but there are still an infinite number of possibilities. The simplest examples with three and four sides are the equilateral triangle and the square, but we can keep on going for ever. If you prescribe any finite number, no matter how large, there is a regular polygon with that number of sides that could be drawn if you had a fine enough pencil. As the number of sides increases the polygon is harder and harder to distinguish from a circle with the naked eye (indeed we might consider a circle to be a polygon with an infinite number of sides). In short, there are an unlimited number of regular polygons.

If we move on from flat polygons to their analogues in three-dimensional space, then we have to consider the collection of convex polyhedra – solid figures with plane polygonal faces that bulge outwards.[3] If we say nothing more about the faces then an infinite number of different ones will be possible. But suppose that we limit the discussion again to those that are regular – that is, whose faces are identical: how many are there?

Remarkably, there are only *five* of these regular solids:[4] the tetrahedron (with 4 triangular faces), the cube (with 6 square faces), the octahedron (with

8 triangular faces), the dodecahedron (with 12 pentagonal faces), and the icosahedron (with 20 triangular faces). The move from two to three dimensions has proved to be infinitely restrictive.[5] Euclid proves that these are the only possibilities at the end of his *Elements* but they were known to the Greeks much earlier. They are usually known as the Platonic solids because Plato describes them in his book *Timaeus*, written in about 350 BC. In this work, Plato started a tradition of ascribing cosmic significance to these five symmetrical figures by equating the tetrahedron with the basic element of fire, the cube with earth, the icosahedron with water, the octahedron with air, and the dodecahedron with the ethereal stuff from which the starry constellations and heavens were made.

Trying to find out who first discovered the regular polyhedra is a bit like trying to find out who invented fire,[6] but Plato (427–347 BC) credits the discovery of the regular polyhedra to Theaetetus of Athens (417–369 BC), who may have been a pupil of Plato's at the Academy. Historians believe that some of the later books of Euclid's *Elements* derive entirely from the discoveries of Theaetetus, and much else that is reported in the works of Eudoxus and Pappus as well. One early source says:

> . . . the five so-called Platonic figures which, however, do not belong to Plato, three of the five being due to the Pythagoreans [about 550 BC], namely the cube, the pyramid, and the dodecahedron, while the octahedron and the icosahedron are due to Theaetetus.[7]

The mystical and astrological associations of the Platonic solids had a powerful grip on Western thinking until the time of Kepler, who, in his *Mysterium Cosmographicum*, sought to accommodate the heavens in the fivefold harmony of the Platonic solids. Kepler's model of the solar system used all the five Platonic solids to describe the orbits of six planets that were known in the sixteenth century. He used the ratio of diameters of the inscribed and circumscribed spheres for each of the Platonic solids to specify the ratio of the greatest distance of one planet from the Sun in its orbit to the least distance of the next outermost planet from the Sun. This produced five ratios for the six known planets. A Platonic solid was introduced in between two adjacent planets, so that the innermost of the pair, when at its farthest distance from the Sun, lies on the inscribed sphere of the Platonic solid, while the outer planet lies on the circumscribed sphere when it is at its closest to the Sun.

The four regular star polyhedra, sometimes known as the Kepler-Poinsot polyhedra. They are the Great Dodecahedron (*top left*), Small Stellated Dodecahedron (*top right*), Great Stellated Dodecahedron (*bottom left*), and the Great Icosahedron (*bottom right*).

When the early Greeks first enumerated the five regular polyhedra that form the Platonic solids, they confined their attention to the convex polyhedra – that is, those that bulge outwards. If we allow polyhedra that bulge inwards, so that any two faces can meet along a common edge at an angle that is *less* than 180 degrees, then there are *four* new members of the club, known together as the regular Star Polyhedra and going by the names of the Small and Great Stellated Dodecahedra, the Great Dodecahedron, and the Great Icosahedron. These were only discovered, one by one, during the Renaissance as craftsmen modified the Platonic figures for ornamental purposes. Kepler also noticed that you could add pyramids of a fixed height to the regular faces of the regular octahedron, dodecahedron, and icosahedron so that the lateral faces of the pyramids fall into the same planes. He introduced the idea of combining polyhedra so that they had intersecting faces, rather like a three-dimensional version of the Star of David. These new possibilities were not systematically understood in the way that the convex polyhedra were, until the work of the French mathematician Louis Poinsot in 1810,[8] and they are sometimes called the Kepler–Poinsot polyhedra, although some of them seem to have been anticipated by the drawings of the remarkable Nuremberg goldsmith Wentzel Jamnitzer, in his 1568 book *Perspectiva Corporum Regularium*. But it took another two years before Augustin Cauchy, in 1812, was able to prove Poinsot's conjecture that his four candidates exhausted all the possible star polyhedra in three dimensions.[9] The slightly curious English names were given to them by the English mathematician Arthur Cayley only later, in 1859.

Wentzel Jamnitzer's designs, beautifully engraved by Jost Amman (1539-1591).

Today, these polyhedra remain objects of aesthetic appeal and geometric fascination to mathematicians.[10] The models that are made of them never cease to amaze us with their combination of beauty, symmetry, and simplicity.[11] It is easy to understand the hold that they once had on the human mind in its search for the numinous and the hidden connections between the things around us here and now and the timeless geometrical harmonies believed to inform the entire Universe about us.

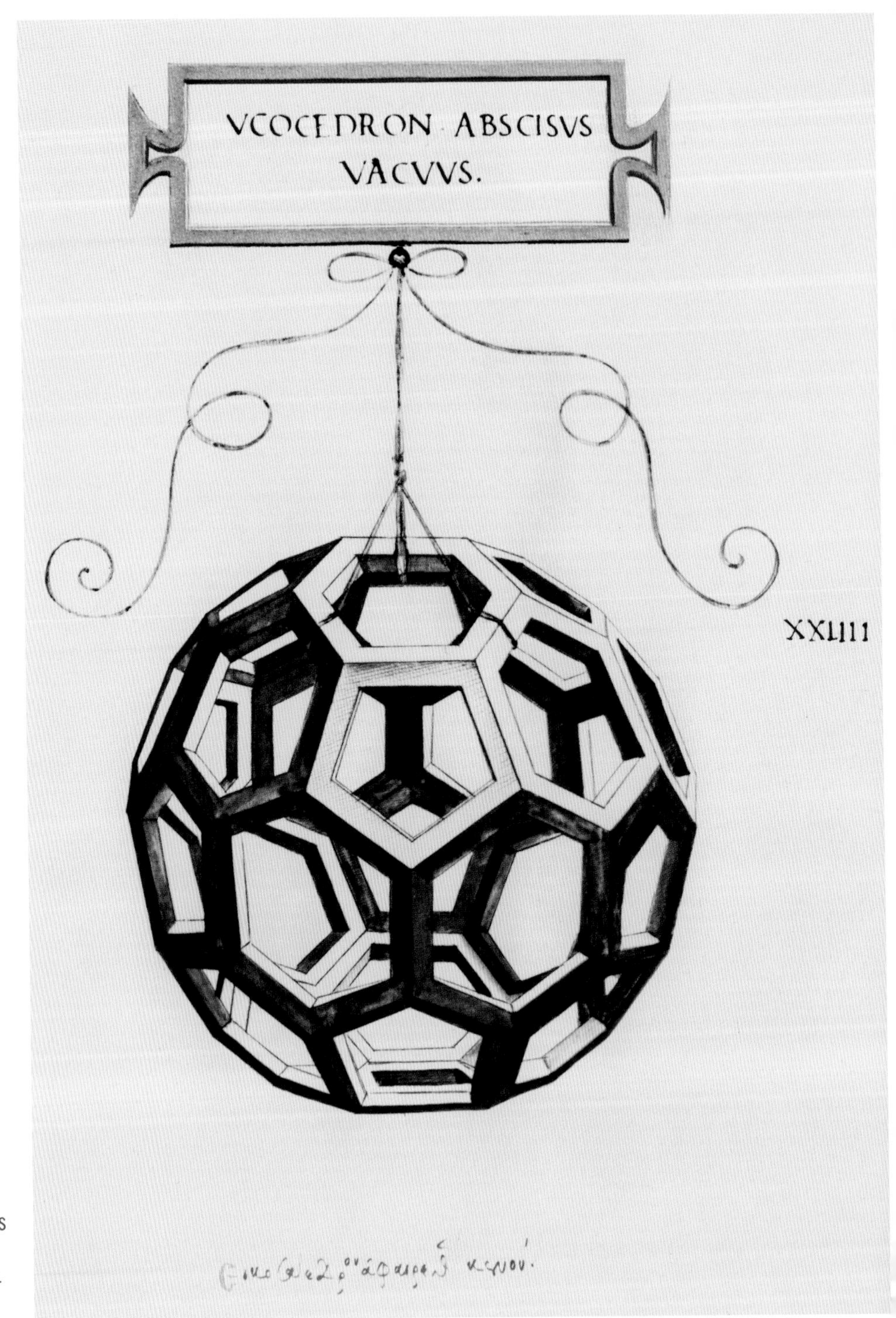

Leonardo da Vinci's Truncated Icosahedron, one of his illustrations for Luca Pacioli's *De Divina Proportione* (1509).

DOES GOD PLAY FOOTBALL?

BUCKY BALLS

God may not play dice but
maybe he plays football.

Harry Kroto

ON contemplating the Platonic solids, Archimedes (*c.* 287–212 BC) soon spotted the possibility of creating thirteen new almost-regular polyhedra. By simply symmetrically sawing off each of the corners of the cube, tetrahedron, octahedron, dodecahedron and icosahedron, he created their five truncated counterparts, which we call the Archimedean polyhedra. These five solids have faces which are still regular polygons – although they are not all the same; their vertices are all alike but their faces are not all the same. Following this prescription, there are a further eight Archimedean polyhedra that can be constructed. We could regard them as the next most symmetrical solids after the Platonic and Star polyhedra.

One of these Archimedean polyhedra has turned out to have a very special significance in the Universe and has been a dominant image in the development of chemistry over the past twenty years. The very special solid is Archimedes' truncated icosahedron. It has 60 vertices, 32 faces, of which 3 meet at every vertex, and 90 edges. The 32 faces consist of 20 hexagons and 12 pentagons, so that 2 hexagons and one pentagon meet at every vertex.[12] This is a beautiful structure and is probably more familiar to you than this catalogue of facts might immediately suggest. A typical football (or 'soccer' ball as it would be called in the USA[13]) had this structure until very recently,[14] with individual pentagonal panels in black and hexagonal panels in white.

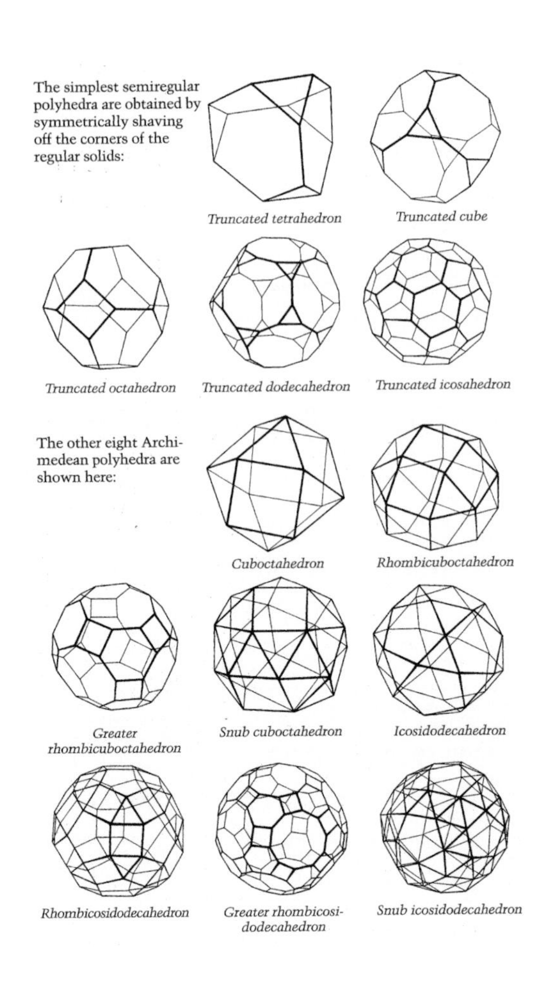

The thirteen Archimedean Polyhedra. All of them have faces which are two or more types of regular polygon.

The architect Richard Buckminster Fuller (1895–1983) made extensive use of icosahedral geometries in the design of geodesic domes, which he invented in 1949. Buckminster Fuller was a self-taught structural engineer who sought to employ mathematical symmetry in the search for building structures that were optimal in various ways, minimising the use of materials or the effort involved in assembly, or in maximising structural strength. He appreciated the way in which some materials that were extremely weak in one context could assume considerable strength if arranged in an appropriate geometric configuration. An eggshell is a familiar example.

Buckminster Fuller's architecture for the American pavilion at the Montreal Expo '67 was a spectacular realisation of a geodesic dome with faces composed of the mixture of hexagons and pentagons that form the truncated icosahedron.

It was a spectacular manifestation of symmetry and function that impressed many scientists and designers with its scale and form. One of those was Harry Kroto, a chemist with a life-long interest in architectural and graphic design. Harry was a colleague of mine at the University of Sussex and even sat on the interview panel that appointed me to my first lectureship. Harry had long been interested in the possibility that carbon molecules might form into long chains in the unusual conditions within molecular clouds in space. Testing an idea like that involves two steps: first, try to create similar chains under extreme artificial conditions in the laboratory, then see if the signature of the spectrum of light from any molecules in space matches those of the artificially created chains. In 1985, Harry pursued the laboratory work with a team at Rice University in Texas led by Richard Smalley and Robert Curl, along with graduate students James Heath and Sean O'Brien. They would smash small clusters of carbon atoms with a laser beam and then explore the vapourised debris after it condensed to see if any interesting new carbon agglomerations had been created. They found that all the new clusters they formed had an even number of atoms, and with a little tuning of the experiment they could create clusters that almost always contained sixty carbon atoms. Harry puzzled over the experimental output as the team tried to understand why carbon was being

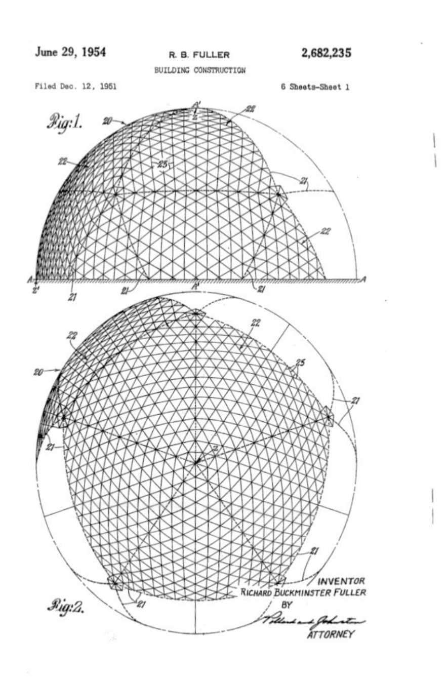

Drawing from Buckminster Fuller's 1954 patent document (no. 2682235).

preferentially created in this carbon-60 form. He remembered the Buckminster Fuller dome and a small cardboard truncated icosahedron that he had made out of cardboard for his children. After phoning home to England to check the geometry of his model, he was convinced that the carbon was forming into a truncated icosahedral structure with one carbon atom at each of the sixty corners. Smalley worked to make a paper model which combined pentagons and hexagons. This work was completed in a frantic eleven-day period beginning on 1 September 1985, and then submitted to the journal *Nature* on 12 September, where it was received on 13 September and published on 14 November with a picture adorning the front cover.

These carbon atoms eventually went by a number of names. At first, it was called 'buckminsterfullerene' in recognition of the insight supplied by Buckminster Fuller's dome. This was soon shortened to 'fullerene' or, more informally, to 'Bucky balls', or even occasionally 'soccerene'.[15]

This new structure for carbon was a great revolution in chemistry,[16] bringing about a unification of inorganic and organic chemistry[17] and offering new ways to engineer materials on the molecular scale. Curl, Smalley and Kroto shared the Nobel Prize for chemistry in 1996. The symmetrical appearance of Bucky balls made them a natural icon for chemistry and there are a host of images on magazine covers celebrating the molecule that only DNA can beat.

Prototype dome by Buckminster Fuller in the form of a rhombic Enneacontahedron, Washington University, St Louis, 1954.

Q
nature
INTERNATIONAL WEEKLY JOURNAL OF SCIENCE
Volume 318 No 6042 14-20 November 1985 £1.90
BETTY & GORDON MOORE
CAMBRIDGE
LIBRARY
14 NOV 1985

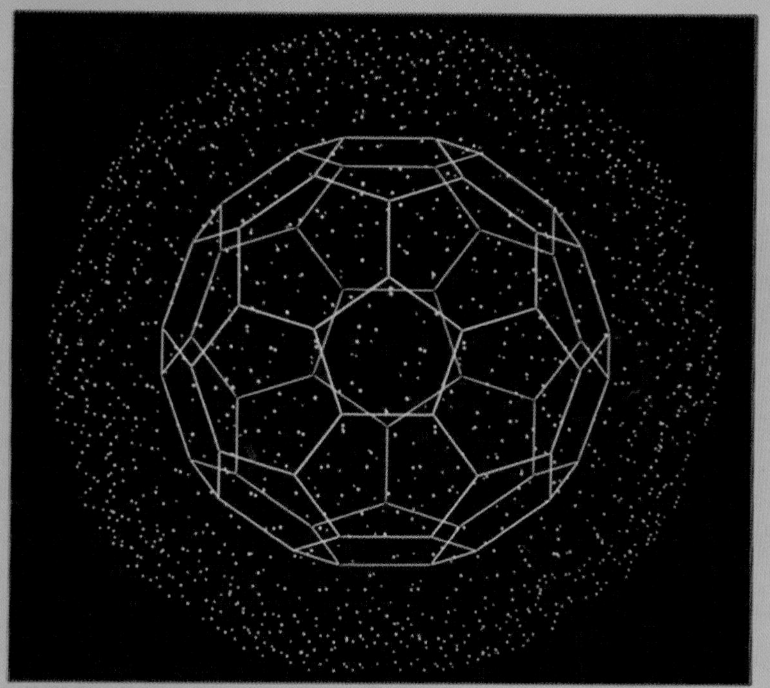

SIXTY-CARBON CLUSTER

AUTUMN BOOKS

Harvey Brooks
(transformation of MIT)

P. N. Johnson-Laird
(brain and mind)

Anthony W. Clare
(psychoanalysis as religion)

A. O. Lucas
(war on disease)

Hendrik B. G. Casimir
(physics and physicists)

Gordon Thompson
(dimensions of nuclear proliferation)

Jacques Ninio
(origins of life)

Edward Harrison
(steps through the cosmos)

The front cover of *Nature* of 14 November 1985, in which the discovery of carbon-60 was announced by Robert Curl, Harold Kroto and Richard Smalley.

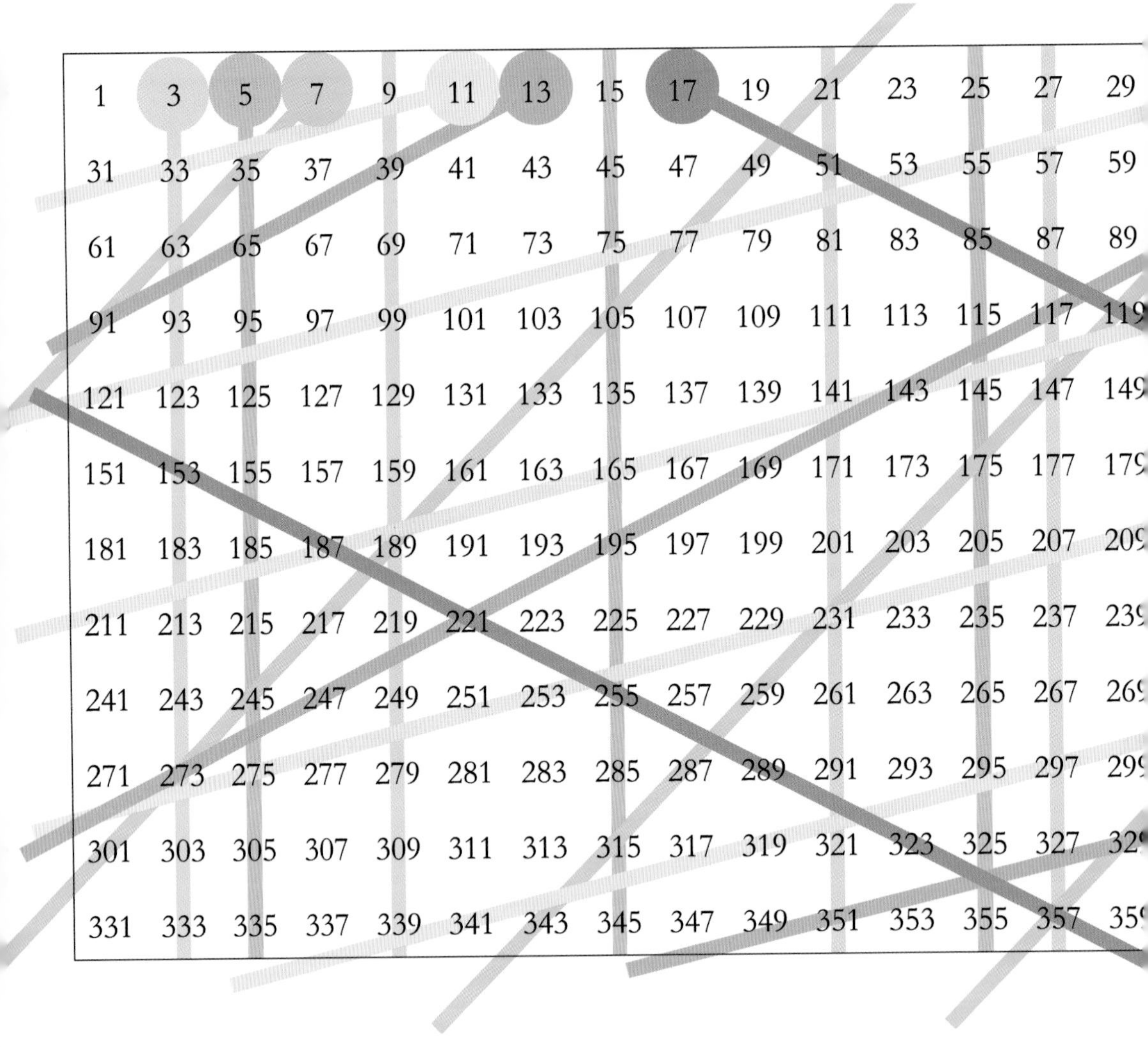

1	3	5	7	9	11	13	15	17	19	21	23	25	27	29
31	33	35	37	39	41	43	45	47	49	51	53	55	57	59
61	63	65	67	69	71	73	75	77	79	81	83	85	87	89
91	93	95	97	99	101	103	105	107	109	111	113	115	117	119
121	123	125	127	129	131	133	135	137	139	141	143	145	147	149
151	153	155	157	159	161	163	165	167	169	171	173	175	177	179
181	183	185	187	189	191	193	195	197	199	201	203	205	207	209
211	213	215	217	219	221	223	225	227	229	231	233	235	237	239
241	243	245	247	249	251	253	255	257	259	261	263	265	267	269
271	273	275	277	279	281	283	285	287	289	291	293	295	297	299
301	303	305	307	309	311	313	315	317	319	321	323	325	327	329
331	333	335	337	339	341	343	345	347	349	351	353	355	357	359

The Operation of the Sieve of Eratosthenes. Lines are drawn through a grid of all numbers to eliminate, in turn, all numbers that are divisible by 3, 5, 7 . . . and their multiples by drawing the sets of parallel bands of different colours. The numbers that are not covered by any coloured band are prime numbers. Prime numbers at the start of the bands on the first row are circled. Even numbers are not included in the grid because they are all divisible by two.

PRIME TIME

THE SIEVE OF ERATOSTHENES

Prime numbers. It was all so neat and elegant. Numbers that refuse to cooperate, that don't change or divide, numbers that remain themselves for all eternity.

Paul Auster[18]

PRIME NUMBERS are the most mysterious quantities that mathematicians have encountered. Divisible by nothing, save themselves and 1, they are the atoms of all numbers. The first few are easy to spot – 2, 3, 5, 7, 11, 13, 17 . . . – and more than 2000 years ago Euclid gave a beautiful proof that they are infinite in number.[19] Yet, there is no simple magic formula that can predict them all. The biggest known primes are much prized by mathematicians and cryptographers because they are used to create the world's most secure codes. At present, the biggest known prime is

$$2^{30402457} - 1$$

which has 9,152,052 digits when it is written out in full. It was found by Curtis Cooper and his colleagues in 2005 within the 'GIMPS' project. This is the mathematician's version of the SETI online project. It uses a central coordin-ating computer server and anyone who adds their own computer power to the prime-number search project will get some huge numbers to test to see if they are prime (just as SETI online gives you radio signals from space to analyse for evidence of intelligent patterns). There is a $100,000 prize, put up by the non-profit Electronic Frontier Foundation, for the first discovery of a prime with 10 million digits. The discovery of the largest primes is very much driven by the power of the biggest computers and proceeds at a rate that reflects the progress in computing power.

Although the search for new primes is now dominated by the capabilities of fast computers, there was a time when all searches were done solely by hand with the aid of human reasoning, which narrows down the number of possible candidates. The first and most influential procedure of this sort was devised by Eratosthenes of Cyrene (276–194 BC), which was then a Greek colony in North Africa but is now Shahhat in Libya. He was the third librarian of the great Library of Alexandria, and devised methods to determine the Earth's circumference and distance from both the Sun and the Moon. His Sieve appears in the *Introduction to Arithmetic* by the Syrian mathematician Nicomedes[20] in about AD 100. It was a handbook for use by students, much easier than Euclid's *Elements*, and so it continued to be widely used until the Middle Ages, both in Europe and in the Arab world. Eratosthenes' systematic process for finding prime numbers by sieving out the others was a useful technique so long as the numbers did not get too large, and it was the first algorithm. Unfortunately, Eratosthenes' works don't survive and we just read about the 'sieve' in the works of other commentators. There we learn that he was recognised as a great polymath[21] by his contemporaries, although he was not the leading authority in any subject. For this reason he was nick-named 'Beta', the second letter of the Greek alphabet, or 'pentathlos', after the pentathletes who excelled at the Games over five events but could not be the individual champion in any one of them.[22] Nicomedes explained the operation of the Sieve in the following way:

> The method of the 'sieve' is as follows. I set forth all the odd numbers in order, beginning with 3, in as long a series as possible, and then starting with the first I observe what ones it can measure, and I find that it can measure the terms **two** places apart, as far as we care to proceed. And I find that it measures not as it chances and at random, but that it will measure the first one, that is, the one two places removed, by the quantity of the one that stands first in the series, that is, by its own quantity, for it measures it 3 times; and the one two places from this by the quantity of the second in order, for this it will measure 5 times; and again the one two places farther on by the quantity of the third in order, or 7 times, and the one two places still farther

> on by the quantity of the fourth in order, or 9 times, and so on *ad infinitum* in the same way.
>
> Then taking a fresh start I come to the second number and observe what it can measure, and find that it measures all the terms **four** places apart, the first by the quantity of the first in order, or 3 times; the second by that of the second, or 5 times; the third by that of the third, or 7 times; and in this order *ad infinitum*.
>
> And analogously throughout, this process will go on without interruption . . . Now these that are not measured at all, but avoid it, are primes and incomposites, sifted out as it were by a sieve.

In other words, the Sieve works like this. List all the positive numbers in rows of ten, up to the largest number you are interested in, call it *N*. Now we cross out all the numbers in this grid which are not prime following Eratosthenes' recipe. First, the number 1 is defined as being not prime so remove it (if you included 1 as a prime number you would end up crossing off every number on the list). Circle the first of the remaining numbers that is not crossed out, 2, and cross out all subsequent multiples of that number. This removes all the even numbers. Circle the next remaining number, 3, and strike out all remaining multiples of that number. Continue like this, circling the first remaining number and crossing out all of its multiples that survive. The numbers that are left circled will be the prime numbers.

You soon find that lots of the numbers scheduled for crossing out because they are multiples of, say, 7 have already been eliminated because they were also multiples of a smaller number[23] (for example, $21 = 3 \times 7$).

The beauty of this representation is that the picture that results contains all sorts of patterns running along columns and diagonals yet there is no systematic way of predicting where the next circled number (a prime) is going to fall.

The most assiduous student of Eratosthenes' Sieve was the American mathematician Derrick Norman Lehmer (1867–1938), who published factor tables for the first 10 million primes extracted from the grid of all numbers by using the Sieve.[24] Lehmer speeded up this tedious process by partially mechanising the sieving. His son built a machine that consisted of a shaft on which thirty gears, each with a hundred teeth, were mounted and meshed with thirty other gears having numbers of teeth determined by one of the thirty prime numbers up to 127. The description of its workings then continues:

> Under each tooth in this second series of gears is a small hole. When the machine is set up and ready for use, some of these holes are plugged and others are open. A beam of light is cast on the side of the machine and then it is set in motion by means of an electric motor. The main shaft gears all revolve at the same speed, but the gears meshing with them revolve at different speeds because of the varying number of teeth. When in the course of perhaps hundreds of thousands of revolutions one hole in each wheel reaches the same point at the same time, when thirty holes are lined up, in other words, the beam of light goes straight through the machine, strikes a sensitive photo-electric plate and stops the machine instantly. A little counter which records the number of revolutions made by the main shaft, gives a number from which the factors of the large number under analysis can readily be obtained.[25]

One of the things that is evident from the picture of the Sieve is that primes become sparser as they get bigger. This is not surprising. As the numbers increase so does the number of factors that are available to divide it: for example, one in four of numbers smaller than 100 are prime, one in six of numbers smaller than 1000, one in 12.7 of those smaller than one million, and only one in 19.8 of those smaller than one billion. So the pattern is that, very roughly, 1 in $2.3N$ of the numbers smaller than 10^N is a prime. There is another way of saying this: of the numbers smaller than N, approximately one in every $\log_e N$ of them is prime, where $\log_e N$ is the natural logarithm of N.[26] Karl Friedrich Gauss made an improved guess, that for numbers near N roughly $1/\log_e N$ of them are prime, that is roughly $N/\log_e N$ of the numbers smaller than N are primes. So there are still plenty of primes getting through Eratosthenes' sieve as N gets bigger. Cryptographers are not going to run out of accessible primes.

This ancient stone shows the square root of two computed to three places in the base 60 system used by the Babylonians, circa 1700 BC (photo by Bill Casselman).

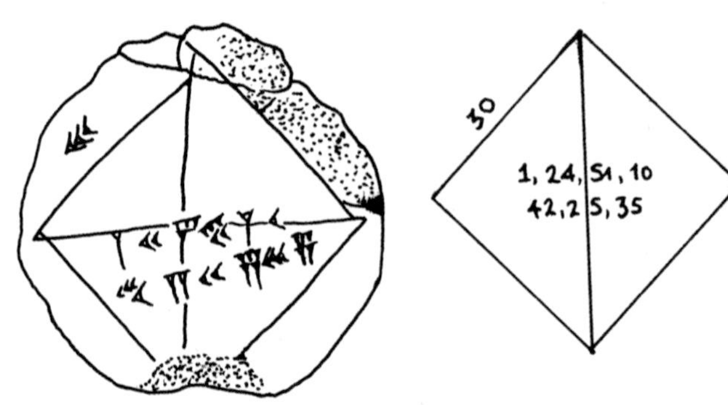

THE SQUARE ON THE HYPOTENUSE

THE BRIDE'S CHAIR

It has been said that the art of geometry is to reason well from false diagrams.

Jean Dieudonné[27]

ONE of the most famous theorems of mathematics, studied by school pupils and college students the world over for 2500 years, is to be found in the first great mathematics book of the Western world. The author, whom we know simply as Euclid, was the senior mathematician at the university and great library in the city of Alexandria, the city founded by, and named after, Alexander the Great in 332 BC. This library, the greatest in the ancient world, with more than 600,000 papyrus scrolls, was opened along with the university in 300 BC.

Euclid gathered the work of the great Pythagorean mathematicians such as Archytas, along with the works of Hippocrates, Eudoxus, and Theaetetus, no doubt adding his own, and then organised the whole into the logical progression of the *Elements*, a collection of thirteen short books that became the most influential teaching textbook in the whole of human history. It is a great treasury of beautiful mathematical argument that shows the power of the axiomatic method at work. Euclid would lay down clear axioms, or assumptions, and rules of reasoning, at the outset, and then use them systematically to prove theorems.[28] This style of rigorous deduction acted as a model for thinkers in many different subjects for thousands of years afterwards. From Thomas Aquinas to Spinoza, we see a style of argumentation that mirrors the layout of

The diagram on the left shows the Babylonian cuneiform numerals on the edge and diagonal of the square. These numbers correspond to the square (*right*) of side 30 whose diagonal is calculated. The two numbers (in base 60) on the diagonal first give 1.414212963, a good approximation to the square root of 2. Below it is the product 30 x 1.414212963, which is the length of the diagonal according to Pythagoras' theorem.

Euclid's propositions and proofs. Indeed, this development went hand in hand with a perspective on mathematics on the part of philosophers and theologians that saw it as part of the absolute truth of things, not a mere description or model of reality.

Until the beginning of the nineteenth century, Euclid's geometry was believed to be the way space truly was. When the non-Euclidean geometries that describe geometry on curved surfaces, such as a sphere or a saddle, were rigorously formulated by Riemann, Lobachevsky, Bolyai, and Gauss, it shocked philosophers to the core. Euclidean geometry was suddenly just one among many: all complete and logically self-consistent, all defined by their own set of water-tight axioms. These geometrical discoveries therefore gave rise to a more general climate of relativistic thinking. Absolute truth was out, whether in geometry, politics, religion, or anthropology.

There is no doubting the breadth and depth of the impact that Euclid's books have had. But there is one result, with its accompanying image, whose impact was stronger than everything else. We find it painstakingly proved in the last but one of the forty-eight propositions in the first book of the *Elements*: it is that property of all right-angled triangles that we know now as Pythagoras' theorem.[29] Take any right-angled triangle whose sides are of length A, B, and C, where C is the length of the longest side (or hypotenuse), then:

$$A^2 + B^2 = C^2$$

One legend has it that Pythagoras (*c.*580 BC–*c.*500 BC) discovered the theorem while waiting to see Polycrates, the tyrannical ruler of the island of Samos. As he waited in the outer hall of the palace, Pythagoras became fascinated by the square tiling pattern of the floor. If he drew a diagonal line which divided a square into two right-triangles, he noticed, then the area of a square drawn above the diagonal was twice the area of the square drawn on an adjacent side. In other words, the square on the hypotenuse is equal to the sum of the squares on the triangle's two other sides. This is the case when $A = B$ in the formula above. Well, it's a nice story!

In the *Elements* Euclid proves the theorem for the more general situation when A does not have to equal B, by means of a diagram whose construction lines bore some resemblance to the cover that was used to drape the chair of honour for the bride at a wedding reception.[30] The theorem states that the area of the square *BCED* equals the sum of the areas of the two squares *AHKC* and *GABF.*

In fact, there are hundreds of different proofs of Pythagoras' theorem and we know that it was known to the ancient Babylonians, Chinese, Indians, and Egyptians long before Pythagoras lived. The Babylonians knew how to construct triangles with choices of *A*, *B* and *C* obeying the Pythagorean relationship earlier than 1600 BC.

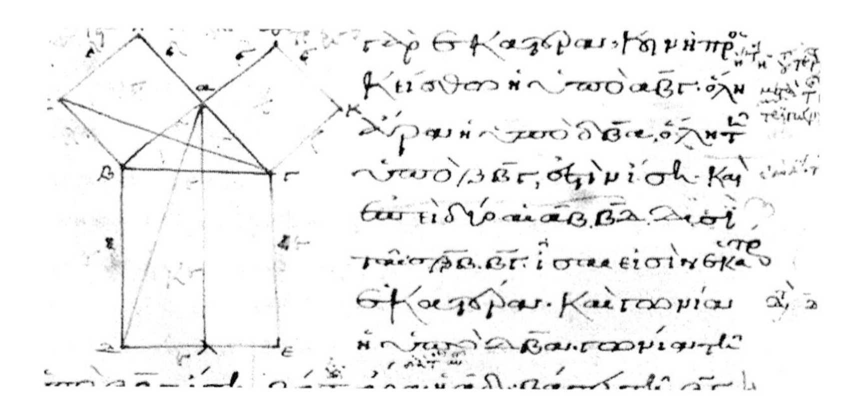

An early Greek edition of Euclid showing the construction of the theorem of Pythagoras.

The small cuneiform tablet with which we opened this chapter[31] shows a square and its diagonals. The inscribed wedge marks give the lengths of the sides in the mixed base-10 and base-60 arithmetic used by the Babylonians.[32] The side of the square is $3 \times 10 = 30$; if you know Pythagoras' theorem you would expect the diagonal to have length equal to $30 \times \sqrt{2}$, so the ratio of the diagonal to the side equals the square root of 2. Here in base-60 (so multiplying by 30 is like dividing by 2), the diagonal is given by 42;25,35 and 1;24,51,10 is equal to the square root of 2. Expanding their base-60 arithmetic notation into our decimal representation, this means its Babylonian value is:

$$1 + 24/60 + (51/60)^2 + (10/60)^3 = 1.41421296$$

which is an excellent approximation good to six decimal places (and four of their sexagesimal places).

The oldest known proof of Pythagoras' theorem is in some ways the most ingenious and beautiful. It is to be found in the oldest text of ancient Chinese mathematics that survives, the *Arithmetic Classic of the Gnomon and the Circular Paths of Heaven*, some of whose parts date back to about 600 BC. It works by drawing four copies of the right-angled triangle with sides A, B and C inside the large square. Then you simply note that the area of the large square $(A + B)^2 = A^2 + B^2 + 2AB$ equals the sum of the areas of the four triangles, that is $4 \times \frac{1}{2}AB$, plus the area of the small central square, C^2. Hence, we see that $A^2 + B^2 = C^2$. Or simpler still, just move the four triangles around inside the big square to make two rectangles and note that the area inside the big square not covered by the triangles must be the same in both pictures and so $A^2 + B^2 = C^2$.

句股冪合以成弦冪

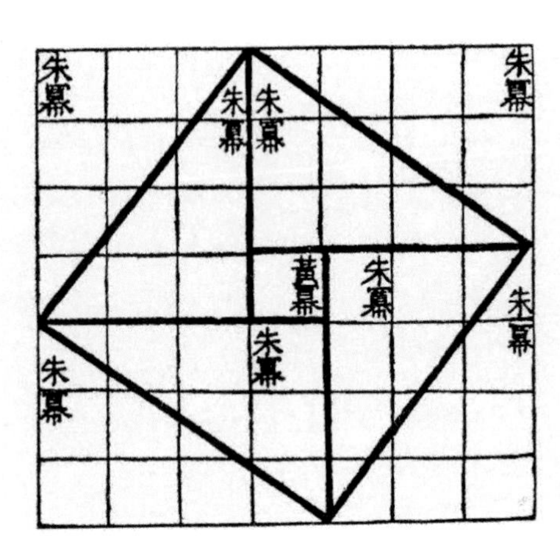

A Chinese diagrammatic proof from 600BC of Pythagoras' theorem which works by noting that the area of the large square equals the sum of the area of the four right-angled triangles and the small square drawn inside it.

Pythagoras' great theorem is still in the news. In 1670, the French mathematician Pierre de Fermat (1601–65) conjectured that the equation

$$A^n + B^n = C^n$$

has no solution if A, B, and C are all positive whole numbers and n is a whole number greater than 2. Fermat's conjecture has a wonderful history because Fermat annotated Problem 11.8 on page 85 of his 1621 edition of *Arithmetica*, a Latin translation of an ancient Greek mathematical text by Diophantus of Alexandria, to the effect that he had a wonderful proof of his conjecture, but the margin was too small to contain it (*Cuius rei demonstrationem mirabilem sane detexi. Hanc marginis exiguitas non caperet*). Alas, no one believed him. Ever since then, Fermat's conjecture has remained the primary challenge problem in mathematics. After hundreds of years of unsuccessful attempts to prove it true or false, the English mathematicians Andrew Wiles and Richard Taylor of Princeton University finally completed a fiendishly difficult proof of Fermat's tantalising conjecture in 1994. Fermat was right. Only when $n = 2$ does the A-B-C equation have a solution for whole numbers A, B, and C.

The pictures we have shown do not just have an impressive and influential history. They, and others like them, have been the source of considerable controversy. Beginning in 1934, a small group of French mathematicians gathered around André Weil began to meet at Capoulade, a café in the Latin Quarter of Paris. Their aim was to reformulate the different parts of mathematics in a new and rigorous fashion that exhibited their shared logical structures. They invented a pseudonym for the 'author' of the project, Nicolas Bourbaki – after a general in the Franco-Prussian war – and set to work. They picked on five central areas of pure mathematics to be expounded and excluded all (impure?) applied mathematics. As Jean Dieudonné remarked:

> Here is my picture of mathematics now. It is a ball of wool, a tangled hank where all mathematics react upon another in an almost unpredictable way. And then in this ball of wool, there are a certain number of threads coming out in all directions and not connecting with anything else. Well the Bourbaki method is very simple – we cut the threads.[33]

The Bourbaki texts were evolved by its members by a process of remorseless criticism and rewriting. They introduced all sorts of pieces of modern mathematical notation and terminology that few now realise they owe to Bourbaki. But Bourbaki had an influence that many mathematicians, especially applied mathematicians, regarded as pernicious. It emphasised mathematical structures at the expense of problems and examples. There were *no* pictures. Mathematical intuition seemed to be flattened out into an axiomatic straitjacket. The emphasis upon common structures behind the developments in different areas of mathematics eventually began to influence the school curriculum in many countries, giving rise to the so-called 'new maths' philosophy. In retrospect this was not successful. It introduced sophisticated concepts

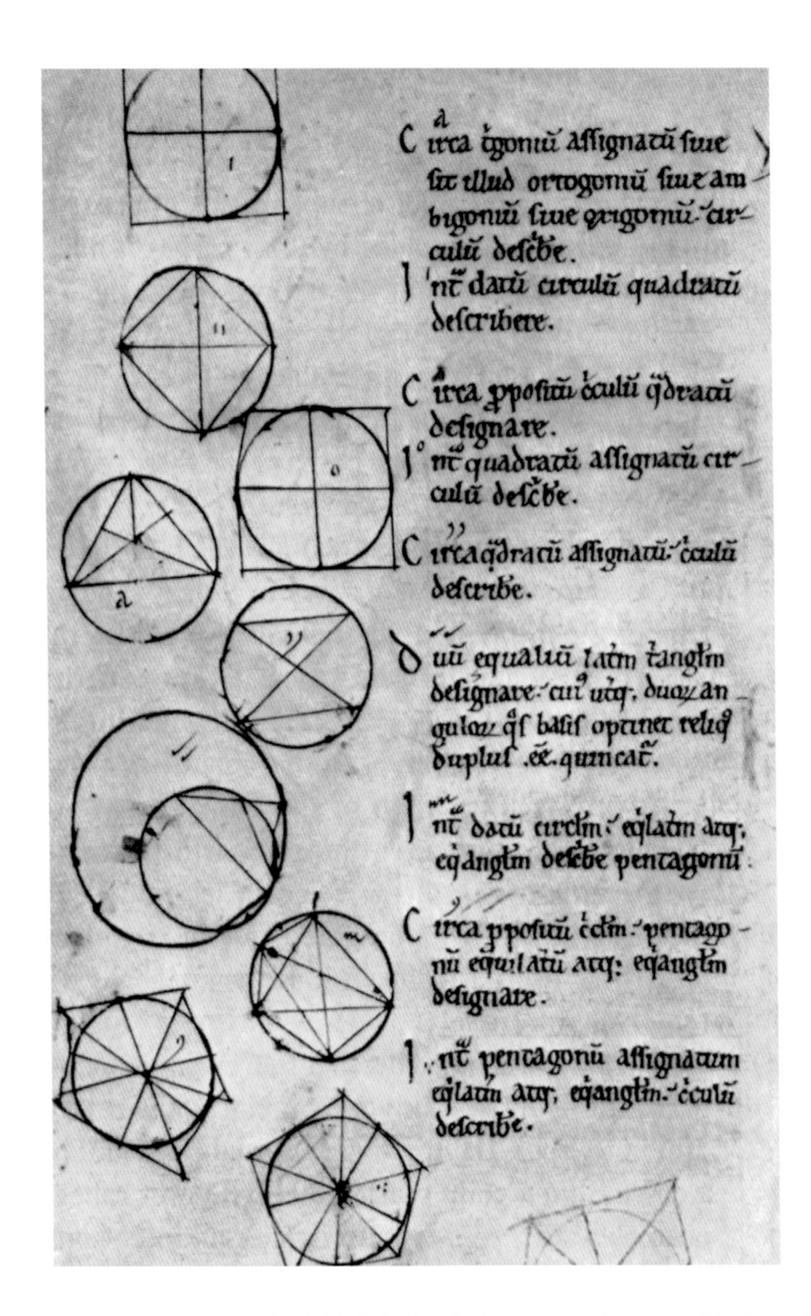

A twelfth-century Latin copy of Book IV of Euclid which states the theorems without proof and provides incomplete diagrams for students to letter and use to provide complete proofs. The original, by Adelard of Bath, from which it was copied, did contain the proofs. Notice how each figure contains labels or checkmarks that tell the reader which statement it is related to. For example, the large circle containing the small circle (fourth from the bottom) contains two 'ticks' to link it to the sixth statement in the list, labelled ∂.

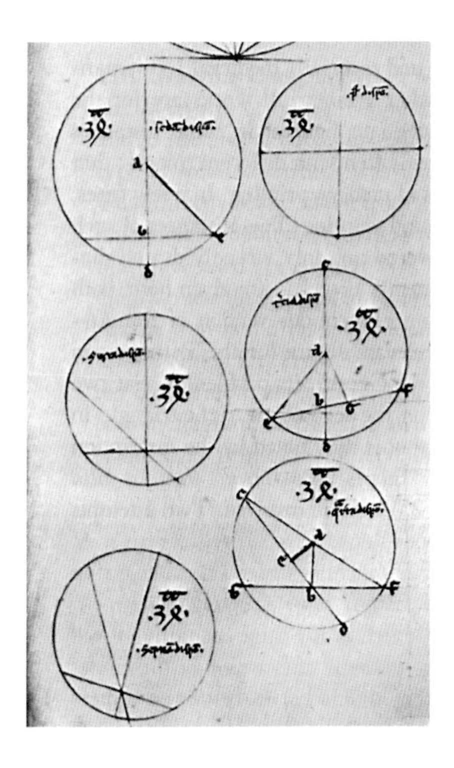

A fourteenth-century copy of an anonymous version of Euclid which cut out the statements and proofs of his theorems but kept the supporting diagrams so that readers could create the proofs for themselves. Here, many different cases are shown of Proposition 3.4 of Book III of Euclid: that, for any intersecting chords in a circle, the segments of one chord from the intersection create a rectangle whose area equals that determined by the segments of the other chord. The different pictures show different cases for the way the chords can be drawn.

too early and students lacked the motivation and interest supplied by problems and failed to develop the skills needed for problem solving. Not least, it prevented parents from helping their children because they simply didn't recognise the type of mathematics being done any more.

The Bourbaki group wanted to banish the practice of 'proof by picture', wherein a diagram played a crucial role in supplying the proof of a mathematical theorem. Such a philosophy was not completely new. Lagrange's famous four-volume work on mechanics, *Mécanique analytique*, of 1788, had been remarkable for its complete absence of diagrams and geometric ideas. Indeed, the author boasted about it in his Preface. His great predecessor, Isaac Newton, had used diagrams wherever they were helpful, just as we do today. For we know that if a proof by picture is a possibility, then there will be a way to formalise the logic that went into that process into something that may apply to a wide range of examples not covered by the picture and not necessarily transparent to the mind's eye. Euclid, for example, was drawn to generalise the original theorem of Pythagoras by replacing the squares on the sides of the right-angled triangle with other shapes.[34]

We might also add that wrongly drawn diagrams could be a source of serious error. In the Middle Ages the term *falsigraphia* was used for wrongly constructed diagrams and arguments. The laborious process of copying ancient manuscripts by hand often led to the figures being omitted or rather carelessly copied. For similar reasons, sometimes the proofs would be omitted and only a selection of diagrams provided to illustrate the statement of the theorems. There are also examples in which the figures appear but are not lettered (a task for someone else). On page 290 we show a twelfth-century Latin copy of Book IV of Euclid that gives statements of theorems but no proofs, together with some incomplete diagrams. Opposite, on page 291, is a fourteenth-century copy which began by giving the statements and proofs and figures for the first six propositions of Book III of Euclid but then decided to finish in a hurry by giving a gallery of figures for the rest of the propositions, but neither their statements nor their proofs. Perhaps they were the analogue of the 'Key facts' type of publications used to aid revision for exams? Or lecture handouts with blank spaces for students to fill in?

After the retirement of its main participants, and an awkward legal battle with publishers, by 1980 Bourbaki was in decline. More than twenty years passed without a significant publication emerging out of the collective and its influence has waned.[35] Perhaps we just like pictures too much? But this would probably be too simple a conclusion to draw. Bourbaki and the opposition to it by many mathematicians showed that there are really 'two cultures' in mathematics. There are those who primarily like to build structures and formalisms of great generality, and there are those who like to pose and solve specific problems (although no mathematician is exclusively in one of these camps all the time). Sometimes the latter focus can lead to the former, and vice versa. The fact that there is a diversity of opinion as to which tendency is (or should be) in the ascendant is probably a good sign that a healthy balance continues to exist between the two.

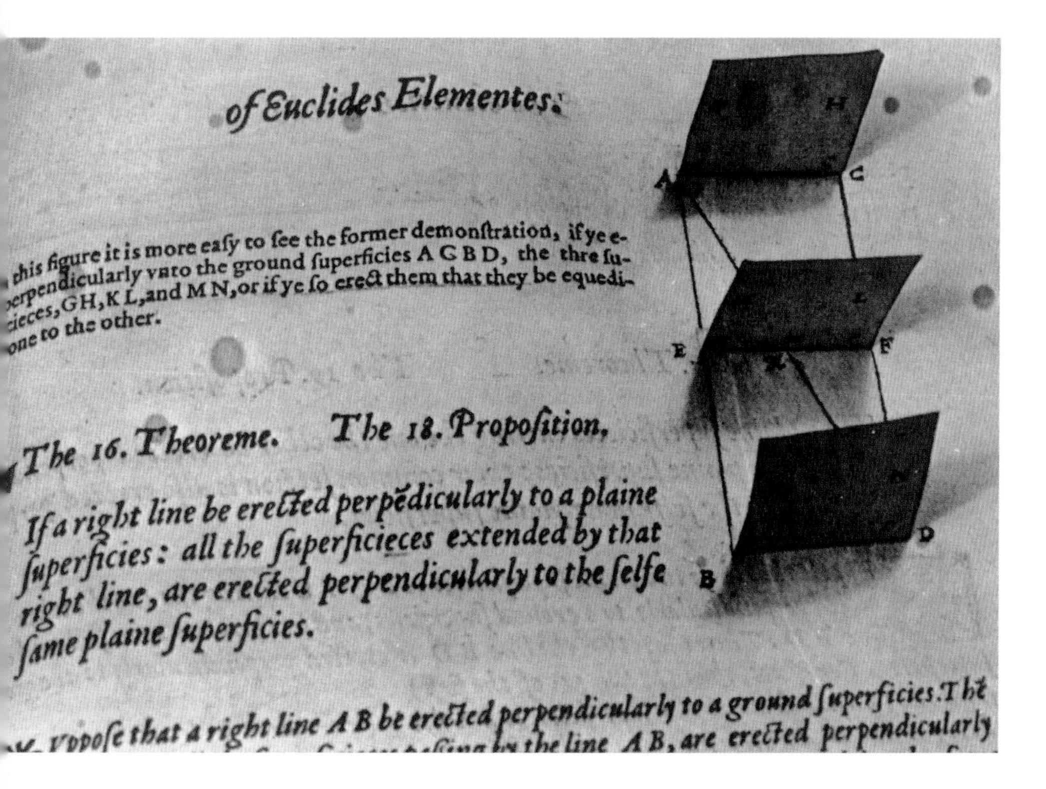
of Euclides Elementes.

...this figure it is more eaſy to ſee the former demonſtration, if ye e-
...erpendicularly vnto the ground ſuperficies A G B D, the thre ſu-
...ieces, G H, K L, and M N, or if ye ſo erect them that they be equedi-
...one to the other.

The 16. Theoreme. The 18. Propoſition.

If a right line be erected perpēdicularly to a plaine ſuperficies: all the ſuperficieces extended by that right line, are erected perpendicularly to the ſelfe ſame plaine ſuperficies.

...ppoſe that a right line A B be erected perpendicularly to a ground ſuperficies. The...

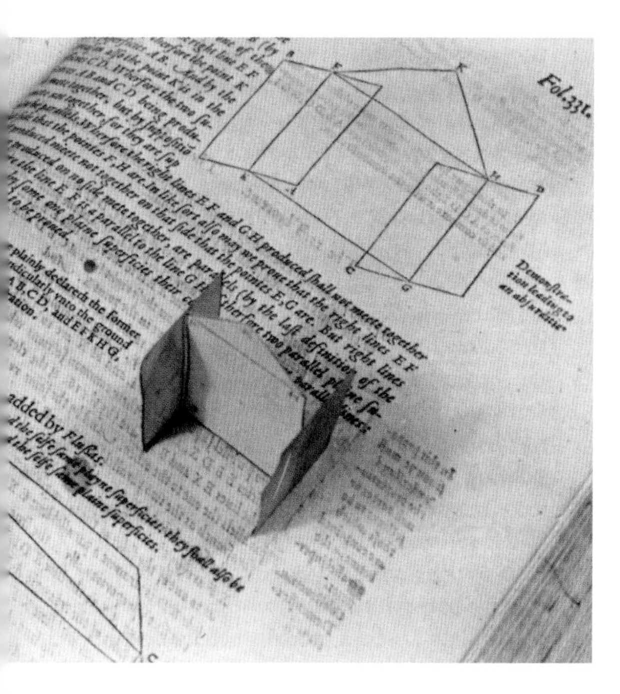

The first 'pop-up' book was about three-dimensional geometry: these two pages from *The Elements of Geometrie of the most auncient Philosopher Euclide of Megara Faithfully (now first) translated into the Englishe toung by Henry Billingsley* (London, 1570) show the clever use of fold-up figures. This was the first English translation of Euclid. There are 34 of these pop-ups in the book and the printing was the work of the renowned English printer John Day.

4 — 17 sen oder deßgley-
3 + 30 chen/ So summier
4 — 19 die zenttner vnd
3 + 44 lb vnnd was auß
3 + 22 — ist/ das ist mi-
Zentner 3 — 11 lb nus dz setz beson-
3 + 50 der vnnd werden
4 — 16 4539 lb (So
3 + 44 du die zendtner
3 + 29 zů lb gemachett
3 — 12 hast vnnd das /
3 + 9 + das ist meer
darzů Addierest) vnd 75 minus. Nun
solt du für hölcz abschlahen allweeg für
ain legel 24 lb. Vnd das ist 13 mal 24.
vnd macht 312 lb darzů addier das —
das ist 75 lb vnd werden 387. Dye sub-
trahier von 4539. Vnd bleyben 4152.
lb. Nun sprich 100 lb das ist ein zentner
pro 4 fl $\frac{1}{8}$ wie kumen 4152 lb vnd kumē
171 fl 5 ß 4 heller $\frac{4}{5}$ Vñ ist recht gmacht

Pfeffer

R

Plus and minus signs in the 1526 edition of Johann Widmann's *Mercantile Arithmetic.*

THE TIMES OF THE SIGNS

$+ \quad - \quad \times \quad \div \quad =$

Biologists think they are Biochemists,
Biochemists think they are Physical Chemists,
Physical Chemists think they are Physicists,
Physicists think they are Gods,
And God thinks he is a Mathematician.
Anon.[36]

Signs and symbols allow us to think without thinking. And the more things that we can do reliably without thinking the better. The higher functions of the brain have evolved to enable many activities to be put on autopilot and performed unconsciously. They are much better done like that – think about where you are putting your feet when you are dancing and you will fall over. Mathematics has become the most elegant example of this streamlining for the mind: by adopting succinct symbols we can save using many words and introduce an unambiguous precision into the reasoning process. It is a language with a built-in logic.

The most ubiquitous mathematical operations are those of addition, subtraction, multiplication, and division. These tasks make the world go round. They have been common operations for mathematical attention in many ancient cultures, but there grew up a vast diversity of instructions and symbols to instruct the reader how to combine quantities together. Their standardisation into a single universal notation solidified in fifteenth-century Europe and its adoption was reinforced by the invention of printing with movable type by Johannes Gutenberg, in 1456.

The earliest known appearance of the arithmetic + and – symbols together in a printed book is in a German text,[37] *Mercantile Arithmetic*, about applicable arithmetic written by Johann Widmann and published in 1489,[38] although the minus sign first appears in a manuscript about algebra written in 1481 and published in 1486.[39]

$$\frac{360x}{144-6x} \qquad 10-x$$

$$x^3+2x^2 \qquad 15-22x$$

Notes taken by a student at Leipzig University, where Widmann lectured, in 1486, show the first use of plus and minus signs. Also shown here, for comparison, are the modern expressions of the algebraic formulae being represented.

The origin of the plus sign itself seems to derive from the multitude of slightly different ways of writing the Latin word *et* ('and'). The origin of the minus sign is less clear.[40] One suggestion is that it derived from the notation used by merchants to distinguish the total weight of a consignment of merchandise from its weight when the container was removed. The latter was called the *tare* or *minus*. Widmann's book contains a host of commercial problems involving adding and subtracting, so he uses + and – signs liberally in the text as well as in equations, instead of the words 'and/plus' and 'subtract/minus'. However, it was not until the latter part of the sixteenth century that these two symbols became widely used in arithmetic and algebra. In Italy, the symbols p̃ and m̃ as abbreviations for 'plus' and 'minus' were used still in the sixteenth century, and also competed with the + and – symbols in Spain and France. There were several varieties of plus sign in use, the Greek (+) which we use today, together with Latin and Maltese crosses with different orientations. This indirect link to religious symbolism brings to mind the story about the great twentieth-century mathematician Paul Erdös, who on being asked if he had enjoyed a tour around a Catholic college in the United States replied that 'there had been too many plus signs.'

Some of the early options for denoting subtraction appear somewhat confusing to our eyes, because subtraction was denoted both by the minus sign, –, and sometimes by the ÷ sign that we use now for division, or simply by a pair of hyphens - -, or a row of dots, so that our sum 7 – 1 = 6 would appear as 7····1 = 6 in some of the mathematical works of Descartes.

The + and – signs were first used in England by the Oxford mathematician Robert Recorde in 1557, in his famous book on algebra *The Whetstone of Witte*, wherein readers could sharpen their wits on all sorts of algebraic manipulations, simply explained.[41] Like all his books, it was written in English, and so was more widely accessible than other books of arithmetic.[42] Recorde studied mathematics and medicine at both Oxford and Cambridge, was probably the first English scholar to support Copernicus' heliocentric cosmology, and was much appreciated as a spirited public lecturer on these subjects. He later became Master of the Mint at Bristol (surviving accusations of treason for not supporting the financing of the King's forces, which saw him imprisoned for eight weeks), then Royal Physician to Edward VI, before ending his career ignominiously in a debtors' prison because of money owed to his long-time enemy Sir William Herbert, the Earl of Pembroke, who had won a libel action against Recorde for the considerable sum of £1000, which he either could not or would not pay. He died in prison in 1558. His *Whetstone of Witte* is also famous for introducing the equals sign, =, into mathematics. He used rather longer horizontal dashes than we do today, more like ═, with the rationale that these were parallel lines and:

> to avoide the tedious repetition of these woords: is equalle to: I will sette as I doe often in woorde use, a pare of paralleles, o: Gemowe of one lengthe, thus: ═ because noe 2 thynges can be moare equalle.

Unfortunately, this notation that Recorde so enthusiastically adopted in his book – which went through many editions – was not universally popular. It didn't appear in print again until 1618, although several mathematicians are known to have used it, from the evidence of their private papers. Some mathematicians used two vertical lines, ||, perhaps as an abbreviation for the first and last letters (iota) of the Greek word for equals, *ἰ σ σ ἰ*, while the alpha symbol

The Arte

as their woorkes doe extende) to distincte it onely into twoo partes. Whereof the firste is, *when one nomber is equalle vnto one other.* And the seconde is, *when one nomber is compared as equalle vnto .2. other nombers.*

Alwaies willyng you to remẽber, that you reduce your nombers, to their leaste denominations, and smalleste formes, before you procede any farther.

And again, if your *equation* be soche, that the greateste denomination *Coßike*, be ioined to any parte of a compounde nomber, you shall tourne it so, that the nomber of the greateste signe alone, maie stande as equalle to the reste.

And this is all that neadeth to be taughte, concernyng this woorke.

Howbeit, for easie alteratiõ of *equations.* I will propounde a fewe exãples, bicause the extraction of their rootes, maie the more aptly bee wroughte. And to auoide the tediouse repetition of these woordes: is equalle to: I will sette as I doe often in woorke vse, a paire of paralleles, or Gemowe lines of one lengthe, thus: ═════, bicause noe .2. thynges, can be moare equalle. And now marke these nombers.

1. 14.ze. —+— .15.ϑ ═════ 71.ϑ.

2. 20.ze. ——— .18.ϑ ═════ .102.ϑ.

3. 26.ʒ —+— 10ze ═════ 9.ʒ ——— 10ze —+— 213.ϑ.

4. 19.ze —+— 192.ϑ ═════ 10ʒ —+— 108ϑ ——— 19ze

5. 18.ze —+— 24.ϑ. ═════ 8.ʒ. —+— 2.ze.

Robert Recorde's *Whetstone of Witte* (1557), showing the first use of the 'equals' sign.

α (which was later used to denote proportionality, rather than equality) was used into the eighteenth century. We even find ⊔ and ⊓ and // and)=(and 3 and æ being used around continental Europe. The use of the latter was a distortion of the first two letters of the Latin word for 'equals', *aequalis*.

There was a risk of serious confusion in the late sixteenth and early seventeenth centuries because some mathematicians on the continent of Europe were using the = sign to mean different things. Some used it as a minus sign, while others, such as Descartes, used it to denote 'plus or minus',[43] which we now denote by ± , while up until 1670 others even used it as the decimal point, so that '134 = 67' in print would denote our decimal 134.67. Then, the geometers wanted to use = to denote parallel lines. So, by 1700, the = sign was being used in five completely different ways in situations that could easily collide in a single mathematical exposition. Under these circumstances it is surprising that it survived at all. It was ripe for replacement by a new symbol that avoided all the possible ambiguities that its different users had invented. In the end, Recorde's = sign became well established in England in the sixteenth century, but it was not until 1660, more than a century after its first English publication, that it would became the equals symbol of choice in the rest of Europe.

Multiplication was well known to the Egyptian, Babylonian, and Indian mathematicians, although it was often viewed as a form of repeated addition. It was signalled by placing numbers alongside one another or, also in India, by placing a dot between the numbers to be multiplied. In Germany, in 1545, we see the first use of the capital (Gothic script) letters M and D used in the roles of multiplication and division signs, by Michael Stifel,[44] and so the algebraic formula $6y$ would be written 6 M y and the fraction one half (½) could be written 1 D 2. Some fifteenth-century writers were careful to note the inconsistency of this notation, in that $5y$ meant 5 multiplied by y but 2½ meant 2 plus ½, and not 2 times ½ as the Indian notation might imply.

The first appearance of our modern 'Saint Andrew's cross' multiplication sign × appears in William Oughtred's 1631 *Clavis Mathematical*, although it appeared as a letter X in 1618 in a special appendix to *Descriptio*, a book by John Napier (the inventor of logarithms) that was probably contributed by Oughtred. Oughtred wrote his multiplication sign slightly differently to ours. It was about half the size of the + and – signs and raised above the line, so $2 \,{}^{\times}\, 2 = 4$. Other seventeenth-century rivals to this multiplication symbol were the asterisk * and the rectangle □, which are still used in mathematics to denote the operation of complex conjugation (rotating through 90 degrees in the x–y plane of a graph) and the completion of a proof, respectively.

Following Stifel's use of D for a division sign, we find the modern ÷ sign appearing first, in 1659, in a book on algebra by the Swiss mathematician Johann Heinrich Rahn.[45] Remember that many authors had used this symbol as a minus sign before.[46] Rahn's work first appeared in an English translation, with additions by John Pell, in 1668. This was fortunate for Rahn because, although his countrymen showed no enthusiasm to adopt his new symbol, the English mathematicians did, and notable mathematicians in Oxford and Cambridge, such as Isaac Barrow and John Wallis, employed it. Alas, in the process, the ÷ sign then tended to become known as 'Pell's symbol' among English mathematicians simply because of Pell's other contributions to Rahn's translated book. It became ubiquitous in mathematical works in English but remained unknown on the continent of Europe. In the seventeenth century in Germany we find an influential mathematician like Leibniz occasionally using a C lying on its side, although more widespread was the use of a colon : or a dot • raised above the line, or one of the forward or back slashes, / or \. The forward slash still persists, as does the colon, the symbol for ratio *a*:*b* being the same as *a*/*b* or a ÷ b. The colon was introduced in this context by the English astronomer Vincent Wing in 1651,[47] but took a long time during the first half of the eighteenth century to dominate over the dot, • , symbol that Oughtred had introduced for the same operation.

The modern symbols for addition, subtraction, multiplication, division, and equality are standard the world over. They are more universally familiar than the letters of any written language. Although their images are simple and aesthetically unimpressive, they are indelibly imprinted upon our conception of the nature of things. They have come to represent the basic changes of Nature, and their appearance in print triggers a reflex in the mind.

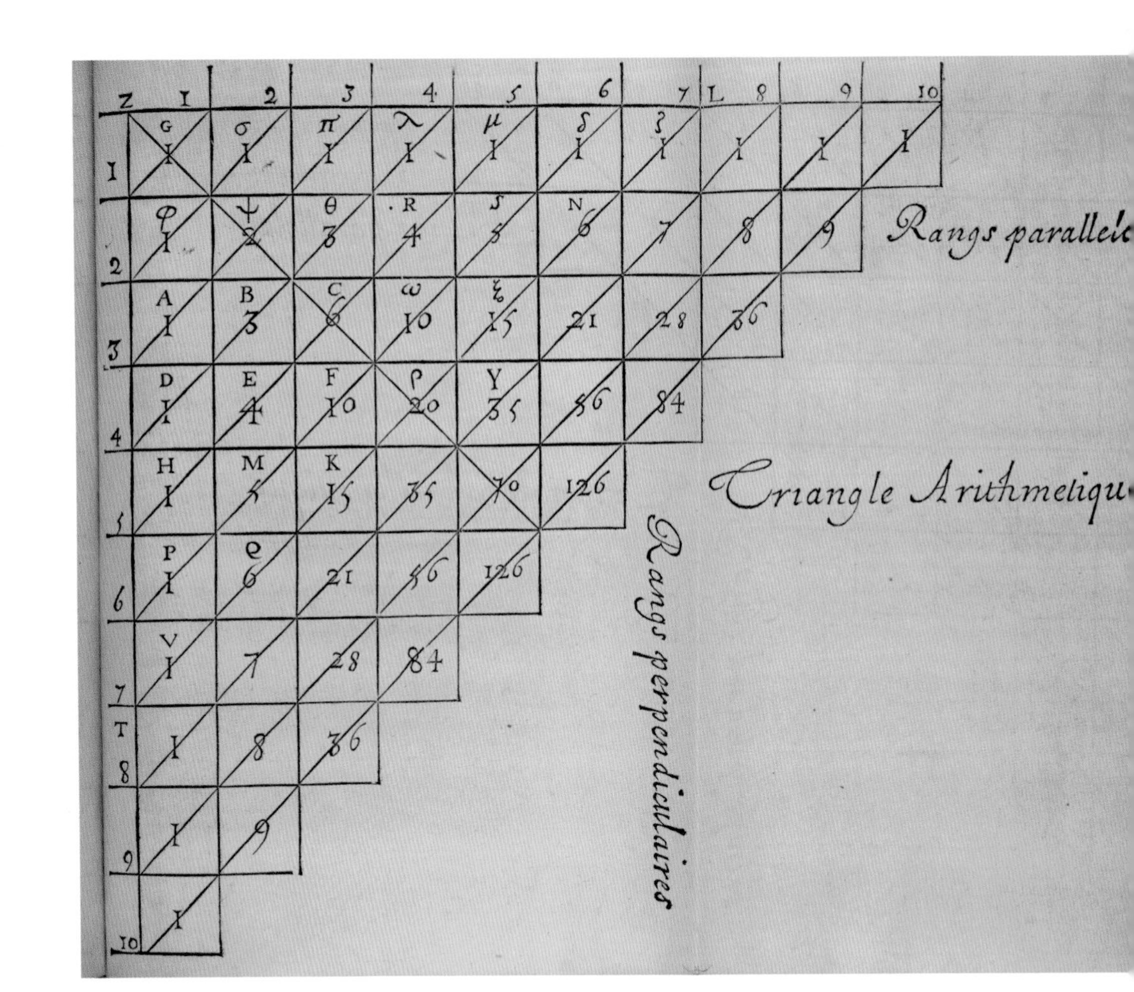

Pascal's version of his triangle of numbers from the front page of his *Traité du triangle arithmétique* (1665).

A PILE OF NUMBERS

PASCAL'S TRIANGLE

Miss Susan: What is algebra exactly; is it those three-cornered things?
Phoebe: It is x minus y plus y and things like that. And all the time you are saying they are equal, you feel in your heart, why should they be?

James M. Barrie[48]

When something is discovered independently several times by people living at different times in completely different cultures it usually means that there is something very significant about it, useful even. This historical confluence is particularly impressive when it comes to the picture that is known as Pascal's triangle. This is shown opposite as a triangle that starts from its apex with three 1's. The rows below are produced by adding together the two numbers immediately above it and to the left of it in the triangle. The two outer sides of the triangle are always populated by 1's.

The triangle can be continued downwards, row upon row, for ever. There are a few obvious patterns to the triangle. The sum of the entries along the nth row down totals 2^n. The outer diagonals are all 1's, the next diagonals in on each side list all the numbers 1, 2, 3, 4, 5, . . . The third diagonal in gives the list of all the 'triangular' numbers 1, 3, 6, 10, 15, 21, 28, 36, . . . , the next gives all the tetrahedral numbers 1, 4, 10, 20, 35, 56, 84, 120, . . . and so on. The second number in each row will divide every other number in the row when it is a prime number, so for example, in row 7 of the modern version we see that

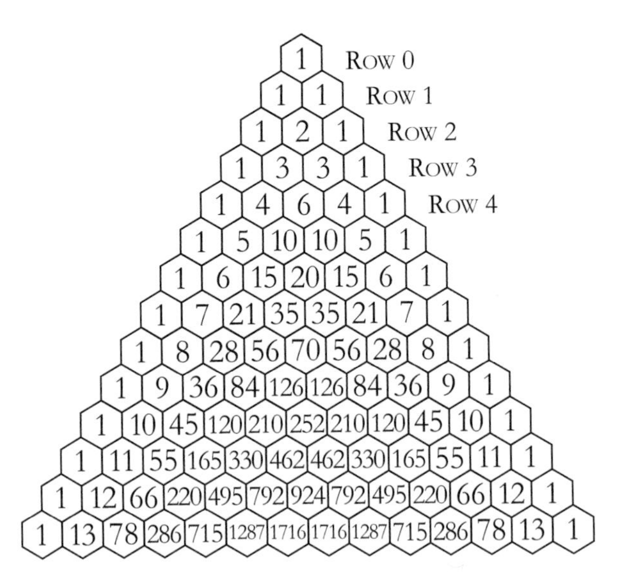

A modern version of the first 14 rows of Pascal's triangle. Each element not forming a side edge is equal to the sum of the two numbers immediately above it to the left and right.

7 divides the other entries 21 and 35. If we take the sums across the shallow diagonals from top right to lower left then each diagonal sums to the numbers of the famous Fibonacci sequence, 1, 1, 2, 3, 5, 8, 13, . . . , where each successive number is given by the sum of the previous two.

The most useful property of the triangle of numbers is that each row gives the quantities that arise when successive powers of the sum of two numbers are multiplied out. If we consider the expansions of powers of the sum of $x + 1$ then we see the numbers in each successive row are the quantities multiplying each term in the algebraic expansion

$(x + 1)^2 = x^2 + 2x + 1$
$(x + 1)^3 = x^3 + 3x^2 + 3x + 1$
$(x + 1)^4 = x^4 + 4x^3 + 6x^2 + 4x + 1$
$(x + 1)^5 = x^5 + 5x^4 + 10x^3 + 10x^2 + 5x + 1$
. . . and so on, for ever.

The great French philosopher, scientist, and mathematician Blaise Pascal (1623–62) got his name attached to this never-ending triangle of numbers because of his book *Traité du triangle arithmétique* which appeared in 1665, although it had been completed and distributed to several mathematical friends of Pascal in 1654. Pascal's version of the 'arithmetical triangle', as he called it, appears in the first pages of the book. It differs only in orientation from the version we see today – rotated clockwise by 45 degrees.

We see Pascal has traced out the diagonals where lines of important entries run. He then proceeded to expound all the remarkable properties of the triangle that he had been able to collect or discover. In the process he devised the method of proof that mathematicians now call 'the method of induction'.[49] One of his special interests was gambling – not so much as a lucrative practice, but as an activity that can be subjected to mathematical study. Pascal was one of the pioneers of the mathematical theory of probability and to make progress he needed to know the number of ways in which certain outcomes could arise in games of chance involving cards and dice.[50] His triangle was an important tool in that quest and, even today, Pascal's triangle continues to reveal unsuspected mathematical properties.[51] Closer to his own day, it inspired a range of generalisations[52] by other mathematicians, such as Leibniz's harmonic triangle, and appears in a symbolic role in Holbein's famous painting of 1533, *The Ambassadors*.[53]

As we cautioned at the outset, Pascal's Triangle certainly did not originate with Pascal. It has very ancient roots in the Indian, Chinese, and Islamic cultures and was devised by a number of European mathematicians interested in gambling and the combinations that were possible when several dice were thrown at once. Since the latter activity tended to be secret, this helpful way of looking at things was not revealed for a long time. Many mathematicians living at different times in various places highlighted many of the triangle's properties. In Italy, it was always known as 'Tartaglia's triangle' (*triangolo di Tartaglia*) after the mathematician Nicolò Tartaglia (1499–1547) who, like Pascal, was interested in developing a theory of (successful) gambling and used the triangle to calculate the number of ways that dice could generate a given total score. He only published the number triangle in 1556 but knew of it much earlier. Indeed, he memorably claimed that he discovered it during the night between Mardi Gras and Ash Wednesday in the year 1523 in Verona. But there was no doubt more to be gained materially from a secret method than from the prestige that might follow from its first publication. Pascal ended up having the triangle named after him because he has the distinction of drawing together all its diverse properties and applications into a grand synthesis.[54]

The oldest picture of the arithmetical triangle can be found in the remarkable civilisation of ancient China.[55] Although the original work no longer survives, we know from references to it by others that the Chinese mathematician Chia Hsien used the triangle between 1050 and 1100 to calculate square and cube roots of numbers, and similar applications were known to Omar Khayyam, the great Persian poet and mathematician, in the early twelfth century. Chia Hsien's work was gradually extended by later Chinese mathematicians, and by 1261 Yang Hui had used a triangle six rows deep to calculate quantities such as $(a + b)^6$ where a and b were any numbers.[56] It was extended to the 8th power expansion by Chu Shih-chieh in his book *Precious Mirror of the Four Elements*, which appeared in 1303. It is referred to there, with no claim to originality, simply as the 'diagram of the old method for finding eighth and lower powers'.

This pattern of numbers has captured the imaginations of many civilisations. It has a symmetry and utility that combine to make it memorable and easy to create. If it were merely encapsulated in a formula then it would have been just one more self-generating sequence of numbers. As a triangle it suggests the operations that are needed to seed the entries that run from the apex downwards – an object lesson in thinking without thinking.

Diagram from the frontispiece of Chu Shi-chieh's Ssu Yuan Yü Chien. It is entitled 'The Old Method Chart of the Seven Multiplying Squares' and tabulates the binomial coefficients up to the eighth power.

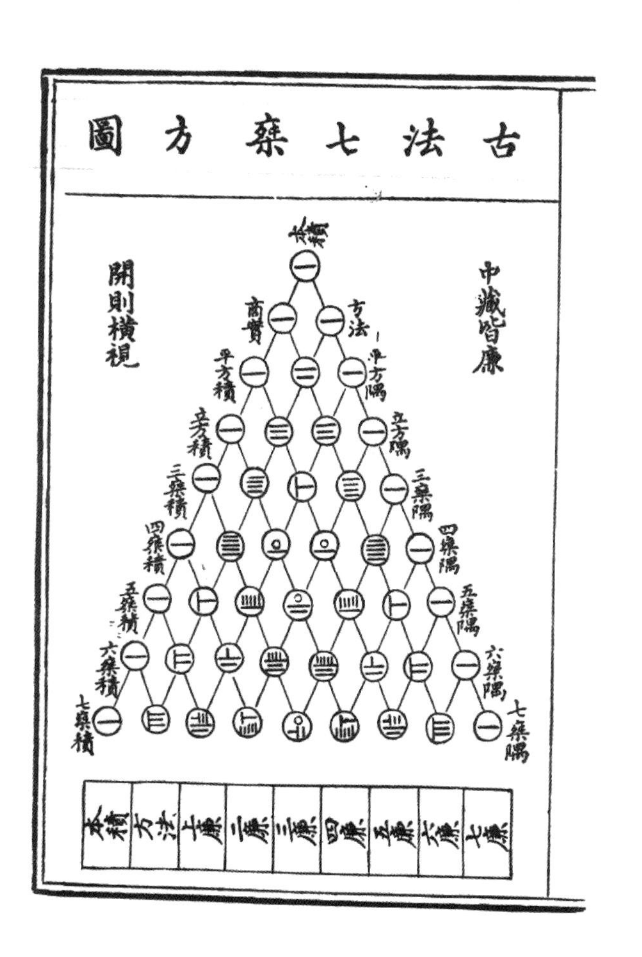
古法七乘方圖
中藏皆廉
開則橫視
本積
商實
方法
平方積
平方隅
立方積
立方隅
三乘積
三乘隅
四乘積
四乘隅
五乘積
五乘隅
六乘積
六乘隅
七乘積
七乘隅
本積 方法 上廉 二廉 三廉 四廉 五廉 六廉 七廉

Ancient dice from
the collection of
Arjan Verweij.

CHANCE AND NECESSITY

DICE

'Come, and let us cast lots, that we may know for whose cause this evil is upon us.' So they cast lots, and the lot fell upon Jonah.

Book of Jonah[57]

DICE are the oldest gaming devices known to man. Today, they can be found in many symmetrical shapes but the most familiar, which became common in the West during the nineteenth century, are cubes, with six faces, numbered so that 1 and 6, 2 and 5, and 3 and 4 are on opposite faces, so that each opposing pair totals 7. This arrangement doesn't uniquely specify how to make dice though. If you are reading this in the West then your die will probably be right-handed in the sense that if you put it on the table with the 'one' spot (or 'pip') upwards and the 'two' spot on your left then the 'three' spot will be on your right. But if you are reading this in China then your local die will be the mirror image of this – left-handed because the 'two' spot will be on your right and the 'three' spot on your left when the one spot is facing upwards. Eastern and Western dice are the mirror images of each other.

Cubes are not the only symmetrical solids that could be used as gaming devices – coins can be viewed as two-sided dice – and they are the simplest. The faces of any of the other Platonic solids could be used for dice (the chances of any face falling down flat on the table most often would be different for each shape) but the cube has the nice property that the upward face that gives the score lies directly opposite the face that lies flat on the ground. If we used a tetrahedron as a die, then we would have to choose the face that was flat on the

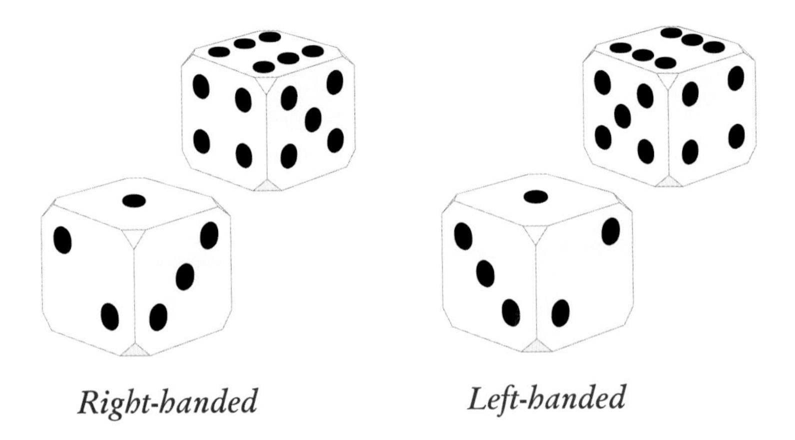

Right- and left-handed dice. The displayed faces of the pair of dice on the left are the mirror images of the pair on the right, and vice-versa.

ground as the scoring face and it would have to be picked up in order to read the score.

The expression 'no dice' that is used in many walks of life to signal that this is not a deal or game that you want to play, derives from the response to dice falling badly so that they don't lie flat, or fall on top of one another, so they need to be thrown again. The most important requirement for a die is to be fair. In order to be fair, dice must be as perfectly cubical as possible. This is important because the behaviour of a die that is not quite symmetrical in shape turns out to be surprisingly biased. Typically, a die will most likely fall with its largest face flat on the ground. Today, machine manufacture and plastic moulding can make dice symmetrical to high precision (typically to 1/5000th of an inch), and casino dice manufacturers even take into account the tiny differences that are introduced to each face by the different number of spots that get printed or embossed upon it. The spots will be drilled out of the acetate surface and replaced by coloured material of the same density until it is flush with the surface of the face. The transparent acetate that is used for casino dice prevents biases being inserted into the interiors to move the centre of gravity closer to one face than the others.[58] By contrast, the cheap dice you buy at the local toy store are likely to have indentations to mark the numbers of 'spots', and will favour the highest numbers because their faces will have the most indentations and so be the lightest. After a very large number of throws, this bias would become noticeable.[59]

Dice symbolise probability and randomness at a glance. They still feature in many games of chance and are hardy perennials in mathematics books concerned with probability and statistics. But there is a fascinating question lurking behind this familiar gaming cube. Why is the mathematical study of probability such a recent creation? Other familiar areas of mathematics that we meet at school, such as arithmetic, algebra, and geometry, go back to the beginning of intellectual history. All the great human civilisations made use of numbers to develop systems of accounting, studying geometry to construct buildings and map the stars. But the study of probability is totally absent from these civilisations. Neither Euclid, nor Plato, nor Aristotle, nor Archimedes had anything to say about the mathematics of chance. For that, we had to wait until about 1660 and the efforts of men such as Blaise Pascal and Christiaan Huygens. This is rather puzzling since games of chance certainly existed in ancient civilisations.

Two possible explanations of this paradox spring to mind. The first is religious in nature. In the ancient world we do not generally find the notion of events being random in the sense of being unpredictable to anyone. This would be tantamount to saying they were irrational. In civilisations where there were gods (or a single God), chance was often the vehicle by which these deities made their will known. The High Priest of Israel had two flat objects which could be drawn from his garment when God's answer was required (two 'heads' = yes, two 'tails' = no, one 'head' and one 'tail' = wait) to an unresolved question.[60] Last year, I saw something similar happening while visiting a Buddhist temple in the old Taiwanese city of Tainen. A man in his early thirties was kneeling at the temple shrine and casting two wooden shapes on the ground, praying aloud as he did so, before going off to draw long sticks on each of which was attached a printed message, in Chinese. After making some inquiries of my hosts, I learned that he wanted the god of the shrine to do something for him, perhaps make him prosperous, or save the life of a relative in bad health, but this required him to promise to do something in return. He was making an offer and then casting the blocks to learn if his offer had been accepted. If the two faces of each die came up the same the answer was 'yes', if they were different it was 'no'. If he got an unfavourable answer then he would go and take another message stick which suggested how he might increase his own offer so as to raise the chance of getting a 'yes' with his next throw. It was fascinating to see this connection between chance and the gods still being acted out. Its significance for the history of probability is that a belief that there is a close link between chance and the will of the gods is likely to stop any systematic study of chance. This would most likely be viewed as a blasphemous meddling with the gods and could result in death.

A second possible explanation (the two are by no means mutually exclusive though) for the late arrival of the theory of probability into the annals of mathematics is the lack of one crucial idea: that of the equally likely outcome.[61] Today, our six-sided die is beautifully made. The corners are square. The faces are of equal size. And the centre of gravity lies in the geometrical centre of the cube. The simple consequence of this symmetrical state of affairs is that the die is equally likely to fall on any face, if fairly thrown. Ancient gaming devices were not like this. Each one was different and extremely asymmetrical. Typically, they were pieces of knucklebone or ankle bones from domesticated animals such as sheep, and throwing dice is still referred to as rolling the bones in some parts of the world – the Arabic for dice is the same as the word for knucklebone, from which we derive our word 'stochastic'. The trouble with knucklebones is that each one is different from all the others and the professional gamester in the marketplace will have learnt from experience about the strong biases of his own device. No general theory of probability is much use to you if you play him because the outcome of the throws are not equally likely. Only when everyone is playing with devices whose biases are known – or which are absent because they are symmetrical and every outcome is equally likely – is it useful or possible to develop a general theory of 'chance' outcomes.

Finally, dice can be designed to have very strange properties. I can make three dice with the property that whichever one of the three you choose to play with, I can choose one of the others which will beat you. There are many of these 'intransitive' dice and the faces of one possible set look like this:

die A: 1, 1, 4, 4, 4, 4
die B: 3, 3, 3, 3, 3, 3
die C: 2, 2, 2, 2, 5, 5

In the long run, die A beats die B, die B beats C, but die C beats A (in each case, the odds of winning are 2 to 1). Things like 'beating' or being 'better than' or 'liking' all have this awkward intransitive property. It's more familiar than you think – if John likes George and George likes Mary, that does not mean that John likes Mary. It is only simpler things, such as 'being taller than', for which relationships are transitive – if John is taller than George and George is taller than Mary then John is necessarily taller than Mary.

Dice symbolise the realm of chance and randomness. They are instantly recognisable all over the world but they hide subtleties of history and complexities of mathematics, and enshrine deep-held beliefs about the nature of reality. They are an image that the electronic era will be hard-pressed to displace into cyberspace.

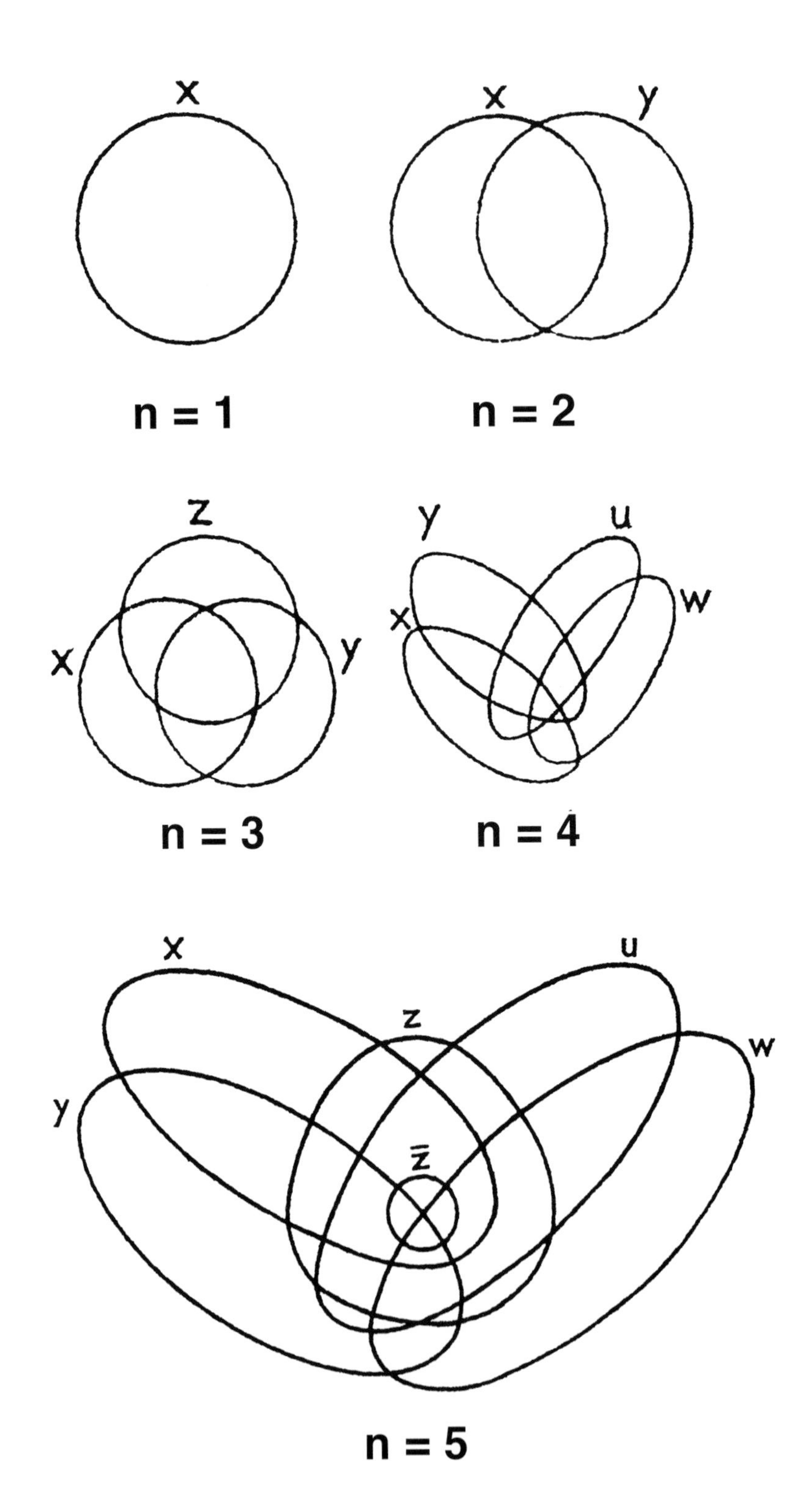

Venn's original diagram for the possible intersections of 1, 2, 3 and 4 regions and his attempt to represent 5 intersecting sets completely.

THE MAN WHO TURNED INTO A DIAGRAM

THE REVEREND JOHN VENN

There is a complication about language . . . namely that words which mean relations are not themselves relations . . . a map is superior to language since the fact that one place is to the west of another is represented by the fact that the corresponding place on the map is to the left of the other; that is to say a relation is represented by a relation.

Bertrand Russell[62]

JOHN VENN came from the east of England, near the fishing port of Hull, received his schooling in London and went – as promising mathematicians did – to Cambridge, where he entered Gonville and Caius College as a student in 1853. Graduating among the top half-dozen students in mathematics, he was elected into a college teaching fellowship. He left the college for four years after being ordained a priest in 1860, following in the line of his distinguished father and grandfather, who were prominent figures in the evangelical wing of the Anglican Church. But, instead of following the ecclesiastical path that had been expected of him, he returned to Caius in 1862 to teach logic and probability. Despite this nexus of chance, logic, and theology, Venn was also a practical man, and rather good at building machines. He constructed a machine for bowling cricket balls which was good enough to clean bowl one of the members of the 1909 Australian touring cricket team on four occasions when they visited Cambridge.

From 1869 onwards, Venn was required to give lectures on logic to students from many colleges and it was during the preparation of those lectures that he devised a way of representing logical alternatives by means of a simple diagram which was to become canonised as the 'Venn diagram', found in every book of logic today. Some years later, in the preface to his book *Symbolic Logic* (1881), he recalls that:

> I now first hit upon the diagrammatical device of representing propositions by inclusive and exclusive circles. Of course, the device was not new then, but it was so obviously representative of the way in which anyone, who approached the subject from the mathematical side, would attempt to visualise propositions, that it was forced upon me almost at once.

In fact, the great Swiss mathematician Leonhard Euler, as well as Lewis Carroll – who, when not writing the adventures of Alice under his nom de plume, was one Charles Dodgson, tutor in logic at Christ Church, Oxford – both invented simpler types of logic diagram.

Venn introduced the diagram that bears his name[63] in a paper of 1880 with the title 'On the Diagrammatic and Mechanical Representation of Propositions and Reasonings'.[64] It allows the common and disjoint members of different sets to be read off from a diagram. Venn shows how to do this for 1, 2, 3, or 4 regions. Consider his case of three regions X, Y, and Z lying inside an all-encompassing region R. Suppose that R is the collection of all cats, while X is the collection of ginger cats, and Y is the collection of male cats, and Z is the collection of lazy cats. The three intersecting sets X, Y, and Z cannot create more than eight separate regions.[65] The overlaps indicate the regions where the cats share the properties of the intersecting sets. The most general situation is where they create eight distinct regions (including the region outside the boundaries of X, Y, and Z which contains those cats which are neither ginger, male, nor lazy) and is shown in Venn's picture for $n = 3$.

In the $n = 4$ case he succeeds in showing all the possible intersections ($2^4 = 16$ in number) of four sets, which are drawn as ellipses. After delineating the tricky case of four regions, he runs into difficulty with the case of five regions to be represented by closed convex curves, and he suggests the diagram shown on page 314. Again, we see that a feature of geometry on a flat surface biases the way in which a diagram can be constructed. If we have more than four sets to represent, then the constructions get more and more complicated and convoluted. Anthony Edwards, like Venn at Caius College, Cambridge, has found ways to create a visually appealing series of Venn diagrams for 3, 4, 5, and 6, or more. Unfortunately, as the number of regions increases it becomes increasingly hard to identify the regions of intersection and interpret their meaning. Edwards helped design the memorial stained-glass window in the chapel at Caius College as a memorial to Venn and his invention.

The Venn diagram for $n = 3$ would have passed for one of Euler's logic diagrams. In 1768,[66] Euler illustrated how four logical relationships could be represented by diagrams:

All A's are B's
No A's are B's
Some A's are B's
Some A's are not B's

However, there are some situations that cannot be represented by a single diagram. For example, if

No A is a B but some C is an A

then there is no single diagram that captures the possible relationship between the sets B and C because B could lie outside C, completely inside C or only partly inside C.

Euler's simple scheme also has some very interesting weaknesses that are a consequence of drawing diagrams on a plane surface. There is a theorem of mathematics, called 'Helly's theorem',[67] after Eduard Helly (1884–1943), who proved it in 1912, which tells us that if we draw four circles on the page, representing the sets A, B, C, and D respectively, then if the common intersection of A, B, and C, of B, C, and D, and of C, D, and A are each not empty, then, geometrically, the intersection of A, B, C, and D cannot be empty, as we can see in the picture overleaf.[68]

However, when A, B, C, and D are collections of things ('sets'), this geometrically enforced conclusion does not necessarily hold. For example, if A, B, C, and D were related like the four faces of a pyramid (including the base), then any face intersects with three others but there is no place where all four

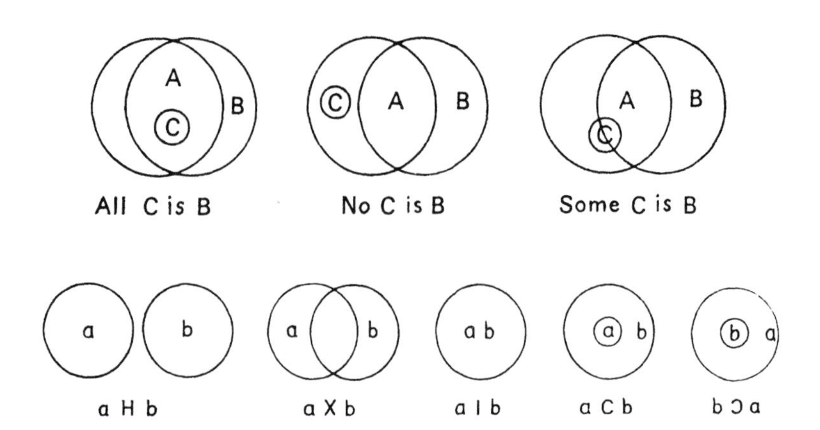

Euler's logic diagrams (1768).

faces intersect. Another example would be provided by having four friends, Alex, Bob, Chris, and Dave, and considering the four different subsets of three friends that can be formed: {Alex, Bob, and Chris}, {Alex, Bob, and Dave}, {Bob, Chris, and Dave}, {Chris, Dave, and Alex}. Then any pair of these subsets has a person in common but there is no person that is common to all four subsets.

Venn saw that the weakness of Euler's earlier proposal was that it failed to capture the incompleteness of our knowledge about the relation of different collections to one another. He solved the problem of removing the ambiguity created by the diagrams by shading out the excluded regions of the diagrams. So if all A's are B's, then the shaded region is excluded. The shading can be seen as a representation of an empty set.

If we also add to the diagram that no A is a B, then the two diagrams exclude any part of the region for A, and together they convey the information that A is nothing.

So far we have drawn the sets A, B, and C as circles, but this is not always possible. When we have four regions, then circles will not be able to include all the possible combinations of overlaps. But ellipses can still do the trick, as we can see in Venn's original n = 4 diagram on page 314.

Venn's influential step was not the first word on diagrammatic representations of logical relations, and nor was it the last word, but it was the most influential.[69] Most of all, we like diagrams that are not merely pictures. They need to convey a visual image, such as a corporate logo, and like any good notation they need to be an aid to thought, or better still something that does the thinking for us. Understanding how such diagrams work is part of the quest to create effective forms of artificial intelligence that can recognise, manipulate, and devise pictures in unambiguous ways.[70]

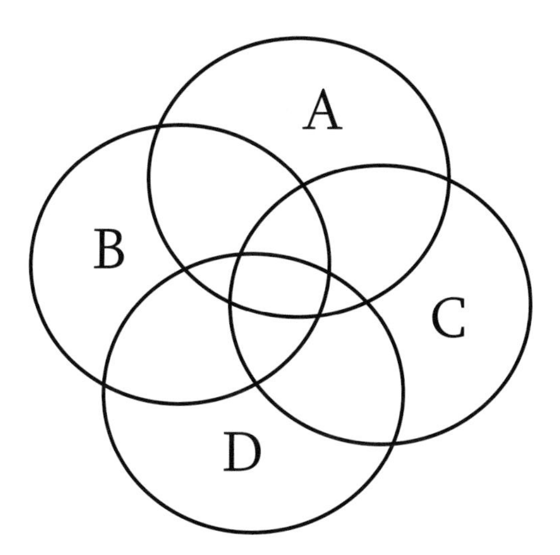

Helly's example (1912). A, B, C and D can be chosen to be collections of things for which this Venn diagram is not true.

A Möbius Strip.

ONE-SIDE STORY

THE MÖBIUS STRIP

'Why did the chicken cross the Möbius strip?'
'To get to the other . . . er, um . . .'

Anon.

TAKE a rectangular strip of paper and pick up the two short ends and stick them together to make a cylinder. At elementary school we did it dozens of times. The cylinder that results has an inside and an outside. But if you repeat the experiment giving the strip a twist before sticking the ends together you will have created something that is strangely different. The resulting loop, which looks like a three-dimensional figure '8', has a surprising property. It doesn't have an inside and an outside: it has only one surface. If you begin colouring the surface with a crayon and keep going, never lifting the crayon from the surface, then you will eventually colour the whole of the surface. Nor is such a property without commercial interest. Factory production systems[71] that use conveyor belts sometimes exploit the one-sided property in order to double the life of the belt before it needs changing. Möbius filmstrip and audiotape were patented in the 1920s to double the length of a continuous loop. The trick is that the twist is just kept away from the rollers.

The first person to notice this curious state of superficial affairs – what mathematicians now call a 'non-orientable' surface – was August Möbius (1790–1868), a German mathematician and astronomer, whose mother's ancestors included Martin Luther. After making many discoveries in mapping and trigonometric astronomy as a young man, Möbius moved from Leipzig, where he had been a student, to the centre of German mathematics at Göttingen, to

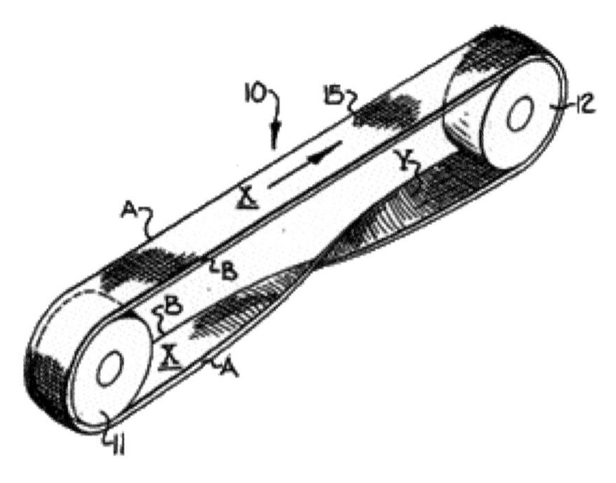

An early patent for a Möbius conveyor belt. Its one-sided form doubles its lifetime compared to that of a traditional two-sided belt which wears on only one side.

study at the Observatory under the direction of the great Karl Friedrich Gauss. From there, he moved to Halle to work with Gauss's own teacher, Johann Pfaff. After many further relocations, the peripatetic astronomer eventually returned, in 1848, to become Professor of Astronomy and Director of the Observatory in Leipzig.[72]

Möbius made many important contributions to astronomy but in the second half of his career began to make new discoveries in mathematics, especially in geometry, and today we still study the Möbius functions and Möbius transformations that resulted from some of this work. As we might expect from a former student of Gauss, Möbius applied very exacting standards to his own work and it took a very long time before any piece of his work was judged finished and suitable for publication. As a result, his paper describing his discovery of the Möbius strip was only found among his papers after his death, although he had discovered it in 1858, while preparing a study of polyhedra for the annual Grand Prix competition of the French Académie des Sciences.[73] It turns out that the Möbius strip was discovered independently in the July of the same year by another German mathematician, Johann Listing,[74] who also worked as one of Gauss's research students in physics, as well as on mathematics. At Gauss's suggestion he began working on problems associated with the structure of spaces and, in the course of his correspondence with his old school teacher about this new subject, he proposed that it be called 'topology', and it is still known by this title today. His co-discovery of the Möbius strip

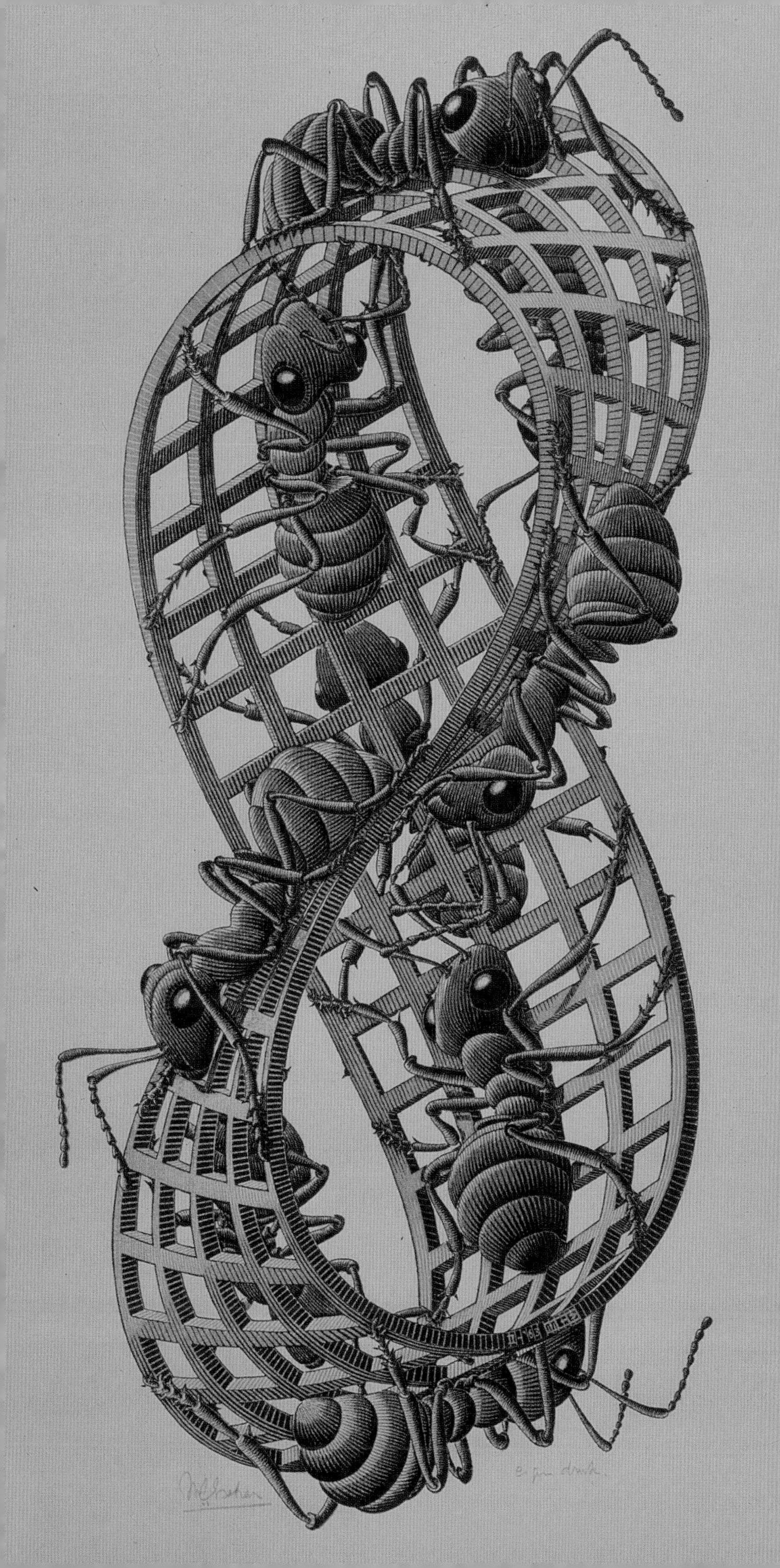

Escher's *Moebius Strip II (Red Ants).* Woodcut in red, black and grey-green, printed from 3 blocks (1963).

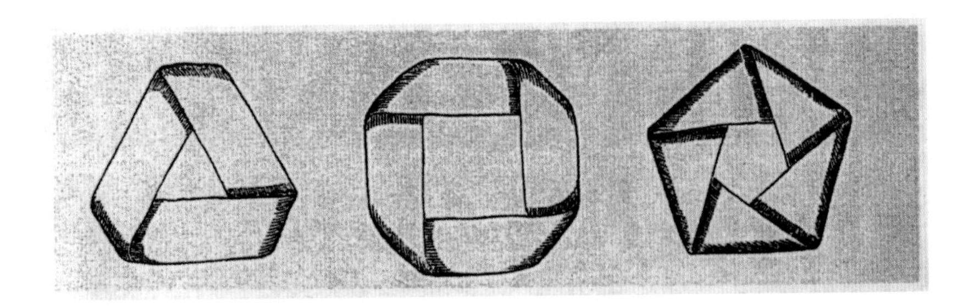

Möbius's original drawing from his unpublished notes (1858).

came at a time when he was facing personal ruin. Despite coming from very poor families, both Listing and his wife lived way beyond their means and were continually facing actions for unpaid debts which left them at the mercy of unscrupulous money lenders. Most of his colleagues seemed to have regarded the couple as irresponsible and had little sympathy for them. Fortunately, one old friend did. When Listing was facing bankruptcy, an old student friend, Sartorius von Walterhausen, came to the rescue. Long ago, Listing had saved his friend's life by nursing him through a serious illness while they were students. Now, thirty years later, von Walterhausen was able to repay him by ensuring that Listing's debt was discharged. A full circle of life that the creators of the Möbius strip would have both found salutary.

The Möbius strip is not only of interest to mathematicians. It has attracted the attentions of artists and designers wanting to make a statement about the infinite and the perfect. The most famous of these is Maurits Escher, whose beautiful drawings of 'living' Möbius strips have become iconic pieces of twentieth-century draughtsmanship. In the first of his two Möbius-inspired works, nine bronzed ants crawl around the never-ending surface of the band.

Often, the appearance of the Möbius strip alongside Escher's gallery[75] of impossible triangles and waterfalls has led viewers to think prematurely that there is something impossible about it. But there is nothing impossible about Möbius's strip. It is simply unexpected.

Escher was not the only prominent artist to exploit the properties of the Möbius strip. In the 1930s, the Swiss sculptor Max Bill believed that new mathematical developments in topology were opening up an unexplored direction for artists to explore and started to create a number of endless ribbon sculptures, in metal and granite.

Bill was developing solid, three-dimensional versions of the Möbius strip, and similar creations were made in stainless steel and in bronze in the 1970s by the American high-energy physicist and sculptor Robert Wilson and the English sculptor John Robinson, whose work *Immortality* uses the Möbius strip to make a gleaming trefoil knot in highly polished bronze seen below floating above a virtual ocean in Nick Mee's computer artwork. Many others have used the Möbius structure in architecture to create exciting buildings and stimulating children's play areas.[76]

Storytellers have also seized upon the strip as a device for fantastic events. In 1949, Arthur C. Clarke used it to describe an entire universe in 'The Wall of Darkness',[77] but the merger of the mundane with the unbelievable is often more amusing, as in Armin Deutsch's short story 'A Subway Named Möbius',[78] about an entanglement in a new line on the Boston subway that turns into a Möbius strip, and trains start to disappear, a Harvard maths professor gets called in . . . Or maybe the problem was that he was called in at the outset!

In new materials and dramatic contemporary contexts, Möbius's strip retains a powerful hold on our imaginations. It is a never-failing source of interest to anyone and one envies the young child who has yet to encounter it for the first time.

Virtual manifestation by Nick Mee of John Robinson's sculpture of a trefoil knot twisted into a Möbius band.

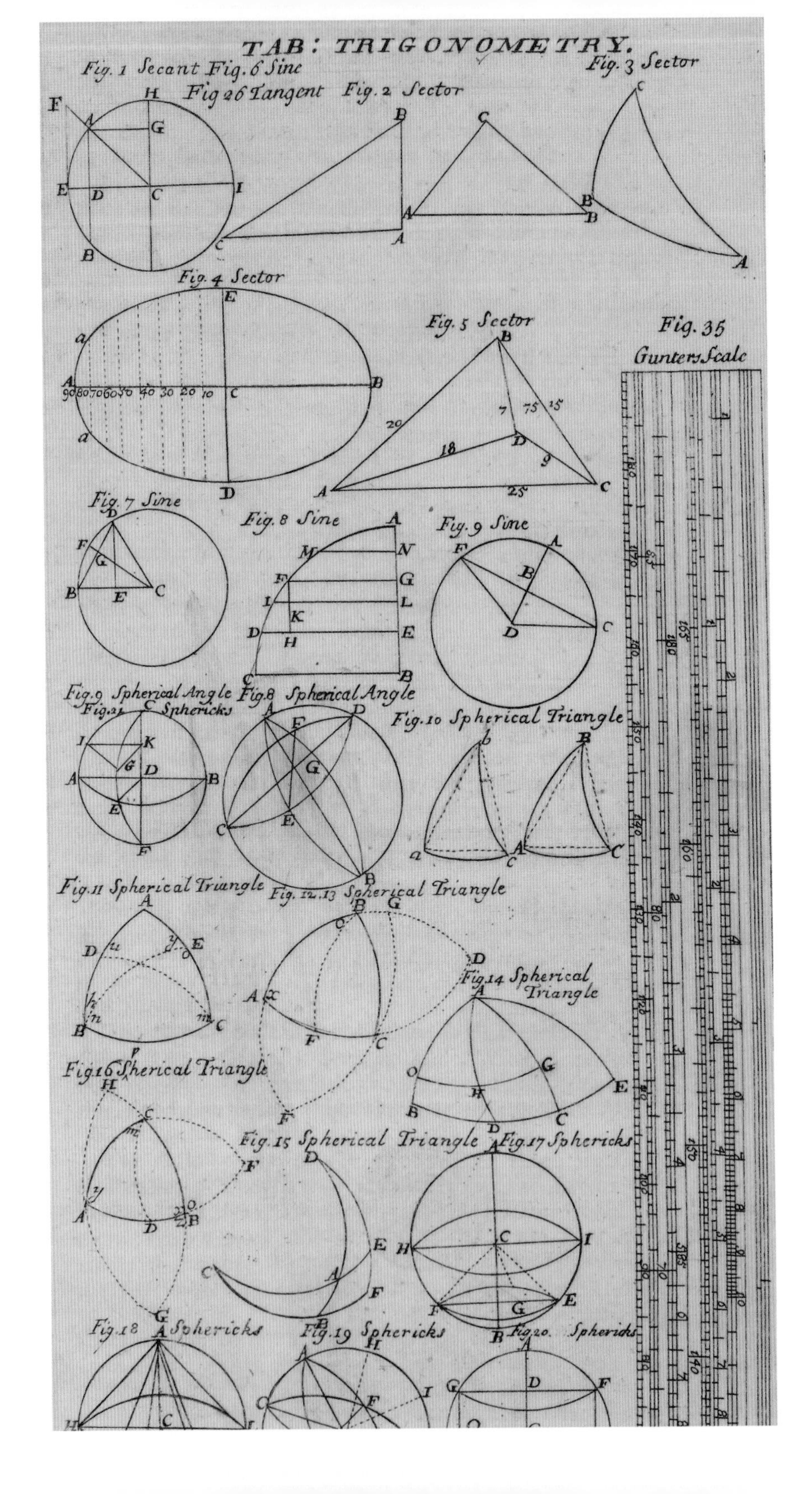

Table of Trigonometry, from E. Chambers, *Cyclopaedia, or, An Universal Dictionary of Arts & Sciences*, an encyclopedia published by Ephraim Chambers in London in 1728 and reprinted in numerous editions in the eighteenth century.

CADMUS AND HARMONIA

SINES AND COSINES

As long as schools continue to teach trigonometry and algebra, there will always be a moment of silence, and indeed prayer, in our public schools.

Senator Scott Howell[79]

TRIGONOMETRY is a modern word for an ancient practice. All ancient civilisations that engaged in building and astronomy needed to carry out surveying, navigation, and accurate recording of the positions of distant points and the angles between different directions. The word 'trigonometry' sounds very Greek. Indeed, it is an amalgam of Greek words.[80] But it is not as ancient as the subject it describes. The word 'trigonometry' was invented in 1595 by a German mathematician and astronomer, Bartholomew Pitiscus (1561–1613), and appears as part of the title of his book *Trigonometria sive de solutione triangularum tractatus brevis et perspicius . . . etc.* The second edition that appeared in 1600 had the more succinct title *Trigonometria sive de dimensione triangulae.*

This ancient subject with a new name revolved around the study of the relations between the three sides and interior angles of right-angled triangles. The most important are the quantities known as the *sine*, *cosine*, and *tangent* of an angle. They are shown in the picture of the triangle overleaf with the three sides labelled, as they are traditionally, opposite (a), adjacent (b), and hypotenuse (c).

The definition of the sine, cosine and tangent of the angle *0* is given by taking the appropriate ratios of the sides:[81]

$$\sin A = a/c,\ \cos A = b/c,\ \tan A = a/b$$

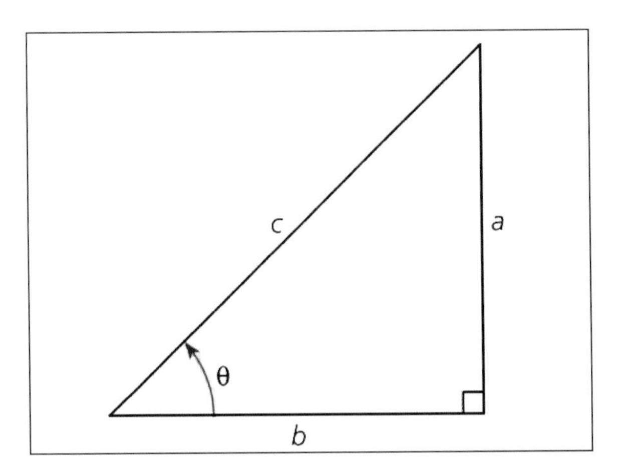

A right-angled triangle.

The origins of these names, sine, cosine, and tangent, are rather convoluted. The first to be named, the 'sine', derives from Indian usage. It was called *ardha-jya* in Sanskrit, which means 'half-chord', in the writings of the Hindu mathematician Aryabhata in AD 499, before being abbreviated simply to *jya*, or 'chord'. When Indian mathematics was translated into Arabic in the eighth century, the letters of this word were just transliterated into a word *jiba* that had no meaning in Arabic.[82] Since vowels are omitted in Arabic scripts, this then turned into the truncated consonants-only form *jb*. Much later, it was translated from Arabic to Latin without any reference to its Indian origins and the translators assumed that *jb* was an abbreviation of *jaib*, the Arabic word for a 'bay', 'cove', 'fold' or a 'pocket' in a garment when they reinstated the vowels.[83] When Gerard of Cremona translated Ptolemy's astronomical classic, the *Almagest*, in the twelfth century, he replaced the Arab *jaib* with the Latin *sinus*, which had the same meaning, from which we have *sine*.[84]

The cosine and tangent have less convoluted origins. The word 'tangent' just comes from the Latin *tangere*, 'to touch'[85] and was introduced by the Danish mathematician Thomas Fincke in 1583,[86] although the mathematical concept appears to originate with the Persian Abu al-Wafa in the tenth century. The cosine – or complementary sine (*complementi-sinus* in Latin) was introduced by the English mathematician Edward Gunter in 1620 simply as a useful counterpart to the sine.

After the appearance of graphs as a means of representing mathematical equations, in the work of Johann Lambert at the end of the eighteenth century, the sine and cosine graphs have become among the most illustrated equations of mathematics. Below is one of Lambert's first graphs of the sine function.

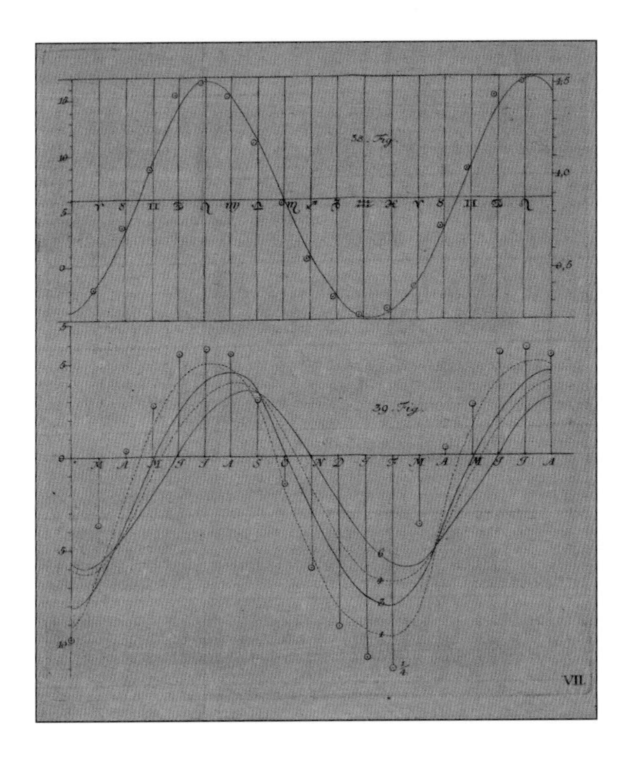

Johann Lambert's first graph of the mathematical sine function (1765).

The sine and cosine curves have turned out to have a special significance for all mathematical curves. Just as it is possible to build up any number by multiplying the numerical signs 1 to 9 together in appropriate ways, or to construct any polygonal figure by joining straight lines in the right way, it is possible to describe almost any curve that repeats at regular intervals by adding together a collection of sine and cosine curves of different wavelengths.[87] This novel method of description is called approximation by 'Fourier series', after the French mathematician Joseph Fourier who developed it to great effect in his

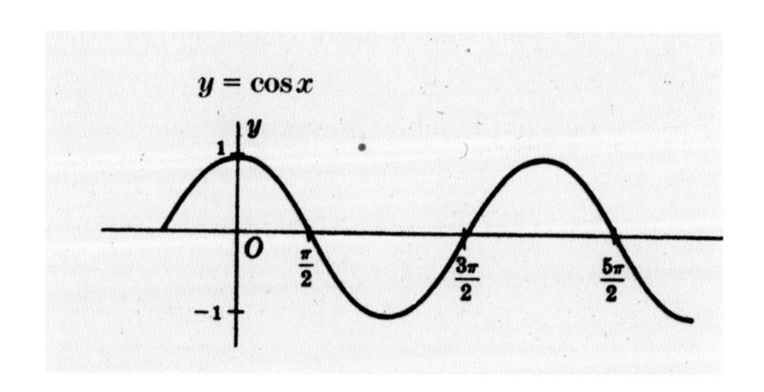

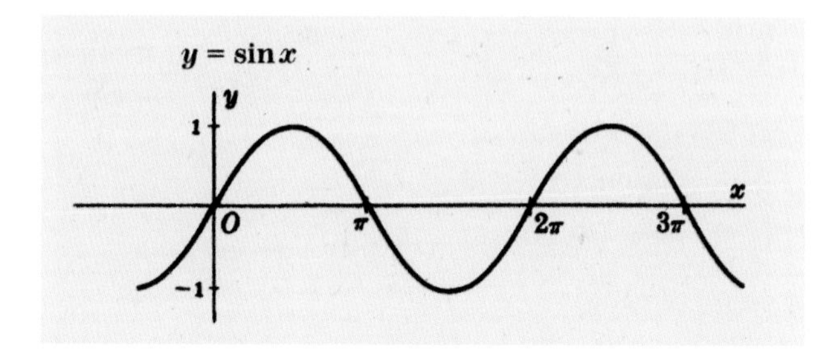

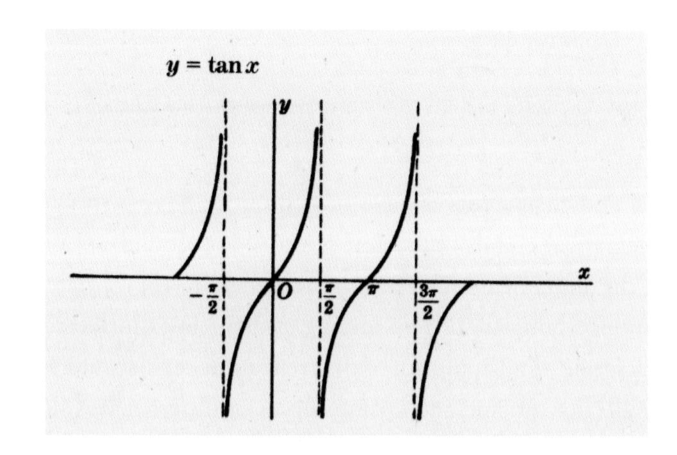

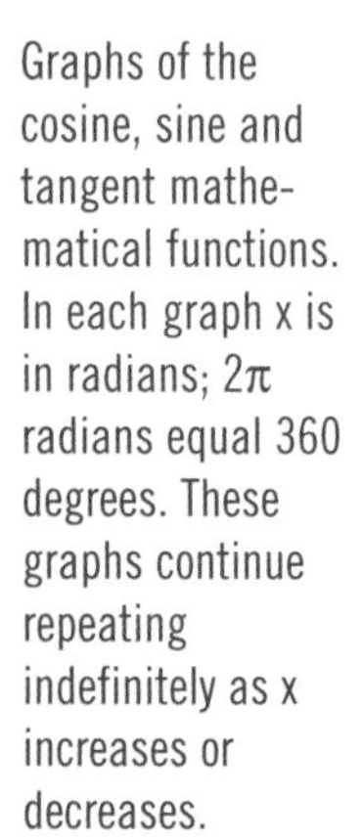
Graphs of the cosine, sine and tangent mathematical functions. In each graph x is in radians; 2π radians equal 360 degrees. These graphs continue repeating indefinitely as x increases or decreases.

book *Théorie analytique de la chaleur* (*The Analytical Theory of Heat*) in 1822. He showed that an infinite collection of different sines and cosines of different periods can match a periodic variation perfectly.[88] In practice, just a few of them produce an extremely good approximation. 'Fourier analysis' remains a major sub-branch of mathematics with countless applications in engineering, electrical circuits, astronomy, physics, and earth sciences.

The sine and the cosine have become the most used graphical functions in all of applied mathematics. There is a very important reason for this. The things that we see persistently in Nature arise so often because they are *stable*. Unstable things – such as needles balancing on their points – are fleeting and rare. Stable things have the comforting property that if they are slightly disturbed or perturbed then they soon recover their equilibrium. Trees behave in this way when wind speeds are not too great. They sway around the equilibrium positions they occupy when there is no wind. There are many other examples – the swing of a pendulum clock, the rolling of a ball in a basin, the rocking of a cradle, the vibration of your lungs when you breathe. All of these disturbances are limited in their level of deviation from rest. They oscillate backwards and forwards about an equilibrium position, and never stray further from it than a particular distance – just like the sine and cosine curves. This is no accident. All of the phenomena in the Universe that exhibit this stability property change like the sum of a sine and a cosine when they are very slightly displaced from their equilibrium positions and their ensuing wobble is very small. No wonder the sine and the cosine are so useful: they describe simply and completely the stability of the world.

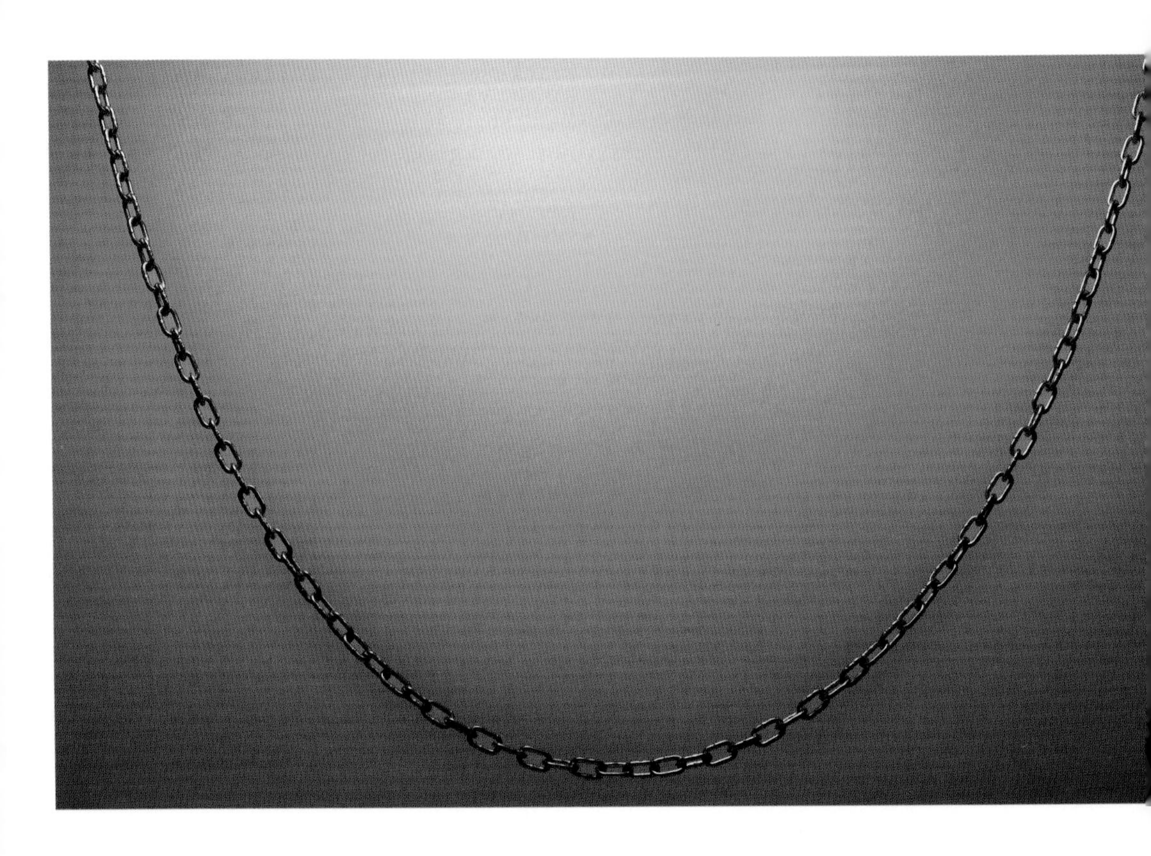

The 'Catenary' shape of a hanging chain.

BRIDGES OVER TROUBLED WATER

CHAINS AND SPANS

A figure with curves always offers
a lot of interesting angles.
Mae West[89]

ONE of the greatest human engineering achievements has been the construction of vast bridges to span rivers and gorges that would otherwise be impassable. These huge construction projects often have an aesthetic quality about them that places them in the first rank of the modern wonders of the world. The elegant Golden Gate Bridge, Brunel's remarkable Clifton Suspension Bridge, and the Ponte Hercilio Luz in Brazil all have spectacular shapes that look smooth and similar. But what are they?

There are two interesting shapes that appear when weights and chains are suspended and they are often confused or simply assumed to be the same. The oldest of these problems was that of describing the shape that is taken up by a chain or a rope whose ends are fixed at two points on the same horizontal level. You can see the shape easily for yourself by holding a piece of string between two ends each at the same level. The first person to claim that he knew what this shape would be was Galileo, who maintained that a chain hanging like this under gravity would take up the shape of a parabola.[90] But, in 1669, Joachim Jungius, a German mathematician who had special interests in the applications of mathematics to physical problems, showed that Galileo was wrong. Unfortunately, he couldn't find the correct answer to the problem himself. It is not an easy one and the equation for the hanging chain was finally discovered sepa-

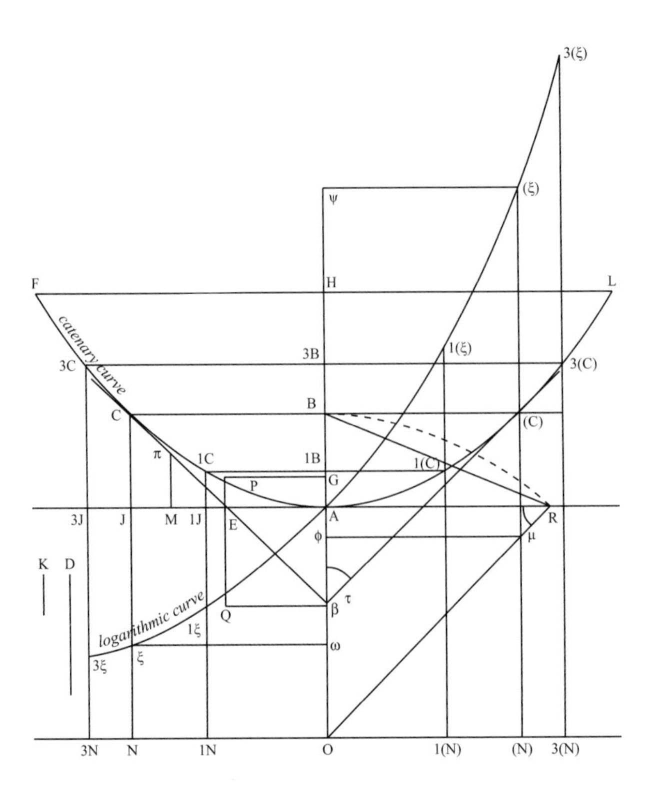

Leibniz's mathematical deduction of the catenary curve (the two curves that he called 'logarithmic' are what we call 'exponential'), which runs from F to A to L.

rately by Gottfried Leibniz, Christiaan Huygens, David Gregory, and Johann Bernoulli, in 1691, after the problem had been publicly announced as a challenge to European mathematicians by Johann Bernoulli, a year earlier. The curve was first called the *catenaria*, a word derived from the Latin word *catena* for 'chain', by Huygens, in a letter to Leibniz (whose explanation of the catenary curve is shown above), but the introduction of the anglicised equivalent, 'catenary', seems to have been due to a future American President, Thomas Jefferson, in a letter dated 15 September 1788 to Thomas Paine, with whom he was corresponding about the design of a bridge arch.[91] Sometimes it was also known as the *chainette* or *funicular* curve.

The shape of a catenary reflects the fact that the tension in the chain supports only part of its weight and the total weight borne at any point is proportional to the total length of chain between that point and the lowest point of the chain.

The equation for the hanging chain in equilibrium that results from this battle between weight and tension has the form:

$$y = B\cosh(x/B) = \tfrac{1}{2}\, B\, \{e^{x/B} + e^{-x/B}\}$$

where *B* is the constant tension of the chain divided by its weight per unit length.[92] If you hold two ends of a piece of hanging string and move them towards each other or apart, then the shape of the string will continue to be described by this formula but with a different value of *B* for each separation. This curve can also be derived by asking for the shape which makes the centre of gravity of the suspended chain as low as possible. This is therefore the most

The St Louis Gateway Arch.

stable and least 'expensive' configuration in terms of the work that must be done against gravity to sustain it.

Catenaries are much used in architecture, often being inverted to create arches. The great seventeenth-century English scientist and architect Robert Hooke made the point in 1671 in a coded statement that translates into English as 'As hangs the flexible chain, so inverted stand the touching pieces of an arch.'[93]

A spectacular example of a man-made catenary arch can be seen in St Louis, Missouri, where the city's iconic Gateway Arch is an upside-down catenary (see previous page). This is the optimal shape for a self-supporting arch: it minimises the shear stresses because the stress is always directed along the line of the arch towards the ground. Its exact mathematical formula is written inside the arch.[94] For these reasons, catenary arches are often used by architects to optimise the strength and stability of structures; another notable example is seen in the soaring high arches of Antoni Gaudí's famous unfinished Sagrada Familia church in Barcelona.

Isambard Kingdom Brunel's Clifton Suspension Bridge (1865).

However, there is a big difference between a hanging chain and a suspension bridge such as the Clifton or the Golden Gate. Suspension bridges don't only have to support the weight of their own cables or chains. The vast bulk of the weight to be supported by the bridge cable is that of the deck of the bridge itself. If the deck is horizontal with a constant density and cross-sectional area all the way along it, then the equation for the shape of the supporting cable is now found to be a parabola $y = x^2/2B$, where B is (as for the hanging chain equation) a constant equal to the tension divided by the weight per unit length of the bridge deck.[95]

In Britain, one of the most remarkable examples from the nineteenth century is the Clifton Suspension Bridge in Bristol, designed by Isambard Kingdom Brunel in 1829, but only completed in 1865, three years after his death. Its beautiful parabolic form remains a fitting visual monument to the greatest engineer since Archimedes.

Infinity sign first introduced by John Wallis in his book *De Sectionibus Conicis* (1655).

INFINITY

WHERE GOD DIVIDES BY ZERO

Infinity converts the possible into the inevitable.

Norman Cousins[96]

THE infinity sign has a dual resonance. It combines the mystic attraction of the great unknown and unknowable with the cold precision of mathematics and the desire to describe the unimaginable. The ribbon like figure-eight on its side is an ancient symbol, a shadow of the ancient ourobos symbol of the snake eating its tail.

It provided the mysterious cross of St Boniface in early Christian tradition, but its entrance into the symbolic world of mathematics did not occur until 1655.[97] That distinction fell to the Oxford mathematician John Wallis. Wallis was both a mathematician of note – the Savilian Professor of Geometry at Oxford University for over fifty years – and a famous code-maker and code-breaker. During the period of the English Civil War (1642–51) Oliver Cromwell's Parliamentarians chose the lawyer John Thurloe to act as their principal spymaster, creating a feared intelligence network that spanned the country and controlled the mail and, thereby, people. Naturally, Thurloe's activities required codes to keep his communications secure and codebreaking to keep the communications of others totally insecure. To this end, he recruited Wallis, the foremost mathematician in England, to deliver these secret services for the Roundheads. Wallis acted as their chief cryptographer from 1643 onwards, and was so effective, it seems, that when Charles II ascended the throne at the end of the Interregnum, he retained Wallis's services for the Royalists. Of course, the links between codes and mathematics are now obvious, and the national security services of countries all over the world are stuffed full of number theorists. The United States' National Security Agency houses the largest collection of mathematicians on Earth.

Wallis wrote little about the general principles of cryptography. He thought that the less others knew about it the better, but he viewed the use of mathematical symbols as tantamount to the creation of a code. The solving of equations was an act of codebreaking: the decrypting of God's secrets.

Wallis was reputed to have created his infinity symbol ∞ from the $\subset | \supset$ symbol sometimes used by the Romans, instead of M, to denote 1000. It was soon used extensively by mathematicians in Europe, but in a rather casual way. Infinity was, understandably, not thought to be like other quantities. The prevailing philosophy, inherited from Aristotle, and widely accepted throughout Europe right up until the mid-nineteenth century, was that actual infinities do not exist, neither in the physical universe nor in mathematics. The infinity sign was just a shorthand – a sort of mathematician's version of 'etc.' – to denote that a series of numbers continued without end – for a 'potential' infinity. The list of all the positive numbers 1, 2, 3, 4, 5, . . . and so on, without end, is the archetypal potential infinity. If the universe is not finite in size then this provides a potential infinity that is physical: we could never reach that infinity in a spaceship any more than we could count to infinity – in fact, you wouldn't even be able to count to a billion in a human lifetime. This denial of actual infinities fitted in comfortably with the medieval synthesis of Aristotle's ideas and Christian doctrine. Infinity was the realm of God alone. Nothing created could rival Him in that respect.[98]

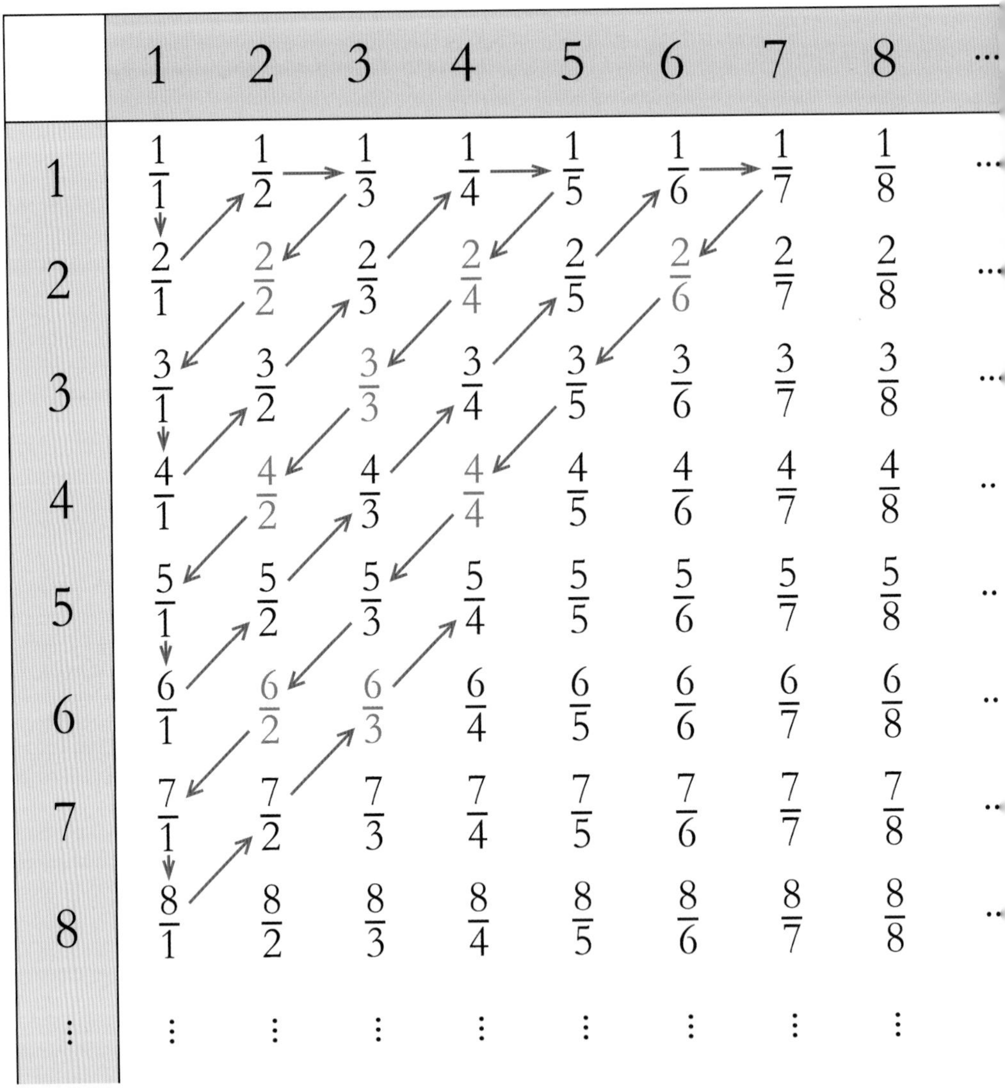

	1	2	3	4	5	6	7	8	…
1	$\frac{1}{1}$	$\frac{1}{2}$	$\frac{1}{3}$	$\frac{1}{4}$	$\frac{1}{5}$	$\frac{1}{6}$	$\frac{1}{7}$	$\frac{1}{8}$	…
2	$\frac{2}{1}$	$\frac{2}{2}$	$\frac{2}{3}$	$\frac{2}{4}$	$\frac{2}{5}$	$\frac{2}{6}$	$\frac{2}{7}$	$\frac{2}{8}$	…
3	$\frac{3}{1}$	$\frac{3}{2}$	$\frac{3}{3}$	$\frac{3}{4}$	$\frac{3}{5}$	$\frac{3}{6}$	$\frac{3}{7}$	$\frac{3}{8}$	…
4	$\frac{4}{1}$	$\frac{4}{2}$	$\frac{4}{3}$	$\frac{4}{4}$	$\frac{4}{5}$	$\frac{4}{6}$	$\frac{4}{7}$	$\frac{4}{8}$	…
5	$\frac{5}{1}$	$\frac{5}{2}$	$\frac{5}{3}$	$\frac{5}{4}$	$\frac{5}{5}$	$\frac{5}{6}$	$\frac{5}{7}$	$\frac{5}{8}$	…
6	$\frac{6}{1}$	$\frac{6}{2}$	$\frac{6}{3}$	$\frac{6}{4}$	$\frac{6}{5}$	$\frac{6}{6}$	$\frac{6}{7}$	$\frac{6}{8}$	…
7	$\frac{7}{1}$	$\frac{7}{2}$	$\frac{7}{3}$	$\frac{7}{4}$	$\frac{7}{5}$	$\frac{7}{6}$	$\frac{7}{7}$	$\frac{7}{8}$	…
8	$\frac{8}{1}$	$\frac{8}{2}$	$\frac{8}{3}$	$\frac{8}{4}$	$\frac{8}{5}$	$\frac{8}{6}$	$\frac{8}{7}$	$\frac{8}{8}$	…
⋮	⋮	⋮	⋮	⋮	⋮	⋮	⋮	⋮	

Cantor's procedure for counting the rational fractions, one by one, so that none are left out. This shows that t finite list of natural numbers, 1, 2, 3…, is the same size as the infinite list of all fractions.

A NEW SLANT ON INFINITY

CANTOR'S DIAGONAL ARGUMENTS

On a clear day you can see forever.

Alan Lerner[99]

GREAT nineteenth-century mathematicians, such as Karl Friedrich Gauss, were still adamant that, even in mathematics, all infinities were merely potential in nature, like the unlimited sequence 1, 2, 3, 4, 5, . . ., whose end is never reached. There could be no physical manifestation or mathematical manipulation of an actual infinity. But a little-known German mathematician, Georg Cantor, changed all that. Cantor stirred up a small hornets' nest of opposition amongst the most conservative ranks of mathematicians who believed that mathematics should have no truck with infinities. Any idea that infinities were in any sense 'real', or could be manipulated, smacked of heresy and threatened the health of mathematics as a whole. By manipulating infinite collections of things, all manner of fallacies might be introduced into mathematics, causing the entire edifice to fall.

Cantor showed how to define 'infinity' precisely, and then proceeded to reveal that there exists a never-ending hierarchy of infinities, each infinitely bigger than the one below it in the hierarchy. The smallest infinity was the one that counts the unending list of positive numbers (1, 2, 3, 4, 5, 6 . . .). All other unending lists that can be put in a one-to-one correspondence with them (that is another way of saying that their members can be counted one by one) are said to be infinities of the same size: 'countable infinities'.

This simple idea has some surprises for our simple intuition. You might think that the infinity of positive numbers 1, 2, 3, 4, 5, 6 . . . and so on, are twice as numerous as the infinite list of all the even numbers 2, 4, 6, 8, 10, 12 . . . and so on. Isaac Newton thought so, arguing that:

> tho' there be an infinite Number of infinite little Parts in an Inch, yet there is twelve times that Number of such Parts in a Foot, that is, the infinite Number of those Parts in a Foot is not equal to, but twelve times bigger than the infinite Number of them in an Inch.[100]

But Cantor revealed that, as infinities, they are the same size, because there is a one-to-one correspondence between the two lists: 1 to 2, 2 to 4, 3 to 6, 4 to 8, 5 to 10, 6 to 12, and so on for ever. They are therefore the same infinite size as each other. This would *not* be true if the lists of numbers were finite, unless they contained the same quantity of numbers in each list. Infinities are different. Cantor captured their essence by defining them as collections which could be put into this one-to-one correspondence with a subset of themselves.[101] In our case, the unending collection of whole numbers can indeed be put in one-to-one correspondence with a part of itself – all the even numbers.

This crucial insight set Cantor off on a spectacular intellectual journey that culminated in a beautiful theory of actual infinities in mathematics. This allowed them to be combined and manipulated rather like ordinary finite quantities, although they obeyed rather different rules of addition ($\infty + 1 = \infty$), multiplication ($\infty \times \infty = \infty$), and division ($\infty \div \infty = \infty$) appropriate for what Cantor called *transfinite* arithmetic.

Cantor went on to show that there are infinities which are not countable, and so cannot be put in one-to-one correspondence with the ordinary numbers: they are infinitely bigger, or 'uncountably infinite'. The collection of all the never-ending decimals falls into this uncountable category. Moreover, there were infinities that were infinitely bigger still than the decimals, which could not be put into one-to one correspondence with them, and so on, for ever. There is a tower of infinities, each infinitely greater than the one below in Cantor's precise sense, and the tower has no pinnacle. There is no biggest infinity.[102]

The hierarchy of different infinities meant that Cantor had to devise some different symbols for each of them because Wallis's simple ∞ couldn't distinguish between them.[103] For all those infinities of the smallest 'size', along with the whole numbers, he used the name 'countable' infinities and labelled them by the Hebrew letter aleph with a subscript zero, $\aleph_0$. Each variety of infinity above $\aleph_0$ that could not be put into a one-to-one correspondence with the infinity below it was labelled by successive alephs of increasing number:

$$\aleph_1, \aleph_2, \aleph_3, \aleph_4, \aleph_5, \ldots$$

Cantor created two famous pictures which revealed two remarkable features of the simple countable infinities. They appear in every university mathematics course in the world in exactly the form that he originally gave.

The first was his demonstration that all the simple fractions, quantities such as 2/3, 3/8, which have the form *P*/*Q* where *P* and *Q* each run through all the positive whole numbers, were countably infinite. Again, at first you might think there are 'more' of these fractions than there are whole numbers because there are infinitely many of them in between any two whole numbers. But this is not the case. As infinite collections they have the *same* size. The trick is to devise a way to count all the fractions systematically so that none are left out. Cantor's picture of the counting route is shown on page 342.

The recipe is simple, when you see how. Count all the fractions in which the top and bottom numbers add to 1, then those that add to 2, then to 3, then to 4, and so on. None will be missed out and you have discovered, as Cantor first did, that the collection of fractions is a countable infinity of size $\aleph_0$. The route through the grid of all possible fractions is a systematic zig-zag up and down the diagonals as Cantor's famous picture reveals.

Cantor then, in 1891, devised another beautiful pictorial argument that first demonstrated that there are other, infinitely bigger, infinities that *cannot* be counted like this. The smallest uncountable infinity is the list of all the never-ending decimals. The picture displays Cantor's famous 'diagonal argument'. Like a chess player offering a piece as a gambit before checkmating the King, Cantor says let's assume that we *can* count all the never-ending decimals that do not end in an infinite string of zeros.[104] List them as we count them one by one. But Cantor can checkmate you because he can always devise a number that cannot appear on the list. He creates it by adding one to each successive digit in the list of decimals. This 'killer' number cannot appear anywhere on the never-ending list because it differs by one digit from every number it contains. The never-ending decimals are therefore uncountably infinite.

3.14159...
1.41421...
1.73205...
2.23606...
2.71828...
0.14285...

Cantor's diagonal argument to show that the list of all unending decimal numbers is uncountable. He assumes that they can be counted and put in a list. A new number is made by taking successive digits from each number in the list, 3.43625... shown in green. A new number is then constructed by subtracting 1 by each of its digits to give 2.32514... . This number must disagree by one digit with every number in the list and so the list cannot contain all numbers as originally assumed.

Ironically, Cantor's ideas were not enthusiastically received by mathematicians of the day in his native Germany. His work was blocked from publication and he was treated by some as a mathematical anarchist, bent upon perverting the structure of mathematics. This led him to withdraw from active mathematics for long periods, but he received encouragement from an unexpected quarter. Catholic theologians realised that his distinction between different orders of infinity was revolutionary. It enabled them to study actual mathematical and physical infinities without threatening the uniqueness of God's infinity.[105] The theological infinity could live at the unreachable top of the never-ending tower of alephs, where it was given the label 'Absolute Infinity', Ω, and enjoyed an infinitely greater status than the mere countable and uncountable infinities that our mathematics and physics had traditionally been challenged to deal with. Cantor's ideas were so simple and clear that there is no reason why they could not have been recognised in outline 2000 years before him. How different the development of scholastic philosophy might so easily have been.

Infinity is also very much a live issue. Since 1975 physicists have been on a quest for the so-called Theory of Everything that unites all the known laws of Nature into a single mathematical statement. That search has been significantly guided by an attitude towards the existence of actual physical infinities. In theories of particle physics, the appearance of an infinite answer to a question about the magnitude of a measurable quantity was always taken as a warning that you had made a wrong turn. For decades its inevitable appearance was managed by a strange subtraction procedure that removed the infinite part from the calculation to leave only a finite residue to compare with observations. Although the results of this so-called 'renormalisation' process gave spectacularly good agreement with experiments, there was always deep unease that this ugliness could not be part of Nature's economy. The true theory must be finite.

This all changed in 1984. Michael Green, now at Cambridge University, and John Schwarz at the California Institute of Technology showed that one particular type of physics theory – a 'superstring' theory – could be totally finite. These theories were based on the idea that the smallest ingredients of the world were not 'points' of energy, but lines, or loops, called 'strings'. They were 'super' because of a symmetry they possessed which allowed matter and radiation to be united.

The upshot was that the things that we call 'particles' are just excited vibrations of the string and might one day have their exact masses and interaction properties precisely determined by calculating the energies of the natural vibrations of the superstring and using '$E = mc^2$' to find the equivalent masses.

The enthusiasm with which the new theories were embraced by physicists was a consequence of their ingenious banishment of infinities, a problem that had plagued their predecessors.

The path towards superstring theories awaits experimental endorsement. But the enthusiasm with which they have been pursued reflects the philosophy of scientists who believe that the appearance of an actual infinity in a physical theory is a signal that it is being stretched beyond its domain of applicability. The solution is to upgrade the theory until the infinities are smoothed into large, but finite, quantities. Engineers, for example, know this well, exorcising the appearance of infinities in simple models of rapid aerodynamic flows by simply including more realism in the description of the air. The crack of a whip is caused by the sonic boom from the tip travelling faster than the speed of sound. A simple calculation that ignored the friction of air would say that this involved something changing infinitely quickly. But a more detailed modelling of the air-flow properties turns this infinity into a very rapid but finite change.

Despite the general adoption of this 'infinities-mean-you-must-try-harder' dictum in relation to physical theories, one area of science has been willing to take predictions of actual infinities more seriously. Cosmologists see a lot of infinities. Many are of the 'potential' variety – the Universe might be infinite in size, face an infinite future, or contain an infinite number of stars. While they pose no local threat to the fabric of reality, we do have to face up to Nietzsche's infinite replication paradox: if the universe is infinite and exhaustively random, then any event that has a finite probability of occurring here and now (such as you reading this book) must be occurring infinitely often elsewhere at this very moment. Moreover, for every history we have pursued here, all possible alternatives are acted out, wrong choices made simultaneously with right choices. This is a grave challenge to ethics and to the theology of almost every religion. Some find it so alarming that they regard it as a powerful argument for a finite Universe. However, it should be remembered that the finiteness of the speed of light insulates us from contact with our doubles. For all practical purposes we can only see and receive signals from a finite part of the Universe.

The challenge to cosmologists does not end there, though. They also have to worry about 'actual' infinities. For decades, cosmologists have been happy to live with the notion that the Universe of space and time began expanding from an initial Big Bang 'singularity', where temperature, density, and just about everything else were infinite at some finite time in the past. Furthermore, when large stars exhaust their nuclear fuel and implode as a result of their own gravity, they appear doomed to reach a state of infinite density in finite time. But this is all neatly kept out of reach. Black holes are believed to be always shrouded by the 'event horizon' – a surface of no return through which things can fall in but not pass out – so that we can neither see the actual infinite density at the centre, nor be reached by its effects.

There are physicists who take a different view about infinities from those engaged in particle physics. Roger Penrose believes actual infinities *do* occur both at the start of the Universe and at the centre of black holes. He has proposed that the laws of Nature provide a form of 'cosmic censorship' that ensures that 'naked' physical infinities never occur: they are always enclosed by event horizons.[106] By contrast, cosmologists with a particle-physics perspective tend to see these cosmological infinities merely as a signal that the theory has overextended itself and needs to be improved. When that is achieved, these infinities will be exorcised just like those that were cured by string theory. As a result we find much interest in the prospect of universes that bounce back into expansion if run forwards in time to contract towards their apparent end. Our presently expanding universe is suspected of having arisen from the rebound, at finite density and temperature, of a previously contracting phase in its history.

From the outside, we cannot see what is happening inside a black hole. But if we fell in, we would be facing an uncertain fate as we approached the centre. Is there a real physical infinity there or does energy slip away into another dimension of space, simply disappear into nothing, or get soaked up by exciting a never-ending sequence of vibrations of the superstrings at the core of all matter and energy? We just do not know. Again, the issue of finite versus infinite is a crucial guiding principle. Do we treat the appearance of an infinity as a signal to update our theory or do we treat these infinities more seriously, as an indication that new types of law govern infinite physical quantities, laws that could dictate how our Universe began and how matter meets its end under the relentless implosion of gravity?

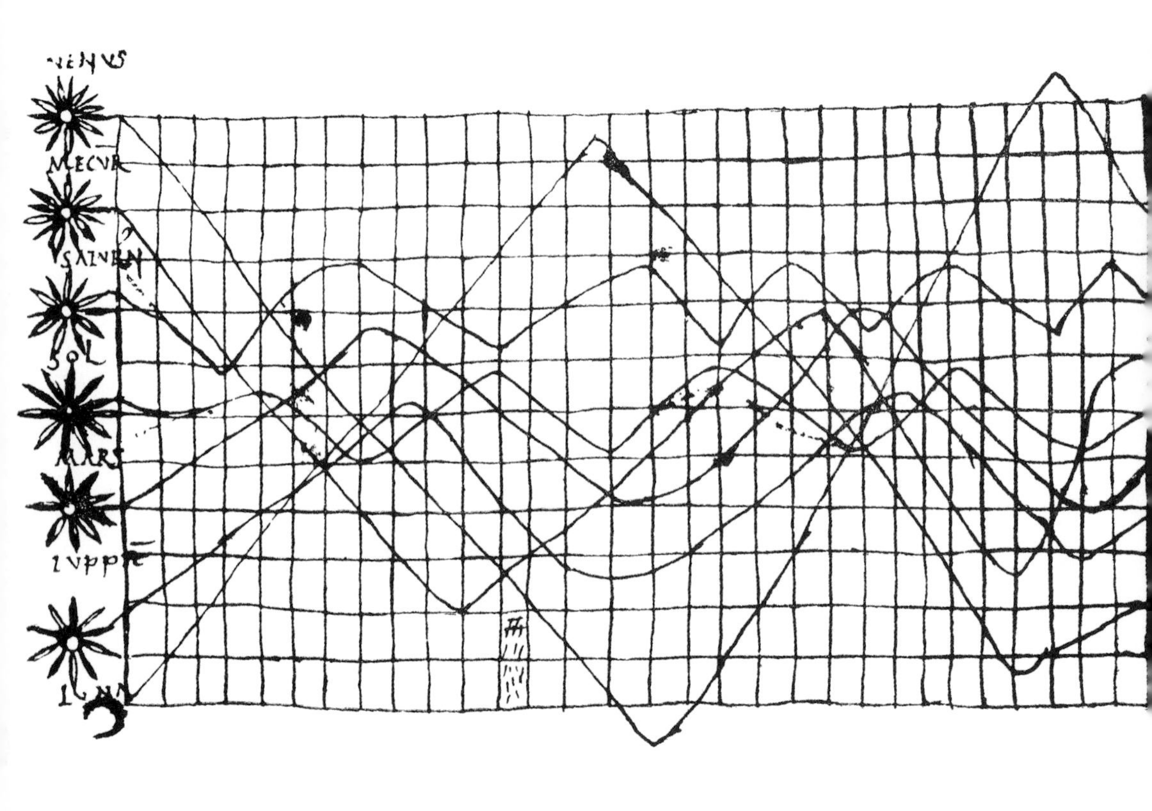

A tenth-century graph representing the apparent positions of the then-known planets and the Sun with time. The scale differs for each planet and the graph may be representing the information presented in a work of Pliny. About one-quarter of the way in from the right, the original author appears to have made some corrections and deleted another curve. This is the first known graph.

THE PLOTTERS
THE ORIGINS OF THE GRAPH

The advantages proposed by [this] mode of representation, are to facilitate the attainment of information, and aid the memory in retaining it . . . Of all the senses, the eye gives the liveliest and most accurate idea of whatever is susceptible of being represented to it; and when proportion between different quantities is the object, then the eye has an incalculable superiority.

William Playfair[107]

OPEN your newspaper, read any maths, science, or economics book, and you won't need to look very far before you find a graph. Graphs are one of the greatest mathematical inventions of all time. Who invented them and why?

The earliest known graphical representation of data dates from the tenth or eleventh century and was intended to show the variation of the inclinations of the orbits of the then-known solar system planets with time. The width of the zodiac was represented vertically by twelve parts, with the horizontal axis divided into thirty units of time. The graph is not easy to interpret as the scale appears to be different for each planet and it is not clear what the undulating line associated with the Sun is representing.[108] However, the interpretation is clearer when it is appreciated that the graph is representing a statement in a passage of Pliny's.[109] The astronomical author is unknown and his graph is contained within a transcription of a commentary by Macrobius on Cicero's

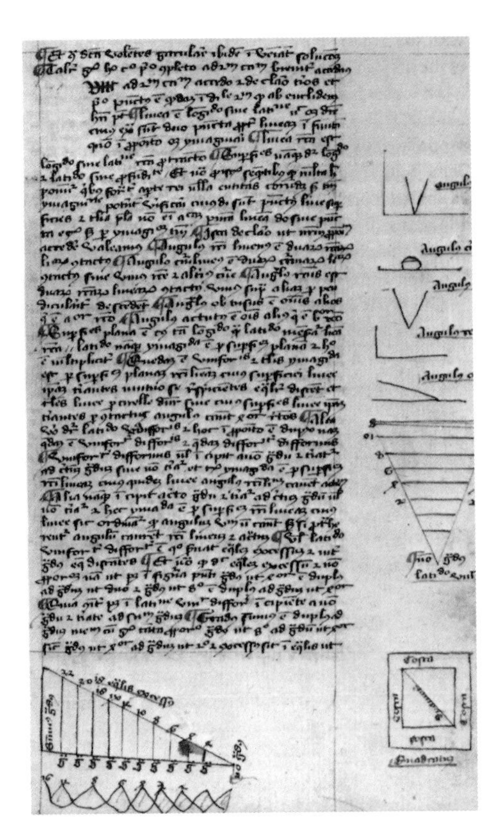

Nicole Oresme used 'latitudes', or diagrams plotting the change of speed of a motion against time, to accompany his discussion of various types of motion. This page is from the *Tractatus de Perfectione Specierum* (*Treatise on the Perfection of Species*, 1486) by Jacobus de Sancto Martino. It is an elementary explanation of the use of latitudes.

An extract from Nicole Oresme's *Treatise on the Configurations of Qualities and Motions.* The small figures are used in the body of the text as a shorthand for descriptions of different types of changing motions. He describes the motion in words and then writes *sicut hic* ('as this') before each little graph that appears as an illustration.

work, *In Somnium Scipionis*, but there appears to be a link to the influential Bishop (and later Pope Sylvester II) Gerbert of Rheims, who was responsible for spreading many mathematical tools – notably the use of the Indo-Arab numerals – throughout monasteries and church schools in Europe.

Graphs seem so pervasive, simple, and useful that you could be forgiven for thinking that they are an extremely old idea, cooked up and used by the Babylonians or the ancient Greeks along with their algebra and geometry. The earliest systematic use of pictures to display trends of change of a quantity with distance or time are found in the work of Nicole Oresme (1320–82), the Bishop of Lisieux and a counsellor to Charles V of France. He was among the first scholars who tried to develop a detailed quantitative theory of motion[110] by introducing the mathematical concept of a functional relationship between two quantities, so if $y = f(x)$ then y depends on the value of x in some way that is defined by the function f. A picture of the relation between y and x will be the graph of the function f. Confusingly for us, Oresme used the word *latitude* to denote the speed of a moving body and *longitude* to mark the time when it was measured. He would represent the longitude as a horizontal line and the corresponding latitude by vertical lines. The resulting pictures had no numerical scales like modern graphs. They simply represented patterns of change against the passage of time: constant acceleration, varying acceleration, constant speed, and so forth.

Oresme would illustrate his work, *Tractatus De Latitudinibus Formarum* (*On The Latitude of Forms*), with a series of miniature figures, resembling bar charts, showing different states of motion. They entered into the text as part of the argument rather as we might use a mathematical symbol, or as illustrative examples. Opposite is an enlarged extract from his most famous work, *Tractatus de configurationibus qualitatum et motuum* (*Treatise on the Configurations of Qualities and Motions*).[111] Instead of elaborating in words the situation where the speed of an object increased steadily, peaked, and then decreased, Oresme would convey this idea in the text by a small version of his graph showing such a motion.

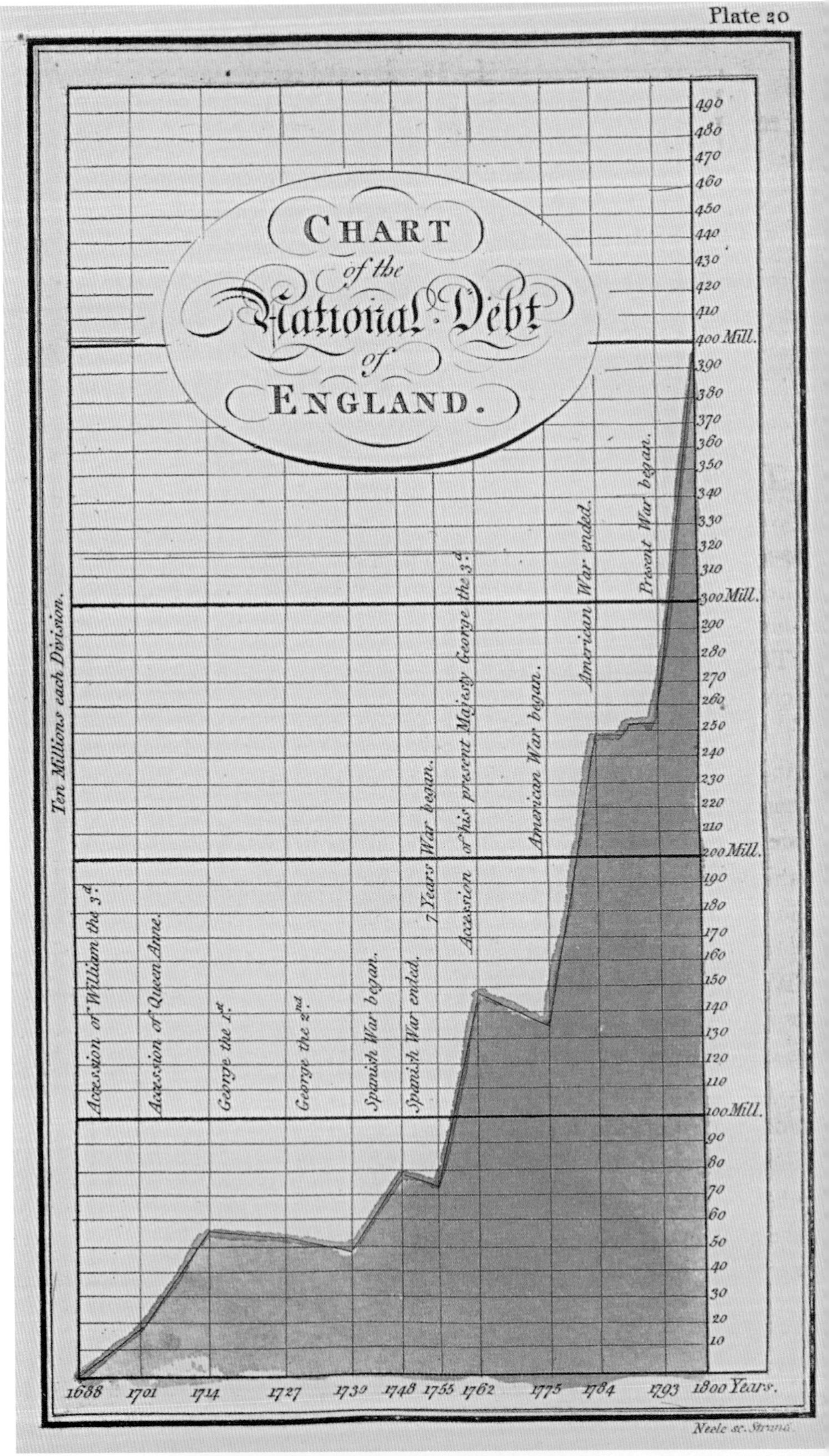

William Playfair's graph of the English national debt from 1699 to 1800, from his *Commercial and Political Atlas* (1786).

THE CHARTISTS

THE ENIGMA OF VARIATIONS

On inspecting any one of these Charts attentively, a sufficiently distinct impression will be made, to remain unimpaired for a considerable time, and the idea which does remain will be simple and complete.

William Playfair[112]

UNFORTUNATELY, Oresme's eye for shorthand explanation didn't evolve into a universal tool. Others used his diagrams when they annotated or copied his works, but there does not seem to have been a systematic development of his simple pictures into useful graphs with numbered scales as we know and use them today. Such graphs are new-fangled things by historical standards. Graphs representing data points and mathematical equations only began to appear in print during the last thirty years of the eighteenth century, nearly a hundred years after the first scientific journals appeared. The word 'graph' was coined later still, by Joseph Sylvester, in 1878, although he applied it to diagrams that displayed structural links between chemical bonds in molecules.[113] When graphs first appeared on paper in their modern forms, they emerged independently in three different contexts, and their originators gave them three different, and equally unmemorable, names.[114]

There are three different common sorts of graph. One plots data against some other quantity, such as time; this is like one of Oresme's diagrams of latitude. Another plots a specified mathematical equation (such as $y = \sin x$) as a

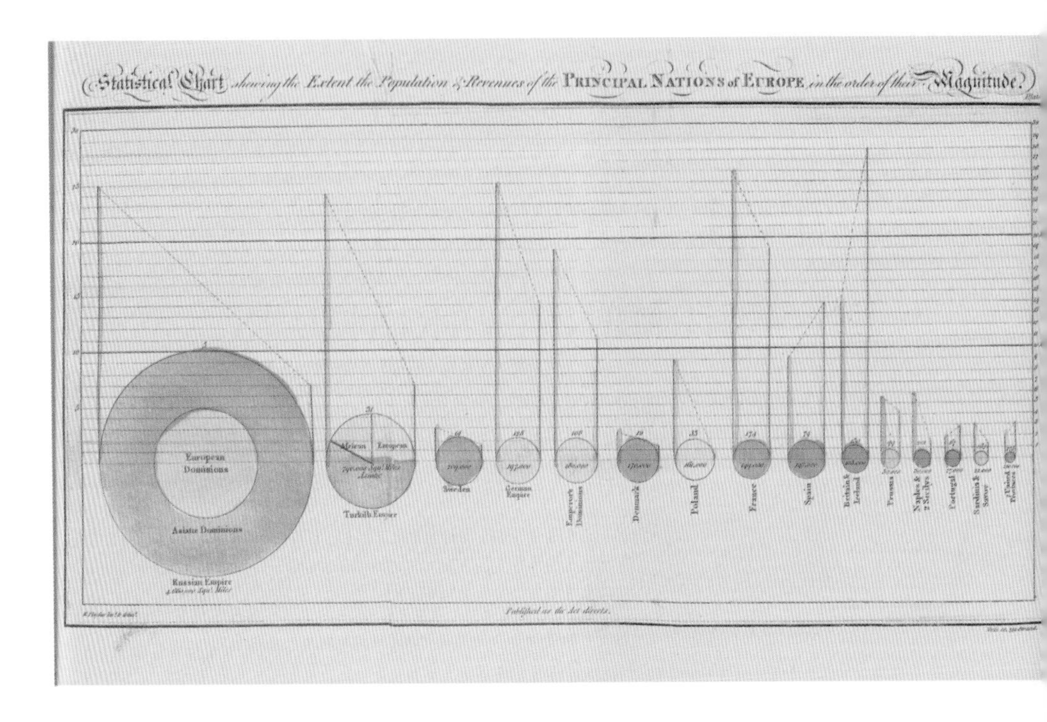

Playfair's invention of the pie chart, used here to compare populations and tax revenues in the principal nati of Europe, was first seen in his *Statistical Breviary* (1801). He uses the areas of the circles to represent surface area of different countries. The vertical red line on the left marks this size scale by giving population si in millions. The tax revenue in millions of pounds is marked on the vertical scale to the right, in yellow. The po lation and revenue levels for each country are then joined by a dotted line. Playfair regarded countries for wh the dotted line sloped upwards to the right, notably Sweden and Britain (sixth from the right), as operating exc sive levels of taxation. Almost all other countries had their dotted lines sloping down to the right.

continuous curve. And another still compares the first two on a single picture, or tries to find the simplest continuous curve that will best approximate the scattering of data points.

In 1785, William Playfair (1759–1823) first created a graphical representation of the change (almost entirely a stepwise growth!) of the English National Debt from 1699 to 1800. He called the pictures examples of 'lineal arithmetic'. His book *The Commercial and Political Atlas*[115] contained forty-three tracers of the development of commercial quantities with time – time series as we would now call them – and a single bar chart. Playfair, a product of the Scottish Enlightenment, was a very wide-ranging thinker and well acquainted with many intellectual developments of the day. His 'Atlas' was not an atlas in the geographical sense. It contained no maps at all. It was a landscape format volume with charts and graphs on fold-out pages that became more than twice the size of the volume itself. It is interesting to see that in the early editions he kept the tables of data from which the graphs were plotted, but by the third edition he was sufficiently confident in the graphical display of the data that the accompanying tables disappear.[116] Later, he invented the 'pie chart' in his book *Statistical Breviary*, which was published in London in 1801. We also find him integrating a bar chart and graphs to create visually arresting correlations, combining, for example, the time trends in the price of a quarter of wheat (bar chart) and in the weekly wages (continuous curve) paid in its production over several centuries.

Playfair was an ingenious mechanic, and from 1777 to 1781 he had worked for James Watt as a draughtsman making drawings of steam engines. He greatly admired Watt's skill and ingenuity. Watt was himself the next graphical pioneer, although he called his graphs merely 'diagrams' or 'indicators'. He devised pen-recorders that would automatically record the level of quantities such as engine pressure by tracing a curve on a paper chart or a piece of smoked glass, but he kept this invention secret until 1823. By this means, Watt created graphs which automatically recorded the time variation of pressure against the volume of steam driving an engine.[117] Later, such automatic self-recording devices were widely used in physiology, medicine, and acoustics.

Graphs were first used to explain things to the general public in France, in 1795, when the revolutionary change to metric units presented educational challenges for the government. The committee charged with educating the public on these changes knew that previous attempts to make changes in weights and measures had produced a mixture of stubborn resistance and apathy. It responded by creating a set of graphs which would enable old units to be converted to new by simply reading off their counterparts. No arithmetic was necessary.[118] An impressive number were created by Louis-Ezéchiel Pouchet, who was a leading figure in the Rouen cotton-spinning industry. He used Playfair's description of them as 'lineal arithmetics', but went one step further than Playfair by representing a number of curves on the same graph, each labelled with a different value of some quantity, and so could represent three pieces of information using only two axes.[119]

The first person after Oresme who seems to have drawn graphs of

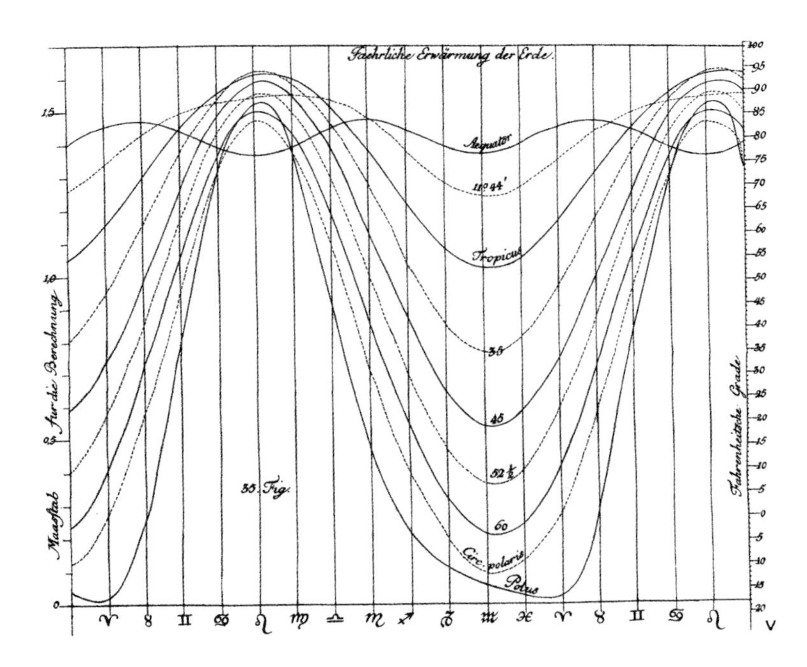

Johann Lambert's 1779 graph showing the effects of solar variation on the Earth's surface with time. The right-hand vertical scale gives the Fahrenheit temperature and each curve is labelled by the latitude where the observations are made.

mathematical equations rather than graphs to represent data taken in time or distributed in space was the talented German mathematician, physicist, and polymath Johann Heinrich Lambert, whom we have already encountered as a pioneer in drawing the sine function. Beginning in the 1760s, Lambert drew a succession of elegant mathematical curves that he called '*Figuren*'. He employed them in many ways that are now familiar but which were then very novel: comparing the curves to the distribution of data points, smoothing out the effects of fluctuations and errors in scattered data points, and drawing the continuous curve that best fits a set of data points. Above is a typical example from his book of 1779, showing the variation of the effects of solar heating on the Earth's surface with latitude. The temperature in degrees Fahrenheit is shown vertically. A time line is shown horizontally and each of the sinuous curves is labelled by a value of latitude.

By the 1830s, graphs were becoming widely used by innovative scientists. The English astronomer John Herschel was among the first astronomers to make good use of them and urged others to use them too. In 1833, he wrote about how he prepared sheets of graph paper with pre-drawn axes ready to be labelled and scaled appropriately, remarking that:

> Such charts are so very useful in a great variety of purposes, that every person engaged in physico-mathematical inquiries of any description will find [benefit] in keeping a stock of them always at hand.

Most graphs have two 'axes' that lie at right angles to one another, a label for each of the axes, together with a clutch of points 'plotted' in different places in the space spanned by the two axes, or a continuous line that marks out a path in the space. In olden days, the two axes used to be known as the 'ordinate' (from Latin for 'an orderly line'), for the vertical one, and the 'abscissa' (from Latin for 'cut off'), for the horizontal one. The quantities marked along the axes are called 'coordinates' – because they enable us to 'coordinate' the positions of different points. They intersect at a point called the 'origin' of the coordinates. Today, the axes tend to be called simply the 'horizontal' and 'vertical' axes, or even the 'x' and 'y' axes because so many graphs use an algebraic letter y to denote the vertical values of the plotted points and x to denote the horizontal points. This procedure requires the notion of a 'coordinate', which was introduced by the French mathematician and philosopher René Descartes, in 1637; allegedly, he had the idea while lying in bed watching a fly walking on the ceiling.

We are most familiar with graphs that have 'Cartesian' axes at right-angles to each other, but there are other ways to specify the position of a point on the page uniquely. Mathematicians have introduced many different coordinate systems that allow them to specify the locations of points more conveniently in some situations. A common system is that of 'polar' coordinates, which locates the position of the point by its distance from the origin of coordinates and by the angle subtended from the horizontal axis. The first graph in polar coordinates was created by the French engineer Léon Lalanne, in 1843, to represent the frequency of occurrence (marked by radial distance from the origin) of different wind directions in his 'windrose' diagram.

The simplest graphs are straight lines. They show that the two quantities being plotted change in direct proportion to one another. Only two points are required in order to draw a straight line that joins them. By contrast, a curve may require many more points to specify it. In 1846, Lalanne realised that by

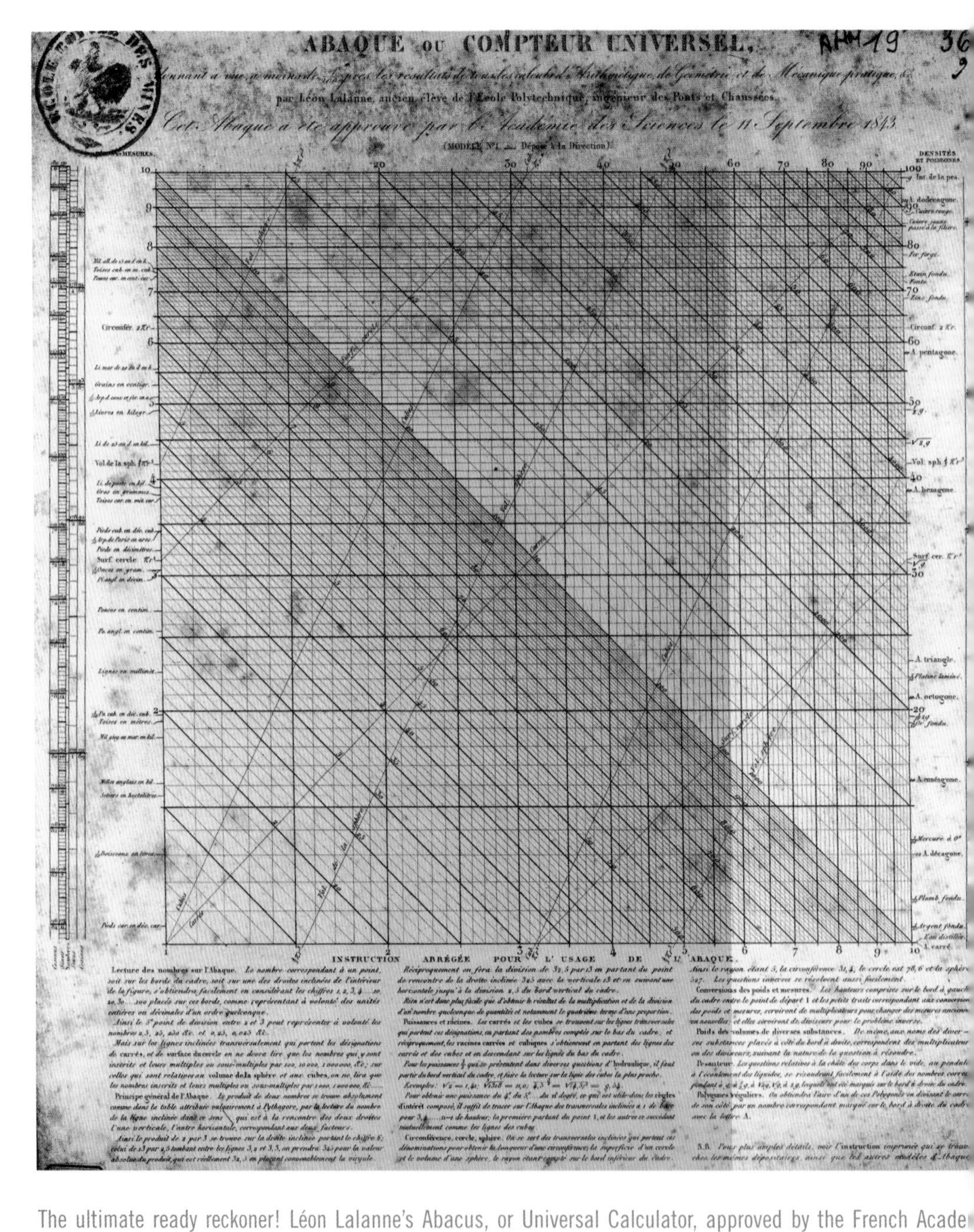

The ultimate ready reckoner! Léon Lalanne's Abacus, or Universal Calculator, approved by the French Acade
Sciences on 11 September 1843. The diagonal lines allow the user to calculate various geometric quantities, su
the volume of a sphere, surface areas and circumferences of geometric shapes, and conversions between dif
systems of weights and measures. The horizontal and vertical scales are both logarithmic.

plotting the logarithms of quantities against each other, rather than the quantities themselves, curved graphs could be changed into straight lines. This device of creating a so called 'log – log' plot is now one of the standard methods for presenting two sets of data that are not linearly related. Lalanne called these 'anamorphic' graphs,[120] after the extreme distortion of perspective that is used in art, and employed them as part of his ambitious diagram to provide a 'universal calculator'.[121]

The use of graphs in seventeenth- and eighteenth-century science and mathematics had strong supporters, who saw it as the ultimate language of science, which they had long awaited. Others ridiculed it as inaccurate and lacking in rigour. The truth is somewhere in between. Graphs allow us to deduce trends quickly and see possible relationships. They allow us to take in far more information at one glance than is possible by merely inspecting equations or lists of numbers. We prefer to receive information in analogue form rather than in digital form. Let me give a personal example. If you are running quickly on a running track and you want to take note of the times you are recording by glancing at a stopwatch that you are carrying, then an old-fashioned sweep-hand watch will allow you to register a split time instantly by a mere glance at the configuration of hands. By contrast, if you are carrying a digital stop-watch, then you will have a slower mental operation to go through to register the time split. You will need to mentally process the numbers.

Finally, it is clear that graphs played a key role in moving many sciences on from the qualitative collection of data towards a quantitative mathematical analysis of data. When data sits in a list or a table there is no immediate stimulus to find a geometrical or algebraic pattern within it. Graphs stimulated a pattern-seeking attitude to data. Simply collecting information is not science – it is a precursor to science. Science only begins when we start seeking for interconnections between different pieces of data, finding rules which reduce the data to something more compact than a mere listing of it.

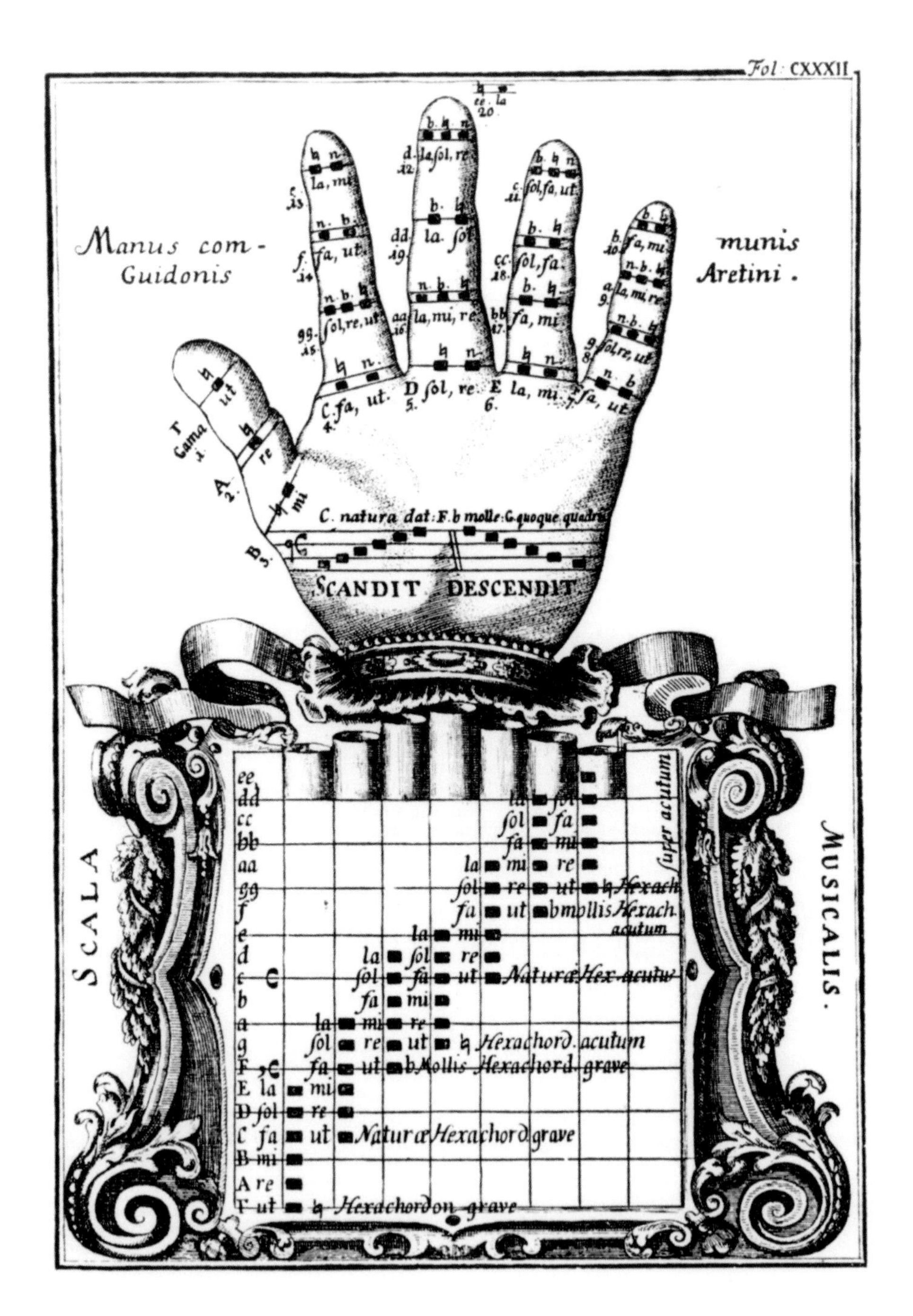

The Guidonian Hand. In the eleventh century, musical theorists started to use the hand as a means of remembering the n musical scale. It was Guido of Arezzo who first developed what we now recognise as the musical staff and systematic not the two are combined in the picture above. It illustrates the sequence of twenty notes on the scale marked by Γ, A, B, C, a, b, c, d, e, f, g, aa, bb, cc, dd, ee (which are our modern G, A, B, c, d, e, f… ć, đ, é). The mnemonic begins with Γ at the thumb, runs down and across the base of the fingers, up the little finger, and then moves in an anti-clockwise spiral th fingertips, ending with dd on the middle finger and, finally, ee, which is specially added above the middle fingertip.

DOE, RAY, ME
THE MUSICAL STAFF

Doe, a deer, a female deer
Ray, a drop of golden sun
Me, a name I call myself,
Far, a long, long way to run
Sew, a needle pulling thread,
La, a note to follow Sew
Tea, a drink with jam and bread,
That will bring us back to Doe . . .
Richard Rodgers and Oscar Hammerstein
The Sound of Music

THE musical staff (or 'stave') is instantly recognisable all over the world. It could even be argued that it was the first 'graph' because it maps the variation in the pitch of sound, vertically, against the passage of time, running horizontally from left to right. It formed an important piece of a medieval student's basic education, along with reading, writing, and arithmetic.[122]

At first, early Western church music was performed from memory but it was essential that everyone was in possession of the same memories. The inevitable discrepancies led to regional styles within the same country, and significant difficulties with coordination and reproducibility. Musical experience had a tendency to remain local knowledge. And when it did travel, purists would complain about distortions of the Gregorian chant by singers elsewhere, and the differing patterns of Latin speaking contributed further unscripted variations.[123]

All this variation did not sit well with the underlying monotheistic Christian theology that all things manifest the character of one God. Variation and non-uniqueness meant that somebody was not doing it right. Unity was undermined. At first, a trace of simple marks, like dots, lines, and hooks, were used informally by monks to guide them through their chants, but the invention of the staff of horizontal lines (four at first, then five), and the systematisation of the musical notation, was the project of an energetic eleventh-century Benedictine choirmaster, now known as Guido of Arezzo (*c.*991–*c.*1033).

Guido introduced his notations so as to facilitate the rapid learning of plainchants among his choristers. His method spread widely through northern Italy, but not without attracting some hostility from fellow monks at his home monastery in Pomposa, near Ferrara in Italy. As a result, he left and moved to Arezzo, a town without an abbey but with many singers wanting to be trained, and so thereafter he was known by the name of that town.

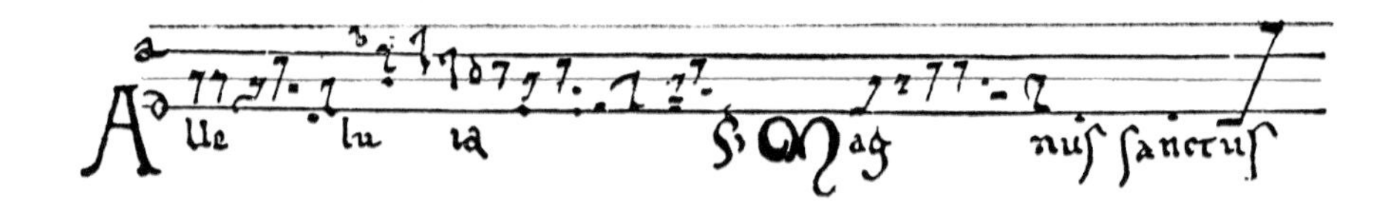

Neume notation of Guido of Arezzo, the Italian monk and musical theorist. The coloured lines fix the relative position of the signs. Red indicates F as the tonic. Yellow indicates C as the tonic.

Guido's further claim to musical fame is that he noticed that the ascending and descending tones marked by the lines of his staff were a match, in the right order, to the first syllables in the words of a very familiar hymn, 'Ut queant laxis', a plainchant hymn to John the Baptist, written in the eighth century by Paulus Diaconus of Lombardy. It is sung in the Roman Catholic Church on 24 June, the Feast of the Nativity of John the Baptist, in three parts at Vespers, Matins, and Lauds. The first Latin stanza reads:

> **Ut** queant laxis **Re**sonare fibris
> **Mi**ra gestorum **Fa**muli tuorum
> **Sol**ve polluti **La**bii reatum
> Sancte Ioannes.[124]

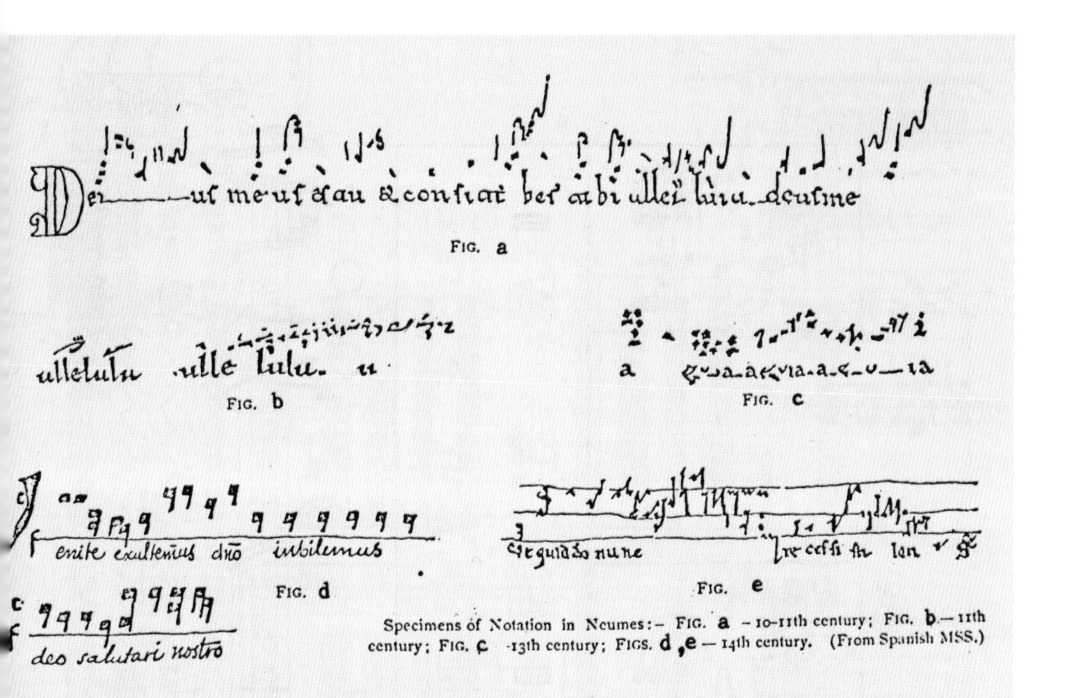

FIG. a
FIG. b
FIG. c
FIG. d
FIG. e
Specimens of Notation in Neumes:– FIG. a – 10–11th century; FIG. b – 11th century; FIG. c –13th century; FIGS. d, e – 14th century. (From Spanish MSS.)

A selection of musical scores from the tenth to fourteenth centuries.

Guido's mnemonic for the ascending scale was Ut, Re, Mi, Fa, Sol, La.[125] If you knew the hymn (and all singers in the Roman Church did) then you would know the notes to sing these syllables at and the eye and the ear were united.[126] The sequence also played the role of solemnising the musical performance.

Some small changes followed. 'Ut' was replaced by a hanging syllable that ended in a vowel, thus providing a nicer open sound for the voice. The choice was 'Do' and was probably taken from the first letters of *Dominus* ('Lord').

Originally, there were only these six notes in the medieval scale, but eventually 'si' was taken from the 'Sancte Ioannes' of the last line and used to create a seventh scale. Later, it was changed to 'ti' so that the last two lines could begin with different letters of the alphabet. And that is how Guido of Arezzo set the scene for *The Sound of Music*.

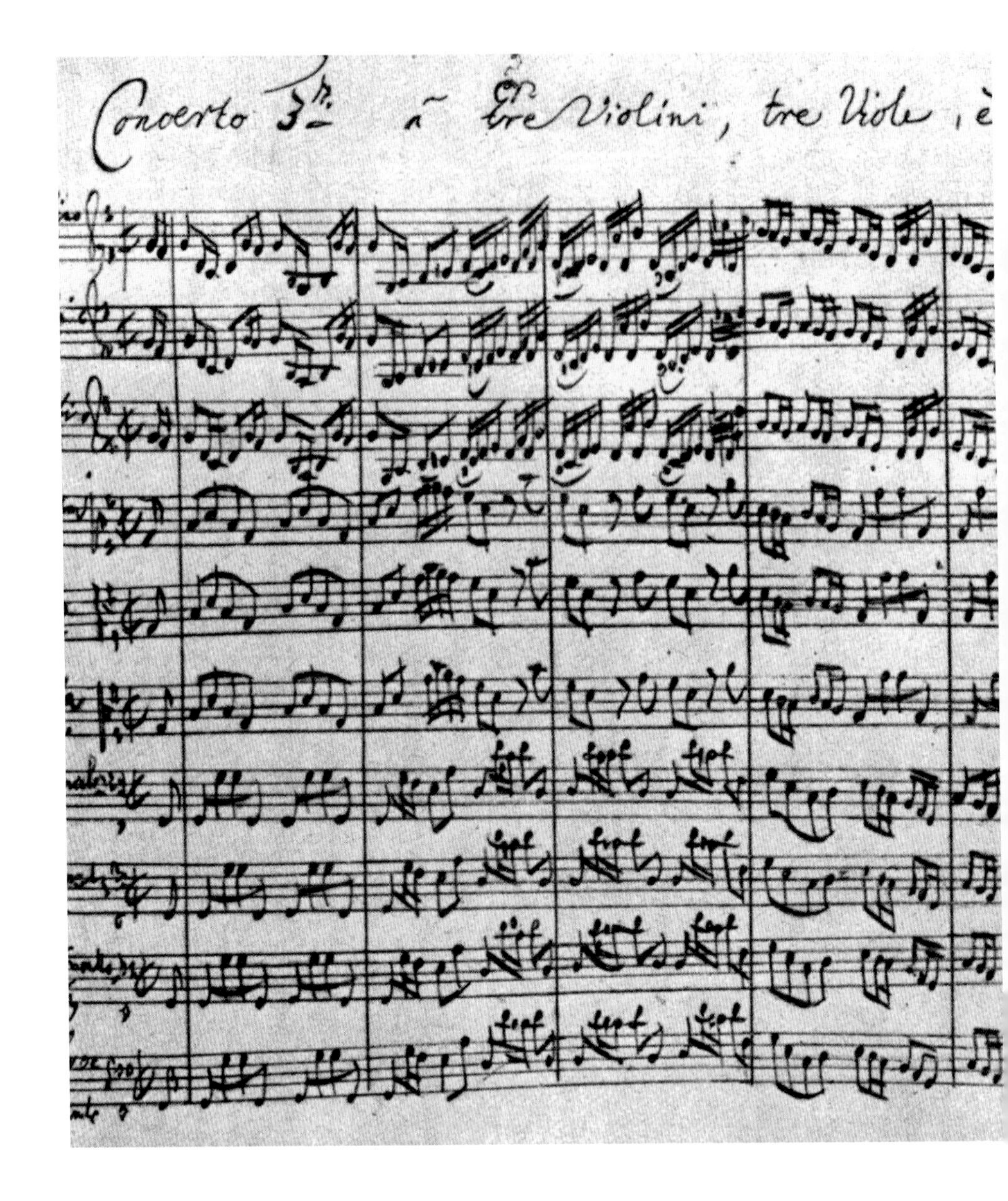
Concerto 3 a tre Violini, tre Viole, è

Ut que-ant la - xis re-so-na-re fi-bris Mi - ra-ge-sto - rum fa-mu-li tu-o - rum,
Sol - ve - pol-lu-ti La-bi-i-re - a-tum San - cte Jo-an-nes

(*Top left*)
Autograph score of J S Bach's Brandenburg Concerto III. *Storia Della Musica*, second volume, A. Della Corte.

(*Bottom left*)
Ut queunt laxis, an eighth-century plainchant hymn to John the Baptist, is a mnemonic for the extended Guido scale.

(*Above*)
A medieval illuminated score.

for himself as to the question of the enclosure of a square, and of a cube.

He would say the square A, in Fig. 96, is completely enclosed by the four squares, A far, A near, A above, A below, or as they are written A*n*, A*f*, A*a*, A*b*.

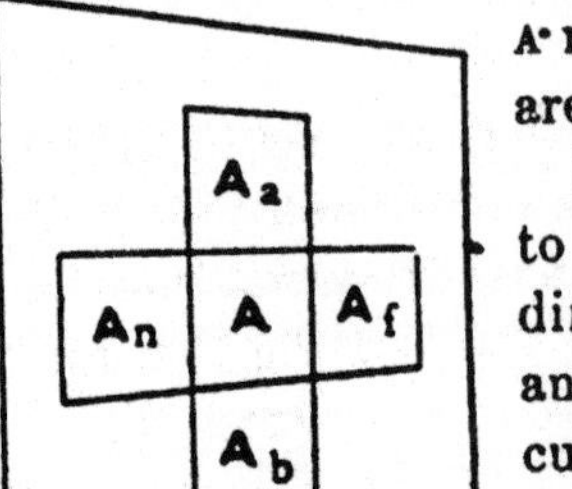

Fig. 96.

If now he conceives the square A to move in the, to him, unknown dimension it will trace out a cube, and the bounding squares will form cubes. Will these completely surround the cube generated by A? No; there will be two faces of the cube made by A left uncovered; the first, that face which coincides with the square A in its first position; the next, that which coincides with the square A in its final position. Against these two faces cubes must be placed in order to completely enclose the cube A. These may be called the cubes left and right or A*l* and A*r*. Thus each of the enclosing squares of the square A becomes a cube and two more cubes are wanted to enclose the cube formed by the movement of A in the third dimension.

The plane being could not see the square A with the squares A*n*, A*f*, etc., placed about it, because they completely hide it from view; and so we, in the analogous case in our three-dimensional world, cannot see a cube A surrounded by six other cubes. These cubes we will call A near A*n*, A far A*f*, A above A*a*, A below A*b*, A left A*l*, A right A*r*, shown in fig. 97. If now the cube A moves in the fourth dimension right out of space, it traces out a higher cube—a tesseract, as it may be called.

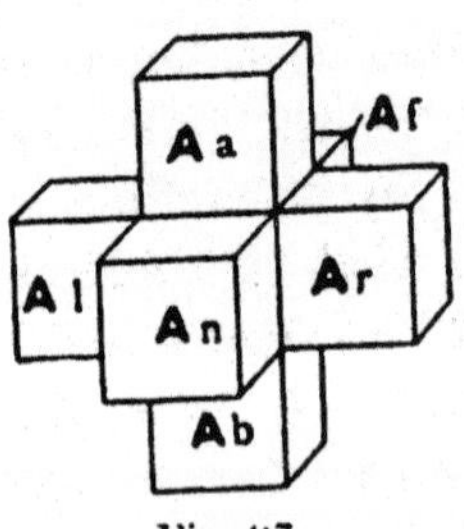

Fig. 97.

Charles Hinton's visual representation of what an unwrapped four-dimensional cube would look like, in three dimensions, by analogy with the two-dimensional appearance of an unwrapped three-dimensional cube, from his *The Fourth Dimension* (1904).

HYPERCUBES

VISUALISING OTHER DIMENSIONS

Hinton married a daughter of the logician George Boole. Unfortunately, having failed to grasp the concept of an either/or proposition, he was convicted of bigamy.

Stephen Eschenbach[127]

HOW can you visualise the fourth dimension? The answer was first supplied by a most unusual English mathematician, Charles Howard Hinton, who worked at the US Patent Office in Washington, DC, at the same time that Einstein worked in the Swiss Patent Office in Bern. His progressive father, James Hinton, had been a surgeon and a charismatic religious philosopher preaching free love and open polygamy; not a recipe for advancement in Victorian England.[128] Alas, young Charles was more interested in polygons than polygamy, and after studying at Rugby school and Oxford, he became a mathematics teacher at Cheltenham Ladies' College and then moved on to Uppingham School. His first published essay, 'What is the fourth dimension?', appeared in 1880 while he was a school teacher.[129] Thereafter, his life became breathlessly exciting. He had clearly listened to the sermons of his father, because in 1885 he was arrested for bigamy. He had married Mary Boole, widow of George Boole, one of the creators of logic and set theory, but then married Maude Weldon as well. Imprisoned for three days, on his release he left for Japan and then travelled on to the USA with Mary, was hired as an instructor at Princeton, and even found time to invent the automatic baseball-pitching machine – a form of high-power rifle that had the effect of petrifying the receiving batter and baking the baseballs with the hot gases.[130] After being fired from this post at Princeton, he moved on to the US Naval Observatory for a period, before coming to rest at the United States Patent Office. His book on the 'Fourth Dimension' appeared in 1904, and probably seemed rather conservative given all that had gone before. But Hinton managed to do something that was visually spectacular: he enabled people to visualise the fourth dimension.

Hinton's memorable contribution to the study of higher dimensions was the series of simple pictures that he created to show how it was possible for us to gain a shadowy impression of what four-dimensional objects would look like. He noticed that the pictures we see in books of real three-dimensional objects are always two-dimensional – flat on the page – and so we should be able to predict what a three- or two-dimensional picture of a four-dimensional object would look like in projection. This image might be its shadow or its projection or an unfolding, just as we would get by cutting along some of the edges of a hollow cube and then laying the unfolded paper as a T shape on a flat two-dimensional surface.

Hinton's ideas for visualising the fourth (and higher) dimensions by extrapolation and analogy were hugely influential, and, in 1909, *Scientific American* magazine offered a $500 prize for the best popular explanation of the fourth dimension. In Europe, a similar fascination with multidimensional perspectives emerged in the world of art. The Cubists seized upon the fourth dimension as a justification for the use of multiple perspectives, most famously in Pablo Picasso's *Portrait of Dora Maar*, although (unsurprisingly) he always denied he had been influenced by developments in mathematics and Einstein's theory of relativity.[131]

Hinton invented the word 'tesseract'[132] to describe a four-dimensional cube viewed in projection. The most famous representation of the unfolded tesseract was used by Salvador Dalí as the Cross in his famous 1954 picture *Corpus Hypercubus*. The American mathematician Thomas Banchoff was contacted by Dalí about the geometrical properties of the unfolding of the hypercube before this image was created.

Salvador Dalí's *Corpus Hypercubus* (1954).

Tesseracts have also found their way into the written arts. In 1940, Robert Heinlein wrote a famous science fiction short story[133], 'And He Built a Crooked House', about a three-dimensional house that was built as a projection of a

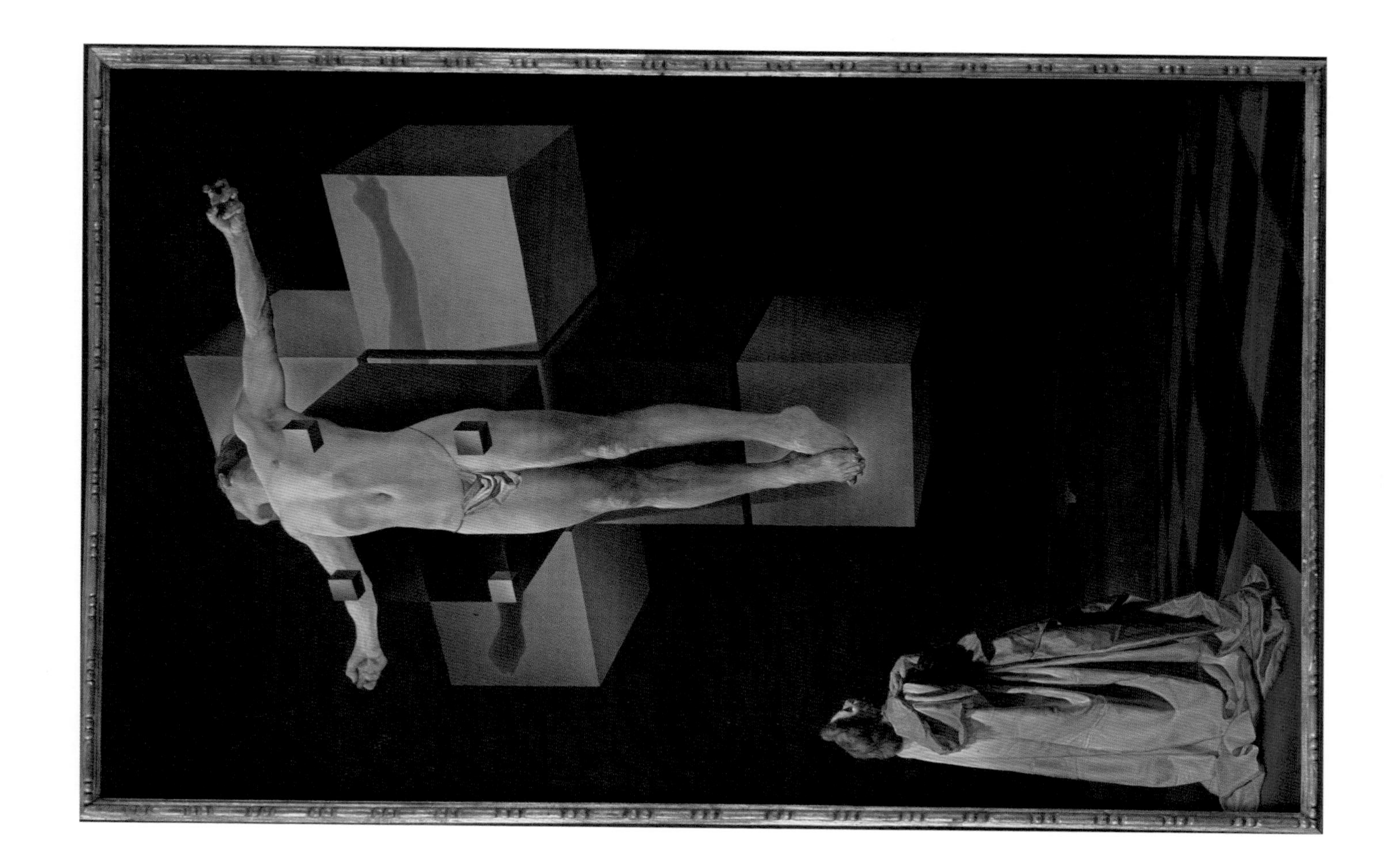

tesseract. Unfortunately, it changes into a real tesseract, with strange results for the occupant.[134]

Hinton's pictures came at just the right time. Before 1900, most talk of higher dimensions was bound up with the strong Victorian interest in spiritism and psychic phenomena. But the ability to visualise higher dimensions had a dampening effect on these occult interests. Higher dimensions became just that: more of what we had already. They were part of geometry and – thanks to Einstein – physics. They were not qualitatively or unimaginably different from the space we inhabited. Remarkably, despite all the attention paid to geometry by the world's leading mathematicians of the time, it was the playful pictures of a comparative amateur that gave us the most vivid images of higher-dimensional worlds. Hinton himself was proud to be a true *amateur*, who could in the end express concern that:

> It seems to me that the subject of higher space is becoming felt as serious . . . It seems also that when we commence to feel the seriousness of any subject we partly lose our faculty of dealing with it.[135]

COSMIC IMAGERY JOHN D. BARROW

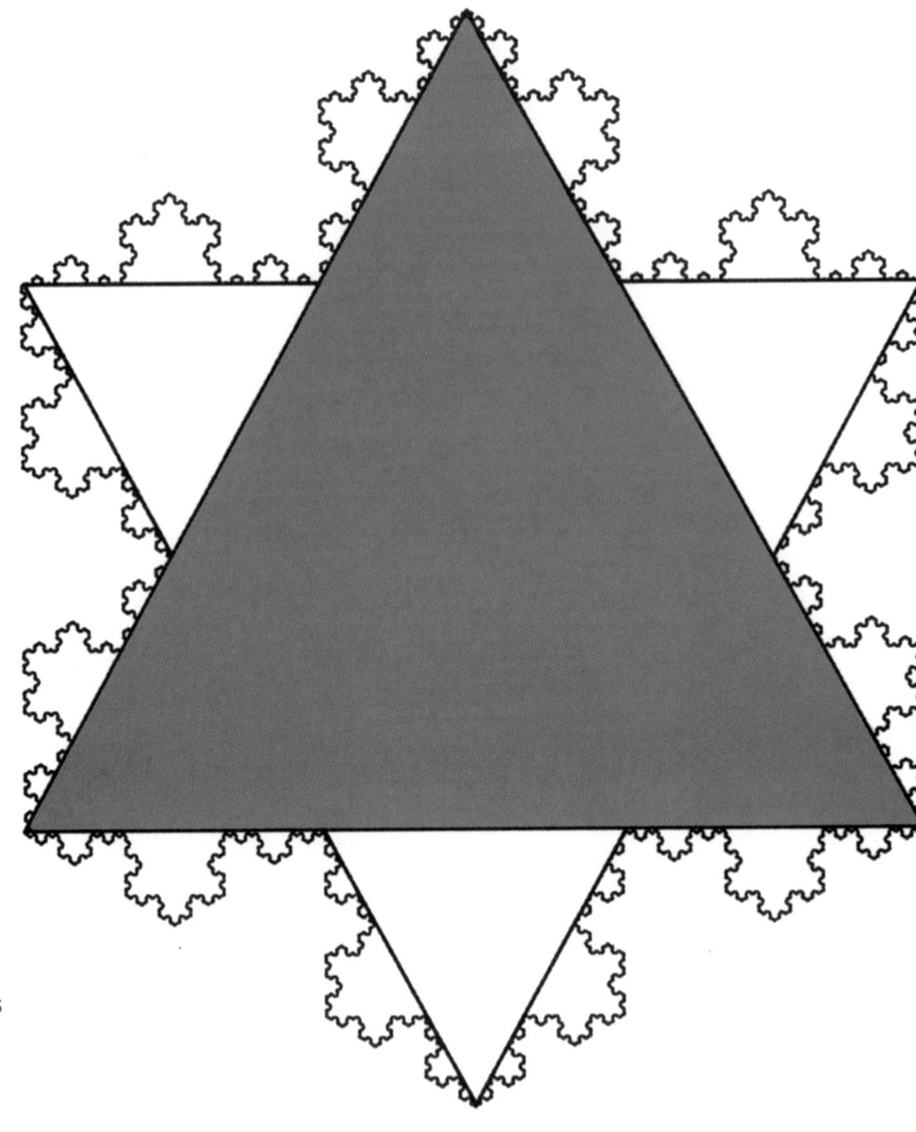

Helge von Koch's triangular 'snowflake' construction erects an equilateral triangle on the central third of every side of every triangle so generated, continuing like this *ad infinitum*.

TAKE IT TO THE LIMIT

KOCH'S SNOWFLAKE, SIERPINSKI'S CARPET, AND MENGER'S SPONGE

The fact that there are some people who are clever at Mathematics but less successful in subjects like Physics, is not due to any defect in their powers of reasoning, but is the result of their having done Mathematics not by reasoning but by imagining – everything they have accomplished has been by means of imagination. Now, in Physics there is no place for imagination, and this explains their signal lack of success in the subject.

René Descartes[136]

ONE of the most extraordinary things about mathematics is that it supplies recipes for creating structures that have completely counterintuitive properties. A beautiful example that has played an important role in our attempts to understand the geometry of Nature is a creation called the Koch snowflake. It was devised by the Swedish mathematician Helge von Koch (1870–1924), who published it first in 1904.[137]

Start with an equilateral triangle and apply a simple rule to each of its three sides: erect another equilateral triangle on the centre one third of each side of the triangle. Now just keep on applying this simple rule to the sides of the three new triangles. At each step of this process each triangle that has already been drawn will produce another three triangles, each with sides one third of the size of the last. Very soon, we have created a finely crenellated structure that

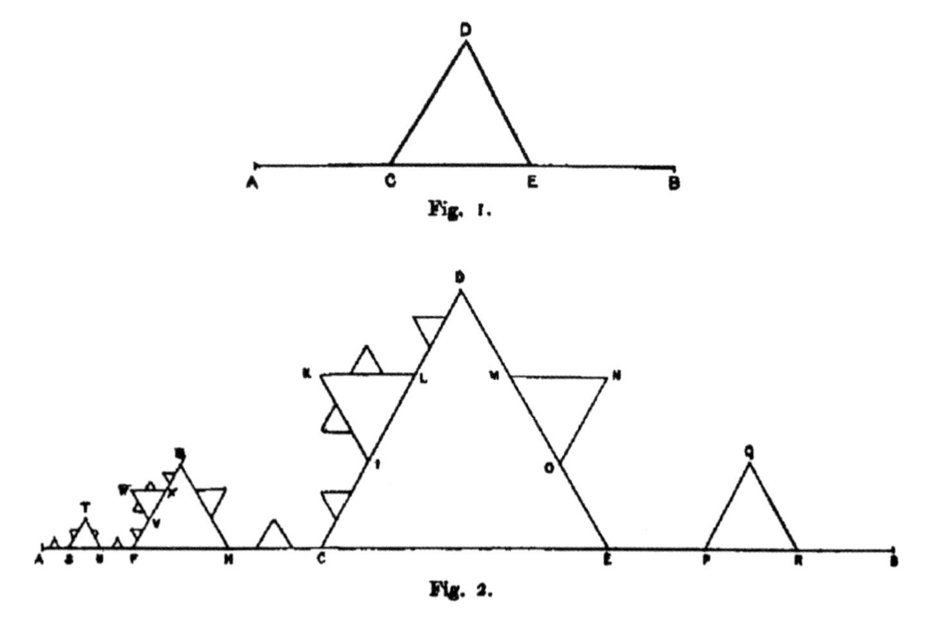

Koch's original description, published in 1904. 'Sur une Courbe sans tangente, obtenue par une construction géometrique élémentaire', *Arkiv för Matematik 1* (1904).

zig-zags around the original triangle. This process of triangular division and multiplication could be continued for ever, making ever smaller and finer zig-zags around the perimeter of the structure. This infinitely continued structure is Koch's 'snowflake'.

The name 'snowflake' derives from its superficial resemblance to the symmetric branches that appear on a snowflake. All snowflakes have a design plan based on six symmetrically placed arms, but those arms continue to branch in novel ways on a smaller and smaller scale as we look more closely at the snowflake. Close scrutiny shows they are not quite symmetrical either. Koch's snowflake lacks the complexity and uniqueness of real snowflakes but, unlike physically real snowflakes, it can continue branching for ever, and in this limiting situation it has some very surprising properties. The perimeter of the Koch snowflake would ultimately possess an infinite number of zig-zags after the triangle trisection had been applied an infinite number of times.[138] In this limiting situation, the area enclosed by the zig-zagging perimeter remains finite, but the length of the perimeter will be infinite.[139] If the area of the initial triangle that starts the process is D and the length of each of its sides is 1, so D equals $\sqrt{3}/4$, then after n trisections of the edge lengths, the total length of the perimeter is:[140]

$$\text{Snowflake perimeter after n steps} = 3 \times (4/3)^n$$

while the total area enclosed is:

$$\text{Snowflake area after n steps} = D \times [1 + 1/3\ \Sigma_r (4/9)^r]$$

where the sum Σ_r is taken over values of r from 0 to n. It is easy to see that as n approaches infinity the perimeter approaches infinity (because 4/3 is bigger than 1), but the area enclosed by it approaches a finite value equal[141] to $8D/5 = 2\sqrt{3}/5$. To get an idea of how fast things grow, after 100 steps the length of each straight-line segment on the perimeter has reduced to 2×10^{-48}, but the total length of the perimeter has grown to 3×10^{12}. If we start with a triangle whose sides are all 1 cm in length, so the perimeter is 3 cm, then the perimeter of the snowflake will have grown to become bigger than the circumference of the Earth after only seventy steps.

If we end the snowflake construction after any number of divisions then the shapes that have been created can be joined together to produce a tiling pattern that will completely cover an infinite plane surface without leaving any gaps if we use two sizes of snowflake tile with areas in the ratio of 3:1.

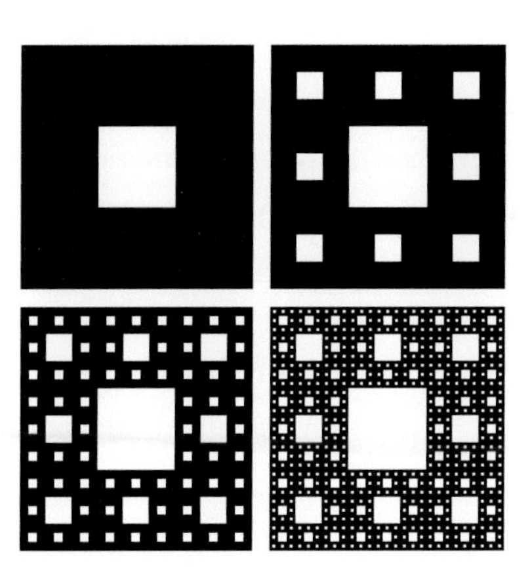

Sierpinski gasket (*left*) and carpet (*right*). A triangle and square are hollowed out by removing small scale copies of themselves on smaller and smaller scales indefinitely.

Menger's Sponge applies Sierpinski's construction to a solid cube.

Koch's creation spawns a host of related infinitely self-reproducing geometric patterns. The Polish mathematician Waclaw Sierpinski (1882–1969) was a world-renowned mathematician who, in 1915–16, contributed some more beasts to the mathematical menagerie that Koch had conceived.[142] 'Sierpinski's carpet' is formed by beginning with a square, divided into nine smaller squares as if marked out for a game of noughts and crosses. Now remove the centre square and continue the process of dividing every square that remains into nine and removing the central part. Carry on for ever. The original square gets chequered on smaller and smaller scales.[143]

Sierpinski devised another striking example that can be seen in some pieces of historic art. It has become known as the 'Sierpinski gasket'. It starts with an equilateral triangle, subdivides it into four equilateral triangles and then removes the middle triangle. The process then continues dividing each of the three remaining triangles in the same way for ever. The Sierpinski gasket is the collection of points on the page which remain if the process is continued infinitely (including the sides of the original triangle).

The last step in these early developments was taken by the Austrian mathematician Karl Menger, in 1926. He provided the recipe for a three-dimensional version of Sierpinski's structures, now called 'Menger's sponge'. We start with a solid cube, subdivide it into 27 sub-cubes, and then drill a square hole through the central square of each face until it emerges from the other end face of the cube. Now subdivide the 8 smaller squares that remain on each outer face into 9 smaller squares and drill a square hole through the centre square to its opposite face. Carry on for ever and what remains after an infinite number of drillings is Menger's sponge.

The Sierpinski carpet and the Menger sponge have a further striking property. Imagine all the most complicated curves that could be drawn on a plane surface. Then, so long as they have a bounded size, so that they fit on a finite piece of paper, they will be found somewhere within the Sierpinski carpet. It is like an infinite catalogue of all possible curves there can be. No matter how complicated you try to make them by adding finer and finer levels of twisting and turning them, so long as you are allowed to stretch and shift the curve without tearing it,[144] so that it fits the angles of change in the Sierpinski carpet, you will eventually come across it. Deformed versions of all possible bounded curves 'live' inside the carpet. If we allow the curves to move around in three dimensions, but remain confined within a finite volume of space, then the same remarkable property is possessed by the Menger sponge: you will eventually find a deformed version of any curve, no matter how complicated, within it.

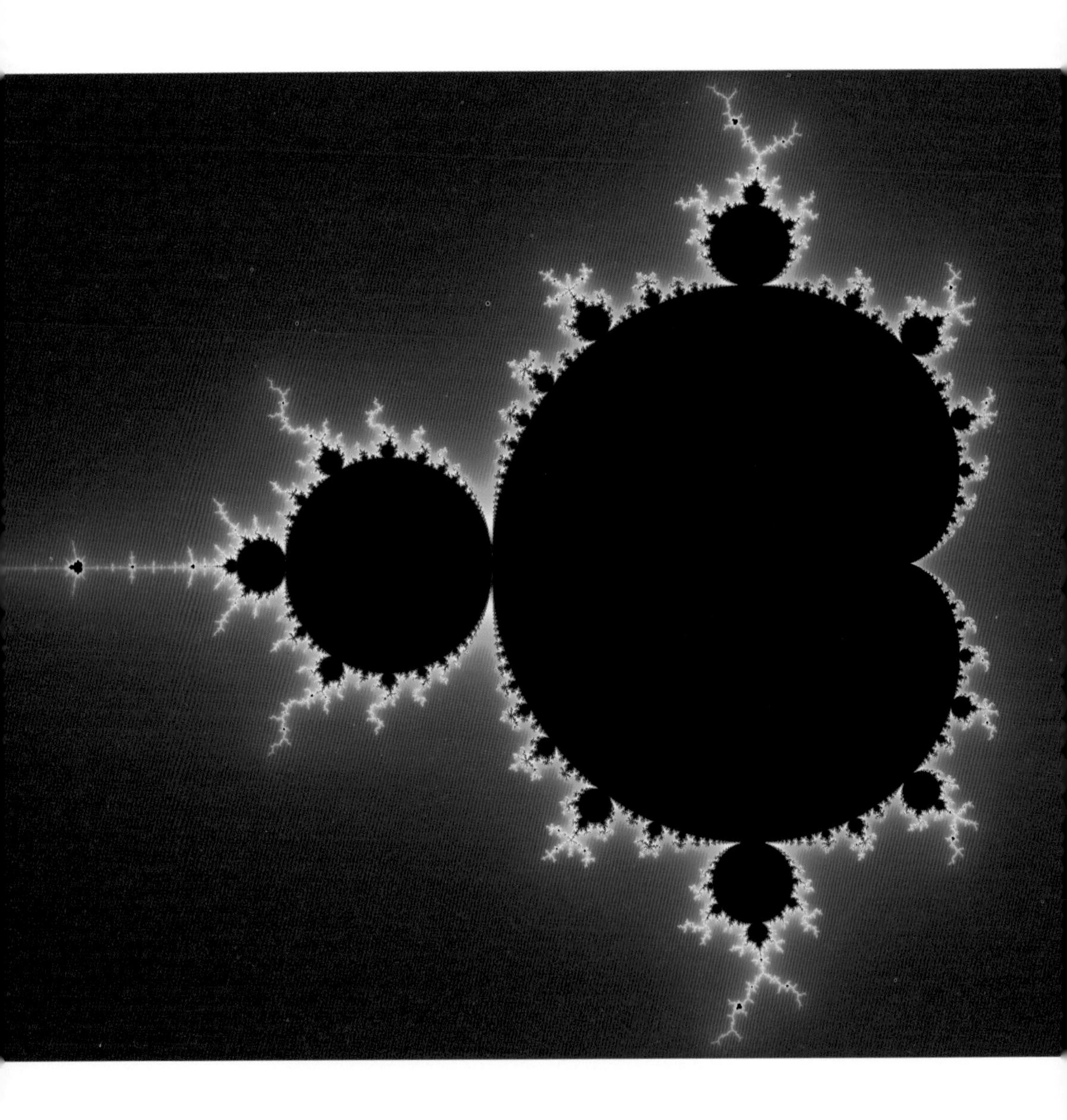

The Mandelbrot set is the black region and has an infinitely intricate boundary. Copies of the whole structure can be found on any scale of the boundary, no matter how small.

WORLD WITHOUT END

THE MANDELBROT SET

Life is complex. This is because it has a real and an imaginary part.

Linux Forum[145]

A REMARKABLE revolution occurred in science and mathematics during the second half of the 1970s. It was not the sort of revolution that philosophers of science ever had in mind. It was not an overthrow of an old paradigm by a new. In fact, it was more like an *industrial* revolution. Previously, computers were the sole preserve of large and well-funded research groups. They were enormously expensive and were used to study 'big' block-busting problems, such as understanding how stars form, forecasting the weather, or simulating nuclear explosions. Gaining access to them was a bit like getting time to observe on a large telescope. Formal applications had to be made from institutional addresses, the proposed use had to be evaluated and rated against other proposals, and if you were eventually successful, you still had a long wait and a pretty frustrating time trying to get the machine to do what you wanted without it crashing.

All this changed rather quickly. The invention of personal computers ('PCs') suddenly gave everyone access to remarkable computing power, at low cost,[146] with easy interactivity, good graphics, and all at desktop size. This is what stimulated the fast growth in the study of what scientists came to call 'complex systems'. Subjects such as 'chaos', 'strange attractors', and 'fractals' became part of the vocabulary of science simply because of the PC revolution.

Some physical problems – for example, the dripping of a tap – can be described by very simple mathematical equations, yet the solutions of those equations display behaviour so complex and unpredictable that they cannot be solved exactly using pencil and paper. The best means of attack is to study how they behave 'experimentally' by using computers. Personal computers fostered the creation of this entire field of experimental mathematics, allowing individuals and small research groups to investigate complex behaviours by watching high-quality computer images in real time. Their easy interactivity allowed for an exploration of different possibilities, altered starting conditions, and generalisations, all at the touch of the keyboard. Visualisation became important. You could look and 'see' what was happening.

These investigations had simply never been practical before. In retrospect, we recognise a turning point in the history of mathematics. The two-fold division of mathematics into 'pure' and 'applied' became a trinity of pure, applied, and 'experimental' mathematics.

The most spectacular discovery that emerged from this changing culture was one of the first. The French mathematician Benoit Mandelbrot had worked for IBM at Yorktown Heights, New York, since 1958. His style of mathematical work was unusual for a pure mathematician – especially a French one – because he was keen to use computation as a guide and a tool in all sorts of mathematical investigations that traditionally lay in the domain of pure thought aided and abetted by heavy algebra.

In the period 1979–80, Mandelbrot began to explore the properties of a seemingly simple mathematical rule and his surprising results began to be announced to other mathematicians in November 1980. The mathematical rule in question takes a point on a plane area of space (labelled by z) and moves it to the location labelled by $z^2 + c$, where c is a constant.[147] Suppose we apply the rule to one point, where $z = 0$ say, then what remains, c, gets moved to $c^2 + c$. Apply the rule again and again and this point gets moved to $(c^2+c)^2+c$, and then to $[(c^2+c)^2+c]^2+ c$, and so on. If we had started at a point other than $z = 0$, then the sequence would have been different. In fact, for a very small change in the starting point we end up with a very different sequence of movements of the point in the plane. This is a familiar feature of a chaotically sensitive system.

As you explore the consequences of these repeated applications of the rule you start to notice for some values of c, and for some starting positions, that the resulting outputs just keep getting bigger and bigger, and the locations cannot fit on the paper any more. For example, with $c = 0$ if the starting value of z had been 3 on the x-axis, then after eight applications of the rule, z would have grown to $3^{2^8} = 3^{256} = 10^{122}$, vastly bigger than the number of particles in the entire visible universe.[148] However, if we had started with a point that was closer to zero than a distance of 1, for example $z = 0.5$ and $c = 0$ still, then any number of successive applications of the rule would have kept the location of z inside the circle of radius 1 centred on $z = 0$. A similar divide occurs as the value of c is chosen differently. For some choices, the values of z get farther and farther away – we say they are *unbounded* because if you first draw *any* sized circle around the starting value of z, then eventually, after enough applications of the transformation rule, the value of z will go outside this circle no matter how big its radius. In contrast, there are other choices of c for which the successive values of z do not wander off to ever-larger values: we say they are *bounded.*

Mandelbrot wanted to know what the collection of c values (or, equivalently, positions on the page) looked like which ensured that values of z stayed bounded. It is a simple question, but it turns out to have an answer that is in one sense infinitely complicated. In general, the collection of points for which the results of repeatedly applying our rule to the starting position $z = 0$ any number of times remain bounded in extent has the remarkable shape shown in the picture.

It is known as the 'Mandelbrot set' and its exquisite structure has been displayed and analysed more than any other mathematical picture of recent times. It has featured on the front cover of many books of mathematics and has inspired computer-generated art and science fiction all of its own. Why is it so special?

The power of computer graphics enables us to map out the frontier of the Mandelbrot set with remarkable precision. It is the black region of the picture. At first glance it appears to consist of a large heart-shaped region with a number of smaller disk-shaped regions attached to it, getting smaller and smaller as we move off to the left. The different colours around the boundary represent the speed with which the nearby points beyond the Mandelbrot frontier become unbounded under repeated application of the mathematical rule. But if we magnify the picture we make a remarkable discovery. Each part of the boundary of the black regions has a structure of unending intricacy that reproduces the same string of heart- and disk-shaped regions in miniature. If we keep looking with greater and greater magnification at smaller and smaller parts of the frontier of the Mandelbrot set, there is always the same structure repeating itself over and over again on finer and finer scales.

This is the hallmark of fractal geometry. The same pattern is reproduced in shape over and over again on smaller and smaller dimensions. But there is more to the Mandelbrot set than this. Other twisting shapes, like sea horses chasing their tails, can be found repeating themselves as well. The overall complexity of the boundary has been shown to be as complicated as any curve on a flat

Part of the top boundary of the Mandelbrot set reveals copies of the whole set.

surface can ever be.[149] These structures contain never-ending small-scale replicas of themselves even though the Mandelbrot set as a whole is not completely reproduced in miniature. Remarkably, the simple mathematical rule we started out with has generated this bottomless sea of complexity.[150] Whereas Koch's snowflake constructed a large-scale pattern by copying a small template over and over again, here the large-scale structure is found to contain reproducing templates, and much else besides, within it.

There is another vein of intricacy to mine out of Mandelbrot's set. Suppose we fix the value of c, but start the point z at a position other than $z = 0$, initially. For what values of the starting points do the locations remain bounded? For each possible choice of c these values form what is known as the interior of the 'Julia set' of c, after Gaston Julia,[151] the French mathematician who, along with Pierre Fatou,[152] separately discovered many of these features of this mathematical transformation rule first in 1918. These Julia sets form the dividing lines between points in the plane that stay bounded and those that disappear off towards infinity as the transformation rule is applied again and again. As the value of c is changed, the variety of Julia sets is stunning. Like the flora of some fictional world, some form spirals and lattices, wisps and spider's webs, while others float like clouds of disconnected points. The Mandelbrot set is made up of all the Julia sets which are single connected pieces.

Mandelbrot's set and the associated Julia sets have become exhibits in art exhibitions extolling the fractal beauty of Nature, while others[153] have displayed them as evidence for the 'reality' of mathematical structures. But, above all, they show us how deep and unexpected the outcomes of very simple instructions can turn out to be. Perhaps the staggering complexity of our minds and bodies might not be such an unapproachable problem after all.

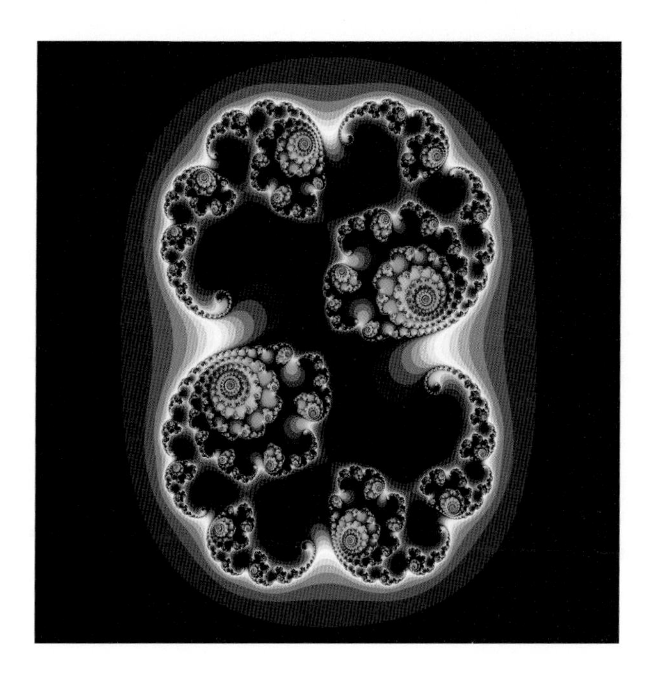

Julia set for the Mandelbrot set generated by the choice of the quantity c in the transformation $z \rightarrow z^2 + c$ as the complex number $c = 0.285 + 0.01i$. The Julia set constitutes the boundary between the starting values for this transformation that escape to infinity and those that remain in a finite region.

Oscar Reutersvärd, Impossible Triangle of *Opus 1 no. 293aa* (1934).

WHERE THINGS HAPPEN THAT DON'T

THE IMPOSSIBLE TRIANGLE

Impossible objects are phenomena which cannot exist but which we can see all the same.

Bruno Ernst[154]

IN 1934, the Swedish artist Oscar Reutersvärd drew the first impossible triangle from an unusual arrangement of nine cubes. He realised that there was an entirely new realm of art to be explored, and enthusiastically set about creating hundreds of impossible figures, or 'illusory bodies' as he called them, of all shapes and complexities, devising impossible cubes and spiral staircases, forks and buildings.

Reutersvärd's impossible triangle was reinvented in 1954 by Roger Penrose and published in an article with his father, the psychologist Lionel Penrose, who had seen the ways in which the Dutch artist Maurits Escher had used 'two-dimensional drawings to convey the impression of three-dimensional objects'.[155] Roger Penrose had been inspired to think in these directions after attending a lecture given by Escher earlier that year, but had not encountered the work of Reutersvärd and others, who had been incorporating impossible constructions into their artworks. The Penroses' article was barely 340 words long and displayed two impossible triangular figures and two versions of an impossible ascending staircase. The solid version of the Reutersvärd triangle was subsequently often called the 'Penrose triangle'. Lionel Penrose was a distinguished psychologist and his co-authorship of the paper ensured that impossible figures attracted the attention of psychologists and researchers interested in the workings of the human visual system.[156] Unlike Reutersvärd's representation, it was drawn in perspective, which added a further paradox of size to the picture.

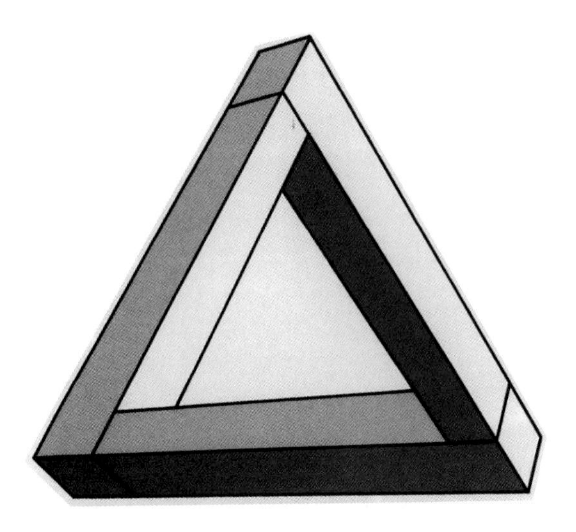

The Penrose Triangle (1954).

In 1961, Escher incorporated the idea drawn from the Penroses' version of the impossible triangle into his famous lithograph *Waterfall*. Here, two impossible triangles have been combined to create a single impossible scene. The waterfall is a closed system, but the water wheel is continuously rotated by it. If this were possible we would have a perpetual motion machine.

Escher wrote to his son, Arthur, in January 1960, about the inspiration he had received from the Penroses' article, but in the end he only drew four of these impossible figures. He devoted his greatest attention to the tiling and tessellations for which he became more famous.

The mystery of the impossible triangle is why the eye persists in trying to interpret it as a single solid object even when it is recognised as being a physical impossibility. The eye first sends the signal that this is a solid triangle, but quickly recognises that in reality this object cannot exist.[157] At this stage we might expect the eye to produce a different interpretation of the scene, perhaps as a collection of planes and lines, but it doesn't. It persists with the interpretation of the triangle as an object, albeit an impossible one.

Maurits Escher, *Waterfall*, lithograph (1961).

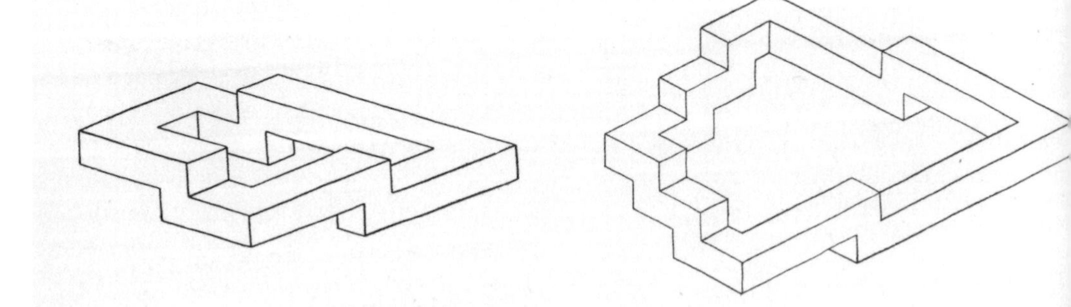

Oscar Reutersvärd's *Travelling Sketches* (1950), drawn on his forty-hour journey from Stockholm to Paris, inspired his creation of the endless staircase.

The impossible triangles (there are actually an infinite variety of them) and all their offspring – impossible cubes and staircases, impossible polyhedra and knots – have become an intriguing interface between the study of human perception and graphic art, joined by a mathematical force. Gradually, artists have gone further and further in the creation of the illusion, simulating illusory models of impossible objects embedded in deceptively realistic landscapes. These enticing creations of impossible realities have ensured that the impossible triangle has become an iconic symbol of the mysterious power of the mind to grasp something that is unreal. The triangle and its offspring are a ubiquitous presence in the world of images, cutting with equal force through art, psychology, and mathematics.

Rotating Snakes 3, by Akiyoshi Kitaoka.

SEEING IS BELIEVING
VISUAL ILLUSIONS

Art has to move you and design does not,
unless it's a good design for a bus.
David Hockney

VISUAL illusions have been a source of perennial fascination for scientists and artists alike over the past hundred years. The Op Art movement of the 1960s cleverly exploited the eye–brain ambiguities and cueing responses by making artworks which appeared to the brain to be dynamic because of the constant mental searching for patterns and lines. The works of artists such as Victor Vasarely and Bridget Riley used these visual responses to create new types of artwork, while biophysicists like Richard Gregory[158] have used them to learn new things about the way the eye and the brain work in unison.

Our visual system is constantly trying to find patterns and the simplest are lines that join up dots to their nearest neighbours. We are notoriously good at doing this, and can invariably 'see' lines in point patterns where no intrinsic linear features exist. Once a linear pattern is detected it will be continued in an anticipatory way, as points that fit an extrapolation are cued by the existing pattern. There are probably good evolutionary reasons why we are not only good at seeing patterns, but even a little too good – because we see them when they are not really there. After all, if you see tigers in the bushes when they aren't there, you will only be thought a little paranoid by your children, but if you don't see tigers in the bushes when they really are there, you may not live to have any children at all.

The dramatic creative power of modern computer graphics has enabled computer artists to create new types of visual illusion that are extremely disconcerting. They do not create movement because of simple visual ambiguity between two perspectives, like Necker's famous cube, an impulse which can always be resisted. Nor are they static impossible figures like those of Reutersvärd and Penrose. Rather, they create continuous movement that cannot be consciously resisted. One of the pre-eminent exponents of this type of mental manipulation with an artistic flourish is the Japanese vision expert Akiyoshi Kitaoka, who exploits the peripheral drift illusion (which makes the snakes appear to rotate) and spiral illusion (which makes grey concentric circles appear to be spirals) to create complex rotations on the page before you.[159] Our peripheral vision has the property of responding differently to regions of different brightness and colour, creating apparent motion from black to dark grey and white to light grey. By carefully arranging the adjacent colours the relative movements can be coordinated to create a range of startling systematic shifts, notably rotation and counter-rotation. Voilà!

TWO EASY PIECES

APERIODIC TILINGS

If we want things to stay as they are,
things will have to change.
Giuseppe Tomasi di Lampedusa[160]

WHEN you set out to tile your bathroom or driveway you will be driven by constraints of cost, time, and the finiteness of human patience to pick a tiling pattern that is periodic. The simplest is to take a pattern of identical squares or rectangles. Any decorative element can be introduced by varying colours or surface textures to overlay a pattern. Again, the one you choose will most likely be a symmetrical one. We like symmetry and it is an easy recipe to follow if you are the tiler. But one possible definition of a mathematician is someone who will *not* be content with a simple periodic bathroom tiling.

The Islamic cultures explored the aesthetic consequences of symmetry more deeply than any other. Their veto on the representation of living things drove their art in a different direction to that found in the Christian cultures of Europe. Elaborate tilings and tessellations cover space as fully as possible, leaving nothing empty. Great buildings, such as the Alhambra, in Granada, are almost overwhelming in their exploration of symmetry and periodicity. In fact, it was the sight of space-filling periodic tilings like these that first inspired the young Maurits Escher to give up the individual style of landscape art that he had begun developing to become a legendary artist of mathematical symmetry in the twentieth century.

Aperiodic tilings of the plane with two tile shapes were first discovered by Robert Ammann and Roger Penrose in 1974.

Despite their beauty, these Islamic tilings have a simple unifying feature. They repeat: they are periodic coverings of surfaces. A long-standing mathematical challenge has been to discover whether it is possible to tile a never-ending plane with a systematic tiling pattern that is *not* periodically repeating. One of the first scientists to investigate these possibilities was the great astronomer Johannes Kepler. He used combinations of polygons and star polygons to tile the plane. However, these tilings were not really aperiodic. The whole local tiling pattern made up of different shapes ends up being repeated periodically if you want to use them to tile the whole plane.

Once the shapes of the tiles are not required to be identical, or extremely symmetrical, it is not at all clear whether or not they will be able to tile an unlimited plane surface. In 1966, the mathematician Robert Berger proved that there is no systematic procedure which can tell whether or not a given set of tile shapes will succeed or fail to aperiodically tile the plane.[161] The problem is one that is too complicated for a mere computer to solve: using the same idea over and over again is not a powerful enough tool for the job. Earlier, it had been shown that there would be a computer program to achieve the tiling if the only tilings that could cover the plane were periodic ones.

Berger also managed to show that sets of aperiodic tiles did indeed exist. The first set that he found was huge: 20,426 different tile shapes would be needed to tile an infinite plane without any periodic repetition of the pattern. Later, he was able to reduce this tiling set to 104 different tiles. In 1968, the great computer scientist Don Knuth reduced the number to 92, before Raphael Robinson reduced the number to 35 and then, in a major advance in 1971, to just 6 tiles. Finally, in 1974, Roger Penrose found examples using 6, and then

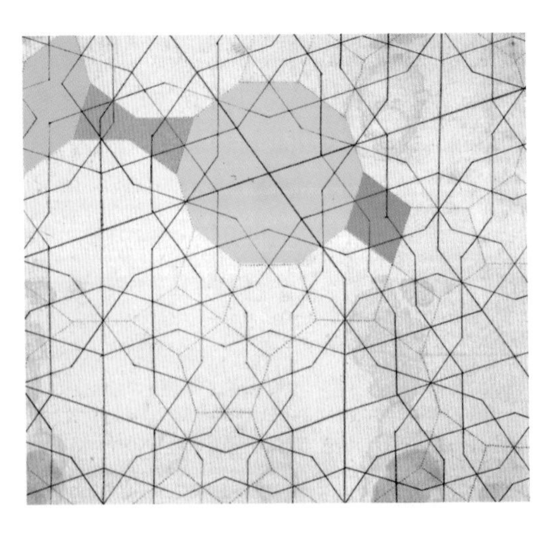

(*Opposite page*) Timurid-Turkman scroll now held by the Topkapi Palace Museum in Istanbul. The faint reddish lines outline the shapes of the underlying tiles. (*This page*) one example of each of the five traditional *girih* shapes has been shaded.

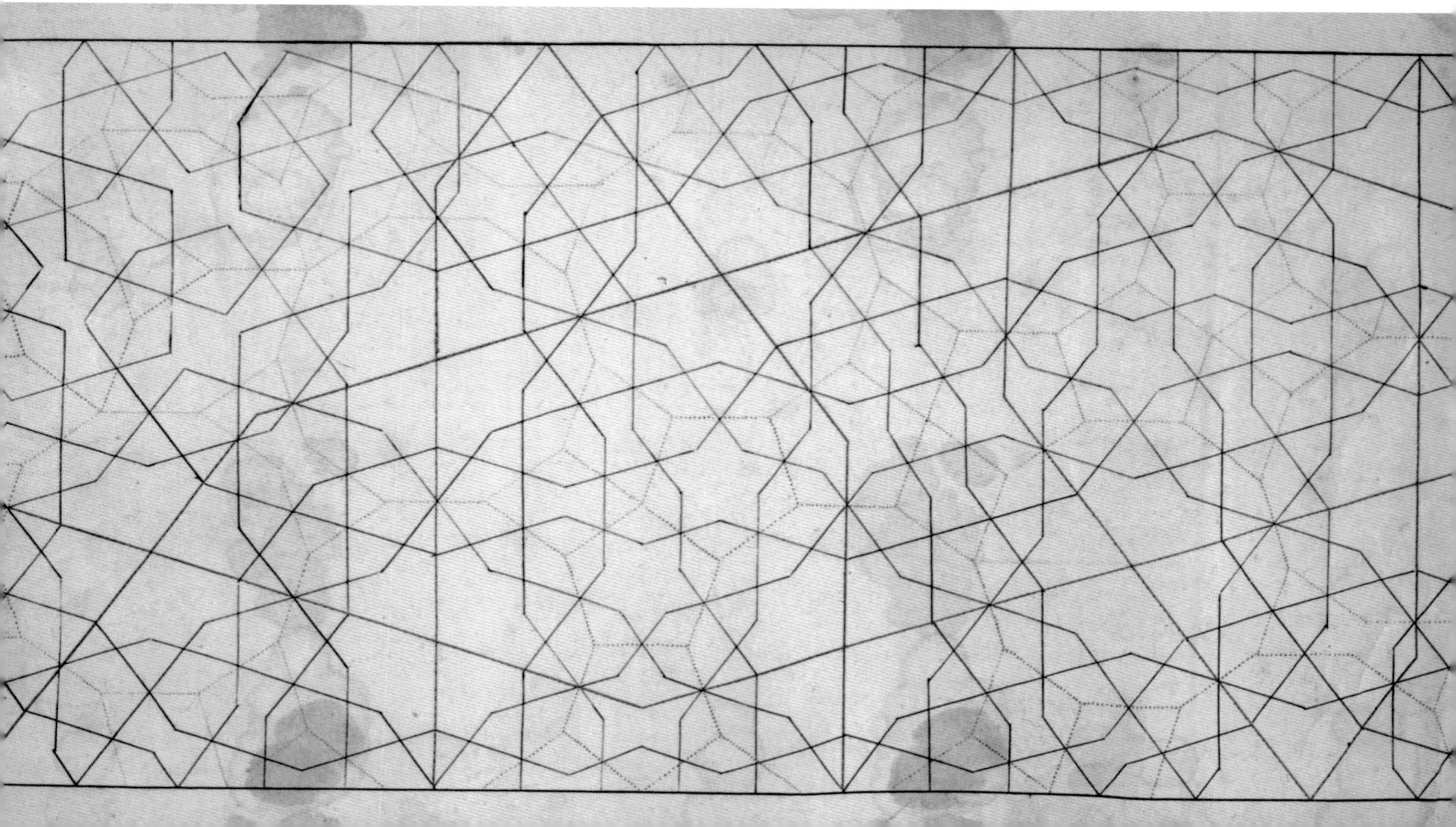

two cases that needed just 2 tile shapes, which could achieve an infinite non-periodic tiling (one pair of pieces were called 'kites' and 'darts', and ended up being commercially manufactured).[162] Robert Ammann, a remarkable American amateur mathematician who had a 'sixth sense' for intuitively seeing these tilings, and who worked sorting mail for the postal service after leaving MIT, found one of Penrose's pairs and a different pair independently.[163] After reading Martin Gardner's column in *Scientific American*, where the Penrose tilings were first announced, he wrote to Gardner telling him of his discoveries.

An infinite number of different tilings of the plane can be made with these two pieces. However, if you take any finite part of any one of these particular tilings, then it will be possible to find it recurring infinitely often inside any of the others. So if you are dropped down in any part of one of the infinite number of Penrose tilings you will not be able to tell which one you are in by just investigating a finite piece of it.[164] They only become different when viewed as infinite wholes. Another consequence of this unusual state of affairs is that no finite part of one of these tilings fixes what the rest of the tiling has to be like. It is clear from this design that the pattern is not a random one. Remarkably, in 1982, the Israeli physicist Dan Shechtman and colleagues discovered[165] that there exist three-dimensional materials which exhibit an internal structure of exactly the sort exhibited by the Penrose tiling. They became known as *quasicrystals*. They possess a microscopic structure that is intermediate between the complete order of, say, diamond and the disorder of glass.

Penrose's tiling has become an important component of many aesthetic designs from Charles Jencks's carpet in Portrack House with its *Garden of Cosmic Speculation*, designed by him and his wife, Maggie Keswick Jencks, to the entrance porch to the library of St John's College, Cambridge, where Penrose was once a research student.[166] It symbolises a whole realm of pattern that contains much that is aesthetically appealing, mathematically unexpected, and yet disarmingly simple to create.

In the spring of 2007, a remarkable discovery was announced by an American physics student, Peter Lu, whose vacation exploring Uzbekistan led him to discover the use of Penrose's non-periodic designs in a number of stunning pieces of fifteenth-century Islamic architecture. He returned to the library in Harvard to seek out other records of Islamic design in Turkey and Iran and recognised the characteristic tiling signature in many examples of traditional

Archway from the Darb-i Imam shrine in Isfahan, Iran, which was built in 1453 CE. The larger pentagons outlined in pale blue were constructed using a large-scale *girih* tile pattern, and the small white pentagons were constructed using a small-scale *girih* tile pattern.

girih tile patterns, where the complex aperiodic patterns would be inscribed onto larger tiles that underlay the design (shown in red in the picture on page 399.[167] Most remarkable was the Topkapi scroll, now held in Istanbul, which functioned as a design guide and tracer for architects, showing them how to inscribe larger pieces with the sub-patterns needed for large-scale non-periodic tilings. Five pieces of different shapes are picked out in different colours in the picture on page 398 and the larger tile is outlined in red.

As an amusing postscript, it is interesting to note that in 1997 Roger Penrose and Pentaplex (the company that manufactured his kite-and-dart tiles for commercial sale) sued a large multinational company, Kimberly-Clark, who had used his tiling pattern as the embossed design on their rolls of Kleenex quilted toilet paper. Apparently, Penrose first noticed it on a packet of the tissue that his wife had bought. It was claimed by the manufacturers that the non-periodicity of the embossed pattern was important in stopping the paper slipping when it was rolled. Presumably, it couldn't have been an exact Penrose tiling if was to be manufactured by mass production of toilet rolls containing 500 sheets. There would inevitably have been a periodicity. The case was eventually settled out of court with both sides sworn to secrecy, but seems to raise interesting questions of mathematical philosophy. As United Kingdom patent law says clearly: discoveries cannot be patented.[168] But inventions can. You cannot patent DNA, or any other discovery that extends human knowledge, rather than human ability. So, if you are a Platonist, and believe that mathematics 'exists', and its truths (like the non-periodic tilings) are discovered, then your mathematical discoveries cannot be patented. But, if you believe that mathematics is a human invention, like chess, then you would be able to patent its structures. It would indeed be interesting to see a legal case turn on the adoption of a particular philosophy of mathematics in this way. Unfortunately the Penrose and Kimberly-Clark case never got that far and seems to have ended up with Pentaplex, Penrose and SCA Hygiene Products, the new owners of the Kleenex range of products, all working in partnership on new products.[169] But who was the unrecognised fifteenth-century genius who got there more than 500 years before?

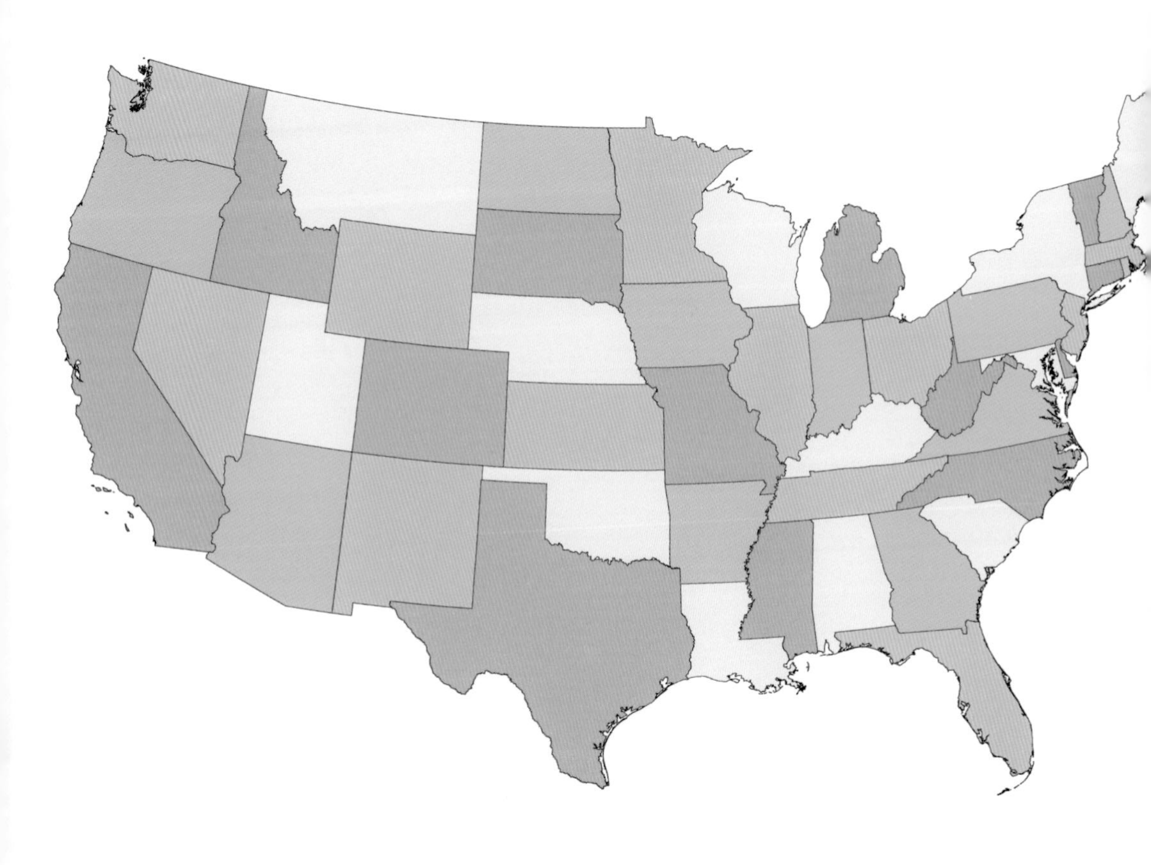

This map of the USA uses four colours to distinguish individual States without any adjacent borders having the same colour.

THE SIGN OF FOUR

THE FOUR-COLOUR THEOREM

Suppose there's a brown calf and a big brown dog, and an artist is making a picture of them . . . He has got to paint them so you can tell them apart the minute you look at them, hain't he? Of course. Well then, do you want him to go and paint both of them brown? Certainly you don't. He paints one of them blue, and then you can't make no mistake. It's just the same with maps. That's why they make every state a different color . . .

Mark Twain *Tom Sawyer Abroad*[170]

THERE is something intriguing about maps. And it doesn't really matter what countries they are charting. Look at the map of the United States. Cartographers have subdivided the map into the separate states in a manner that makes them simple to distinguish. No two adjacent states have the same colour meeting along a border. If you count up how many different colours have been used to create such a variegated map, then you will find that there are just four.

The question is, could you colour all possible maps in a similar way, avoiding a collision of colours, with fewer than four colours? The first person to conjecture that four was the smallest number of colours needed to colour all possible maps was Francis Guthrie in the 1850s, who had noticed that four colours sufficed to colour the map of the counties of England and wrote about the question to his brother Frederick, who was then a student at University College, London. But neither of them could prove what they suspected, that four colours always suffice to colour a map. In these circumstances it is as well to 'phone a friend', or at least ask your professor, so Frederick asked Augustus De Morgan at University College. He made some interesting observations about the problem, but couldn't prove or disprove it either, so he asked another friend, the formidable William Rowan Hamilton at Trinity College, Dublin, who had been a professor there since he was just sixteen years old, and proceeded to make a succession of remarkable discoveries in mathematics and physics. Alas, even Hamilton could offer no new insight as to how the conjecture could be proven, and it appears that he didn't spend much time trying. Mathematicians seemed to lose interest in the problem and the only person who showed any spark in the decade to follow was the American philosopher Charles Sanders Peirce, who failed in his attempt to prove the four-colour conjecture in the 1860s.

It was twenty-five years before the problem attracted the attention of mathematicians again. In June 1878, Arthur Cayley presented the problem to a meeting of the London Mathematical Society and then published an article (with no pictures) entitled 'On the Colouring of Maps' in the *Proceedings of the Royal Geographical Society*, the following April, admitting that while it was a theorem known to map-makers, of its truth 'I have failed to find a proof'.[171]

Cayley was a remarkable person. He graduated as the top mathematics student in his year at Cambridge in 1842, but, a few years later and without private means, needed to seek employment in the legal profession at Lincoln's Inn in London. He worked full-time as a lawyer but pursued his mathematical interests in his spare time. One wonders whether he ever slept, because during the fourteen-year period that he worked at the bar he published 250 outstanding research papers in mathematics. Finally, he gladly took a salary cut, and accepted the Sadlerian professorship of pure mathematics at Oxford. By the time he died, aged seventy-four, in 1895, he had written over 900 mathematical papers covering every area of mathematics.

It was his connections to Cayley, both as a student in Cambridge and as a lawyer, that led Arthur Kempe, a keen amateur mathematician, to attempt to prove the conjecture. The result was a short paper sent on 17 July 1879 to the journal *Nature*, and a longer one to the *American Journal of Mathematics*,[172]

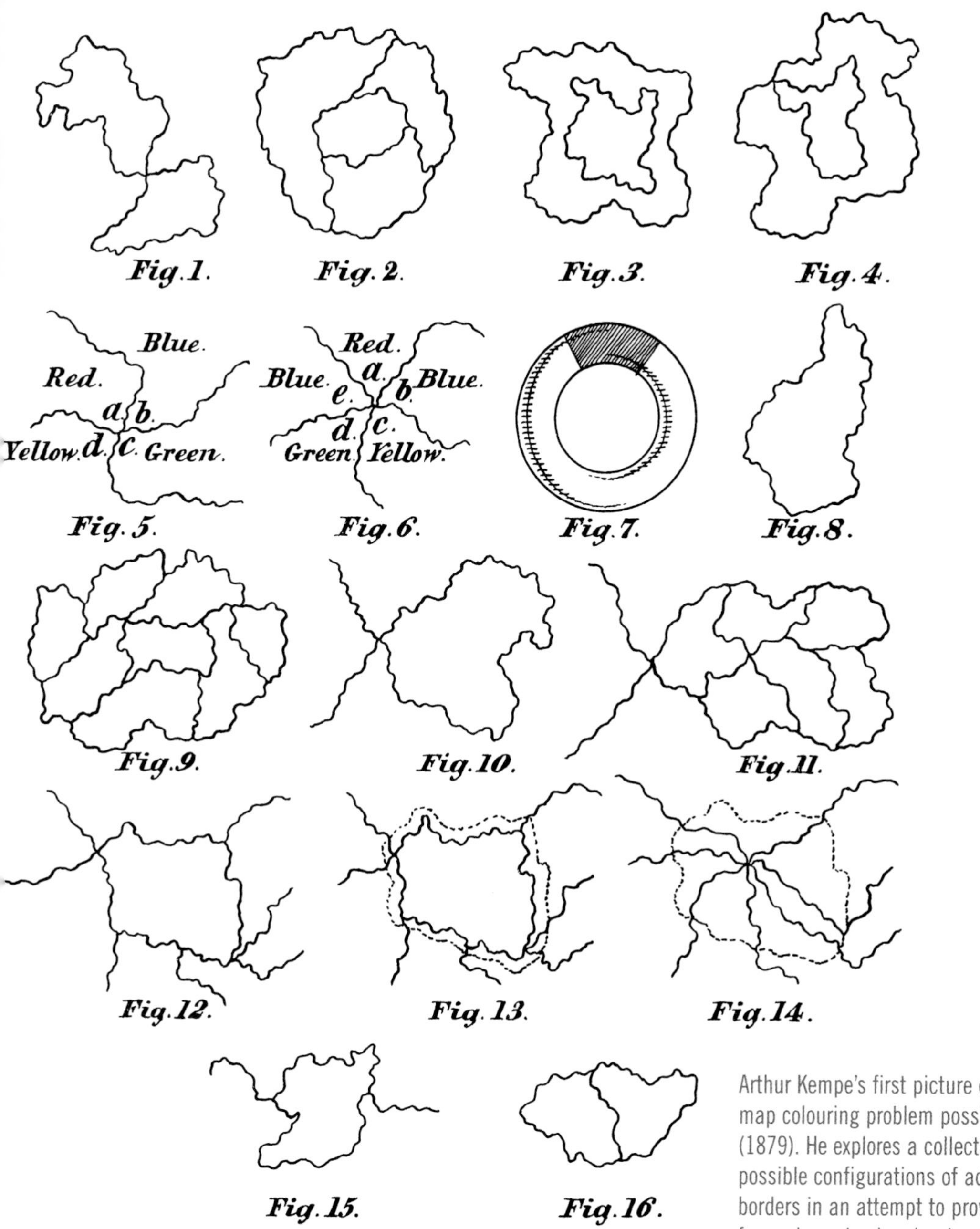

KEMPE, Geographical Problem.

Arthur Kempe's first picture of the map colouring problem possibilities (1879). He explores a collection of possible configurations of adjacent borders in an attempt to prove that four colours (and no less) could colour all possible maps without adjacent regions having the same colour. Kempe's 'proof' was found to be flawed ten years later.

in which he claimed to have proved that four colours suffice for colouring any flat or spherical map and displayed the first picture illustrating the problem, which we show here. His arguments were ingenious and novel, and for a decade mathematicians were convinced by them. It was not until 1890 that Percy Heawood, a mathematics lecturer in Durham, found a gap in Kempe's proof. When the flaw was mended, alas all that could be proved was that every map can be coloured by at most five colours.[173]

This was a step backwards. In the years that followed, many mathematicians attacked the problem by starting small and building up the size and complexity of the maps. By 1922, it was known that every map with twenty-five or fewer regions ('states') could be four-coloured. Gradually, this critical size was raised from twenty-five to ninety-five. But there are an awful lot of numbers bigger than ninety-five. Things really began to change in the 1970s, by which time the problem had become quite famous and mathematicians working on it all hoped that they might be immortalised by finding the new trick that would give a proof for all possible maps. The problem was simple to state and visually obvious, and so attracted the attention of mathematicians all over the world. Martin Gardner even teased them by publishing a spoof counter-example that could not be four-coloured in his Mathematical Games column in *Scientific American* – although it turned out to be incorrect.

Something dramatic happened in 1976. The four-colour conjecture turned into the four-colour *theorem*. Then the controversy really started. Kenneth Appel and Wolfgang Haken,[174] from the University of Illinois, announced a proof which was completed with the assistance of a computer that had been programmed to check whether a very large number of possible exceptions to the conjecture were really counter-examples. The checking took 1200 hours of computer time because the list of difficult cases was so long.

Why was this so controversial? For the first time, mathematicians were confronted with the proof of a theorem – rather than just a piece of numerics – that had not passed through any single human mind. The computer's operations were so voluminous that no one could ever check by hand what had been done. Different computer programs could be written[175] to repeat the checks, or different machines run to check the integrity of the checking program itself, but the overall proof was a super-human task.

The computer-assisted proof of the four-colour conjecture was a disappointment to many mathematicians. One famous mathematician, on being told how the proof was obtained, was reported to have remarked: 'so it wasn't a good problem after all.' Even the first public announcement of the proof and its methodology at a major conference of mathematicians evoked only a muted response. The reason is interesting. Although the popular image of mathematics imagines that the subject expands its realm by enlarging the library of mathematical truths and continually adding new theorems to its catalogue, this is not the whole story, or even its main theme. Mathematicians are looking primarily for new types of arguments, new methods of proof, new tricks that can be used in different places to extend what is proved into new parts of the unknown. The computer-assisted proofs of the four-colour theorem revealed no new type of argument. There was no useful legacy from Appel and Haken's proof and no surprises – just the satisfaction of knowing that it really was true after all.

Very soon after the announcement of the Appel–Haken proof there sprang up a serious debate about the role of computers in mathematics.[176] The new proof was not traditional. Some people didn't regard it as 'real' mathematics at all. Others drew parallels with drug-assisted world records in sport. But, with time, this storm has abated, and the problems raised have been dissolved, if not resolved.

For a long time there had been examples of very long proofs, hundreds of pages in length, which only a very small number of mathematicians would be qualified to check in any given case. The four-colour theorem was the natural extension of this trend. Its length was too great for any one mathematician to survey and feel convinced by. We began to see that the infinite sea of mathematical truths that sit waiting for us to discover or construct are hidden from reach by a series of horizons. Close by is the horizon of all the truths which can be established by 'short' proofs that can be deduced, written out, and read by one mathematician, or by a small number of collaborators. Go further out and we enter the realm of longer and longer proofs, until we reach a horizon beyond which we cannot go without the assistance of computers. Almost all the mathematics that we have found so far lies in the realm of short proofs. It is a remarkable feature of the physical Universe and the 'universe' of mathematics which describes all possible patterns that so many deep truths reside there, just a few short steps from base. Perhaps there are amazing structures and truths that lie far beyond the reach of humanly surveyable proofs. The most fascinating feature of this realm of super-human proofs is that we believe it is possible for conjectures which are very short to state to have enormously long proofs.[177]

In recent years this debate has re-emerged with the appearance of a proof of 'Kepler's packing problem', which asks for the most efficient way to pack many spheres into a volume of space – you can see some intuitive and correct solutions to this problem by looking at any greengrocer's display of stacked oranges. Yet a proof eluded mathematicians from 1661 until 1998, when Thomas Hales completed a proof that Kepler's conjectured maximal packing density (the so-called face-centred cubic packing) was correct. Again, it was computer-assisted. The proof consisted of 250 pages of human mathematical analysis together with 3 Gigabytes of computer code and stored output. Hales used the computer to check exhaustively the many possible contrary cases one by one in order to ascertain that Kepler's conjecture was correct. The paper describing the proof was examined by experts over a very long period of time before being published in the leading mathematical journal. However, the journal *Annals of Mathematics* declined to publish the computer programs that completed the proof, judging it to be not mathematics in the purist sense that the journal required. The computer materials would be published in the *Journal of Discrete and Computational Geometry*. In October 2004, stimulated by these developments, the Royal Society of London hosted a two-day conference of mathematicians, computer scientists, and experts on artificial intelligence to discuss the implications of these developments for the nature of mathematical proof.[178]

Of course, in both these famous cases of long computer-assisted proofs there is no reason why a short, entirely human, proof might not eventually be found.

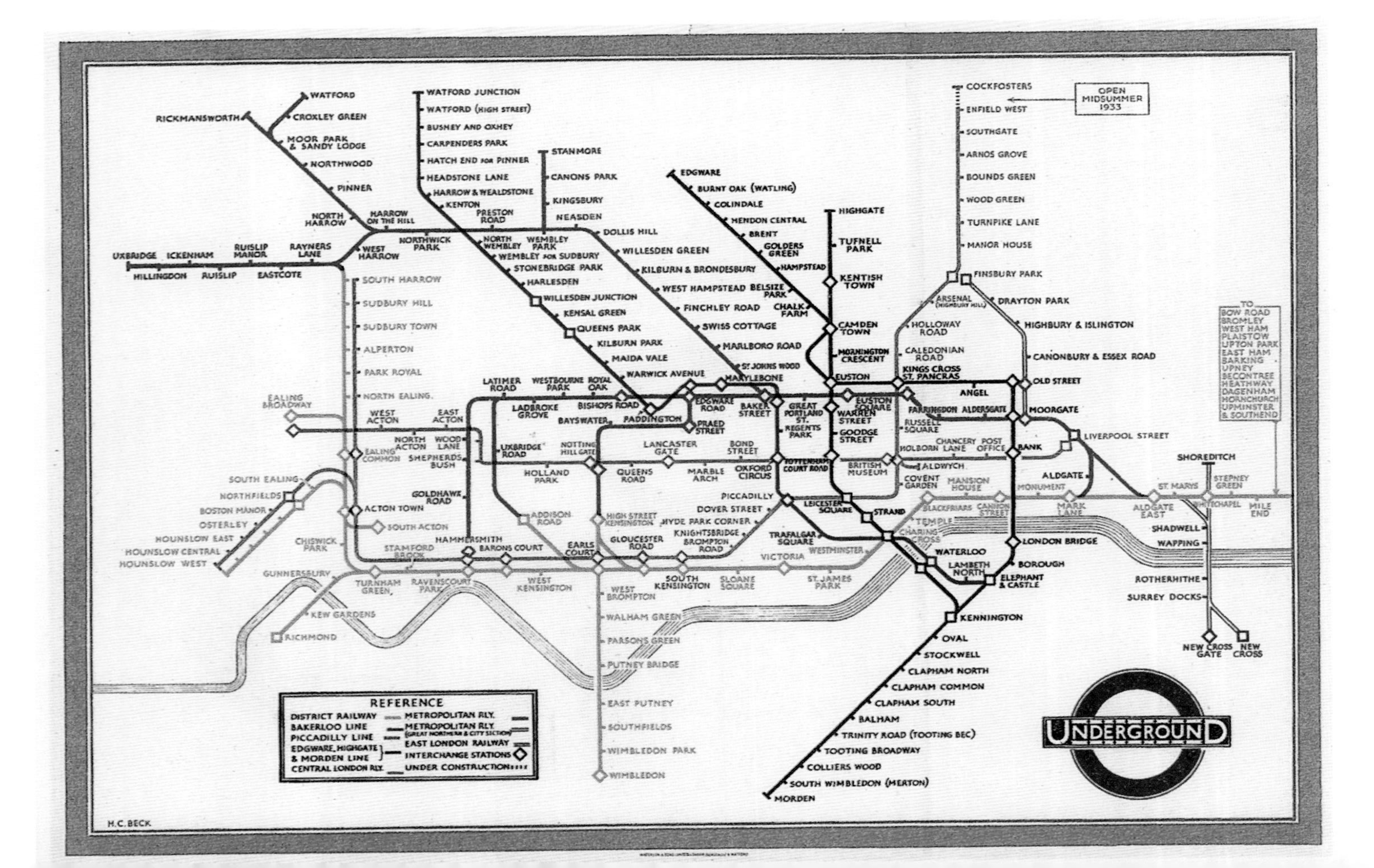
UNDERGROUND
REFERENCE
DISTRICT RAILWAY
BAKERLOO LINE
PICCADILLY LINE
EDGWARE, HIGHGATE & MORDEN LINE
CENTRAL LONDON RLY.
METROPOLITAN RLY.
METROPOLITAN RLY. (GREAT NORTHERN & CITY SECTION)
EAST LONDON RAILWAY
INTERCHANGE STATIONS
UNDER CONSTRUCTION
OPEN MIDSUMMER 1933
COCKFOSTERS
ENFIELD WEST
SOUTHGATE
ARNOS GROVE
BOUNDS GREEN
WOOD GREEN
TURNPIKE LANE
MANOR HOUSE
FINSBURY PARK
ARSENAL (HIGHBURY HILL)
DRAYTON PARK
HIGHBURY & ISLINGTON
CANONBURY & ESSEX ROAD
TO BOW ROAD BROMLEY WEST HAM PLAISTOW UPTON PARK EAST HAM BARKING UPNEY BECONTREE HEATHWAY DAGENHAM HORNCHURCH UPMINSTER & SOUTHEND
WATFORD
RICKMANSWORTH
CROXLEY GREEN
MOOR PARK & SANDY LODGE
NORTHWOOD
PINNER
NORTH HARROW
HARROW ON THE HILL
WATFORD JUNCTION
WATFORD (HIGH STREET)
BUSHEY AND OXHEY
CARPENDERS PARK
HATCH END FOR PINNER
HEADSTONE LANE
HARROW & WEALDSTONE
KENTON
PRESTON ROAD
STANMORE
CANONS PARK
KINGSBURY
NEASDEN
DOLLIS HILL
WILLESDEN GREEN
KILBURN & BRONDESBURY
WEST HAMPSTEAD
FINCHLEY ROAD
SWISS COTTAGE
MARLBORO ROAD
ST. JOHNS WOOD
EDGWARE
BURNT OAK (WATLING)
COLINDALE
HENDON CENTRAL
BRENT
GOLDERS GREEN
HAMPSTEAD
BELSIZE PARK
CHALK FARM
HIGHGATE
TUFNELL PARK
KENTISH TOWN
CAMDEN TOWN
MORNINGTON CRESCENT
UXBRIDGE
ICKENHAM
RUISLIP MANOR
RAYNERS LANE
HILLINGDON
RUISLIP
EASTCOTE
WEST HARROW
NORTHWICK PARK
NORTH WEMBLEY
WEMBLEY PARK
WEMBLEY FOR SUDBURY
STONEBRIDGE PARK
HARLESDEN
WILLESDEN JUNCTION
KENSAL GREEN
QUEENS PARK
KILBURN PARK
MAIDA VALE
WARWICK AVENUE
SOUTH HARROW
SUDBURY HILL
SUDBURY TOWN
ALPERTON
PARK ROYAL
NORTH EALING.
EALING BROADWAY
WEST ACTON
EAST ACTON
NORTH ACTON
WOOD LANE
EALING COMMON
SHEPHERDS BUSH
LATIMER ROAD
WESTBOURNE PARK
ROYAL OAK
LADBROKE GROVE
BISHOPS ROAD
BAYSWATER
PADDINGTON
UXBRIDGE ROAD
NOTTING HILL GATE
HOLLAND PARK
LANCASTER GATE
QUEENS ROAD
MARBLE ARCH
BOND STREET
OXFORD CIRCUS
MARYLEBONE
EDGWARE ROAD
PRAED STREET
BAKER STREET
GREAT PORTLAND ST.
REGENTS PARK
EUSTON
EUSTON SQUARE
WARREN STREET
GOODGE STREET
TOTTENHAM COURT ROAD
BRITISH MUSEUM
KINGS CROSS ST. PANCRAS
ANGEL
FARRINGDON
ALDERSGATE
RUSSELL SQUARE
HOLBORN LANE
CHANCERY LANE
POST OFFICE
ALDWYCH
COVENT GARDEN
OLD STREET
MOORGATE
LIVERPOOL STREET
BANK
MANSION HOUSE
MONUMENT
ALDGATE
SHOREDITCH
ST. MARYS
STEPNEY GREEN
WHITECHAPEL
MILE END
ALDGATE EAST
MARK LANE
SHADWELL
WAPPING
ROTHERHITHE
SURREY DOCKS
NEW CROSS GATE
NEW CROSS
SOUTH EALING
NORTHFIELDS
BOSTON MANOR
OSTERLEY
HOUNSLOW EAST
HOUNSLOW CENTRAL
HOUNSLOW WEST
ACTON TOWN
SOUTH ACTON
CHISWICK PARK
GOLDHAWK ROAD
HAMMERSMITH
STAMFORD BROOK
BARONS COURT
ADDISON ROAD
EARLS COURT
HIGH STREET KENSINGTON
GLOUCESTER ROAD
PICCADILLY
DOVER STREET
HYDE PARK CORNER
KNIGHTSBRIDGE
BROMPTON ROAD
LEICESTER SQUARE
TRAFALGAR SQUARE
STRAND
BLACKFRIARS
CANNON STREET
TEMPLE
CHARING CROSS
VICTORIA
WESTMINSTER
ST. JAMES PARK
WATERLOO
LAMBETH NORTH
LONDON BRIDGE
BOROUGH
ELEPHANT & CASTLE
GUNNERSBURY
TURNHAM GREEN
RAVENSCOURT PARK
WEST KENSINGTON
SOUTH KENSINGTON
SLOANE SQUARE
WEST BROMPTON
WALHAM GREEN
PARSONS GREEN
PUTNEY BRIDGE
EAST PUTNEY
SOUTHFIELDS
WIMBLEDON PARK
WIMBLEDON
KEW GARDENS
RICHMOND
KENNINGTON
OVAL
STOCKWELL
CLAPHAM NORTH
CLAPHAM COMMON
CLAPHAM SOUTH
BALHAM
TRINITY ROAD (TOOTING BEC)
TOOTING BROADWAY
COLLIERS WOOD
SOUTH WIMBLEDON (MERTON)
MORDEN
H.C. BECK

YOUR TUBE
THE LONDON UNDERGROUND MAP

The Central and the Circle
When they are both full-known,
Of all the lines that are on the Tube,
The Central bears the crown.

Roger Tagholm 'The Central and the Circle'[179]

ONE of the most instantly recognisable icons in the United Kingdom is the map of the London Underground train system. It is a beautiful piece of design that has endured with very few changes, other than the updates necessary to keep pace with the appearance of new lines and stations, since it was first created by Harry Beck in 1931. Indeed, so familiar is its colourful web of London stations that you may well have never recognised what a revolutionary piece of draughtsmanship it was, and why it is a candidate for inclusion in our gallery of influential mathematical images.

The London Underground was the first train system of its type in the world. It began to issue maps for its passengers in 1906, when the four competing train companies which ran separate routes came together to form the single 'Underground' company and their four routes were renamed the District, Piccadilly, Bakerloo, and Northern lines, which still exist in extended forms today. The first maps of the system were conventional and geographically accurate. This looked a bit of a mess. London is an ancient city that has grown in a ragged and largely unregulated fashion of aggregation. Unlike the rectangular grid system of streets of New York, or the radial geometry of Paris, London is asymmetrical and complicated, with no readily discernible centre. As

Harry Beck's 1933 map of the London Underground.

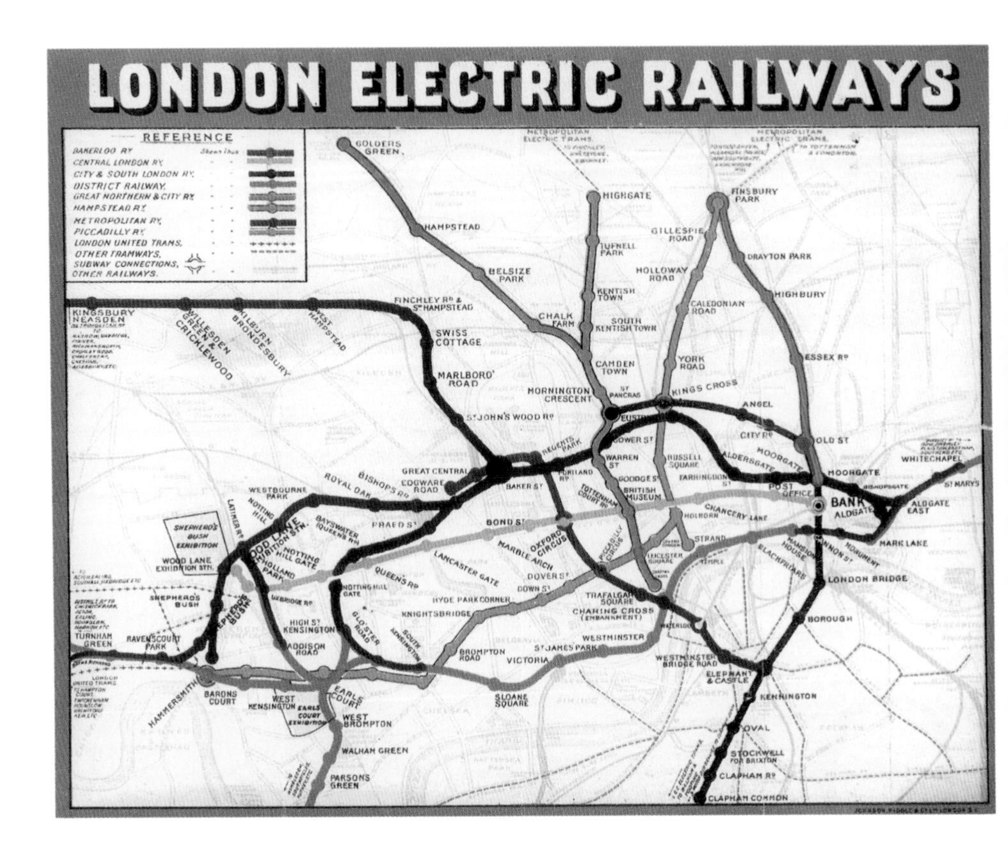

First generation London Electric Railways map, 1908, which shows the true geographical positions of lines and stations, except for a distortion to fit in the key.

a result the Underground routes meandered around in crowded paths and the purely geographical maps looked ugly and complex.

Things changed in 1925, when Harry Beck, a young draughtsman and designer with a background in electronics, joined the company as a junior draughtsman in the Signal Engineer's Office, but he was still able to contribute pieces of freelance work. He set about creating a revolutionary style of map that was, at first, rejected by the Company, but then began to appeal to the hard-pressed managers of the Underground railway. Revenue was low and Londoners had not taken to travelling on the system – not least because the existing maps made it look so complicated. Working in his spare time, Beck departed from the geographical tradition of transport maps accurately showing where lines were going with respect to the street system, and adopted a

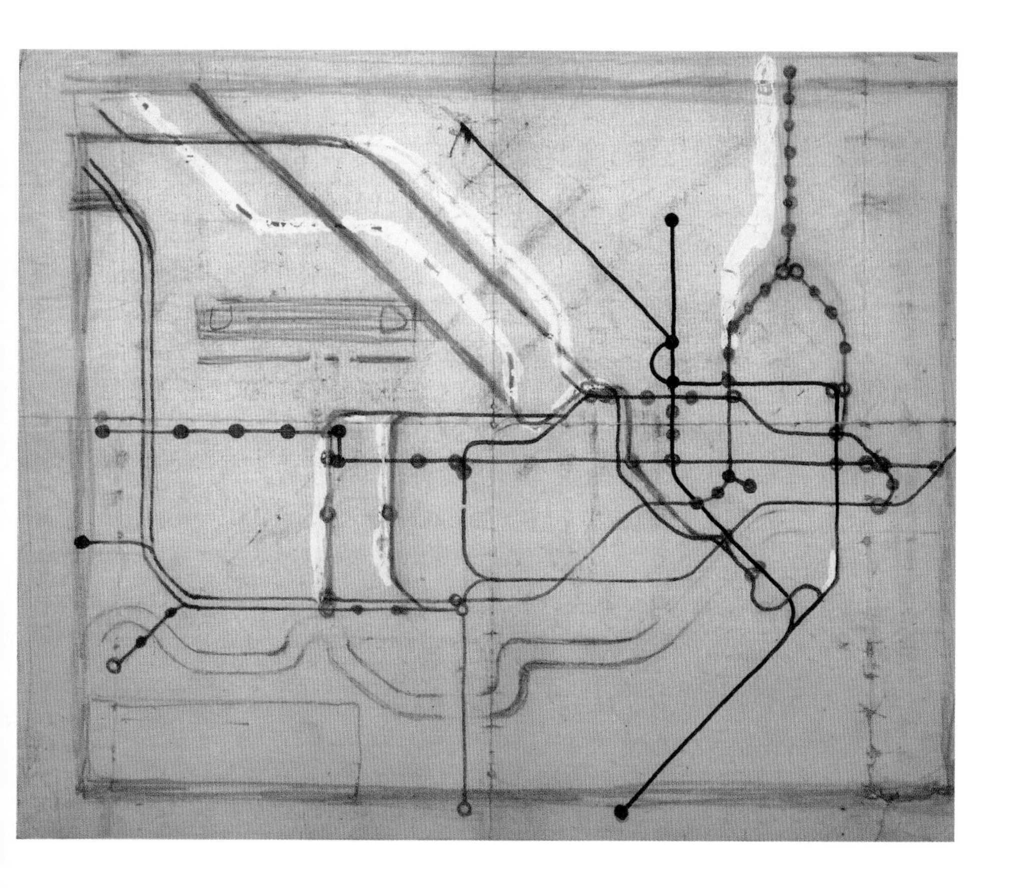

Harry Beck's original sketch for his Underground map, on two pages of an exercise book. The River Thames was added later, but the key features of his simple use of horizontals, verticals and diagonals with colour coding and station markers are already in evidence. Beck said that he 'tried to imagine that I was using a convex lens or mirror, so as to present the central area on a larger scale. This, I thought, would give a needed clarity to interchange information.'

topological structure which emphasised only the connectivities created by the train lines. The relative positions of stations on neighbouring lines were distorted where necessary in the interests of clean design and visual simplicity.

Beck moved towards a picture in which all lines were drawn straight, either horizontal, vertical, or at 45 degrees. This is the easy part, and mirrors the layout of an electronic circuit board, which was so familiar to Beck. The challenge is to fit in the lettering of all the stations in the crowded central region, while clearly showing the exchanges that are possible between different lines at the cross-over stations. Beck succeeded brilliantly. Although it was initially rejected by the Publicity Department in 1931, his disappointment changed to delight a year later, when he was summoned to hear that the Department had decided to print 750,000 copies of his map to be distributed free to commuters. The first Beck maps were released to the public in January 1933. The Central Line was orange and the Bakerloo Line was red. They were given their present colours of red and brown, respectively, in 1934.

This was the start of his lifelong passion. Over the years he evolved the representation of the tricky exchange stations where passengers could switch from one line to another, chose good colours to distinguish the different lines, refined the representation of the River Thames, and tried different symbols to mark the ordinary stations, the intersections, and the British Rail mainline stations. The result was an iconic piece of seamless design that helped the London Underground become a popular and successful means of transport.[180] Beck's maps were displayed in large format on the walls at all stations and distributed freely to the public in the form of handy pocket-sized booklets.

Beck's topological maps, or diagrams, as they became known, also had a subtle sociological effect upon the city of London. A true geographical map would have reflected accurately just how far from the centre of London places like Morden, Uxbridge, and Cockfosters really are. They were never really 'in' London at all. Indeed, in the pre-Beck era these extreme stations were actually excluded from the railway maps, appearing in a list of stations down the line at the edge of the map. But Beck's maps made the extremities of the Tube lines appear quite close to the heart of London, and thereby encouraged the concept of Greater London and the commuting culture of Underground train users. Actually, it takes longer to get into the centre of London on the Underground from some of the distant stations than it does to travel into London from Reading or Cambridge on overground trains, yet the remote Underground stations still seemed to be 'in' London because of their apparent nearness on the Underground map.[181] Commenting on Beck's 'diagram' in a BBC television series about design classics, the historian Adrian Forty put it like this:

> The prospect of making a journey to Cockfosters or Ruislip, if one had looked at a geographically correct map, would have seemed rather formidable. Looking at the Underground map, it looks reasonably simple.

In 1951, Beck created a topological map for the Paris Metro system in the same style as that for the Underground. It does not seem to have been an official commission and was never used by the Metro authority, but it shows what might have been. Other systems did not enjoy the success that Beck brought to the London Underground. The New York Subway system introduced a new map in 1972, claiming that it would do for New York what Beck's diagram had done for London. Alas, it was not to be. It was confused and unclear and was replaced after a few years by a plainer but more useful alternative.

The last part of Beck's career was difficult. London Transport introduced other designers to refine the map in the 1960s and his name ceased to appear as a credit on the map. Beck left their employment, feeling cheated of the rights to his design, and became a lecturer at the London School of Printing, where he taught the history of type and typefaces. He embarked on a long correspondence with London Transport in an attempt to convince them that his proposals for the evolution of the map were superior to those of Paul Garbutt, another London Transport employee, which they had chosen to adopt. Yet today's Underground map is still just an extension of Beck's classic, including many new routes. Sadly, his name no longer appears in the bottom left-hand corner, but the map still bears his unmistakable stamp.[182]

Beck's map has also begun to inspire other artistic creations. Simon Patterson's 1992 artwork *The Great Bear*, now hanging in the Tate Modern in London, replaces Beck's station names with famous writers, sporting figures, scientists, and cultural figures. David Booth's *The Tate Gallery by Tube* (1986) was one of the most popular posters created for the Underground.[183] It showed the coloured inner Underground lines squeezed from a tube of oil paint labelled with the word 'Pimlico' – the closest Tube station to the Tate Gallery. Last, but not least, Professor Albus Dumbledore, the headmaster of Hogwarts School of Witchcraft and Wizardry, claims to have a scar on his left knee which is an exact map of the London Underground.

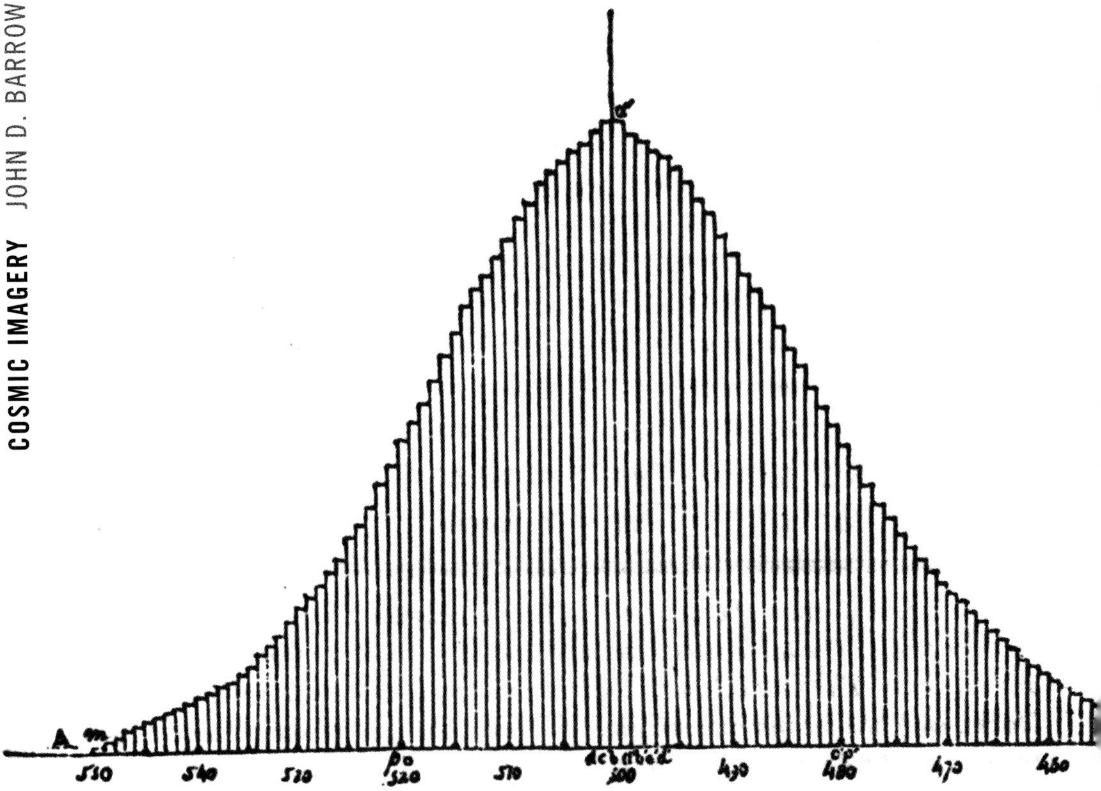

Quetelet's representation of the normal curve as the limit of a binomial distribution (1846).

LIES, DAMNED LIES, AND STATISTICS

THE IMPORTANCE OF BEING NORMAL

The
Normal
Law of Error
Stands out in the
Experience of mankind
As one of the broadest
Generalisations of natural
Philosophy. It serves as the
Guiding instrument in researches
In the physical and social sciences and
In medicine, agriculture and engineering.
It is an indispensable tool for the analysis and the
nterpretation of the basic data obtained by observation and experiment.

William J. Youden[184]

THE study of statistical trends began in the late eighteenth century, when many mathematicians and economists tried to derive a formula for the likelihood of occurrence of random fluctuations or errors in a set of data. One branch of this inquiry, which focused upon the different political systems of states and countries, was given the name *Statistik* by the German author Gottfried Achenfeld. Eventually, it merged with the quantitative study of political and economic data related to deaths and population sizes that were of special interest to those costing insurance premiums. Alongside these pragmatic developments in actuarial science, there continued a study of probability, largely motivated by the evaluation of gambling odds led by famous mathematicians such as Pierre de Fermat, Blaise Pascal and James Bernoulli. From this more mathematical thread there was spun a formula for the distribution of probabilities that would later be called the 'normal' distribution of probabilities. It was found first by Abraham De Moivre, a French Huguenot mathematician who in 1685 had sought sanctuary in London from religious persecution. He first published the formula in an article written in 1733. This paper was the first to determine the probability of occurrence of an error of any given size under particular assumptions. He included it in the second edition of his book *The Doctrine of Chances* in 1738 as an approximation to the binomial distribution of probabilities for a large number of outcomes.[185] Alas, this work went largely unnoticed at the time and its presentation contained no graphs or pictures.[186] Several eighteenth-century mathematicians then rediscovered the crucial formula by approaching the problem in different ways, but still no one took the simple step of drawing a picture. Who did it first?

The standard histories of statistics are not too helpful about this and some detective work is needed to find the answer. The first picture of the 'normal' distribution of probabilities seems to be found in a book by Augustus De Morgan, written in 1838, with the title *An Essay on Probabilities and on Their Application to Life Contingencies and Insurance Offices.*[187] De Morgan was a great mathematician who made contributions to many areas of mathematics, statistics, and logic. He became the first professor of mathematics at University College, London, his career path in Cambridge having been blocked by his refusal to take the theological test needed to receive his MA degree (although he was actually a member of the Church of England – he was to resign from his chair in London not once, but twice, on matters of principle, refused to be nominated for a Fellowship of the Royal Society, and declined an honorary degree from Edinburgh – obviously not an easy person to deal with). His book contained chapters on the laws governing the occurrence of random errors with several pictures of the characteristically bell-shaped curve which he called the 'standard law of facility of error'.[188]

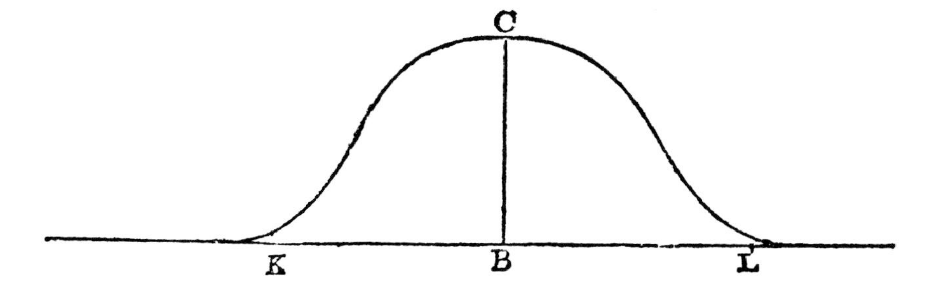

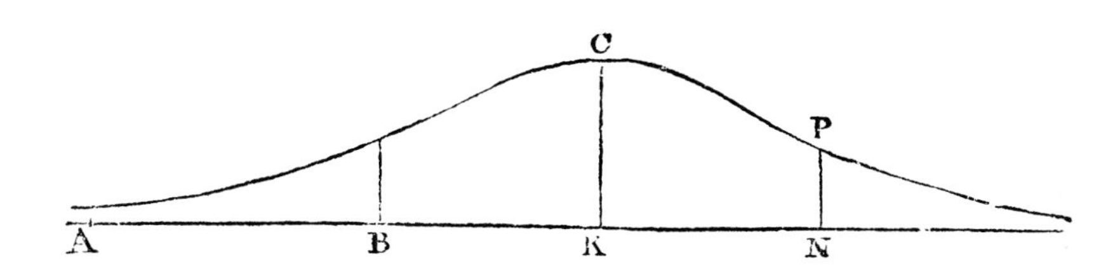

The first pictures of the normal probability curve by Augustus de Morgan in his *Essay on Probabilities* (1838).

The application of statistics to the study of social trends was promoted most effectively by an energetic and multi-talented Belgian, Adolph Quetelet (1796–1874), who developed quantitative and comparative studies of statistical trends in a remarkable fifty-year period of research activity beginning in 1824. His work was characterised by its pioneering use of graphs and charts to represent data in vivid ways across a wide range of sciences. He called the normal distribution of probabilities 'the curve of possibility' – in his original French texts '*la loi de possibilité*' or '*la courbe de possibilité*'.[189] The picture shown on page 418 from his book of 1846 shows this distribution being built up as a limiting form from a large number of binomial events. This distribution is characterised by two quantities – its mean or central value – and its degree of concentration about its central value, which we call its variance. Distributions with small variance are more sharply peaked around the central value than those with larger variance, which are flatter and the probability is more evenly spread among the outlying possibilities. The distribution extends all the way to infinity in both positive and negative directions, but in practice there will eventually be a largest and smallest value which is too improbable to be of practical relevance. This is why some of the curves of De Morgan and Quetelet seem to end at some large positive and negative values along the horizontal axis.

Quetelet and his followers, such as the English scientist Francis Galton, a cousin of Charles Darwin, who we met earlier in connection with the first weather maps, and who introduced the term 'normal distribution' for this pattern of probabilities, seized upon its ubiquity and used it as a basis for the study of society and its sub-sections en masse. Quetelet introduced the concept of 'the average man' (later to become hackneyed as 'the average (wo)man in the street' so beloved of political commentators). The reason why this characteristic curve of frequencies of occurrence arises so often in our study of the natural and human worlds is simple. It is the inevitable result of the statistics for the sum of a number of different, but independent, random processes. As the number of processes being added gets larger, so the normal distribution is approached more closely as a description of the likelihood of each possible different outcome arising.[190] Many of the things that we see in the world look as if they depend upon a perplexingly large merger of different things in ways that are lost in history. But so long as the different processes are independent and additive the outcome is a normal distribution with some mean value and some variance which will be determined by the characteristics of the actual process that created it.[191]

Quetelet invented the idea of 'social science', and called it 'social physics' (*physique sociale* or *mécanique sociale*, echoing Pierre Laplace's *mécanique céleste*) to reflect his strong belief in the predictable and deterministic nature of social behaviour in the large; others called it 'moral statistics'. His interpretation of the significance of social averages was a little extreme. Having identified the possibility of defining the 'average man' for one nationality or another, he imagined using him to represent that nation's characteristics, and even those of the human race as a whole, so as to determine variances from the average:

> If the *average* man were ascertained for one nation, he could represent the type of that nation. If he could be ascertained according to the mass of men, he would represent the type of human species altogether.[192]

His most famous work, from 1842 and translated as *A Treatise on Man*, arrived at a statistical construction of his average man. This also enabled him to predict the average number of suicides, murders, marriages, and crimes to be expected in different societies each year. Curiously, he doesn't place great emphasis on 'chance' in these statistically predictable events, but takes a rather fatalistic view of humanity, its members at the mercy of inevitable statistical laws.[193] This is why he sees his inquiry as a form of social 'physics' in which there are 'laws' of human behaviour, just as there are laws for gases and moving bodies.

In the century to follow, this approach would become a snare to many who wished to pursue the study of eugenics, begun by Francis Galton (who also invented the name), in order to assert the superiority of one class, sex, or race over others. The normal distribution often played a key role in that argument, known by its new title of the bell-shaped curve, which had been introduced by Esprit Jouffret[194] in 1872 to reflect its characteristic shape. The assumption that some human attribute was normally distributed was used as a means of determining its mean and thereby characterising sub-populations as systematically above or below the mean. Interestingly, even in the time of Quetelet, one voice resolutely opposed the development of social statistics and its use by his influential supporters, who included the nursing and statistical pioneer Florence Nightingale, an admiring student of Quetelet, and a significant contributor to the development of new statistical methods. That opponent was the great Victorian novelist and social commentator Charles Dickens.

Dickens was implacably opposed to statistics. To his mind, it reduced everything to averages and failed to recognise the needs of individuals. Governments could appeal to statistics to argue that the lot of the average man was quite good, even getting better, or that the average number of industrial accidents in a given industry was small, so as to avoid having to enact legislation that would help the poorest members of society or improve hazardous working conditions. Statisticians were 'representatives of the wickedest and most enormous vice of this time', he wrote in 1864.[195] Dickens's novel *Hard Times* was published in 1854 and attacks the status quo for its glaring economic inequalities. One of its memorable curmudgeons is Mr Thomas Gradgrind, a retired merchant and schoolmaster, who wants nothing but facts and figures. Even his pupils are known by number rather than by name. He always carries on his person a ruler and a pair of scales so as to be ready to 'measure any parcel of human nature, and tell you exactly what it comes to'. When tragedy falls upon his family and his son Tom is found to be a thief, Tom defends himself by appealing to Quetelet's discoveries that the law of averages means that there must always be some members of society who are dishonest: it is a law of Nature. Then his daughter Louisa enters into a doomed marriage that Gradgrind assents to because he places his faith in the favourable statistics regarding successful matches rather than whether the couple love each other. Louisa's life has been ruined by her father's failure to balance numbers and facts with love and emotions; his life has been impoverished by an addiction to averages and statistical laws, like the economist who drowned in a lake of average depth six inches.

In more recent times, the 'bell curve' has continued to be at the centre of controversial claims concerning the social and racial stratifications of intelligence and economic achievement. The notorious book[196] *The Bell Curve* by Richard Herrnstein and Charles Murray used the image of the normal distribution as its title and tried to characterise complex qualities such as 'intelligence' as single (unchanging) numbers that could be ranked on a single scale. Not surprisingly, the book attracted more criticism and controversy than almost any other published in modern times.[197] Ever since the appearance of Herrnstein and Murray's book the term 'bell curve' has meant something more, and something more sinister, to many social commentators than the simple picture of the distribution that De Morgan and Quetelet first drew. It shows at once the advantages and the dangers of simple pictures. As Einstein was once reputed to have remarked, 'things should be made as simple as possible, but no simpler.'

Mind Over Matter

World of R&D 2005

Size of circle reflects relative amount of annual R&D spending by coun

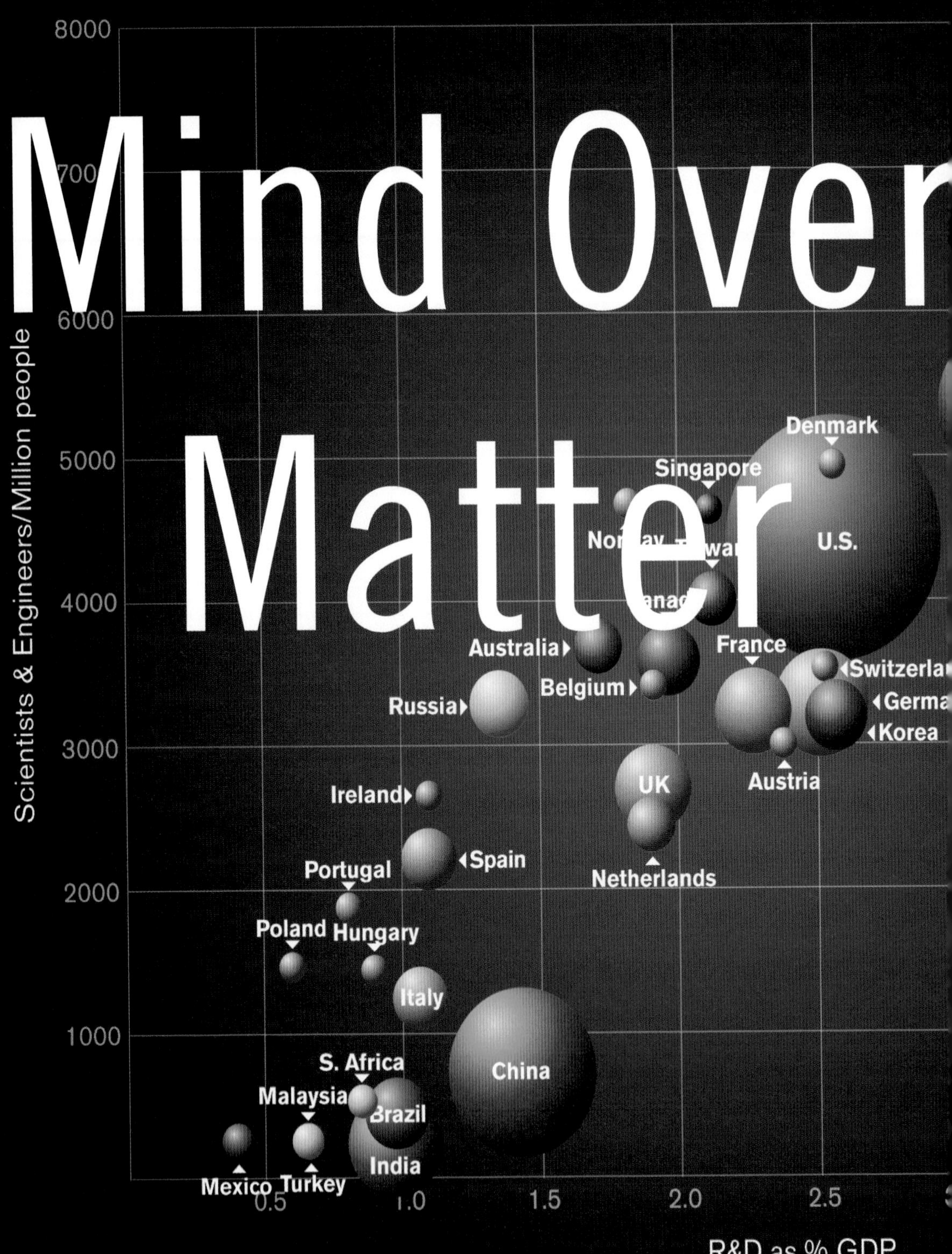

Asia Americas
Europe Other

Source: R&D

Work is of two kinds: first, altering the position of matter at or near the Earth's surface relatively to other such matter; second, telling other people to do so. The first kind is unpleasant and ill paid; the second is pleasant and highly paid.

Bertrand Russell[1]

The number of scientists per million of population vs percentage of Gross National Product (GNP) spent on Research and Development for 2005, produced by the World Bank.

THE physicist Lord Rutherford once said that all science is either physics or stamp collecting (well, he would, wouldn't he?). What he meant was that at the core of chemistry and biochemistry is the physics of atoms and molecules. At the heart of that physics lies Schrödinger's equation of quantum mechanics, from whose solutions flow all that we know about the detailed structure of matter and energy. In this last part, we look at some of the pictures that physics and chemistry frame for humanity. We see the images of atoms and molecules themselves, and the spectacular representations of complex molecules that chemists use to display their intricate geometries. We look back at the vehicles for discovery, such as Newton's prism, that have become symbols of the entire spectrum of discoveries that the Newtonian philosophy of experiment could achieve.

In chemistry, the periodic table of the elements is the great organising diagram which finally made sense of the patterns observed in the properties of the elements. Mendeleyev made some leaps of faith to create the table, but he got it right and only later would it be understood in terms of atomic properties. Remarkably, he resisted the urge to explain only what was known, and so he was able to use his table to predict the existence of undiscovered elements, along with their expected properties. The periodic table was conceived as a picture. Its influence has survived the full understanding that quantum theory supplied, and it still adorns the wall of every chemistry laboratory, like the roster of the players on Nature's teamsheet.

The canonical image of physics is neither of a table nor of an equation – it is a person. The picture of Albert Einstein has become the trademark of human intellectual achievement, deep thinking, and imagination. It created the public image of the scientist as an elderly, somewhat eccentric, and other-worldly gentleman, but there is an interesting history to all this imagery. Einstein did not always look like this and the change in his style of practising physics mirrors the change in his image and status in the world at large.

The quest to understand the atomic nucleus and the energy locked up within it, which Einstein's work made feasible, gives rise to several of our most dramatic images. The 'mushroom cloud', seen first at the Trinity nuclear test, has become the image of the destructive capability of nuclear weapons. Behind it is the simple graph of the variation of binding energy against nuclear size that reveals the two ways in which atomic energy can be extracted. The quantum basis of matter that underwrites these patterns also enables us to understand why things are as they are, why atomic matter, planets, and stars have the sizes and masses that they do. A simple picture of the masses and sizes of everything reveals a pleasing simplicity that is readily understood – the nearest thing to a periodic table in astrophysics.

The second half of the twentieth century saw new probes of the world of sub-atomic particles by means of devices which rendered the invisible manifest. Cloud chambers and bubble chambers track transient elementary particles of matter and watch their interactions with one another. Unfortunately, understanding these interactions is not always so easy. The addition of quantum uncertainty to the classical theories of electromagnetism and light produces a theory that is simple to state but almost impossible to solve. The extraction of its predictions required a new calculus of pictures to be invented by Richard Feynman. Feynman's diagrams are the principal tool by which high-energy physicists do the mathematical book-keeping necessary to calculate what we should see in high-energy experiments.

The quantum strangeness that makes the interaction of matter with light so complicated has its own iconic images. They confound our attempts to interpret the meaning of a theory that makes correct predictions with such dazzling accuracy. Schrödinger's cat is an apparent paradox for the interpretation of quantum reality, which seemingly requires us to regard true reality at some moments to be a mix of different everyday realities. But what does it mean for a cat to be present in a mixture, half-dead and half-alive? Physicists have produced dramatic pictures of these strange entangled states of reality. The glimpses that these images give us of the realm of the very small show what might lie ahead. Already we can manipulate individual atoms to engineer on the scale of atoms.

The exploitation of the small seems to be the road to the future. It may also show us how narrow-minded we have been about extraterrestrial forms of intelligence. We have always imagined advanced civilisations in the Universe doing spectacularly big things, such as engineering planets and stars, which require huge amounts of energy. Astronomers even fashioned searches for the waste products from such energetically extravagant activities. Now we can appreciate that this is unlikely to be a sustainable path of technological development. Advanced civilisations are likely to engineer smaller and smaller, using less and less raw material, generating a minimum of waste energy and pollutants. Their space probes are likely to be imperceptibly small.

Last, we take a look at some of the larger consequences of developments in materials science. The legendary picture of Moore's law of progress in the computer business tells us something about computers, but it also revolutionised the business of their production because it facilitated the coordinated development of hardware and software. The world wouldn't look the same without the 'Xerox' machine, but who made the first Xerox copy, and how?

Our final chapter examines the sandpile, a simple image with a deep message. It has become a paradigm for the growth of organised complexity in the face of chaotic unpredictability. Its creators hoped that it might explain the growth of all complexity. Alas, it doesn't, but it restores our faith in the usefulness of simple pictures. You don't need multi-million-dollar equipment to make important discoveries about how Nature works. Simple things, rightly seen, fully interpreted, can provide a new perspective on reality. A picture is all you need.

Yousuf Karsh's
1950 portrait of
Albert Einstein.

THE ACCEPTABLE FACE OF SCIENCE

EINSTEIN AS ICON

A genius! For thirty-seven years I've practised fourteen hours a day, and now they call me a genius!

Pablo Sarasate

SCIENTIFIC celebrity has a relativity all of its own. Some scientists are celebrated by their peers, some are treasured by their students, whereas others are lionised by the public at large. But very few are given the burden of being a celebrity to everyone, everywhere, all the time. Albert Einstein achieved that universality. Publicists must despair that he did so in the pre-TV era, without an agent or the help of a public relations company. He didn't even have a website. More striking still, I doubt that more than a handful of readers have ever seen a colour photograph or a film of Einstein, and even fewer know what his voice sounded like. Yet his face has become an icon for wisdom, imagination, creativity, and concentrated mental power; his brand name so pervasive a synonym for genius that he once had to confess that 'I am no Einstein.'

There have always been scientific celebrities. Isaac Newton was the ornament of his age: people talked about him, made fun of him, and even devised Newtonian models of government and morals. But Newton never became an icon. Instead, he remained severe and aloof. In the nineteenth century, Charles Darwin became, with Thomas Huxley's assistance, a reluctant public intellectual. Although he kept to the wings, his ideas were centre-stage in long-running controversies about science and religion, the origins of humankind, and our relationship to hairier members of the animal kingdom. In retrospect, what is interesting about Darwin's work is that at some popular level it was rather *too* well known. People *did* think that they could understand what he was saying. Indeed, that was the problem – they understood too well.

Einstein restored faith in the unintelligibility of science. Everyone knew that Einstein had done something important in 1905 (and again in 1915), but almost nobody could tell you exactly what it was. When Einstein was interviewed for a Dutch newspaper in 1921, he attributed his mass appeal to the mystery of his work for the ordinary person:

> Does it make a silly impression on me, here and yonder, about my theories of which they cannot understand a word? I think it is funny and also interesting to observe. I am sure that it is the mystery of non-understanding that appeals to them . . . it impresses them, it has the colour and the appeal of the mysterious.[2]

Relativity was a fashionable notion. It promised to sweep away old absolutist notions and refurbish science with modern ideas. In art and literature, too, revolutionary changes were doing away with old conventions and standards. All things were being made new. Einstein's relativity suited the mood. Nobody got very excited about Einstein's theory of Brownian motion or the photoelectric effect (for which he was awarded the Nobel Prize for physics), but relativity promised to turn the world inside out.

The fact that it was supposedly understood by a mere handful of people on the planet was a comfort and added cachet. You didn't need to try to understand and no one would think you stupid for not knowing. Especially appealing was the fact that Einstein's ideas about space and time used familiar words like mass and energy but endowed them with new and richer meanings. Everyone knew what these words meant, but sentences such as 'everything is relative' or 'moving clocks run slow' did not by themselves convey useful information. This irony was beautifully captured by Rea Irvin's cartoon for the *New Yorker* Magazine in 1929, which showed a New York street scene composed of a baffled road sweeper scratching his head, surrounded by tradesmen and passing shoppers in various poses of puzzlement. The caption quotes Einstein: 'People slowly accustomed themselves to the idea that the physical states of space itself were the final physical reality.'[3]

A younger, dapper Einstein, *c.* 1905.

Einstein's celebrity began in 1919 when the British eclipse expeditions confirmed his prediction of the extent to which light was bent by the gravity of the Sun (see pages 169–71). This was headline news and in the years following the misery of the First World War it re-created a sense of order and hope about the Universe and about human enquiry. Pictures of Einstein were then quite different. In 1905, he was a dapper young man working in a Swiss patent office. He was usually shown in a smart European-style suit and bow tie, and would be instantly recognisable as an archetypal European professor.

After the Second World War, Einstein moved across the Atlantic to Princeton, where he became increasingly isolated from the main currents of development in theoretical physics because of his aversion to quantum theory. With no research group or school of students, he pursued an isolated search for an extension of the general theory of relativity which would unite it with the theory of electromagnetism to produce a 'unified field theory' – the forerunner of today's 'Theories of Everything'. It is from this latter part of his life that his iconic legacy principally derives. The slightly bohemian-looking, dishevelled old gentleman with the sad eyes – captured so beautifully by Yousuf Karsh's camera in 1950 – became the symbol of the power of the mind.

Several things about Einstein's delayed celebrity are interesting. First, Einstein's image in America contrasts sharply with that of the smart young man from Switzerland who made the acclaimed discoveries. The ubiquity of the later Einstein image meant that the concept of the scientific genius became associated with the image of an old fatherly figure.

The other great Einstein transformation, which occurred as the icon was emerging, is in some ways more interesting. This was in the style of his scientific work. In the brilliant ideas of 1905, we see the hallmarks of the young Einstein: penetrating ideas, motivated by crucial thought-experiments, and finally fashioned into perfect form by mathematics that is as simple as possible but no simpler. Yet the development of the general theory of relativity introduced Einstein to the power of abstract mathematical formalisms, notably that of the tensor calculus. A deep physical insight orchestrated the mathematics of general relativity, but in the years that followed the balance tipped the other way. Einstein's search for a unified theory was characterised by a fascination with the abstract formalisms themselves. Rather than using physical arguments or thought-experiments, the strength and degrees of completeness of the formalisms themselves were used to sift their appropriateness as physical theories.

After his death in 1955, Einstein became the epitome of intelligence in general, not just of scientific genius. His image appeared on postage stamps and magazine covers. He became the subject of poetry, sculpture, and painting. 'Einstein advertisements' implied that buyers were smart to buy the product, or likely to become smart if they did. Conversely, the simplicity and ease of use of a product was endorsed by the assurance that you did not need to be an Einstein to use it. The most elaborate advertising campaign that exploited Einstein's image was the 1979 Pennaco Hosiery catalogue, intriguingly entitled *The Theory of Relativity–Fall–Winter 79*. Women were exhorted to:

> 'Believe in the theory of relativity'. Fashion relates beautifully – making each part relative to the next – and thus creating fantastic total outfits. Thank you, Mr Einstein, for the theory – (sort of) . . . on to the facts – and practice – and happy 100th Birthday! What would we have done without you!

Today, the Einstein industry is as strong as ever. I was surprised to find that I owned an Einstein suit, an Einstein toy, and inevitably an Einstein calculator.

The most amazing thing of all is that – despite the hullabaloo and the inevitable cynicism about celebrity in our age, especially in response to media-created icons – Einstein's scientific legacy is greater than ever. His predictions about gravity have steadily been confirmed with just a few remaining out of reach of our experiments. His scientific work was an unqualified success and his personal demeanour and response to fame an object lesson to all. His picture will continue to act as an image of physics and the quest to understand the Universe.

Medio, vel tenebris terminabatur.

Cæterum non opus est ut Hypotheses ejusmodi refutem, quæ, ex inventâ tandem veritate, suâ sponte corruent. Satius itaque fecero, si proferam experimentum tandem, quo omnia quæ de genesi colorum hactenus explicui, non modò probari possunt, sed etiam videri. Quamobrem sit ABCac prisma quod radios per foramen F in obscuratum cubiculum transmissos

40

refringat versus lentem MN ut colores quos efficit in p, q, r, s, t per lentem deinde trajiciantur ad X et ibidem commisceantur in albedinem, ~~pro~~ sicut in præcedentibus ostendi. Deinde aliud prisma DEGdg priori parallelum ad locum X, ubi albedo redintegrata est, statuatur, quod lucem versus Y refringat. Hujus autem prismatis verticalis angulus Gg sit æqualis angulo verticali Cc prismatis anterioris, aut eo

93. Instrumentum scribitur quocum omnia de coloribus hactenus dita dilucidissime probantur.

Numb 54

Extract from Newton's notebook recording his experimental investigations into the splitting of white light into colours by a prism of glass. Here he shows the effects of a sequence of prisms.

UNWEAVING THE RAINBOW

NEWTON'S PRISM

Of Newton, with his prism and silent face,
The marble index of a mind forever
Voyaging through strange seas of Thought alone.
William Wordsworth[4]

NEAR the end of his life, Sir Isaac Newton recalled the day on which he bought a simple glass prism at the Stourbridge Fair in August 1665.[5] He had long been interested in the effects of glass on sunlight and the origin of coloured light. The experiments he performed with that prism led to our modern understanding of the refraction of light, the design of reflecting telescopes, and a full understanding of why rainbows form. His work on optics and light culminated in the publication of his second great book, *Opticks*, in 1704.

Newton did many deep and difficult things. He was multitalented. One of the most powerful mathematicians who has ever lived, he was also a gifted experimentalist who not only built high-precision instruments, such as the first reflecting telescope, from scratch, but even made the tools with which to build them. He was the first scientific celebrity in England, 'the ornament of the age', and was knighted by Queen Anne for his 'political service'. His books were only accessible to the mathematically trained, but everyone liked to talk about them, and he suffered the occasional lampoons of the popular press as a result. He was aloof and severe, autocratic and impatient with rivals, spending the later part of his career as President of the Royal Society and Master of the

Royal Mint. At the Mint, he made a number of important innovations, including the introduction of milled edges to coins to show if silver had been illicitly clipped from the edges. When Newton began his work at the Mint it was under great pressure. The country was running out of coins; there was threat of war with the French and no coins to pay the army. 'Clippers' were systematically stealing silver from the edges of coins before exchanging them for new unclipped coins. A new coinage was scheduled, but French agents were keen to sabotage it and the Mint, then located in the Tower of London, was an uneasy mix of minters and the military, each suspicious of the other, and anxious to be in the ascendant. Accidents happened. Not surprisingly, Newton did not think it safe to live in the rooms allocated to him in the Mint and took rooms in Jermyn Street instead. Yet Newton saved the day, successfully supervising the new coinage and foiling the clippers.

Perhaps because of the mathematical difficulty of so much of Newton's scientific work, the part that is most widely known and appreciated is the understanding of colour that his prism made possible. Unlike the work on motion and gravity, it was published in English rather than Latin. To the popular mind – and even to many scholarly minds – this was part of the mystique of Newton. For thousands of years, knowledge about the world was believed to be attainable only by study of the ancient Greek philosophers, especially Aristotle, who produced the first natural philosophy. Understanding grew out of linguistic skill, scholarly interpretation, and respect for ancient authorities. But Newton's work exemplified a new 'experimental' philosophy. To understand the way the world worked, you didn't need to read old books. You could observe and experiment and hypothesise and test for yourself. Newton did all these things better than anyone, and he often did it using extraordinarily mundane objects. Who would have thought that the nature of light could be understood by means of a simple glass prism that you could buy in the market? And, most important of all, each of us could easily acquire a prism, and do Newton's experiment for ourselves. This was science as public knowledge at its best.

Before Newton began to think about the prism, the general view of light was that it was white, but it was possible to add colours to it. Newton demonstrated that white light was just a mixture of different colours of light, which he first called the 'spectrum' of light in 1671. Different colours of light are bent, or 'refracted', by different angles when they pass through a glass prism into air.

This creates the vivid splitting of the white light into the spectrum of red, orange, yellow, green, blue, indigo and violet (memorised by countless generations of English schoolchildren by means of the mnemonic Richard Of York Gave Battle In Vain[6]).

Newton's sketch of his 'laboratory' and optical set-up. Light enters the room through a small hole in his window shutter on the right and either passes only through the lens to focus at a small spot on the bottom of the board on the left, or it passes through the lens and prism. Light passing through the prism forms a spectrum of colours at the top of the board. Newton has made small holes in the board so that only one of the refracted colours passes through each of them. He then passes each single colour of light through a second prism behind the board. This enables him to show that its path is bent by the prism, but following the second refraction its colour is unchanged. In the top corner Newton has written 'Nec variant lux fracta colorem', or 'nor does the refracted light change colour'.

By adding a second prism, Newton was able to feed coloured light back through the prism to reproduce the original white light. This experiment enabled Newton to understand why images from refracting telescopes, which used lenses, were blurred by colours (so called 'chromatic aberration'); although he wasn't able to remove it completely, this led to his invention of the reflecting telescope, which used mirrors rather than lenses to magnify images without causing aberration. All major astronomical telescopes are now based on Newton's idea. However, it was Newton's explanation of the rainbow that was most impressive for his contemporaries.[7] Some, like the poet John Keats, saw it as a subtraction of some of the world's magic and mystery, by which Newton did 'unweave a rainbow'.[8]

He showed that the refraction of sunlight by water droplets was responsible for the production of the rainbow of colours. The same effect can be produced on a smaller scale by light passing through the spray of water in the air near a waterfall.[9] If the water droplets are very fine, like those in a fog, the bow will be almost white and this phenomenon was accurately captured in Turner's famous painting of *Buttermere Lake*, in 1798.

Water droplets are almost spherical. Sunlight is refracted by the water and then reflected back off the opposite face of the droplet, and refracts again as it passes back into the air with its greatest intensity at an angle about 42 degrees to the incoming path.

The dispersed colours are focused through a point inside the droplet, and so the spectrum crosses itself before exiting into the air, and the rainbow forms in the sky opposite the Sun, where it is easy to see. This means that, although blue light is bent most by the refraction, it is the red light that ends up highest in the sky and forms the outer colour of the rainbow.[10] If the Sun is higher than 42 degrees in the sky, the rainbow cannot be seen from ground level because it is below the horizon. The whole arc of the rainbow spans 84 degrees on the sky and requires a wide-angle lens to capture it well.

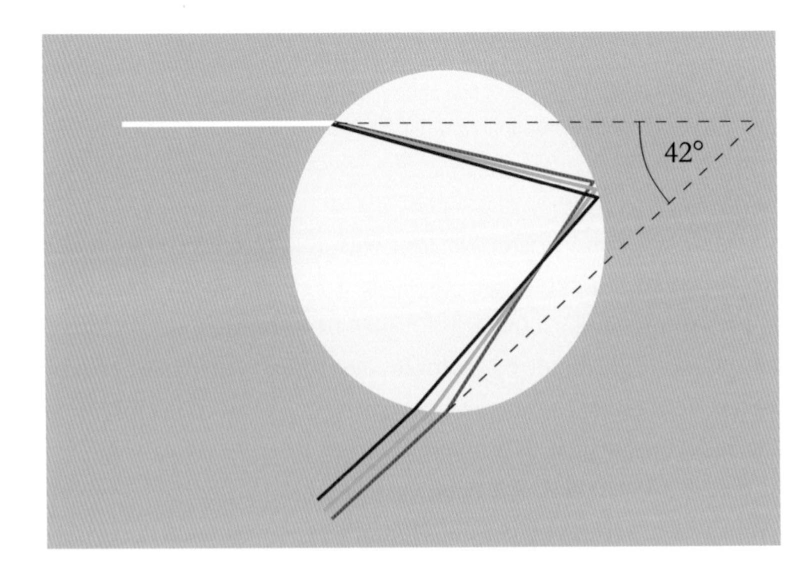

Rainbow formation by the refraction of sunlight by water drops. The rainbow can only be seen above the horizon when the sun is lower than 42 degrees on the sky.

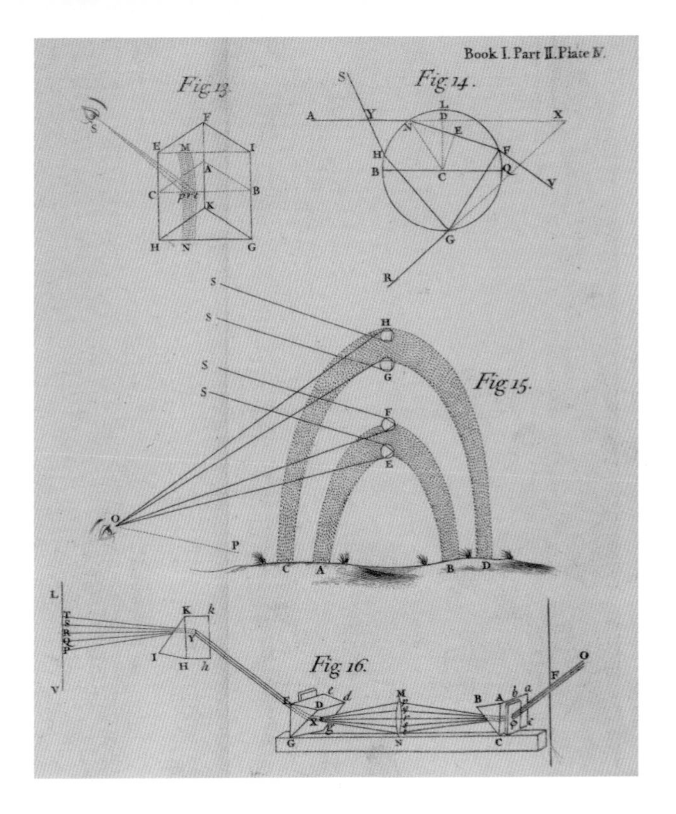

The seven colours of the rainbow have a curious history. In reality, all wavelengths (that is, colours) of light are present in the make-up of white light. Newton chose the colours that make the biggest impression on the eye to define the parts of the spectrum. The choice is fairly clear at the yellow–orange end of the spectrum, but at the blue end it is not at all obvious how you distinguish blue from indigo and violet. Many of these colours change their appearance as their brightness alters. The yellow–orange[11] begins to appear brown at low intensity. The existence of indigo[12] has always been controversial. We do not really see it as a separate colour (although denim blue jeans try hard to make it a fact of life) and it is often argued that Newton wished the spectrum to have seven colours for reasons of religious or Pythagorean harmony, and introduced an innocuous indigo just to make the six distinct colours into seven.[13]

A plate from Newton's *Opticks* (1704) shows the optical path of light that leads to a rainbow and the basics of the refraction process when light passes through prisms and a lens.

Today, Newton's decomposition of white light into its spectrum of composite colours by his simple prism is a symbol of the Newtonian analysis of the world. It was the basis for the development of our theory of light in all its complex manifestations, behaving as though it is both a wave and a particle. Newton did unweave the rainbow, but the threads that he unravelled would twist and turn themselves into an even more fantastic tapestry.

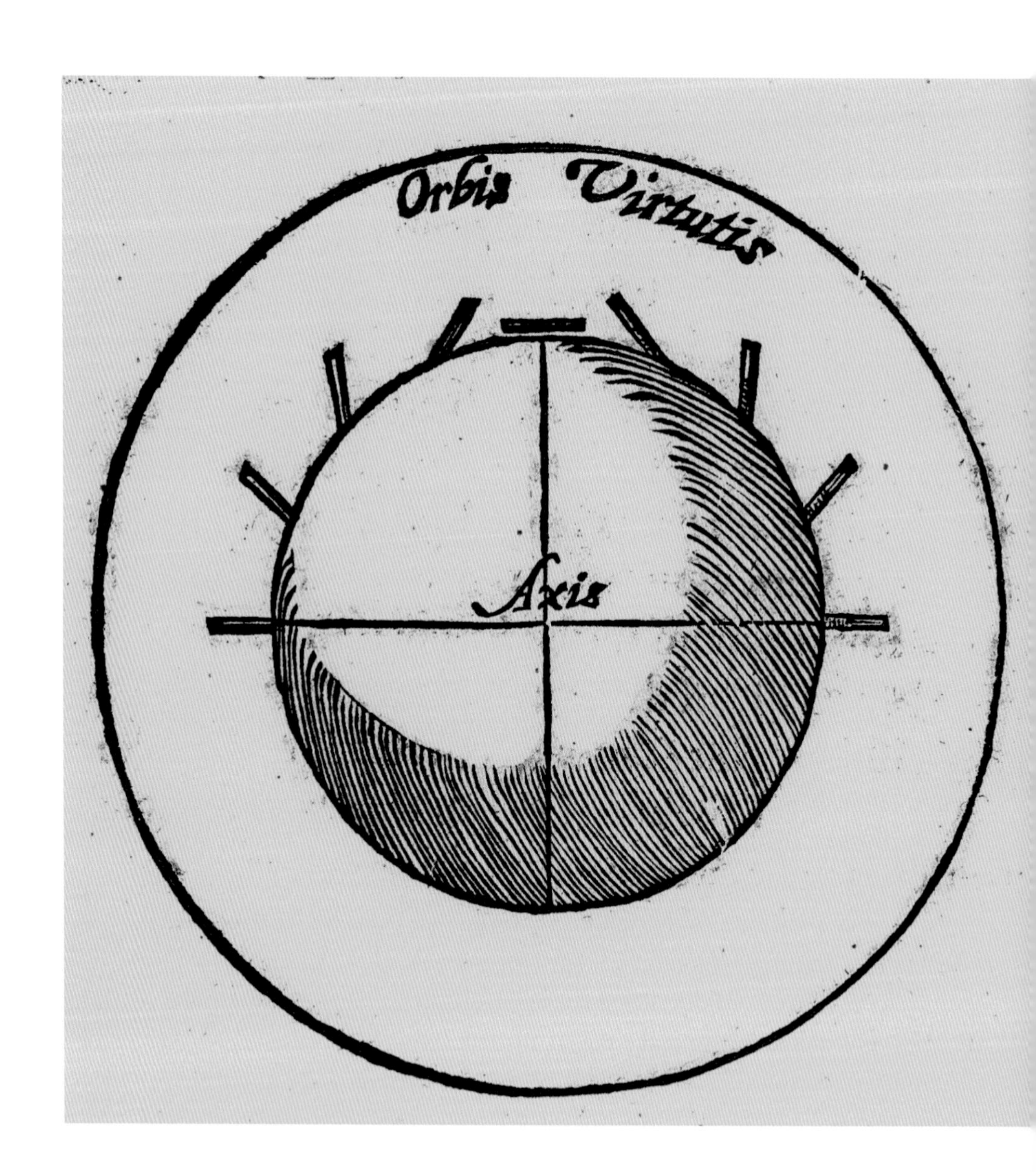

William Gilbert's picture of compass needles dipping in the magnetic field of the Earth, from his classic 1600 book on the nature of Earth's magnetism, *De Magnete*. Gilbert was the first to establish that the Earth acts like a large magnet with north and south magnetic poles.

FOETAL ATTRACTION

EARTH'S MAGNETISM IS BORN

We may see how far from unproductive magnetick philosophy is, how agreeable, how helpful, how divine! Sailors when tossed about on the waves with continuous cloudy weather, and unable by means of the celestial luminaries to learn anything about the place or the region in which they are, with a very slight effort and with a small instrument are comforted, and learn the latitude of the place.

William Gilbert[14]

WHEN scientists are asked to nominate who they think did most to create the modern practice of science, they might be tempted to pick the Italian pioneer Galileo Galilei, the philosopher of scientific method Francis Bacon, the astronomer Johannes Kepler, the great Isaac Newton, or one of the founders of the Royal Society, such as the 'Roberts' Boyle or Hooke. But most of these illustrious candidates made their marks after the ground-breaking work of the remarkable William Gilbert (1544–1603) of Colchester and St John's College, Cambridge, eldest of eleven children of a good family, physician to Queen Elizabeth I, President of the Royal College of Physicians, and the leading medical authority of his day before his death, a victim of the Plague, in 1603.

Gilbert sounds as though his scientific claim to fame would be for advances in medicine. While he was conspicuously successful in that sphere, his fame rests upon a life-long interest that had nothing to do with medicine. His intellectual passion was to understand *magnetism*. He pursued it with a single-mindedness and straightforwardness that was decades ahead of its time. He took no notice of fables and unfounded reports from the past. Instead, he carried out a series of critical experiments to test simple ideas about the action and source of magnetism. Crucially, in order to understand the behaviour of the magnetic field of the Earth, he built a model Earth, a small magnetic sphere of lodestone, or 'terrella' as he called it, and carried out a series of experiments to determine the behaviour of metal needles and other materials near its surface at different positions between its two magnetic poles. He then wrote an account of all these investigations in a clearly illustrated book, published in Latin in 1600, which leads the reader, clearly and simply, from ignorance to natural knowledge.

In many ways, magnetism typifies the ineffectiveness of many strands of ancient science. Knowledge was accumulated, but there was little attempt to reduce it to the action of simple laws, whose properties could be interrogated by creating artificial circumstances. The ancient Greeks knew about the attractive properties of minerals, such as lodestone, which could attract iron, and the Chinese developed working compasses in the tenth century by noticing the constant orientation taken up by a piece of lodestone or magnetised iron placed on a piece of wood and floated on water. Navigators documented the idiosyncrasies of their compasses over the next 500 years but no one understood why they pointed north, and what was the nature of the 'force' that directed them there. All kinds of superstitions remained commonplace – Gilbert blames Ptolemy for originating the crazy idea that garlic cloves nullify the action of a compass and so should never be taken on board ship! So, for all the centuries of navigation with compass needles, all Gilbert knew at the outset was that compass needles point northwards for some mysterious reason.

Gilbert's experimental investigations culminated in a book, *De Magnete,* that was divided into six sub-books. It was revolutionary in style as well as in content. He begins by surveying the beliefs and explanations of the past, not for want of authorities to bolster his own arguments but because he was going to subject them to critical analysis. He introduces the basic facts about lodestone before describing his barrage of simple experiments and their results. The overall presentation amounts to a textbook of the sort that had not been seen before. There were special asterisks in the margins to highlight key points, speculations, summaries of unsolved problems for future work, and many clear diagrams to ease the reader's task. His deductive work, focus upon experimental knowledge, meticulous observations, style of presentation and lack of adherence to ancient authorities provided the model that would inform the creators of the Royal Society of London sixty years later.

The most memorable and influential part of Gilbert's classic unravelling of the nature of magnetism comes in the latter part of the first book, where he explains his momentous discovery that the Earth behaves as if it is a great spherical magnet with a North to South Polar axis running through it. By means of experiments with his magnetised model of the Earth, he could move a small magnetic needle around its surface and show that the model reproduced the behaviour of compass needles at different latitudes on the Earth's surface, pointing always to the Poles. In effect the 'South Pole' of the little bar magnet created by the magnetised compass needle is attracted to the North magnetic Pole of the Earth. And so it was that Gilbert solved a problem that had stared every navigator in the face for 600 years. He saw how this property of the Earth could be used to determine latitude (although the accuracy was not good enough to make this a practical proposition).

His deduction of the magnetic Earth is the beautiful outcome of logical reasoning, small-scale simulation of a large-scale problem, and wonderful physical intuition. On page 444 is his classic picture of compass needles dipping in the magnetic field of the Earth (or 'sphere of power', *Orbis Virtutis*). Having noticed that the magnetic axis of the Earth was close to the axis of rotation, he moved a small compass needle around his small magnetised sphere and saw that it always pointed north–south when placed in a horizontal plane tangential to the sphere, and also dipped downwards at an angle towards the Earth's axis when allowed to pivot on a horizontal axis.[15]

Today, we know that the core of the Earth is surrounded by a mixture of molten iron and nickel that makes it behave as a giant magnet because of the flow of electric current in the molten metal. The powerful magnetic field in the core weakens as it passes out through the surface of the Earth into space. The fluid nature of the magnetic core means that the Earth's magnetic poles are not fixed for all time but slowly wander in direction; nor are they exactly aligned along the Earth's rotation axis. In fact, the North and South Poles of the Earth's magnetic field have swapped over 170 times during the past 76 million years. The last reversal happened about 770,000 years ago. At the moment the Earth's magnetic field is steadily weakening, seemingly heading towards another switch-over in a few thousand years' time.

The Earth's magnetic field is a crucial feature in making the Earth a possible abode for the evolution of life. The Sun emits a steady 'wind' of electrically charged particles, called the 'solar wind'. If the Earth had no magnetic field then over time the impacts of these fast-moving particles would have blown away the Earth's life-supporting atmosphere of nitrogen and oxygen gas molecules. This is what happened to the ancient atmosphere that once existed around the planet Mars. Mars has no significant magnetic field and is defenceless against the electrically charged particles in the solar wind. Earth is different. Its magnetic field deflects the solar wind particles around the Earth's atmosphere leaving it virtually intact.

Go towards the North Pole and you will witness one of the most spectacular of all natural phenomena: incoming solar wind particles exciting atoms in the Earth's atmosphere which subsequently emit their extra energy in the form of light to create the fantastic light-show in the sky that we call the Aurora Borealis, or the Northern Lights. Some of the electrically charged protons and electrons in the solar wind collide with the atoms and molecules in our upper atmosphere and excite them to emit light with different energies, and colours, according to the types of atom being hit. Oxygen collisions produce greens and reds, while nitrogen creates mauves, blues, and pinks.

The terrestrial magnetic field that Gilbert first understood and displayed in his simple picture more than 400 years ago is the reason life is possible on Earth and the source of so much that is predictable, intriguing, and beautiful about the environment in which its inhabitants now live.

The Northern Lights.

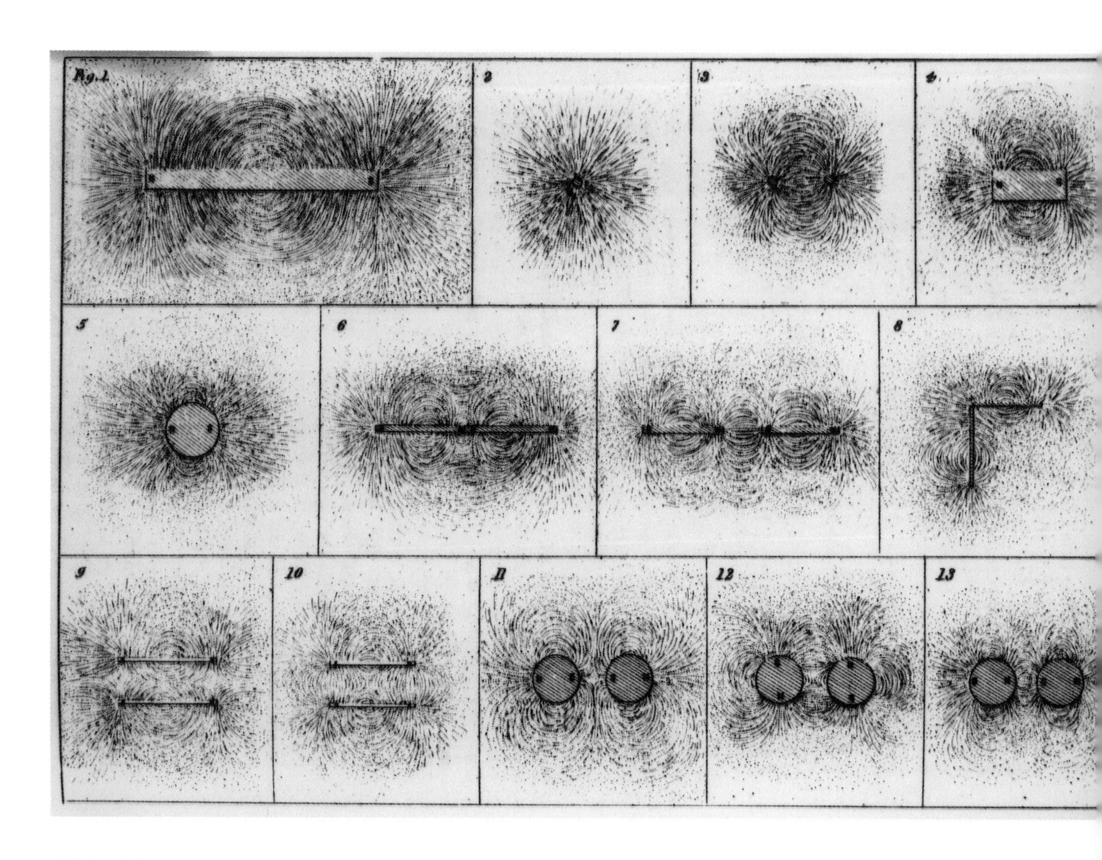

Michael Faraday's pictures from *Electricity* (1852) show the pattern of iron filings near a magnet. Fig. 1 shows the p
around a bar magnet; fig. 2, the pattern around one pole of the magnet; fig. 3, around two opposite poles; fig. 4, ar
small magnet; fig. 5, around a magnetic disc. Figs. 6-13 show other configurations of magnets with different stren

MAY THE FORCE FIELD BE WITH YOU

MAGNETIC FIELD LINES

Faraday, in his mind's eye, saw lines of force traversing all space, where the mathematicians saw centres of force attracting at a distance. Faraday saw a medium where they saw nothing but distance.

James Clerk Maxwell[16]

MICHAEL FARADAY (1791–1867) rose from humble, uneducated beginnings to become the foremost experimental physicist of his times. Even then, he lacked the mathematical knowledge and skill of even the weakest graduating student of his day. Yet, by acute observation and careful creation of decisive experiments, he unveiled a previously unseen world of electrical and magnetic phenomena and made it known both to physicists and to the general public. His view of the world was strongly informed by his non-Conformist religious views and he was a member of the Sandemanian Church formed in 1730 by John Glas as a breakaway from the Scottish Presbyterian Church. Glas was later joined by Robert Sandeman (whose family are still known to us through Sandeman port), and the sect became known as the Sandemanians, while Glas spread the church to the USA. Sandemanians allowed no interpretation of Holy Scripture and therefore no preaching. This would have been viewed as a human distortion of the divine, or 'plain style', of the words of God. There were no clergy and the Church sought its inspiration via direct revelation to its members and elders, of which Faraday was the most eminent.[17] The Church had unusual beliefs about some Christian doctrines, especially the meaning of faith, and very rigid church discipline. Sunday attendance was compulsory and Faraday was excluded from the Church for a period because he had missed a Sunday service – in fact, he had been dining with Queen Victoria.

Faraday's attitude to mathematics and 'theory' was similar to the Sandemanian attitude towards theological exposition: it was an unwanted embellishment of the plain truth.[18] The only way to read the book of God was literally and simply; the only way to read the book of Nature was by experimental inquiry. All additions were subtractions. The use of mathematics was a transformation of God-given reality into a fallen representation: a mere human interpretation.

This literal approach to God's action in Nature led Faraday to take very seriously the concept of magnetic lines of force. Whereas others familiar with this concept of the action of magnetic fields through space regarded it as a useful (but some had said 'senseless') device for visualising what was occurring when a magnet imposed forces of attraction or repulsion on other magnets, for Faraday it was more than a helpful picture. He regarded the lines of force exerted by magnetic fields as entirely real and was able to trace them beautifully by sprinkling iron filings on a piece of paper lying over a bar magnet. This simple experiment has been done millions of times since by school students all over the world. It supported a view of magnetic force that was subtly different to what had gone before. After Newton, scientists began to consider forces acting instantaneously over large distances, a conception generally referred to as 'action at a distance'.[19] No mechanism was suggested by which this action operated. Faraday's picture of force fields saw the influence of the magnetic field filling space with its tentacles, changing in complicated ways when it encountered the effects of other magnets. The simple patterns shown in Faraday's beautiful pictures convinced him that they were really there.

In retrospect, Faraday's simple visual pictures were a watershed in the development of physics. The understanding of electricity and magnetism was about to be revolutionised by James Clerk Maxwell, the greatest British physicist after Newton. His approach was highly mathematical and, where necessary, quite abstract. For Maxwell, visualisation was fine if it was possible, but it was not a requirement for our picture of the world to be correct. Indeed, one might even argue that it might make you a little suspicious. The world is not constructed for our convenience and it might be thought something of a coincidence that the ultimate structure of the Universe and its laws readily admit descriptions in terms of little billiard balls, lines of force, or ropes and pulleys.

After Maxwell, physics moved towards the new insights of quantum theory and relativity that revealed a microscopic and macroscopic world that was nothing like what our everyday intuition about things had led us to expect. The flow of time could be changed; matter could behave like a particle or a wave and be in two places at once. Lines of magnetic force had their place but they also had their limitations. Faraday's genius was to see that they were sufficient to solve the problems that needed to be solved first. His natural ability as an explainer of difficult concepts to non-specialists, even those who like him had no formal education, meant that these simple pictures of magnetic forces were used to explain the discoveries of the moment to the general public at the Royal Institution and other venues, including the Surrey Institution in Blackfriar's Road, where Faraday would regularly talk to the public about his work. His images of fields of force formed a lasting impression upon the minds of scientists, and they remain the first explanatory tool that they reach for.

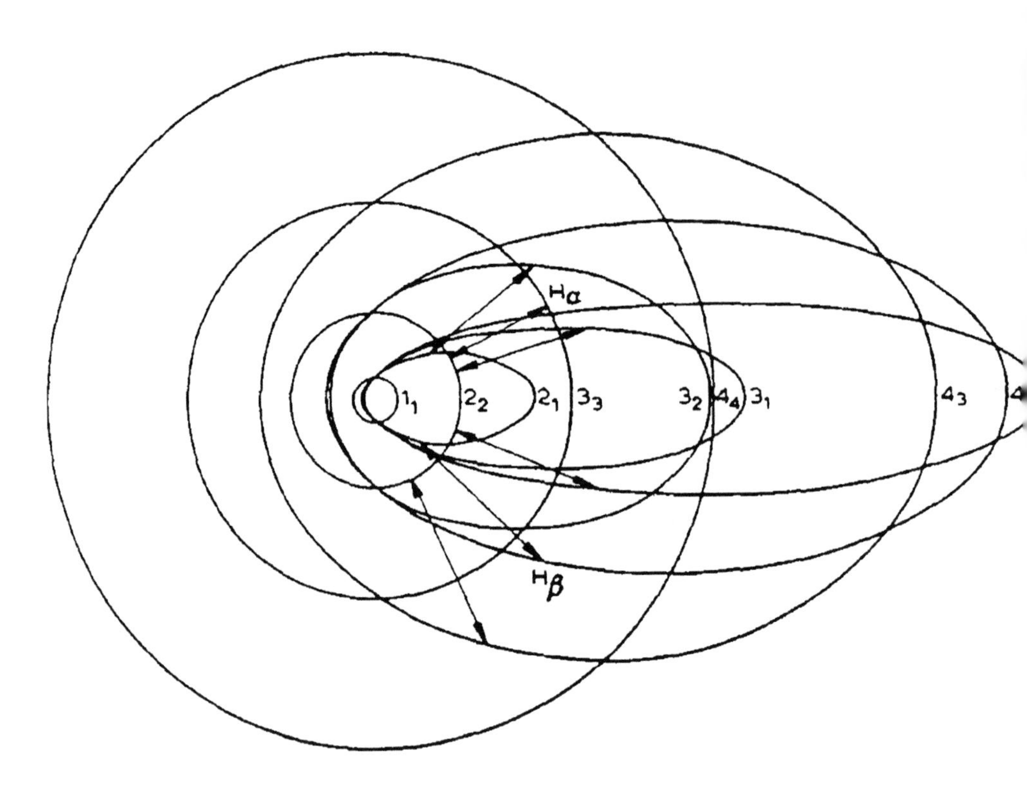

Niels Bohr's diagram of electron orbits and quantum energy levels in hydrogen, from his 1922 Nobel Lecture.

GREAT DANE

BOHR'S ATOM

You see, the chemists have a complicated way of counting: instead of saying 'one, two, three, four, five protons', they say, 'hydrogen, helium, lithium, beryllium, boron.'

Richard Feynman[20]

THE image that many people still have of the atom is that of a mini-solar system: electrons orbit around a nucleus composed of protons and neutrons. The picture seems natural enough. In the real solar system it is the attractive force of gravity that keeps the moving planets along elliptical paths around the massive Sun, providing the inward force needed for orbital motion. Similarly, in the atomic realm, it was first imagined that the electrostatic force of attraction between the negative electric charges of the orbiting electrons and the positively charged protons in the nucleus would perform an analogous function. Alas, such a model fails miserably. Orbiting electrons are accelerating – constantly changing their direction of motion as they orbit – and accelerated charges radiate energy. The electrons would very quickly lose all their energy of motion and spiral into the nucleus. And, even if a cure could be found for that sickness of the model, there is another awkward problem: *every atom would be different.* Take one proton and one electron. This is the recipe to make an atom of hydrogen. But the electron can orbit in a circle around the proton at *any* radius for some speed. In practice, this means that the random formation of hydrogen atoms would result

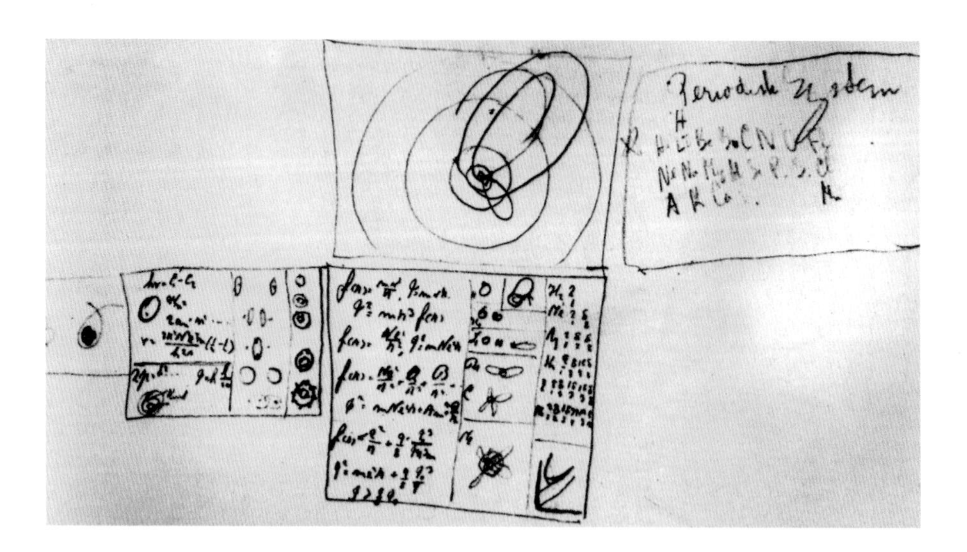

Bohr's notes, from 1921, of the arrangement of electrons in quantised orbits around the nucleus of an atom. The box at the top right, headed *Periodische Zystem* (Periodic System) is the beginning of a full understanding of the Periodic Table of the Elements.

in electrons at different radii moving at different speeds in different atoms:[21] all hydrogen atoms would be different. There would be no reproducibility or stability in the world. Even if they all started out being the same size and energy, they would gradually become different as each of them suffered different perturbations from collisions with other particles and rays of light.

The resolution of this deep problem about the world required a bold and brilliant idea: an idea that changed the nature of physics for ever. That idea was the 'quantum'. In 1900, Max Planck proposed that changes in energy of all magnitudes could not occur. There was a smallest possible change (a 'quantum' of energy) and all changes in energy had to be equal to some whole number of these quantum units. This simple, but radical, idea of 'quantisation' allowed Planck to provide a beautiful explanation of the behaviour of heat radiation, from which flowed a new understanding of the interactions between light and matter. We saw its universal presence on pages 92–98.

The next step up the staircase towards an understanding of the quantum hypothesis was taken by the young Danish physicist Niels Bohr in 1913.[22] Following Planck's idea, Bohr changed the conception of the mini-solar system model of the atom by proposing that the energy of an atom can only change by multiples of a whole quantum.[23] This produced predictions for the wavelengths of light emitted when electrons moved from high to low energy levels which perfectly matched the observations of real atoms. It also provided an elegant solution to the problem of the stability of matter. If a single electron was put in orbit around a single proton, held by the electrostatic force of attraction between their opposite electric charges, then the energy of the orbit could only take particular values – multiples of the basic energy quantum – and the electrons could only reside in one possible orbit nearest to the proton. Once set in that orbit the electrons do not radiate and cannot drift steadily off in energy and orbital radius as a result of the buffeting of incoming waves of radiation. In order to change their energies they have to be hit by multiples of a whole quantum of energy in order to shift them into another allowed orbit farther from the nucleus, or remove them completely from the atom, leaving it in a positively charged (or 'ionised') state. As a result, the quantisation of energy formed the basis for the whole stability of matter and the reproducibility of those atomic and molecular properties that makes our existence possible.

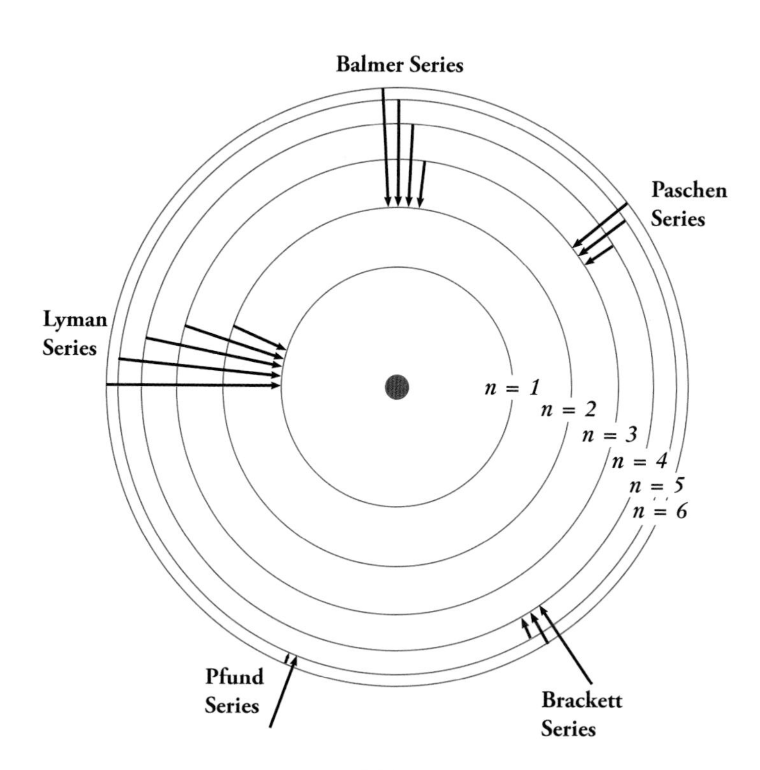

When electrons move from one quantised orbit to another with lower energy, light is emitted. The series of transitions that end at each electron orbit are named after Lyman, Balmer, Paschen, Brackett and Pfund. Only six energy levels are shown.

If the number of quanta of energy that are emitted when an electron comes in from far away to reside in an orbit around the atomic nucleus is denoted by n, then the allowed stable orbits that the electron can reside in are labelled by $n = 1$ (which is closest to the nucleus), 2, 3, . . . and so on. When an electron jumps from a distant orbit (with large n) to one nearer the nucleus (smaller n) then the change in radius of the orbit can be calculated along with the change in the total energy required to move orbits. The energy change makes energy available so that light can be emitted when there is a transition to a lower energy orbit.[24] This produced an energy-level diagram that logged all the transitions that occur from the orbits with quantum numbers $n = 2, 3, 4, 5, \ldots$ etc., down to the lowest (or 'ground state') energy level with quantum number $n = 1$, and was called the *Lyman series* (after an American, Theodore Lyman (1874–1954)); those from orbits with quantum numbers $n = 3, 4, 5, \ldots$ to the $n = 2$ level, called the *Balmer series* (after a Swiss, Johann Balmer (1825–98)); from orbits with $n = 4, 5, \ldots$ to the $n = 3$ level, called the *Paschen series* (after a German, Friedrich Paschen (1865–1947)); from orbits with $n = 5, 6, \ldots$ to the $n = 4$ level, called the *Brackett series* (after an American, Frederick Brackett (1896–1980)); and from those with $n = 6, 7, \ldots$ to the $n = 5$ level, called the *Pfund series* (after an American, August Pfund (1879–1949)).

Bohr's beautiful conception of the hydrogen atom – which could be extended to describe all other types of atom – gave rise to a picture which shows the scheme of transitions between quantised outer and inner orbits ('Bohr orbits') occupied by electrons around the nucleus, corresponding to the different spectral series.

— 70 —

но въ ней, мнѣ кажется, уже ясно выражается примѣнимость выставляемаго мною начала ко всей совокупности элементовъ, пай которыхъ извѣстенъ съ достовѣрностію. На этотъ разъ я и желалъ преимущественно найдти общую систему элементовъ. Вотъ этотъ опытъ:

			Ti=50	Zr=90	?=180.
			V=51	Nb=94	Ta=182.
			Cr=52	Mo=96	W=186.
			Mn=55	Rh=104,4	Pt=197,4
			Fe=56	Ru=104,4	Ir=198.
			Ni=Co=59	Pl=106,6,	Os=199.
H=1			Cu=63,4	Ag=108	Hg=200.
	Be=9,4	Mg=24	Zn=65,2	Cd=112	
	B=11	Al=27,4	?=68	Ur=116	Au=197?
	C=12	Si=28	?=70	Sn=118	
	N=14	P=31	As=75	Sb=122	Bi=210
	O=16	S=32	Se=79,4	Te=128?	
	F=19	Cl=35,5	Br=80	I=127	
Li=7	Na=23	K=39	Rb=85,4	Cs=133	Tl=204
		Ca=40	Sr=87,6	Ba=137	Pb=207.
		?=45	Ce=92		
		?Er=56	La=94		
		?Yt=60	Di=95		
		?In=75,6	Th=118?		

а потому приходится въ разныхъ рядахъ имѣть различное измѣненіе разностей, чего нѣтъ въ главныхъ числахъ предлагаемой таблицы. Или же придется предполагать при составленіи системы очень много недостающихъ членовъ. То и другое мало выгодно. Мнѣ кажется притомъ, наиболѣе естественнымъ составить кубическую систему (предлагаемая есть плоскостная), но и попытки для ея образованія не повели къ надлежащимъ результатамъ. Слѣдующія двѣ попытки могутъ показать то разнообразіе сопоставленій, какое возможно при допущеніи основнаго начала, высказаннаго въ этой статьѣ.

Li	Na	K	Cu	Rb	Ag	Cs	—	Tl
7	23	39	63,4	85,4	108	133		204
Be	Mg	Ca	Zn	Sr	Cd	Ba	—	Pb
B	Al	—	—	—	Ur	—	—	Bi?
C	Si	Ti	—	Zr	Sn	—	—	—
N	P	V	As	Nb	Sb	—	Ta	—
O	S	—	Se	—	Te	—	W	—
F	Cl	—	Br	—	J	—	—	—
19	35,5	58	80	190	127	160	190	220.

Mendeleyev's periodic table. Printed in 1869, this is the first version of the periodic table drawn up by the Russian chemist Dmitri Ivanovich Mendeleyev (1834-1907). He left gaps in the table for new elements, which were indeed later discovered, vindicating his theory. This version has the elements listed using chemical symbols, and ordered by atomic weight, but the fully correct chemical arrangement has not yet emerged. The final version (1871) had the elements arranged in the familiar vertical columns or groups.

ALL HUMAN LIFE IS HERE

THE PERIODIC TABLE

As long as chemistry is studied there will be a periodic table. Even if someday we communicate with another part of the Universe, we can be sure that one thing both cultures will have in common is an ordered system of the elements that will be instantly recognizable by both intelligent life forms.

John Emsley[25]

THE ancient Greek philosophers, following Aristotle's lead, believed that the material world, and all the transformations of its constituents that we see around us, could be explained in terms of just four basic substances: earth, fire, air, and water. This simple-minded view held sway until the seventeenth century, when chemistry, born of alchemy, began to identify other elemental ingredients. 'Earth' was not one single substance and nor was every gas simply air. The eighteenth century witnessed a dramatic growth in the resulting chemical spectrum. Many new metals were discovered – notably cobalt, nickel, manganese, tungsten, chromium, magnesium, and uranium – and gases such as hydrogen, nitrogen, oxygen, and chlorine were isolated for the first time.

The notion of a chemical 'element' was first defined by the French chemist Antoine Laurent de Lavoisier,[26] in 1789, following Robert Boyle, as something that could not be decomposed any further by physical processes.[27] Lavoisier picked on thirty-three candidates to endow with this elemental status and divided them into four groups: metals, non-metals, earths, and gases. Some of these candidates are now known to be compounds and a few, such as heat and light, are not chemicals at all. Here is his list of elements. Only those in italics are now known to be chemical elements:

Gases: heat, light, *hydrogen*, *nitrogen*, *oxygen*.
Earths: alumina, barytes, lime, magnesia, silica.
Metals: *antimony*, *arsenic*, *bismuth*, *cobalt*, *copper*, *gold*, *iron*, *lead*, *manganese*, *mercury*, *molybdenum*, *nickel*, *platinum*, *silver*, *tin*, *tungsten*, *zinc*.
Non-metals: *sulphur*, *phosphorus*, *carbon*, chloride, fluoride, borate.

The 'earths' are actually all oxides (for example, lime is calcium oxide and silica is silicon dioxide), but it was not possible at that time to remove the oxygen atoms to leave the associated elements free. The other non-elemental entries are in the non-metals group, and again they could not be separated into the basic elements of chlorine, fluorine, and boron with the techniques available to Lavoisier. Lavoisier, who made an enemy of Jean-Paul Marat, was guillotined during the reign of terror, in 1794, because he was involved in the collection of taxes for the state. The judge announced that 'the Republic has no need of geniuses', but just 18 months later the government announced that he had been wrongly executed.

A further step was taken by an English science teacher working in Manchester. In a talk and a paper presented to the Manchester Literary and Philosophical Society in 1805, John Dalton sought to explain the various ways in which elements combined with one another in terms of basic ingredients possessing specific weights. Most chemists of the day regarded 'atoms' as being too small to be studied, but Dalton was more adventurous. He proposed a table of twenty elements, together with their weights, with a symbolic notation for representing their rules of composition. Substances were represented by little pictures composed from their elemental diagrams. From this basic list of elements, further compounds could be arrived at: compound 21 was water and was described as HO; compound 22, NH, was ammonia. These representations provided the germ of the idea for the chemical formulae that we are so familiar with today.

Dalton's notation was a little too complicated to be appealing and the modern chemical symbols that we use are those introduced in 1813 by the

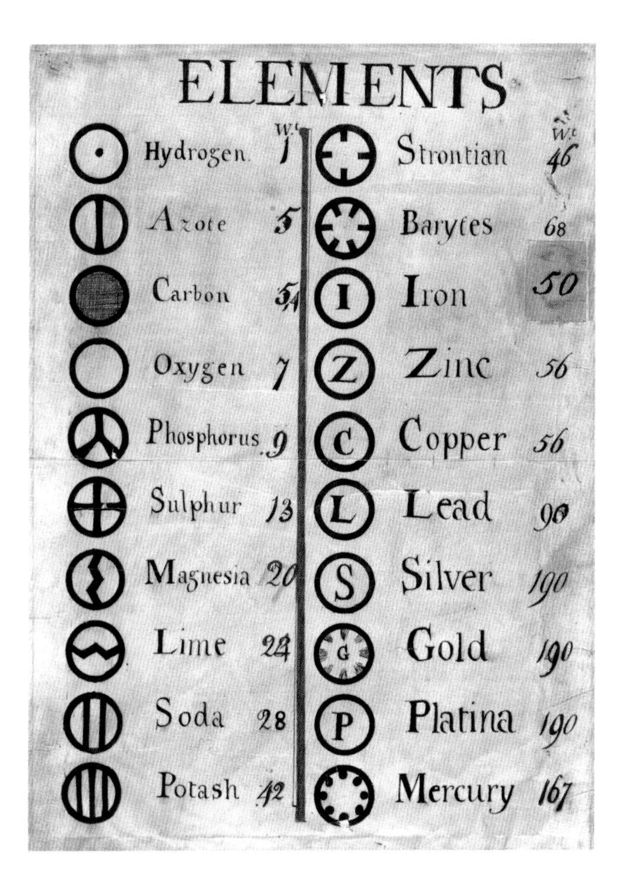

ELEMENTS
W.t
Hydrogen 1
Azote 5
Carbon 5,4
Oxygen 7
Phosphorus 9
Sulphur 13
Magnesia 20
Lime 24
Soda 28
Potash 42
W.t
Strontian 46
Barytes 68
Iron 50
Zinc 56
Copper 56
Lead 90
Silver 190
Gold 190
Platina 190
Mercury 167

John Dalton's 1805 classification of 20 chemical elements, together with their weights.

Swedish chemist Jöns Berzelius,[28] an admirer of Dalton's work.[29] He used the simpler notation of an initial letter for the name of the element (sometimes from its Latin or even, in the case of potassium, its Arabic name), or two letters if there would otherwise be ambiguity, as with C for carbon but Co for cobalt. These symbols could be strung together to represent compounds, such as H_2O, very easily and were in general use after about 1835, eventually being used as components of equations representing chemical reactions; for example:[30]

$$CuSO_4 + 2HCl \rightarrow H_2SO_4 + CuCl_2$$

Dalton was horrified by the new complexities of chemical language – 'A young student in chemistry might as soon learn Hebrew,' he said despairingly, on seeing the new scheme.

New elements were being discovered rapidly, and Lavoisier's original earths were decomposed into their true elemental constituents by Humphry Davy using electrolysis. By 1863 there were more than sixty elements, and it was tantalising to ask if there was a limit to the population explosion – if so, what might it be, and what factors could conceivably determine it?

Valiant attempts were made during the nineteenth century to produce a classification that predicted the properties of elements in terms of their weights. Most of the world's leading chemists[31] had a go at developing something of this sort.[32] Yet they were all scooped by an almost unknown Russian chemistry professor from Siberia.

Dmitri Mendeleyev, one of fourteen children, born in 1822 to the local schoolmaster and his wife in Tobolsk, Siberia, was educated in St Petersburg as a result of his mother's fierce belief that her son was especially talented and should have every possible educational advantage. She was right. He was a top student at the university and moved on to work in France and then to Germany, where he assisted the famous Robert Bunsen – of 'burner' fame – in Heidelberg. Eventually, he returned to St Petersburg to became professor of chemistry at the university in 1867.[33]

One spring day in 1867, he was prevented from going out by bad weather and so worked unplanned on his new textbook, *The Principles of Chemistry*. He encountered the problem of how to present and organise the burgeoning number of chemical elements and their properties. So he wrote the name of each element on a piece of card, along with a list of some of its properties, oxides, and hydrides. Then he set about arranging the cards in all sorts of different ways – rows having the same valency, columns in descending atomic weights – looking for patterns. Suddenly, he found an especially impressive arrangement of his cards. He wrote it down on the back of an old envelope that can still be seen in St Petersburg.[34]

A tidier version soon followed. Mendeleyev selected the first seven elements from lithium to fluorine in increasing weight,[35] and then the next seven, from sodium to chlorine in the same way, horizontally. A periodicity appears: in the vertical columns there are now two chemically similar elements and the main valency for the first of the seven column entries is one, for the next is two, then three, then four, three, and one respectively. Soon, Mendeleyev realised that his table would be clearer if it was turned around by switching the

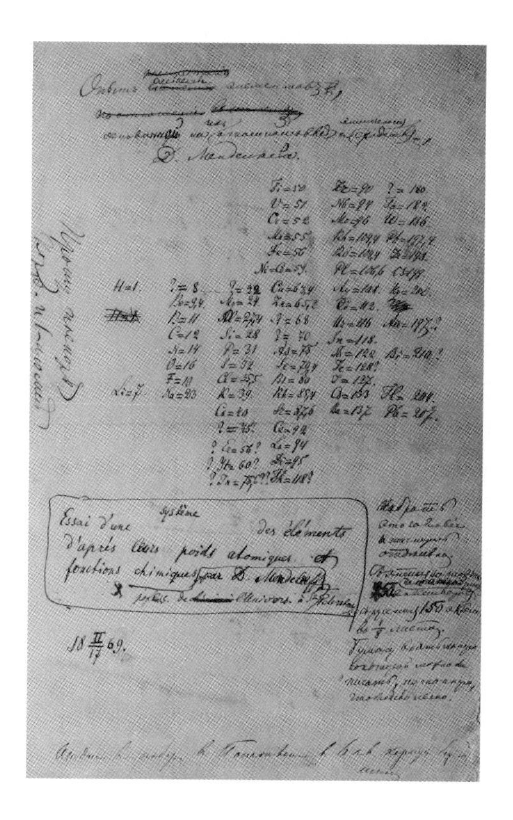

The original periodic table, sketched by Mendeleyev in 1867.

rows and columns. The result is the form that we still recognise today, despite its enlargement as new elements have been discovered.

There are eight columns, or periods. In his next major refinement in 1870 Mendeleyev started by distributing the sixty-three known elements in twelve rows, starting with hydrogen and ending with uranium, in order of increasing atomic weight so that each element is in a column featuring elements that are chemically similar.

One of the most remarkable feats of intuition that Mendeleyev displayed using his table was to predict the existence of new elements. He resisted the urge to force all the known elements into a *complete* periodic table. That's the sort of thing that Aristotle would have done. Instead, he believed that the logical organisation meant there could be gaps in the table, and he predicted that new elements would be found to fill them, using the periodicity properties of the table to predict the details of their expected atomic weights and densities. Below boron, aluminium, and silicon he identified three 'missing' elements which he named 'eka-boron', 'eka-aluminium', and 'eka-silicon'.[36] They were subsequently discovered with the atomic weights and densities he had predicted: eka-aluminium was discovered in Paris in 1875 and called gallium (Latin for France); eka-boron was discovered in Uppsala in 1879 and called scandium (Latin for Scandinavia); eka-silicon was discovered in Freiberg in 1886 and called germanium (Latin for Germany). Mendeleyev also predicted a new member of the 4th (titanium) group of elements with an atomic weight of about 180. It was found in 1923 at the University of Copenhagen, with a weight of 178.5, and named hafnium (Latin for Copenhagen).

Mendeleyev became the Director of the Russian Bureau of Weights and Measures in 1893 and made a memorable contribution. He defined the official alcoholic composition of vodka as one molecule of ethyl alcohol for two molecules of water. The molecular weights imply 38% alcohol versus 62% water, by volume. The legal standard introduced in Russia in 1894 rounded this up slightly to 40% alcohol to 60% water, which is defined as '80% proof' (i.e. proof = 2 x alcohol volume per cent) in America.

A nice analogy is drawn by Gerald Holton which captures the nature of Mendeleyev's success and the big impact it must have had on his contemporaries:

> It is as if a librarian were to put all his books in one heap, weigh them individually, and then place them on a set of shelves according to increasing weight – and find that on each shelf the first book is on art, the second on philosophy, the third on science, the fourth on economics, and so on. Our librarian may not understand in the least what the underlying regularity is, but if he now discovers on one of these shelves a sequence art – science – economics, he will perhaps be very tempted indeed to leave a gap between the two books on art and science and look about for a missing philosophy book of the correct weight to fill the gap.[37]

The periodicity in the table, and hence the reason for its name, can be seen if we plot a property of an element, such as the atomic volume,[38] against its atomic weight, as was first done by Julius Meyer in 1870. The alkali metals appear at the peaks of the graph.

Mendeleyev made no claims to understand the reasons for the structure and periodicity of his table. It was a brilliant intuitive leap. It was his belief that there existed an underlying symmetrical structure into which the elements fitted, rather than simply a desire to provide a useful cataloguing device, that led to his dramatic discoveries and predictions. Although he couldn't explain the patterns present, he knew that their existence would lead others to do so.

In its modern completed[39] form, the periodic table of the elements has seven rows, or periods, along which there are found 2, 8, 8, 18, 32, and 32 different elements respectively. This pattern was fully understood after the quantum theory of the atom was found. The quantum wavelike nature of the electron means that only a whole number of wavelengths can 'fit' around the circumference of each orbit. The numbers arising in the row periods of the table reflect the numbers of electrons in 'orbit' around the central nucleus of each atom. Quantum mechanics allows the innermost orbit (or 'shell', as it is called) to contain 2 electrons, the next to hold 6, then 10, and then 14. The number of entries in the rows of the table determined from these are the numbers of electrons that are in full orbits, so $8 = 2 + 6$, $18 = 2 + 6 + 10$, and $32 = 2 + 6 + 10 + 14$. When we arrange the elements in rows of increasing atomic number with the columns having the same outer electron orbit, we arrive at the modern form of the periodic table. Across each row we are steadily adding electrons to an orbit until it is full, whereupon we end up at one of the noble (that is, chemically inert) gases at the right-hand end of the row. Then we start again in the next row by filling up the next orbit. Remarkably, Mendeleyev was able to see this pattern long before either the electron or the proton was discovered. By fixing on atomic weight (which is determined by the number of protons in the nucleus of the element) and on the valency (which is determined by the extent to which the electron orbits are complete), he captured the essence of their role in a simple way. Today, a picture of his periodic table can be found on the wall of every chemistry laboratory in the world.[40] His mother really was right.

An East German stamp from 1979 celebrating Kekulé's discovery of the structure of benzene.

THE LORD OF THE RINGS
BENZENE CHAINS

I remember ordering Perrier in Boulogne after the benzene scandal and specifying that I wanted my water 'sans benzene'. The joke was not appreciated.

'Everyday life in Hong Kong'[41]

BENZENE was first discovered in 1825 by the English physicist Michael Faraday, who extracted it from gas produced from oil. He gave it the rather dull name 'bicarburet of hydrogen' but it didn't stick. The German chemist Eilhard Mitscherlich called it *Benzin* after distilling it from benzoic acid, made from the resin of the Benzoin tree that is found in Asia and used for a range of aromatic and medicinal purposes, and so, after anglicisation, 'benzene' it remained. A natural component of crude oil, today benzene is one of the most commercially important chemicals. It is contained in petrol, rubber products, dyes, drugs, solvents, and plastics.

For a long time after its discovery, benzene presented chemists with an intriguing mystery. Its chemical make-up of six carbon atoms and six hydrogen atoms (displayed in its simple chemical formula C_6H_6) seemed to present them with an impossible set of combinations between the twelve atoms. Usually, carbon formed four bonds and hydrogen one bond, so how did six of each join together? The key insight that solved the mystery was made by a remarkable German chemist, Friedrich Kekulé, and it created the whole modern field of structural organic chemistry.

As a student, Kekulé was a lively mind with many interests, including an aptitude for languages, of which he spoke many, and any activity that required good spatial awareness. He was an able artist, dancer, and gymnast who set his sights on becoming an architect, and it was with that intent that he enrolled as a student at the University of Giessen, in 1847. But by the time he graduated four years later, he had been seduced away from civic architecture by the discovery of molecular architecture. He became so fascinated by chemistry after enrolling for a minor course that he changed his major course of study completely. After graduating, he did some dull work for a time, testing water samples, before moving to London to work as a research assistant at St Bartholomew's Hospital, in 1853. He quickly made new friends and began taking an interest in their attempts to classify organic chemical compounds according to their structure. He spent many hours discussing chemistry late into the night with his friend Hugo Mueller. After one of these socio-chemical evenings, the bonding patterns of atoms, and the architectures that would make their final complex structures possible, were turning over and over in his mind as he made his journey back to his lodgings late one night. Wearied after a long day, he fell asleep on the open-top bus ride home to Clapham Road. Here is Kekulé's account of what he began to dream:

> One fine summer evening I was returning by the last bus, riding outside as usual, through the deserted streets of the city . . . I fell into a reverie, and lo, the atoms were gamboling before my eyes. Whenever, hitherto, these diminutive beings had appeared to me, they had always been in motion. Now, however, I saw how, frequently, two smaller atoms united to form a pair: how a larger one embraced the two smaller ones; how still larger ones kept hold of three or even four of the smaller: whilst the whole kept whirling in a giddy dance. I saw how the larger ones formed a chain, dragging the smaller ones after them but only at the ends of the chains . . . The cry of the conductor: 'Clapham Road,' awakened me from my dreaming; but I spent a part of the night in putting on paper at least sketches of these dream forms. This was the origin of the 'Structural Theory'.[42]

Kekulé spent the next part of his career developing this dreamy picture of closed chains of atomic bonds into a solid theory of how atoms form chains of atomic bonds, establishing the four-fold valency of carbon in 1857. By now, he had become a well-known figure in theoretical chemistry, and was invited to become professor of chemistry at the University of Ghent the following year. But personal tragedy struck in 1864. His wife Stephanie died just two years after their marriage. He found that he could not work like he used to and just seemed to think aimlessly about chemical compounds such as benzene, unable to concentrate and work on them with any purpose. In these unpromising circumstances, Kekulé had another extraordinary dream, about the structure of benzene, that would change the course of chemistry. Later in life, he recalled that:

> I was sitting writing on my textbook, but the work did not progress; my thoughts were elsewhere. I turned my chair to the fire and dozed. Again the atoms were gamboling before my eyes. This time the smaller groups kept modestly in the background. My mental eye, rendered more acute by the repeated visions of the kind, could now distinguish larger structures of manifold conformation; long rows sometimes more closely fitted together all twining and twisting in snake-like motion. But look! What was that? One of the snakes had seized hold of its own tail, and the form whirled mockingly before my eyes. As if by a flash of lightning I awoke; and this time also I spent the rest of the night in working out the consequences of the hypothesis.[43]

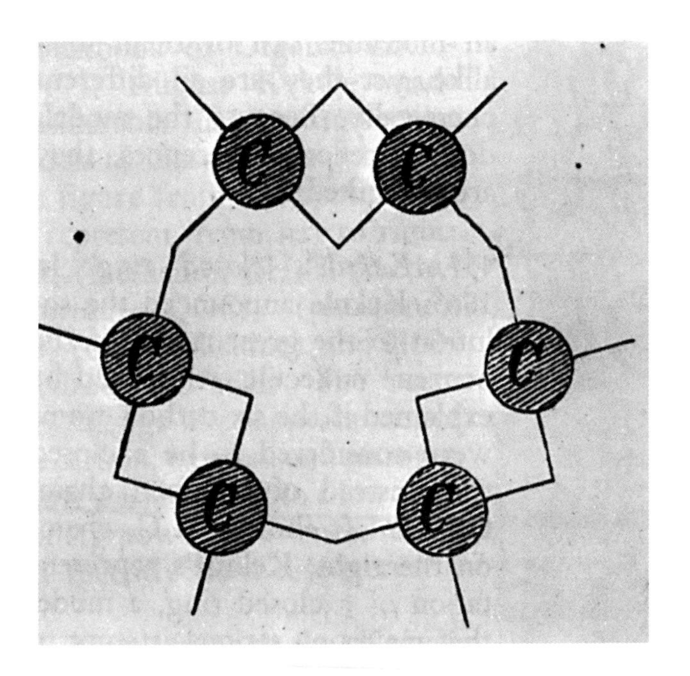

A representation of the benzene ring from Kekulé's *Lehrbuch der organischen Chemie* (1861–1867).

Kekulé soon put flesh on the bare bones of the idea in his dream. In 1865, he attempted to represent benzene by a simple hexagon with one carbon atom at each of its corners. It was the first structure to be proposed with alternating single and double bonds.

A year later he published a more intricate model in which there were alternating single (|) and double (||) bonds in a cyclic structure, like the snake eating its tail in his dream.[44]

In arriving at this picture in 1872, Kekulé contributed a brilliant idea that was to guide the whole field thereafter. He suggested that the atoms in the benzene molecule are oscillating rapidly, and this causes collisions between different neighbouring atoms. The number of collisions that any atom makes in a given period of time determines its valency – the number of bonds it can make with other atoms. So, in its first oscillation, the carbon C1 atom might hit the neighbouring C2, C6, H, or C2 atoms, but in the next oscillation the order of collisions might be reversed to give C6, C2, H or C6 instead. As a result, there would be a double bond linking atoms C1 and C2 during the first oscillation but a double bond linking C1 and C6 during the second. In this way, Kekulé showed how the valency of atoms was related to collisions and hence to the modern picture of resonant oscillations. In 1997, an American Madison Area juggling troupe even created a routine called the 'Benzene Ring for Four Jugglers' which plays out the way the hydrogen and carbon atoms change their bonding partners.[45]

Kekulé was greatly honoured for his successes in chemistry, and when he looked back in later life to reflect on the things that guided him, he liked to highlight the architectural inclinations of his youth that had proved so useful in building another type of structure, and his propensity for being a dreamer – a 'beautiful dreamer' of pictures.

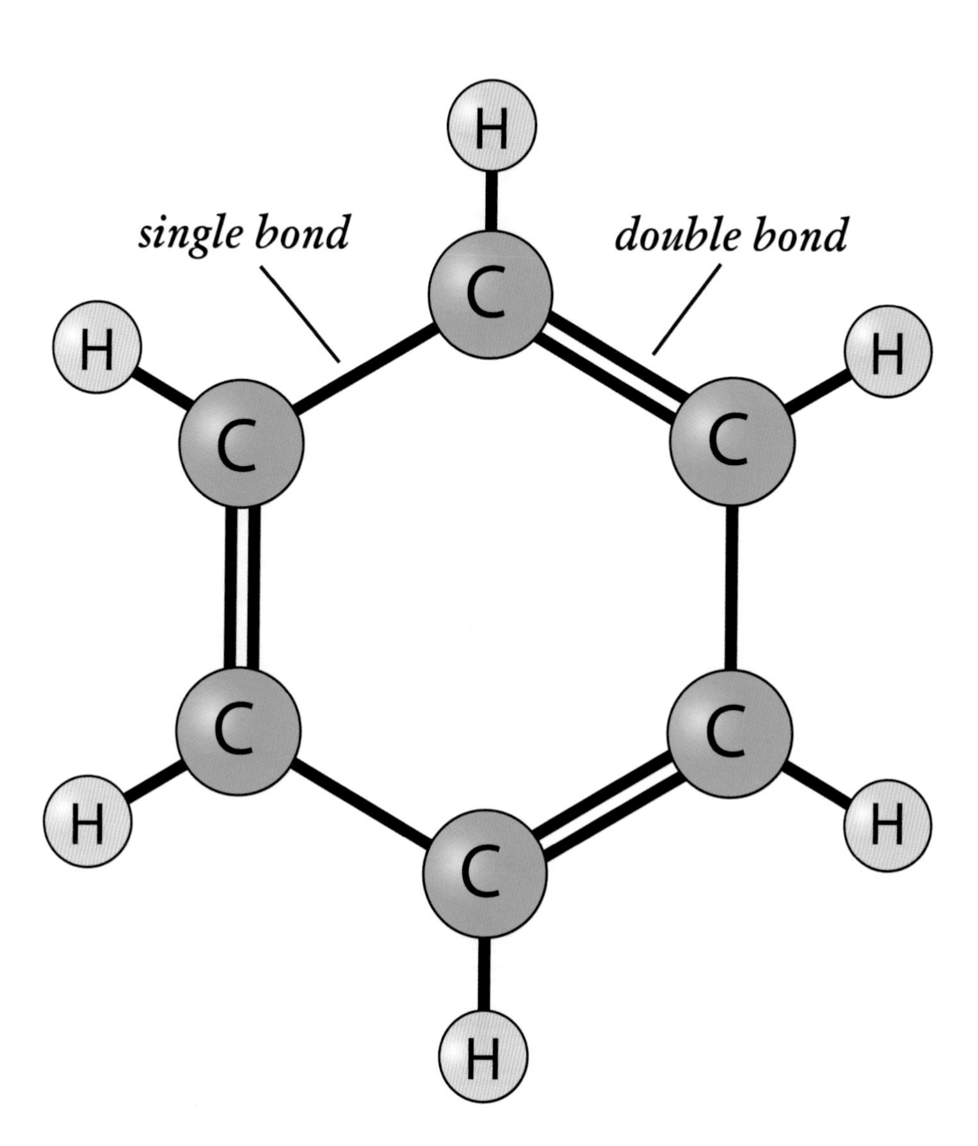

A modern picture of the benzene structure showing the alternate single and double atomic bonds.

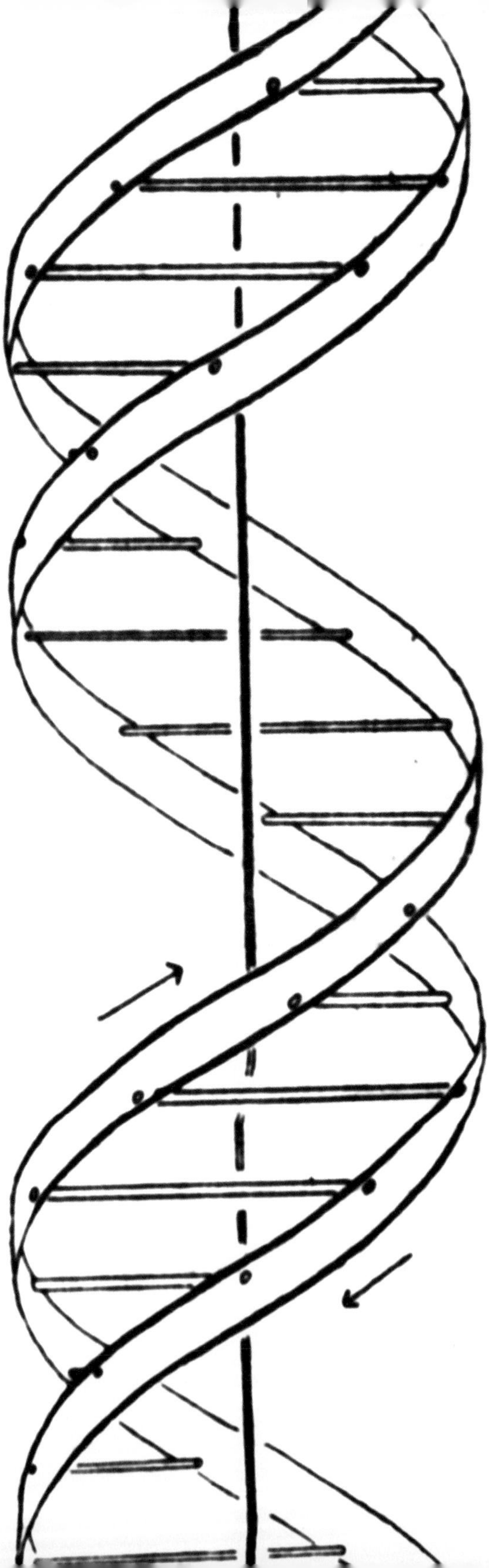

Watson and Crick's 'double helix' from their original *Nature* 'discovery' paper of 25 April 1953.

LIKE A CIRCLE IN A SPIRAL

DNA, THE COIL OF LIFE

1953 became the annus mirabilis. The Queen was crowned. Everest was climbed. DNA was solved.

Max Perutz

WHEN it came to deciding the person of the last millennium there was a good deal of competition. In the UK, the wartime Prime Minister, Winston Churchill, eventually won out over Isambard Kingdom Brunel, Isaac Newton, William Shakespeare, and Princess Diana. In the USA, Albert Einstein, Elvis Presley, and Thomas Jefferson figured highly. In China, it was a straight fight between Chairman Mao and Confucius. There is no hope of a global consensus. But ask scientists for the molecule of the millennium and there will be no uncertainties at all. There is only one winner. Deoxyribonucleic acid – or 'DNA', as it is always known in public – is the most important molecule of all. For without it there would simply be no one to pose the question.

At the heart of every cell in your body is a tiny kernel of sticklike chromosomes inside which genetic information is encoded by the molecules of DNA. DNA looks like a stepladder that has been twisted around; the sides of the spiral ladder are made of sugar and phosphate molecules, while the rungs of the ladder are composed of the nitrogen bases of adenine (A), thymine (T), guanine (G) and cytosine (C) which attach to one another in pairs so that A always pairs with T, and G always pairs with C. This means there are four possible rungs, or base pairings: AT, TA, GC, and CG. When a cell divides into two, it is possible for this structure to be copied to each daughter cell that results. DNA divides like a zip. Each half takes with it one of the sides of the ladder and half of the rungs. Later, they rearrange themselves to produce two sets of paired rungs again, and two identical new sets of the original DNA pattern are created.

In 1953, Francis Crick and James Watson made history by deducing the double helix structure of DNA from images that Maurice Wilkins and Rosalind Franklin had taken of the molecule, using the new technique of x-ray diffraction, and Erwin Chargaff's earlier discovery of the A, C, G, and T components.[46] Chargaff had found that the amount of A equalled that of T and the amount of G equalled that of C, facts which suggested to Crick and Watson that DNA possessed two strands, one linking G to C and the other linking A to T. The helical geometry was hit upon by careful interpretation of the x-ray images, and first appears in a schematic diagram in Crick and Watson's short letter to the journal *Nature* announcing their proposed structure for DNA.[47]

The DNA is very tightly coiled into our cells; each pair is barely a millionth of a millimetre across. Remarkably, a single cell in our bodies contains about 170 cm of tightly wound DNA and so the total length of DNA entwined inside us would stretch to the Moon and back more than 6000 times.

Crick and Watson announced their discovery in the 25 April 1953 issue of the journal *Nature*. Its significance was immediately evident, and ever since

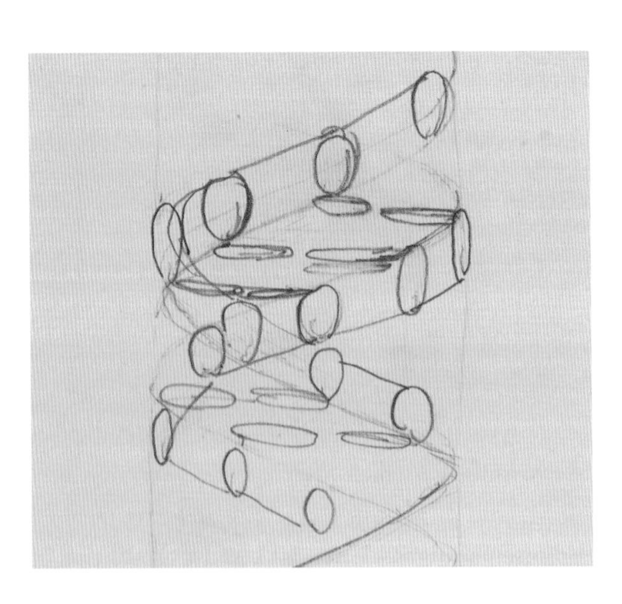

Crick's first sketch of the helical structure of DNA.

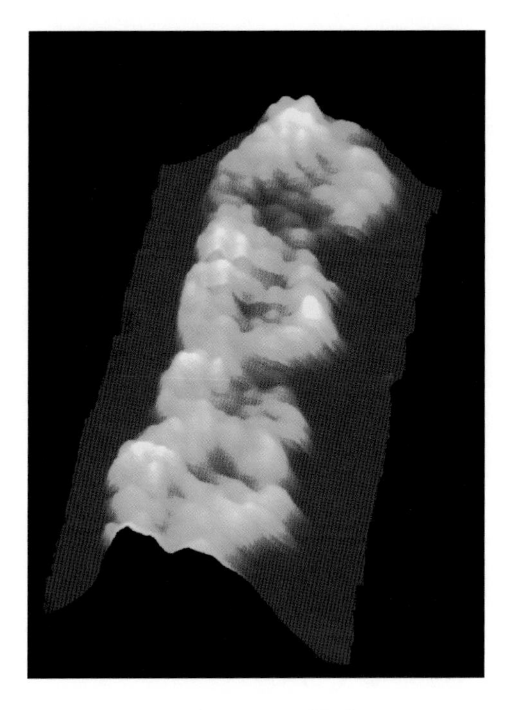

An actual image of DNA.

then the 'double helix' has been a shorthand for the signature of human life.[48] Biologists then cracked the code to determine how the sequence of A, C, G, and T is used to pass information into the new unwinding strands of DNA.

The impact of DNA on our conception of life has been incalculable. Around the world there are sculptures and artworks in different media that celebrate its discovery and symbolic meaning. It showed us how elegant the self-replication process can be at a molecular level, and how much information and diversity can be packed into something so small. Over the past half-century, we have seen DNA become an almost foolproof way of identifying individuals with very high probability of uniqueness. DNA 'fingerprinting' is now routinely used by law-enforcement agencies and courts around the world. Yet, despite the extraordinary degree of individuality that our DNA gives us, its geometry is the greatest of all shared human properties. More than 99 per cent of the 3 billion base patterns present in each person's DNA are shared by all people. As long as there is life on Earth, the double helix will be its unique and instantly recognisable signature.

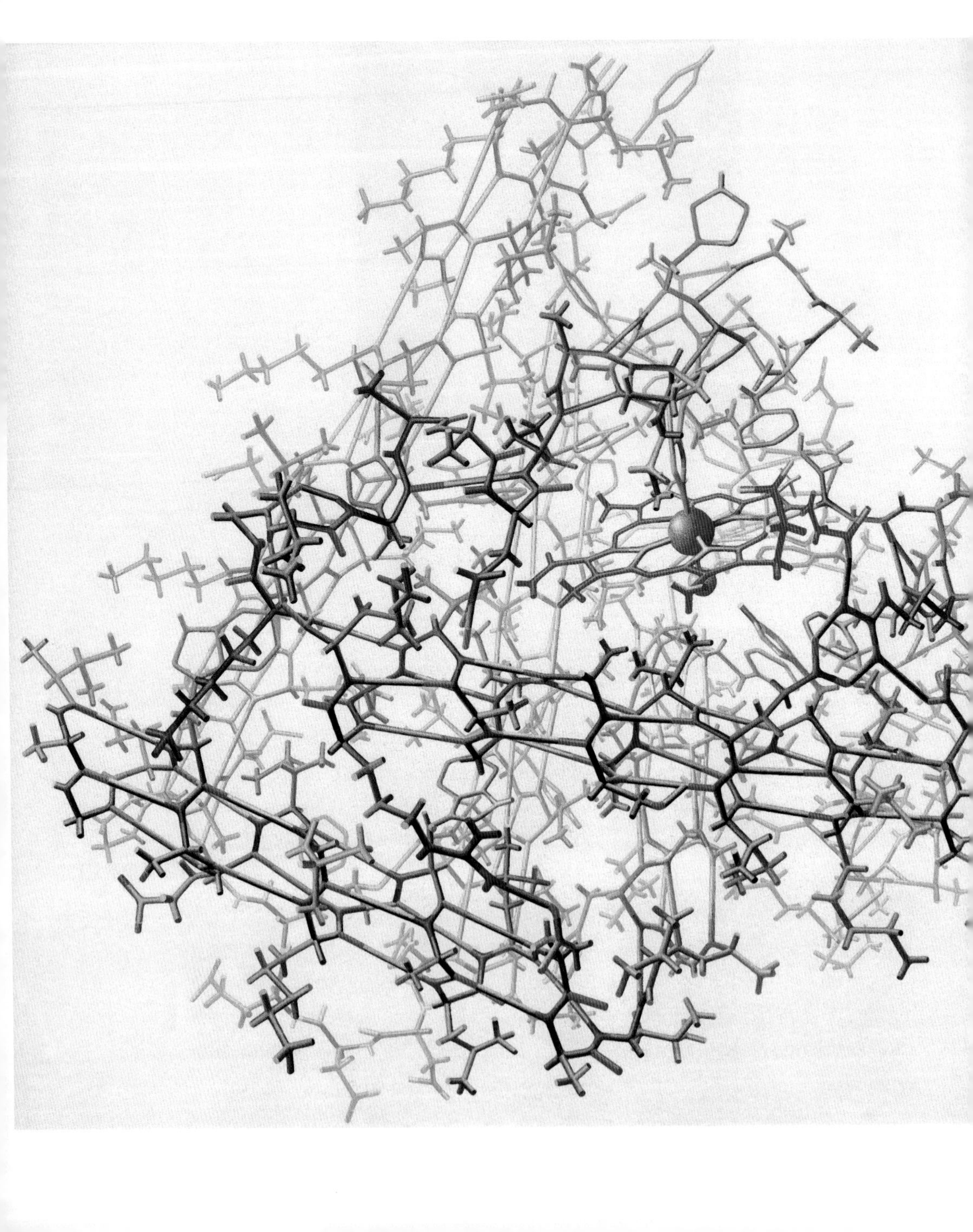

ALL JOIN HANDS

MOLECULAR ARCHITECTURES

'Working with keys is always meaningful', Holmberg said, and smiled, 'Locking and opening is, in a sense, man's very purpose on this earth. Key rings rattle throughout history. Each key, each lock, has its tale. And now I have yet another to tell.'

Henning Mankell[49]

SOMETIMES scientists want to represent things in three (or even more) dimensions in ways that require the skills of a professional artist. A few scientists and mathematicians are well known for their artistic abilities, but most are quickly defeated by the challenges of perspective and layout. Fortunately, there are now useful software packages that make even a rank amateur look moderately competent. However, until this software began to appear in the 1990s, scientific illustrations were generally done in a university's drawing offices, where black-and-white line drawings were the norm. There was little scope for colour publication in professional scientific journals because of cost, and even 35 mm slide presentations at conferences tended to be in black and white, or two other primary colours. Scientific work lacked colour. You might even say it lacked perspective. One of the consequences was that magazines like *Scientific American* then had a special appeal for working scientists that has now been somewhat diminished by the prevalence of good computer graphics. These magazines had first-rate illustrators to hand who could transform your rough amateur efforts at illustration into beautiful multi-

Opposite: Irving Geis's 1961 drawing of the 2600 atoms in the sperm whale myoglobin protein, published in *Scientific American.*

Irving Geis working on his famous watercolour of the myoglobin protein.

toned colour pictures that you could reuse for teaching or at research conferences – and it was all free. Of all the illustrators that played this important role, there is one who stands out.

Irving ('Irv') Geis (1908–97) first studied fine art and architecture, but became famous for his pre-eminent skill in representing complex molecular architectures. After Gerald Piel and Denis Flanagan acquired the ailing *Scientific American* magazine in 1948, they resuscitated it by implementing a new style of high-level popularisation in words and pictures. During the Second World War, Geis had been head of graphics in the Office of Strategic Services, the forerunner of the CIA. After the war, he became a technical artist, whose skills were soon noticed by Piel and Flanagan. Geis was used extensively for key articles that established the growing reputation of the magazine for visual appeal and technical accuracy.

Geis's greatest piece of work for *Scientific American* became of immense importance for the representation and appreciation of molecular structure. In 1960, Geis was asked to prepare a picture for an article by John Kendrew.[50] After years of meticulous work, Kendrew had managed to establish the first detailed structure of a protein crystal by an extraordinary piece of three-dimensional reconstruction using a huge number of two-dimensional x-ray images of the three-dimensional molecular structure – work for which he received the Nobel Prize in 1962. Kendrew took the first step towards understanding the structure of human haemoglobin (blood) by picking upon its smaller relative, the protein myoglobin from the sperm whale, and completing an x-ray map of its structure at a resolution of 2 angstroms. This molecule is composed of a complex three-dimensional lattice work of 2600 atoms. By protein standards it is rather small. It is red in colour and is present in muscle, where its role is to carry oxygen, and it is responsible for the familiar red colour of meat. The oxygen-carrying demands of deep-diving mammals such as whales make the role of myoglobin crucial for their bodily functions. Picturing it was a challenge that Kendrew's remarkable geometric intuition finally mastered in 1957. Drawing it for all to see was the challenge for Geis. The result of more than six months of painstaking draughtsmanship was one of the most beautiful scientific illustrations ever made for *Scientific American*, and appeared in the June 1961 issue. Looking at it now it is hard to believe that it is not a modern computer-generated image; but it was completely coloured by hand in 1961. Geis produced his image after careful study of Kendrew's real three-dimensional model of the molecule constructed from rods, clips, and coloured balls, along with his three-dimensional contour map of the electron distribution, built up from a stack of fifty clear sheets of printed Lucite. At times he worked with microscopic glasses in order to see the effects of single brush hairs.

Geis's image contains accurate representations of interatomic spacings, constituent atoms, and bond angles and types, and manages to create a three-dimensional feel by variations in colour and clever use of distortion. Today, this would be done by a moving computer graphic which would create three-dimensional perspective by having the viewer orbit around the molecule. The eye and the brain do the rest.

A number of important features of myoglobin can be identified from Geis's picture. There is one main protein chain threading its way through the architecture in blue. At the centre, we can see the large orange iron atom surrounded by four nitrogen atoms and a water molecule. The smaller orange oxygen molecules are gripped tightly by the iron atom. This picture is actually just a snapshot of a structure that oscillates in time. Some of the time, the oxygen atoms are enclosed by the protein chain, but at other moments the chain relaxes outwards, and allows the oxygen to pass in and out.

Geis's success was the inspiration for a generation of molecular artists to follow in his footsteps, and facilitate new ranges of visualisation in chemistry and biochemistry. Eventually, special imaging software packages were created for use with computers which enabled scientists to 'paint' their own molecular structures in impressive styles. The most notable of these is *Molscript*, a graphics program produced by Per Kraulis in 1991.[51] The accompanying picture, derived from nuclear magnetic resonance spectroscopy, shows one of Kraulis's three-dimensional images[52] of the Ras p21 protein.[53]

A host of other exotic protein structures provide an entire genre of *Molscript* artworks that have a compelling abstract style even if you know no chemistry at all.

The modern study of macromolecules presented a major challenge for image-makers. The graphics that are used to display their architecture produce images of great beauty and, as vividly as a Lowry townscape, its figures seem to be alive and moving. They deserve their place in any exhibition of the art of science.

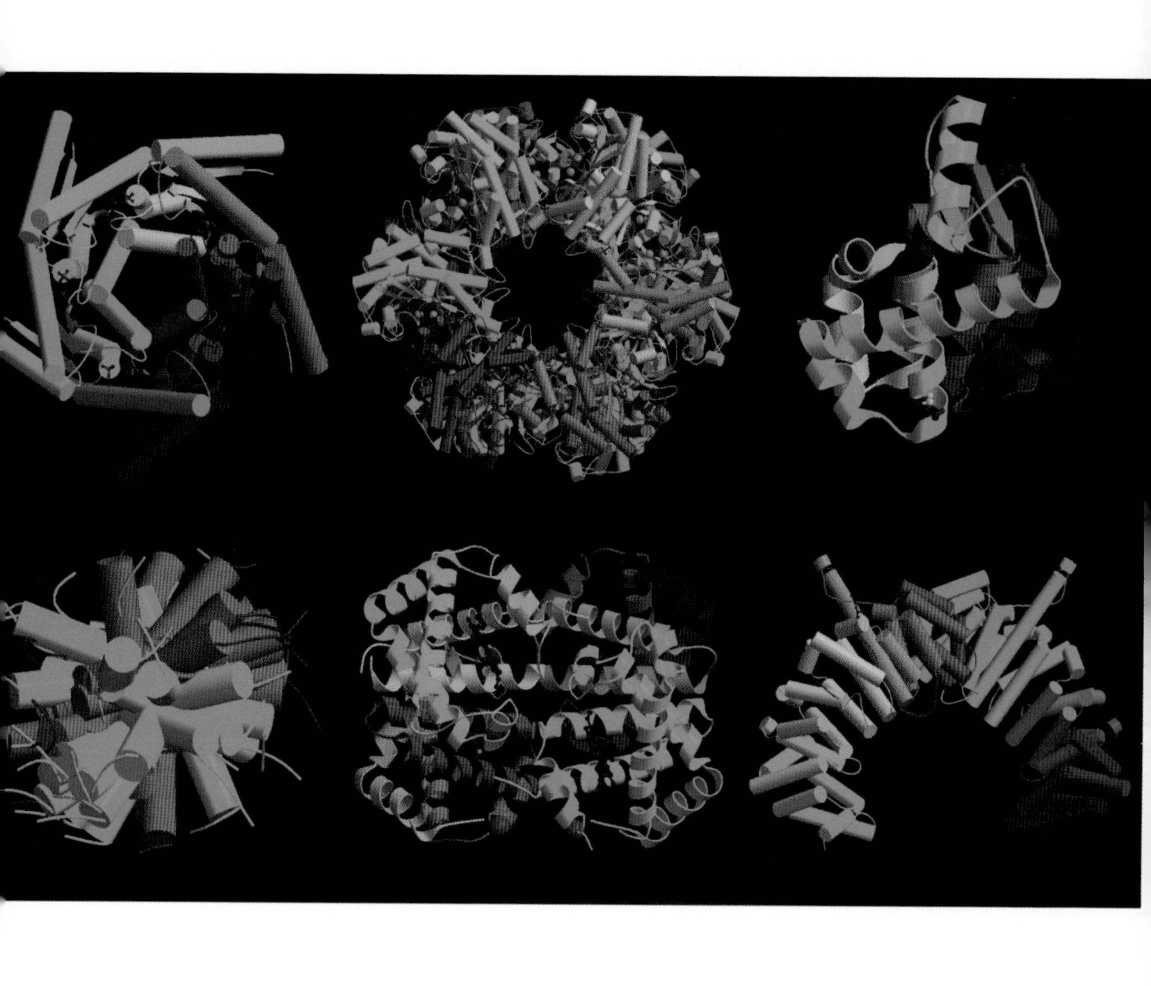

Molscript images
of the Ras p21
protein molecule.

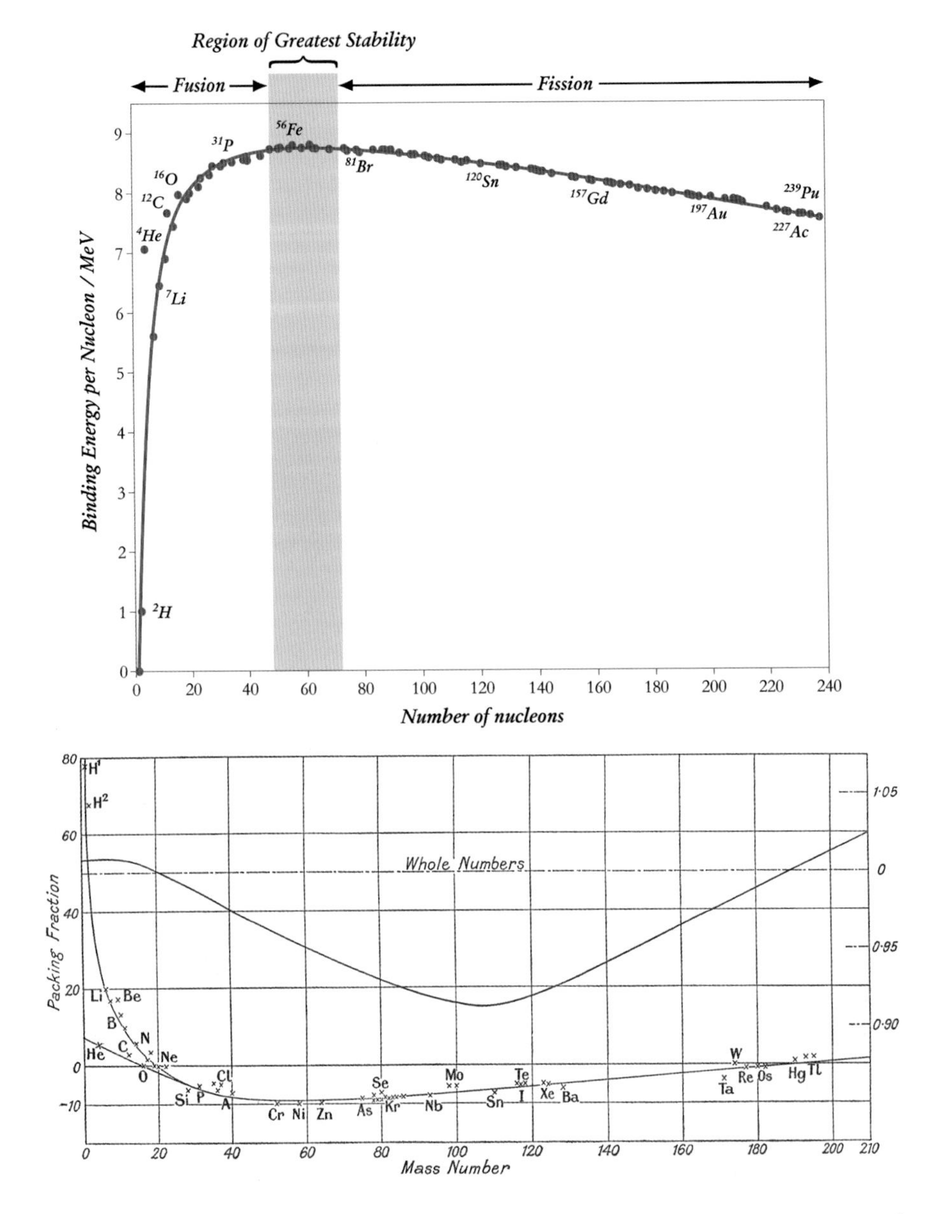

(*top*) The Variation of the Binding Energy per constituent nucleon plotted against the number of nucleons (protons and neutrons) in an atom's nucleus. It increases with the number of nucleons for small nuclei, reaches a maximum in the neighbourhood of iron (with 56 nucleons) where nuclei are most stable, and then falls as the number of nucleons increases. The pattern means that energy will be released when two small nuclei (such as hydrogen) are joined together by nuclear fusion, or when a heavy nucleus (such as uranium) is split into smaller nuclei by nuclear fission. (*bottom*). The first version of this diagram was drawn by the British Nobel Prize-winning physicist Francis Aston, in 1922. He draws the diagram the other way up, identifying the 'packing fraction' as the negative of the binding energy available plotted against the Mass Number, which is the number of nucleons in the nucleus.

LET IT BE

THE BINDING ENERGY CURVE

Very dangerous things, theories.

Dorothy L. Sayers[54]

THE first quarter of the twentieth century brought about a dramatic growth in our understanding of atoms. We learnt that they are not indivisible. We learnt that they are not immutable. And we learnt that they are storehouses of unimaginable energy. They possess a nucleus of protons and neutrons surrounded by a cloud of electrons that can be thought of as 'orbiting' the nucleus, bound in their orbits by the electrostatic forces of attraction between the negative electric charge of the electrons and the positive electric charge of the heavier protons. If you can hit an atom strongly enough, then one or more of the orbiting electrons can be removed from the grip of the nucleus to leave a positively charged ion. But if you hit the atom with a few million times more energy still, then it is possible to shatter the nucleus itself and liberate its protons and neutrons.[55]

Imagine that you make a ball – just like the core of a golf ball – by winding lots of long rubber strips around a common centre. If you don't hold them tight, the strips will spring back and the ball will break up. When you do let go, the elastic energy that is binding the stretched rubber strips tightly around the centre will be released. The nucleus is similar in some respects. If you can liberate the protons and nucleons, then it should be possible to release the energy that is binding them together. This is the secret on which the manufacture of nuclear weapons and controlled nuclear power is based. Take the example of the helium nucleus, which is sometimes called the 'alpha' particle. It is composed of two protons and two neutrons. The mass of the two protons is 2.01456 mass units[56] and the mass of the two neutrons a little less, at 2.01732. In total, the four constituents therefore weigh in at 4.03188. But if you measure the mass of a bound helium nucleus it is significantly lighter: 4.00153 mass units. There is a mass deficit of 0.03035 units.[57] If we were to use Einstein's famous $E = mc^2$ formula linking energy (E) with mass (m) via the square of the speed of light (c), then any mass deficit is equivalent to an energy deficit. This is the energy that binds the protons and neutrons together into the stable nucleus of helium. We call it the *binding energy* of helium.[58] It appears to have been drawn first by the English Nobel Laureate Francis Aston in 1922.[59]

As the number of particles in the nucleus increases, the total binding energy of the nucleus will increase because the 'mc^2' contributions from the additional protons and neutrons will grow. As a result, it is more interesting to know the average binding energy per nucleon[60] when we divide the total binding energy by the total number of protons and neutrons. This is the binding energy per nucleon.

A fascinating trend emerges if a graph is plotted of this average binding energy per nucleon (with energy measured in units of MeV) against the total number of nucleons, the mass number (denoted by A). The values $A = 1, 4, 7, 9, 11, 12, \ldots$ correspond to the lightest nuclei, of hydrogen, helium, lithium, beryllium, boron, and carbon, respectively.

We see that, as the number of nucleons in the nucleus increases, the average binding energy increases sharply, but eventually levels out to produce a fairly flat plateau at a value around 8 MeV,[61] with a maximum at the mass number 56 corresponding to iron at 8.79 MeV, before it falls off gradually towards larger values of A, falling to about 7.6 MeV at uranium with $A = 238$.

What does this mean? There are two dominant effects at work. Short-range nuclear forces pull the nucleons together and bind them tightly. But they are countered by longer-range electric forces that repel particles (the protons) with the same electric charge. In order for the nuclei to remain bound as the number of nucleons increases and the repulsive forces become more significant, they need to have relatively more neutrons present because, being electrically neutral, they do not contribute to the electric repulsion. The higher the binding energy per nucleon that results, so the more stable is the nucleus.

The characteristic shape of the graph has a momentous consequence.

If two small nuclei are joined together ('fusion'), the heavier nucleus that results will have larger average binding energy per nucleon and energy will be given out in the fusion process. By contrast, if large nuclei are split into smaller nuclei ('fission') then energy will also be given out in the fission process. Heavy nuclei need only a small amount of extra energy (usually achieved by adding a neutron to the nucleus) to be added in order to enable the electric forces of repulsion to overcome the attractive nuclear forces and split the nucleus into two pieces. The first atomic bomb was based upon the nuclear physics of fission – splitting large nuclei to release their huge binding energies. Subsequently, the creation of the so-called hydrogen (H) bomb exploited the energy release from fusing together small nuclei. Although the fission of uranium ($A = 235$) by the addition of a neutron releases about 200 MeV of energy in total, a relatively high mass of uranium fuel is being used. By contrast, the fusion of two isotopes of hydrogen – for example deuterium (one proton and one neutron) and tritium (one proton and two neutrons) – may only release about 17.6 MeV, but the amount of energy obtained per unit mass of fuel is much greater. These efficient fusion reactions involving hydrogen and its isotopes play a key role in powering the Sun and other stars and it has long been hoped that they could provide safe and efficient sources of energy on Earth.

The simple binding energy curve with its single peak and steep and shallow gradients on either side is the reason we have atomic and nuclear weapons, the possibility of atomic and nuclear power, stable stars, and stable nuclei that make life a chemical possibility. It is a picture that says it all.

Life magazine cover of February 1950 showing the 'Baker Day' underwater atomic bomb test of 25 July 1946 – pa
of Operation 'CROSSROADS'.

THAT HIDEOUS STRENGTH

THE MUSHROOM CLOUD

Batter my heart, three-personed God

John Donne[62]

THE first explosion of an atomic bomb occurred at the Trinity test site in New Mexico, 210 miles south of Los Alamos, on 16 July 1945. It was a watershed in human history. The creation of this device for the first time gave human beings the ability to destroy all human life: to tap into energy sources of almost unimaginable power and destructive consequences that remained radioactive for hundreds of years after the devices were used. Subsequently, an arms race saw the escalation in the energetic yield of these bombs as the United States and the Soviet Union sought to demonstrate their ability to produce increasingly devastating explosions. Although only two of these devices were ever used in conflict,[63] the ecological and medical consequences of these early tests in the atmosphere, on the ground, below ground, and underwater are still with us.

The explosions were much photographed and produced a characteristic fireball and cloud of debris that came to symbolise the consequences of nuclear war. The book *100 Suns 1945–1962* by Michael Light contains one hundred astonishing images of these explosions, without comment, in colour and black and white.[64] The name 'mushroom cloud' became commonplace in the early 1950s, but the comparison between bomb debris patterns and 'mushrooms' dates back at least to newspaper headlines in 1937.

The familiar mushroom shape forms for a reason. A huge volume of very hot gas with a low density is created at high pressure near ground level – nuclear explosions are generally detonated above ground to maximise the effects of the blast wave in all directions. Just like the bubbles rising in boiling water, the gas accelerates up into the denser air above, creating turbulent eddies curving downwards at their edges while additional debris and smoke streams up the centre in a rising column. The material at the core of the detonation is vapourised and heated to tens of millions of degrees, producing copious x-rays, which excite the atoms and molecules of the air above, creating a flash of white light whose duration depends upon the magnitude of the initial explosion. As the front of the column rises it rotates and draws in material from the ground to form the 'stem' of the growing mushroom shape; its density falls as it spreads, and eventually it finds itself with the same density as the air above it. At that moment, it stops rising and disperses laterally so that all the material drawn up from the ground scatters backwards and descends downwards to create a wide 'mushrooming' area of radioactive fall-out.[65]

Ordinary explosions of dynamite, TNT, or other non-nuclear weapons have a rather different appearance to the atomic bomb's mushroom cloud because of the lower temperatures created at the start of a chemical explosion. This results in a turbulent mix of exploding gases rather than the organised stem and umbrella-like mushroom cloud.

One of the pioneers of the studies of the shape and character of large explosions was the remarkable Cambridge mathematician Geoffrey ('G.I.') Taylor. Taylor wrote the classified report on the expected character of a nuclear explosion in June 1941.[66] It was declassified in 1950, and then appeared as a research paper[67] later that year. Taylor had become a legendary figure in the wider public mind when, a little earlier, *Life* magazine in the USA published a sequence of time-lapsed photographs of the 1945 Trinity test in New Mexico. The energy yield from this and other American nuclear explosions was still top secret. But Taylor showed how a few lines of algebra enabled him (and hence anyone else with a smattering of simple mathematics) to work out the approximate energy of an explosion just by looking at photographs like these. This was the type of simple deduction about physical problems that seemed bewilderingly complex at first sight for which Taylor was renowned.

Taylor was able to work out the expected radius of the explosion after any lapse of time after its detonation by noting that it will depend significantly on

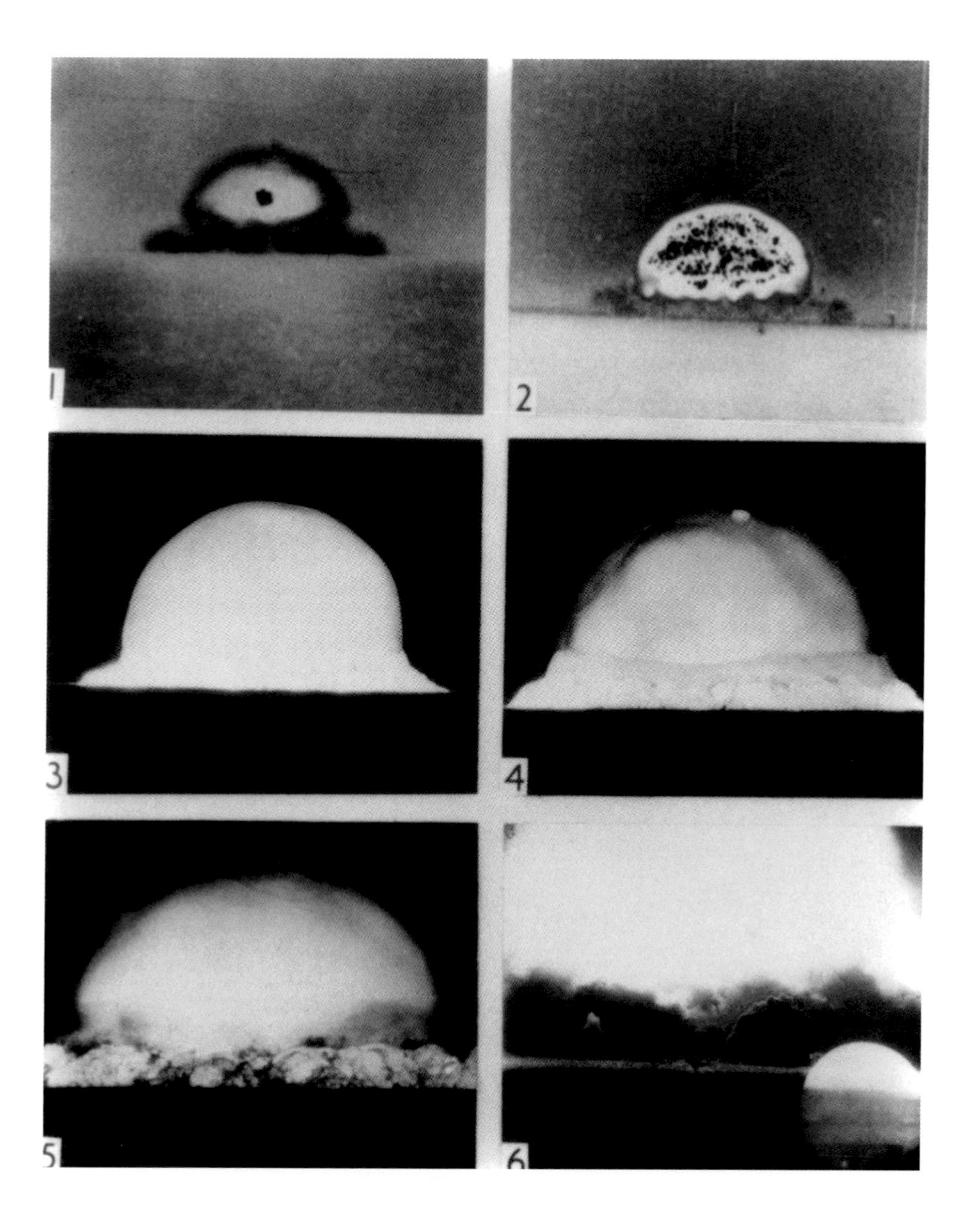

Six images recording the detonation of the world's first atomic bomb. Photographed by U.S. Army motion picture camera at distance of 8 miles from blast. At 5:30 a.m., July 16, 1945, in a desert near Alamogordo Air Base, New Mexico. Codenamed 'Trinity', the explosion illuminated the night sky 'with an intensity many times that of the midday sun', shattered windows 125 miles away and produced a mushroom cloud between 50,000 and 70,000 feet in height. 'Trinity' was the result of over two years' work by a group of engineers and scientists based at Los Alamos, New Mexico, led by J. Robert Oppenheimer. Three weeks after the 'Trinity' explosion, a second atomic bomb was detonated over Hiroshima, Japan.

only two things: the energy of the explosion and the density of the surrounding air that it is ploughing into.[68] There is only one way that such a dependence can look, so approximately:

$$\text{Energy of bomb} = \frac{\text{air density} \times (\text{radius})^5}{(\text{time})^2}$$

From the first frame of the photo we see that at $t = 0.009$ seconds the blast wave of the explosion has a radius of roughly 94 m. We know that the density of air is 1.2 kg per cubic metre and so the equation then tells us that the energy released was about 10^{14} J, which is the equivalent[69] of about 25 kilotons of TNT. For comparison, the 2004 Indian earthquake released the energy equivalent to 475 megatons of TNT.

The mushroom cloud is an ambiguous image. It represents the atmospheric response to the release of stupendous energy bound up inside the nucleus of the atom; it is strangely beautiful yet immensely destructive, its fearful symmetry pregnant with information about the battered atoms at its heart.

Bubble chamber tracks. Straight-line tracks are made by uncharged particles; spiral tracks are left by charged particles responding to a magnetic field applied across the chamber.

WRITING ON AIR
BUBBLE CHAMBER TRACKS

I'm forever blowing bubbles,
Pretty bubbles in the air
Jaan Kenbrovin[70]

CLOUD chambers were conceived by Charles Wilson in 1911 in order to study the formation of rain clouds. His original chamber, made in 1912, is displayed in the Cavendish Laboratory Museum in Cambridge. Wilson exploited the fact that when a neutral atom loses one of its electrons, and becomes a charged ion, then water vapour droplets will condense around it if it is placed in an atmosphere saturated with water vapour that is suddenly expanded. More dramatic still, if a charged alpha particle[71] moves through a gas it leaves a trail of ions in its path, and the water vapour condenses around them marking out the path of the alpha particle, just like the vapour trail of a fighter jet across the sky. Wilson also observed the tracks formed by beta radiation (electrons), gamma rays, and x-rays. These observations gave us the first concrete look at the smallest particles of matter and the influx of cosmic rays from space. The cloud chamber led to many dramatic discoveries, including the first antiparticle – the positron[72] – in cosmic rays, and Wilson received the Nobel Prize for physics in 1927 for the 'discovery of a method of rendering discernible the paths of electrically charged particles'. Yet Wilson was a modest fellow, of whom Lord Blackett wrote, after his death in 1959, that 'Of all the scientists of this age, he was perhaps the most gentle and serene, and the most indifferent to prestige and honour. His absorption in his work arose from his intense love of the natural world and from his delight in its beauties.'[73]

Unfortunately, the cloud chamber has one significant drawback: the density of gas in the chamber is very low, so there are rather few events possible in the chamber volume. The next development found a way to increase the density in the chamber a thousand-fold.

The Wilson cloud chamber was eventually superseded in 1952 by the bubble chamber, which was invented by the American physicist Don Glaser, who received the 1960 Nobel Prize for his work. The detection chamber was filled with liquid hydrogen squeezed by a piston to very high pressure (typically at least ten times normal atmospheric pressure), so that it could remain at a temperature well above its boiling point under normal atmospheric pressure, without boiling away. If anything is introduced into this superheated liquid then it will act as a local hot spot that causes the liquid to boil there, and the path of any moving interloper will be revealed by a track of bubbles when the pressure is lowered, causing them to expand. The tracks are then photographed to reveal the paths of the particles that have passed through the chamber. By using a different liquid, such as xenon, the chamber can be optimised to detect different particles. The application of a magnetic field will cause electrically charged particles to move in spiral paths (positive and negative charges with different handedness) but leaves neutral particles unaffected. Unfortunately, there is a pot-luck element to this type of detector. It can't be set up so that it is triggered by incoming particles, so it is not an ideal tool in the search for very rare events.

Glaser decided to change research fields to biology once his invention was seen to be a success. His Berkeley colleague Luis Alvarez saw new possibilities and set about building larger and larger chambers, up to 2 m in length, rather like a large wardrobe. The key to success, however, was the seamless wedding of this technology to the computer. Computer-control enabled pictures to be taken rapidly and the whole output of the chamber became a three-dimensional record that could be studied and restudied by different research groups all over the world.

These bubble chamber photos that dominated physics imagery for so long now seem like the old tinted photographs that you find hidden away in a drawer. Just as our old black-and-white family photos have been superseded by digital cameras, high-resolution colour, zoom facilities, instant movies, and video recordings, so the detection of elementary-particle collisions and decays

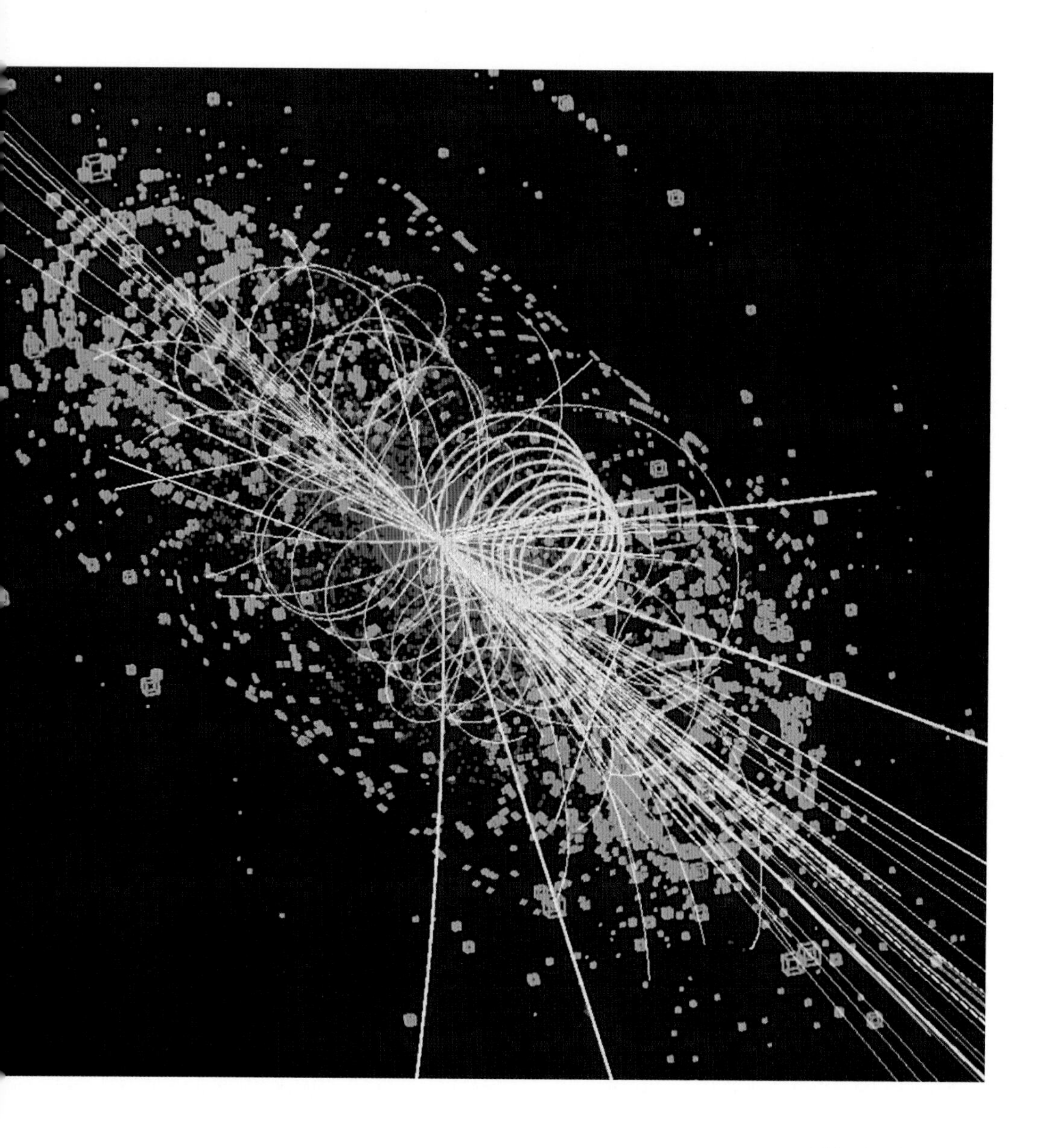

This simulation shows the decay of a Higgs particle following a collision of two protons which might be seen at the Large Hadron Collider at CERN in the future.

is now conducted using multi-layered cylindrical chambers. Each can detect a certain type of particle, and some measure velocity, some momentum, some just energy. Computers can quickly reconstruct the sequence of events occurring in the chamber from the information deposited in the different layers of the detector. The Stanford Linear Accelerator (SLAC) detector weighs 4000 tons and stands six storeys high.

The old bubble chamber pictures define a key era in the development of human knowledge about the Universe. They revealed the movements and commotion in the world of elementary particles that had been going on all around us – and within us – unnoticed. They made the most important discoveries in sub-nuclear physics and allowed physicists to feel that they were actually 'seeing' sub-nuclear particles as they passed silently through their laboratories. The transition from these old black-and-white pictures to new multi-colour interaction pictures is a step away from a direct feeling of 'seeing' particles. You know the reconstructions are artificial in the sense that they are colour-coded and edited to show only the things you are interested in watching. But this development parallels what has happened in the theoretical view of elementary particles as well. Many of the particles that were discovered and studied by bubble chamber photographs – kaons, pions, and protons – are now known to have sub-components. They are not the most elementary pieces of matter; their constituents are quarks and they have a strange property: they cannot be seen. They are 'confined' inside other larger particles.

Particles like the kaon and the pion are made of quark–antiquark pairs, while protons and neutrons are each made of three quarks. You might think that by breaking a pion you could liberate the quark and the antiquark and see them individually. But they are bound so strongly that the energy released on separating them is enough to make another quark–antiquark pair and you end up with two pions and no free quarks. It is rather like trying to obtain a single north magnetic pole by cutting a bar magnet in half: you just end up with two bar magnets, each with a north and south pole.

The modern study of elementary particles therefore pushes us increasingly towards things that cannot be 'seen' in the conventional sense. At present, it is suspected that the most elementary states of matter are excitations of strings that may only reveal their special features at energies far beyond those we are able to create in any terrestrial laboratory. The pictures we see here may one day be viewed nostalgically as the last glimpses we had of a world that behaved like a small-scale version of the large-scale everyday world that we inhabit, and from which we gain our first intuitions about the nature of things.

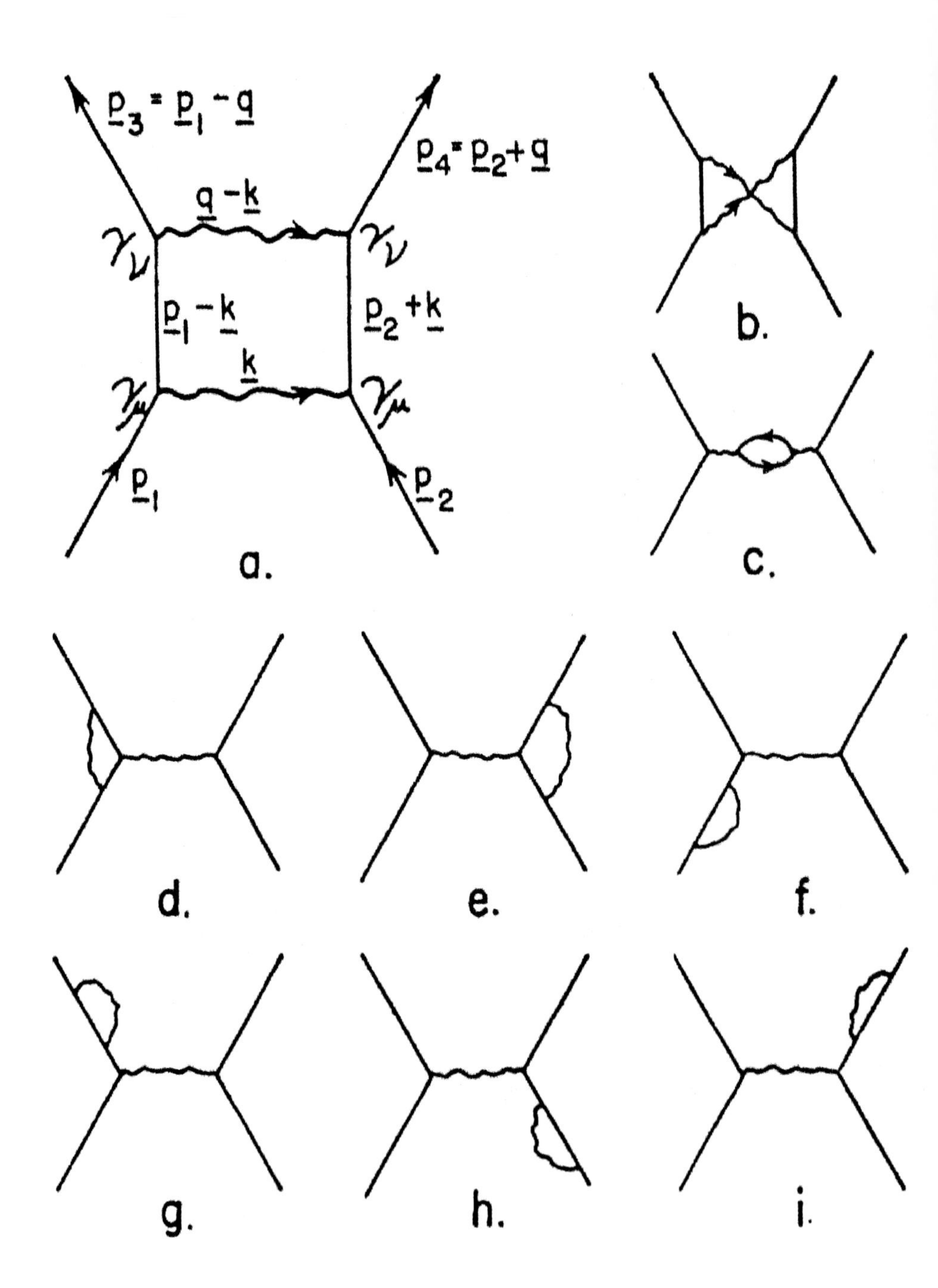

Richard Feynman's diagrams showing the interaction between incoming elementary particles. These are space–time diagrams in which time runs vertically, space horizontally. Straight lines display the space–time paths of real (observable) particles; wavy lines show virtual (unobservable) particles that mediate the incoming particles.

‘DICK’S FUNNY LITTLE PICTURES’

FEYNMAN DIAGRAMS

Einstein’s great work had sprung from physical intuition and when Einstein stopped creating it was because he stopped thinking in concrete physical images and became a manipulator of equations.

Richard Feynman[74]

THE quantum picture of reality that emerged from the work of Planck, Bohr, Born, Heisenberg, Dirac, and Schrödinger in the first quarter of the twentieth century provided a means of calculating all the properties of atoms and molecules with unparalleled accuracy. But, as is often the case in physics, finding a theory is much easier than extracting its outcomes: formulating equations is much simpler than solving them. This is a consequence of a deep feature of the universe and reflects the way in which symmetry and simplicity seem to dictate the forms of the laws of Nature without necessarily being manifested in the complex outcomes of those laws. Look at Nature at the level of its laws and it looks amazingly simple; look at the level of the states that those laws produce and it looks like a higgledy-piggledy mess of complications. As a result of this dichotomy, the great challenge for physicists is to find procedures for finding solutions, either exactly or by a series of ever-improving approximations, to their mathematically beautiful equations.

In the late 1940s, the world's leading theoretical physicists were preoccupied with the problems of the very small. Following the technical achievements of the Manhattan Project led by J. Robert Oppenheimer, they sought to extend their understanding of the atomic nucleus and the interactions between its constituents. But they were also deeply engaged with the quest to understand the nature of light and its interactions with particles of matter, such as electrons, which carried electric charge. Prior to the discovery of quantum mechanics, a united theory of electricity and magnetism had been spectacularly conceived by James Clerk Maxwell. The startling successes of the new quantum theory meant that Maxwell's classic theory had to be upgraded to incorporate a probabilistic quantum interpretation, and new physical effects created by elementary particles of matter. Nineteenth-century electrodynamics plus the quantum complexion of light and matter required a new 'quantum electrodynamics', or 'QED', as it soon became known.[75]

The overwhelming problem faced by those first seekers after a quantum theory of light and its interactions with matter in the 1940s was the sheer complexity of it all. Heisenberg's uncertainty principle leads us to a picture of the 'vacuum' in which any single particle moves as if in a sea of particles and antiparticles appearing and disappearing over intervals of time so brief that they are unobservable directly. Their indirect effects, however, are still felt and have measurable consequences. These unobservable or 'virtual' particles, as they became known, created an enormous complication in what used to be very simple processes such as the scattering of an electron by a photon. Virtual particles mean that we have to consider the probability of the incoming electron emitting a virtual photon, then interacting with that photon, before subsequently reabsorbing the virtual photon. There is then a hierarchy of increasingly complicated possibilities of this sort to be evaluated. For example, the incoming electron could emit a virtual photon that produces a virtual electron–positron pair, which then annihilated to make a virtual photon, which is reabsorbed by the electron after scattering from the photon. These more convoluted histories have lower probabilities of occurring, because they involve more interactions occurring in a chain, but the correct numerical calculation of the rate of occurrence of the interaction to a specified number of decimal places needs to take a significant number of them into account. Mathematically, this is a formidable book-keeping and computational challenge. A number of the world's greatest physicists have made mistakes calculating the contributions by this hierarchy of virtual interactions.

In the spring of 1948, twenty-eight of the world's leading theoretical physicists met at the Pocono Manor Inn, a small inn in the Pennsylvania countryside.[76] Their goal over the few days of their stay was to come to grips with the problem of calculating the observable consequences of interactions between electrons and light in the emerging theory of QED. The upcoming Harvard star, Julian Schwinger, lectured all day, with a few breaks for meals and coffee, expounding the complicated mathematical procedures he had developed for carrying out these calculations. Only at the end of the long day did the young Richard Feynman get his chance to present his way of attacking the same problems. Feynman's approach was quite different; it was not immediately transparent to his audience, but in time it would become the way in which all their calculations were going to be done.

Feynman introduced a new pictorial way of thinking. By representing interactions between particles and light by means of diagrams that showed their motions in space and time, he could draw in all the possible virtual processes that might be making smaller and smaller contributions to the overall probability of a particular interaction taking place. These 'Feynman diagrams', as they became known, enabled physicists to 'see' what the possible interactions were, just by joining up all the possible links between initial and final situations for an interaction. It made book-keeping easy. But Feynman could do much more with his diagrams. His brilliant intuition enabled him to transform the graphics into the detailed formulae needed to compute the numerical answers that could be compared with experimental results. It wasn't long before they were producing predictions that agreed with experiments to spectacular levels of accuracy. Eventually, Freeman Dyson, a young British physicist working at the Institute for Advanced Study at Princeton, showed that the mathematical methods of Schwinger and those of a Japanese physicist, Sin-Itiro Tomonaga (who had solved the problem of QED independently, working in Japan, by yet another method), produced the same answers as the diagrammatic methods of Feynman. Schwinger, Tomonaga, and Feynman shared the Nobel Prize for physics in 1965. In fact, Dyson played a major role in promoting the use of Feynman's diagrams by physicists all over the world. He meticulously demonstrated the equivalence of the different calculational techniques and produced a rigorous step-by-step guide to the construction of Feynman diagrams and their use in performing calculations. However, such guides are not so easy to use just by following the rules – it's a bit like learning to use a computer by reading the manual – and Dyson also trained a large number of young physicists in their effective use; they, in turn, spread the word to others, and the use of Feynman diagrams came to dominate the major physics centres of the world.

Feynman's diagrams and the theory of their use first appeared in a classic paper published in 1949.[77] Ever since then, the blackboards in theoretical physics departments have been filled with them, and PhD students in particle physics have been trained to use them to compute the rates of countless physical processes. The largest of these calculations involve calculating and combining the contributions from nearly 900 different Feynman diagrams to yield a numerical prediction good to thirteen decimal places that can be compared with experiment. When Feynman's methods were first employed, unaided by computers, they were used to predict answers to about five decimal places.

The visual importance of Feynman diagrams is the greater because they are vivid representations of physical processes that cannot be pictured directly by cameras. Some physicists, such as Schwinger, were content to calculate the results they wanted without having a visual picture of the processes involved. Although Feynman was perfectly capable of doing the same, he was troubled by the fact that we could write down mathematical expressions for processes occurring in the world that we could not visualise. Like all the best formalisms, his diagrams avoided the need for thought. They had a structure that naturally implied where the next pieces had to fit. If you want to know what comes next, as Feynman put it, 'it's like asking a centipede which leg comes after the other.'

I know of one other famous physicist who has a private pictorial notation for carrying out computations in mathematical physics. Roger Penrose uses a diagrammatic system of lines, curves and links instead of traditional tensor analysis to carry out calculations in tensor analysis. It hasn't caught on like Feynman's because it doesn't do anything that can't be done conventionally. It is just handy shorthand which has the nice property that the pictures guide the logic and expand the range of things that you can do without thinking.[78]

Richard Feynman went on to become one of the greatest twentieth-century physicists and an iconic figure known all around the world as a result of his

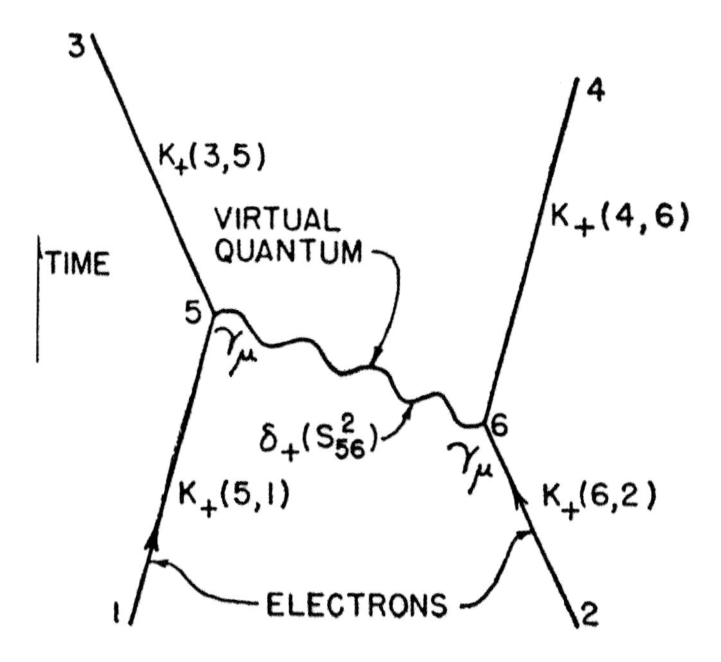

One of Richard Feynman's diagrams from his first 1949 paper, showing the scattering of two electrons mediated by virtual photon exchange. The incoming electrons are labelled 1 and 2; after their interaction they are labelled 3 and 4.

eccentric book of memoirs[79] and his contribution to unveiling the problems that caused the Challenger Space Shuttle disaster. Appropriately, he owned an old van that was painted with Feynman diagrams and carried the number plate QUANTUM. Those diagrams remain his most enduring legacy: the new toolkit of the physicist.

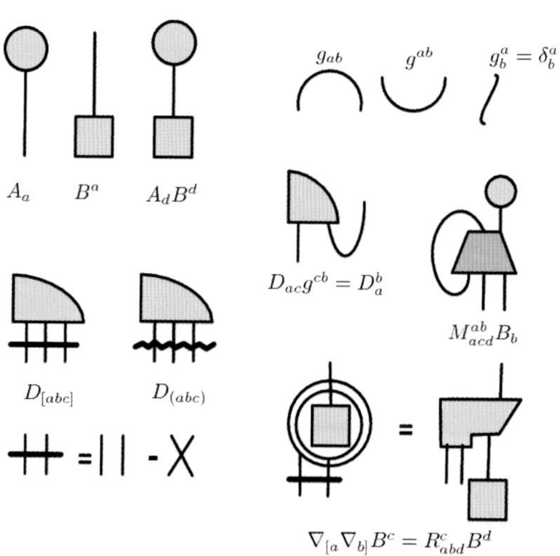

Some of Roger Penrose's picture calculus for carrying out calculations in space–time geometries.

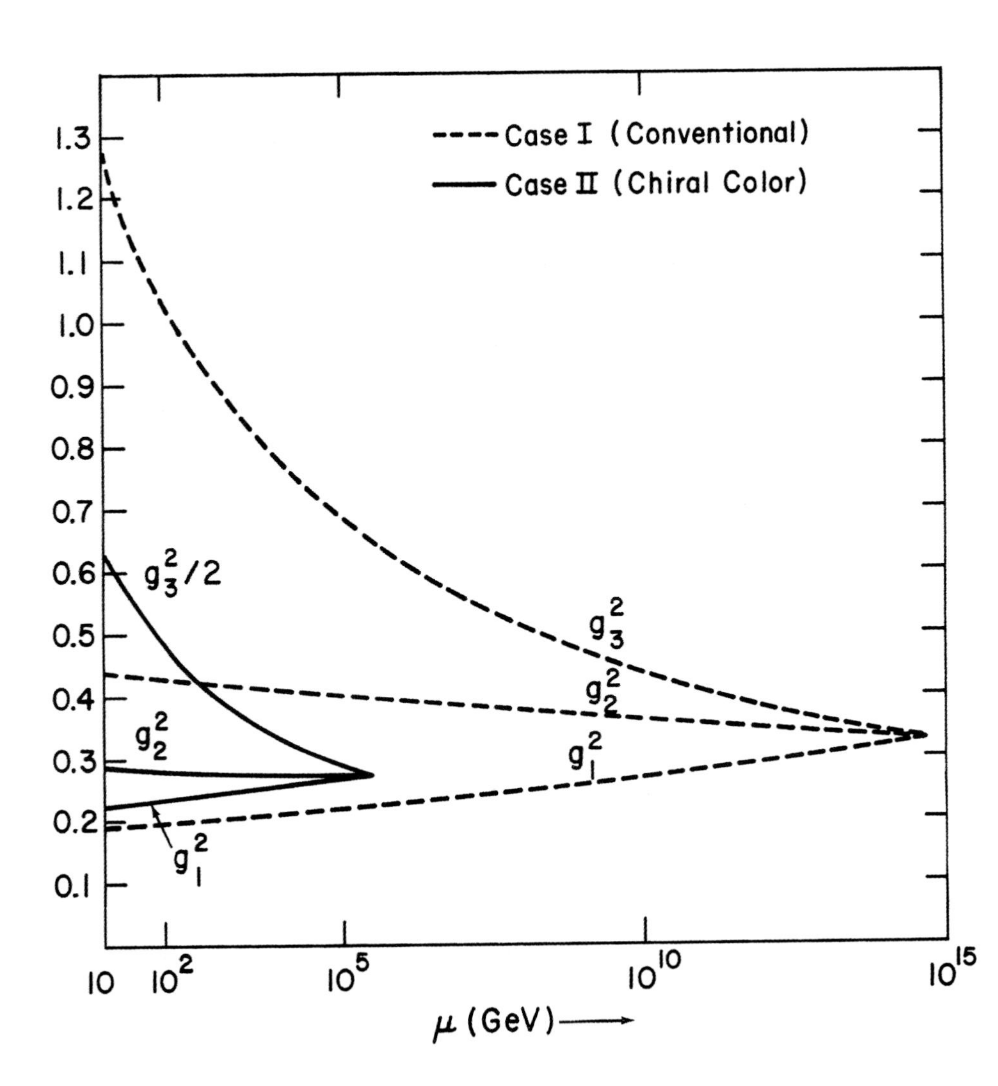

Jogesh Pati's first picture of the grand unification near 10^{15} GeV of the effective strengths (dashed) of the electromagnetic (g_1), weak (g_2) and strong (g_3) forces of Nature. The force strength is shown vertically at the energy, shown horizontally, at which it is measured. An alternative scenario of low-energy unification was included (solid curves) for comparison. It produces a cross-over at 10^5 GeV but this theory did not gain support.

BAND ON THE RUN

A UNIVERSAL TRIPLE POINT

And if a kingdom be divided against itself
that kingdom cannot stand.

St Mark[80]

ONE of the dreams of physicists has always been that there would be a single theory to unite all the four basic forces of Nature into a single universal law. In the early eighteenth century, Isaac Newton first realised that gravity needed to be supplemented by other, far stronger, attractive forces of Nature that could explain why small pieces of matter held together. The force of gravity alone was not enough to explain the world. Inspired by Newton's work, a Croatian Jesuit, Roger Boscovich, was the first to attempt a universal theory in his book *A Theory of Natural Philosophy*, published in Venice in 1763. He assumed that Newton's theory of gravity, with its inverse square law of attraction between separate masses, would control things at very large distances of separation, but when they were close it would be modified in form. He provided a picture of the form that he expected, with the force alternating between attraction and repulsion in an oscillatory fashion. The places where the net force is zero were expected to result in stable structures such as atoms or solids.

How far have we got in realising Boscovich's dream of a universal theory? During the twentieth century, physicists developed wonderfully accurate mathematical theories of the forces of gravity and electromagnetism, along with the so called 'weak' and 'strong' forces that govern radioactivity and nuclear processes. Each of these four theories gives extremely accurate explanations and predictions about phenomena within its own domain, but there is something disjointed about a world governed by four separate pieces of legislation. The pragmatist might complain that this viewpoint is just a projection of a religious belief in one God whose decrees govern all – why shouldn't there be more than one law? Indeed, monotheistic beliefs were a source of the very concept of laws of Nature, whether they are single or plural in number.[81] Equally, it might be argued that we have an evolved sense of symmetry and harmony that has its origins in our need to recognise living things and distinguish them from inanimate objects.[82] But experience has steadily revealed Nature's liking for symmetry and the remarkable properties of the symmetries that lie at the base of natural laws which seem to have the special properties needed for a unification of their patterns to take place.

Electricity and magnetism seemed at first like two different things, but nineteenth-century investigations revealed that the flow of electricity could create magnetism and moving magnets make electric currents flow – as in the dynamo that powers your bicycle lights. Thereafter, electricity and magnetism were replaced by the single force of 'electromagnetism'. In the late 1960s, it was then predicted independently by Abdus Salam and Steven Weinberg that electromagnetism and the weak force of radioactivity were also just different aspects of a single underlying 'electroweak' force.

In any of these quests to join together what the physicists of the past had put asunder, we face two big problems that seem at first sight insuperable. The different forces act on different collections of particles and they have enormously different strengths. Electromagnetism is a hundred times weaker than electromagnetism under laboratory conditions. How then can they be the same force in disguise?

The answer was supplied by the unusual properties of the quantum vacuum. As Maxwell was the first to stress, the physicists' vacuum[83] is not the everyday notion of 'nothing at all'. It is merely what remains when everything that can be removed has been removed: it is the lowest available energy state. It is also a seething mass of activity. Particles and antiparticles are continually appearing and annihilating one other, so quickly that they cannot be observed directly because of Heisenberg's quantum uncertainty principle. These are therefore virtual processes[84] like those accounted for by Feynman's diagrams.

Let's look at the interaction between two electrons. They each have one unit of negative electric charge and so will repel if fired towards each other. The strength of the recoil is a measure of the strength of the electromagnetic force of Nature. The quantum vacuum changes that simple picture in unsuspected ways. Think of one electron being introduced into a vacuum sea of continually appearing and annihilating electrons and positrons (which have positive electric charge). Since opposite electric charges attract, the virtual positrons (+) will be attracted towards the electron (-). The result is a segregation, or 'polarisation', of the vacuum. The electron is surrounded by a cloud of virtual positive charges which shield its own central 'bare' negative charge. Another, slowly approaching electron will not feel the undiluted repulsive effect of the bare negative charge of the first electron sitting in the vacuum. Rather, it will feel the weaker repulsive effect of the shielded charge and hence will recoil more weakly than expected.

Things become more interesting still if we increase the energy of the incoming electron. The higher its energy, the further it will penetrate into the shielding cloud of virtual positive charges and the more it will feel of the full bare negative charge of the target electron. So, it will be deflected more strongly than an incoming electron of lower energy that doesn't get so close to the unshielded charge.

This has dramatic consequences. It means that the effective strength of the electromagnetic force of Nature *increases* with the energy (or temperature) of the environment in which it acts. As the energy rises, incoming electrons move faster and get a closer 'look' at the bare unshielded charge of the target electron. It is rather like wrapping two billiard balls in cotton wool. If you collide them very slowly they will rebound very gently because only the cotton wool layers touch. But collide them at high speed and the cotton wool will have little effect. The hard surfaces will hit and a stronger rebound will occur.

The same considerations can be brought to bear on the strong force of Nature that governs the interactions between gluons and quarks, the elementary constituents of protons and neutrons. Whereas the electromagnetic force acts on all particles that possess electric charge, the strong force acts on particles that carry the so called 'colour' charge attribute and is sometimes called the 'colour force'.

The situation for the interaction between two quarks is more complicated than that of two electrons. When we consider the effects of the cloud of virtual electrons and positrons we could neglect any photons of light which mediate the electromagnetic interactions between them because they carry no electric charge themselves. Their presence doesn't affect the charge distribution around the target electron. However, in interactions between quarks, which carry the colour charge, the role of the mediating photon is played by the gluons; but gluons carry the colour charge themselves and so their presence affects the colour charge distribution significantly.

As a result, there are *two* different effects on the interaction between two quarks as a result of the quantum vacuum. As before, there will be a cloud of oppositely colour-charged virtual quarks surrounding our first target quark and they serve to shield its colour charge and so weaken the interaction with any incoming quark of the same colour. Again, the effect would be to make the strong interaction between them become effectively stronger as the energy of an incoming quark rises. But the cloud of virtual gluons introduces a significant new effect. Virtual gluons have the opposite effect to the virtual quarks. They tend to smear out the central colour charge with the same type of colour charge and make it less pointlike. When scattering occurs from this bloated colour charge, it occurs more weakly. The winner in this battle between two opposing effects depends on how many different varieties of quark there are to appear in the virtual pairings. If the number is as low as the six we see in Nature then it is the gluon smearing that wins out and the measured strong interaction between quarks will get effectively *weaker* when we observe it at higher and higher energies.

This unusual property is called *asymptotic freedom* because as energies get ever higher the quarks are predicted to become more and more like 'free' particles, feeling no forces at all. It is now observed in experiments spanning a wide range of energies and was first predicted to occur in 1973 by the American physicists, David Politzer, David Gross, and Frank Wilczek,[85] who received the 2005 Nobel Prize for physics for this discovery.

This insight into the way strong forces behave was a very important breakthrough. Before 1973, the study of the strong interaction in Nature was beginning to look like a lost cause. Many physicists expected that strong interactions would get stronger and increasingly complicated as the energies of interaction increased. Understanding what was happening at very high energies would be intractable. The discovery that things got simpler and easier to calculate at high energies was a godsend to the development of our understanding of high-energy physics and for our quest to understand the early history of the Universe. The discovery and acceptance of asymptotic freedom, and the theories that encapsulated it, led to the first believable investigations of the physical processes occurring in the very early history of the Universe and the beginning of the new interdisciplinary study of 'particle-cosmology', out of which the idea of the 'inflationary universe' emerged.

The changes in the effective strengths of the electroweak and strong forces of Nature at high energies solve the problem of uniting them posed by their very different strengths. The differences we see are just a reflection of the very low energies where we make our laboratory measurements, which is a consequence of the low-temperature environment that is needed for physicists to exist. Life could not exist in the high-temperature regime where the true symmetry of the forces of Nature is revealed.

As we go to ultra-high energies, the strong force weakens and the electromagnetic and weak forces are predicted to strengthen. Remarkably, they all appear able to cross at an energy corresponding to a temperature of about 10^{27} degrees Kelvin. Above the crossover energy, they are expected to look the same.

In 1974, the first prescription for a Grand Unification of the electroweak and strong forces of Nature along these lines was suggested by Howard Georgi and Sheldon Glashow.[86] It proposed the first 'Grand Unified Theory' (GUT) of three forces of Nature (no one knew at that time how to go about including gravity in the game). As we have seen, the change, or 'running', of the force strength allows the problems of their very different low-energy strengths to be overcome. But what about the other problem – the fact that they act on different types of elementary particle? This was solved by the appearance of new particles in these unified theories that mediate interactions between particles which would otherwise, in the separate non-unified descriptions, not interact with one another.[87]

In fact, a closer scrutiny of this cross-over in the three force strengths proved to be very revealing. In 1991, Ugo Amaldi, Wim der Boer and Hermann Fürstenau[88] showed that the cross-over just failed to happen unless a special 'supersymmetry', that had long been suspected to exist in Nature, existed. Its effect was to double the number of elementary-particle types in existence, and so slightly alter the way the force strengths change as energies increased. The result was a more-or-less exact cross-over at high-energy. This cross-over picture brought about huge interest in supersymmetric theories that continues unabated to the present day.

The simple suggestive picture of the three-fold intersection of the strengths of the electromagnetic, weak, and strong forces of Nature, which was first drawn by Jogesh Pati in 1978,[89] has played an inspirational role in the search for a unified description of Nature.[90] The convergence of the running force strengths suggests that unification does exist and led to the exploration of the early history of the Universe, an understanding of the preponderance of matter over antimatter within it today, and the search for the right way to include gravity in the unification scheme. It is a simple symbol of the Universe's deep unity in the face of superficial diversity, which is what we mean by beauty.

Opposite: a calculation of the change in the strengths of the electromagnetic, weak and strong forces' strengths versus increasing energy in the standard theory of grand unification in 1991 (*top*) shows that there is a significant 'miss' of the target of a single cross-over at one energy. By contrast, if the property of supersymmetry exists in Nature, then it changes the number of elementary particles that must exist and produces a convincing single cross-over at high energy (*bottom*). These pictures, produced by Amaldi, de Boer and Fürstenau, created an explosion of interest in theories of supersymmetry that continues to this day.

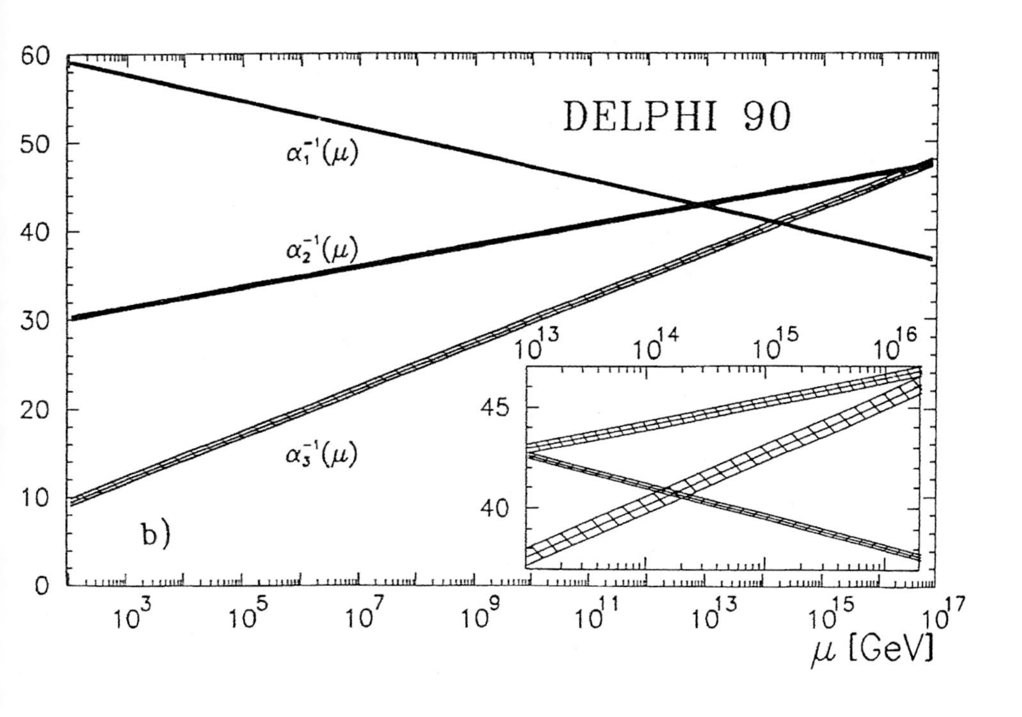
DELPHI 90
$\alpha_1^{-1}(\mu)$
$\alpha_2^{-1}(\mu)$
$\alpha_3^{-1}(\mu)$
b)
60
50
40
30
20
10
0
10^{13} 10^{14} 10^{15} 10^{16}
45
40
10^3 10^5 10^7 10^9 10^{11} 10^{13} 10^{15} 10^{17}
μ [GeV]

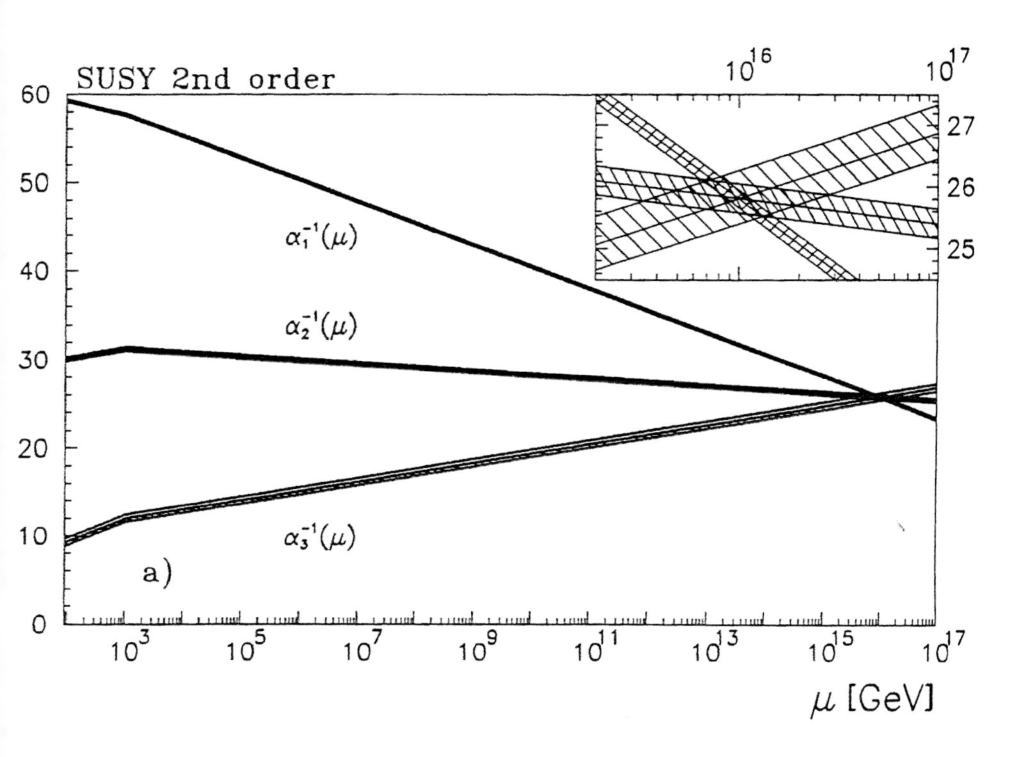
SUSY 2nd order
$\alpha_1^{-1}(\mu)$
$\alpha_2^{-1}(\mu)$
$\alpha_3^{-1}(\mu)$
a)
60
50
40
30
20
10
0
10^{16} 10^{17}
27
26
25
10^3 10^5 10^7 10^9 10^{11} 10^{13} 10^{15} 10^{17}
μ [GeV]

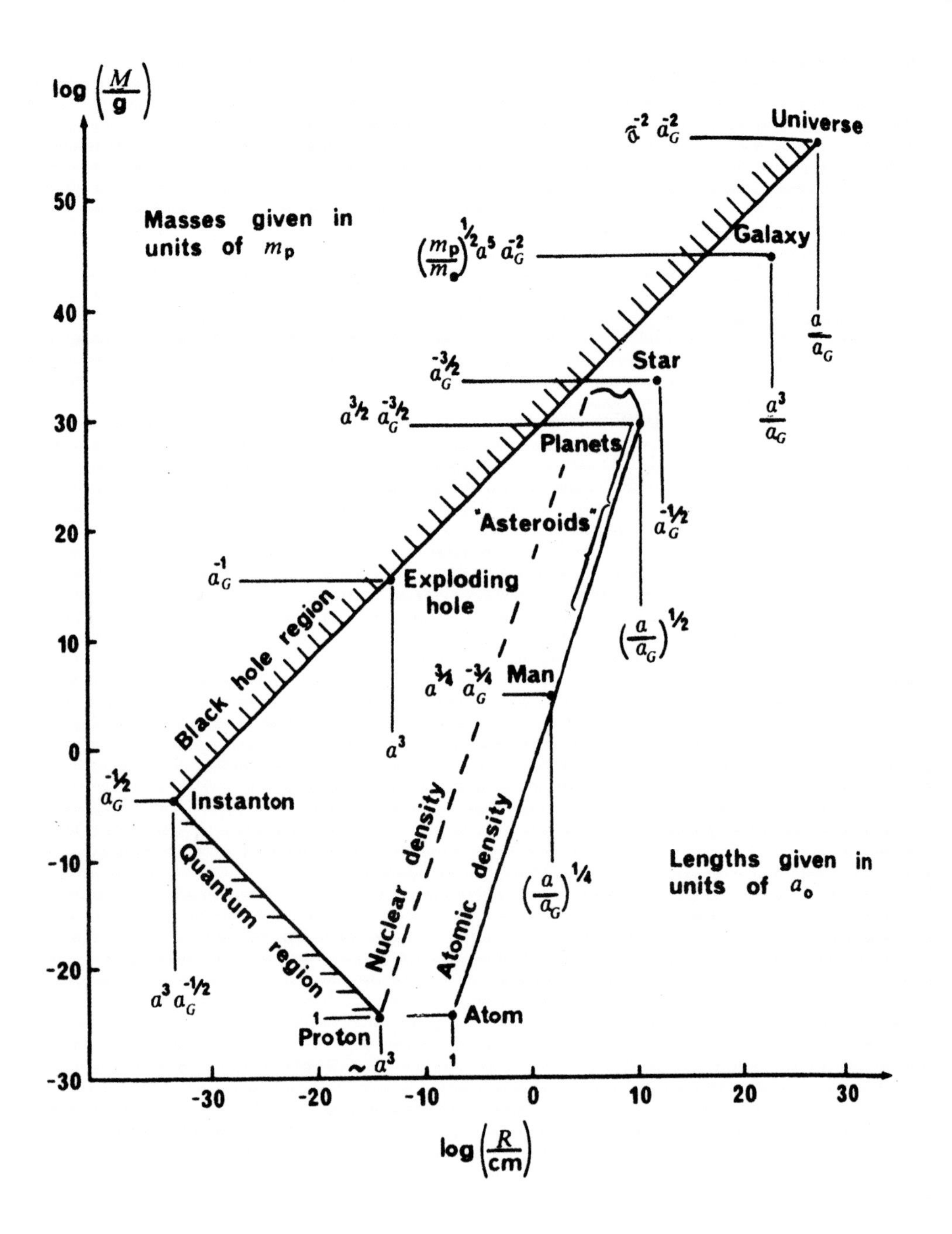

An encapsulation of the sizes (in cm) and masses (in grams) of the significant structures of the Universe, b Carr and Rees. All sizes and masses are determined in units of the radius of the hydrogen atom ($a_0 = 5.3x10^{-}$ cm) and all masses in units of the Planck mass ($m_p = 2.2 \times 10^{-5}$ gm). The line marked 'Atomic Density' is line of constant density (M/R^3) equal to the atomic density (about 1gm/cc). Regions in the top left corner i the 'black hole region' are inside black holes and unobservable from the outside. Regions in the bottom lef corner in the 'Quantum region', cannot be observed directly because of the constraints on measuremen imposed by the Heisenberg Uncertainty Principle. The sizes and masses of objects are also given in terms o the fine structure constant ($\alpha = 1/137$) and the gravitational structure constant ($\alpha_G = 10^{-39}$).[91]

THE GREAT UNIVERSAL CATALOGUE

THE SIZES OF EVERYTHING

'Neutrinos have mass?' Langden shot her a stunned look. 'I didn't even know they were Catholic!'

Dan Brown[92]

SUPPOSE you commissioned a survey of all the masses and sizes of the different things to be found in the Universe – stars, planets, atoms, molecules, comets, rocks, people, and so on. Plot the answers on a graph showing their masses and their sizes. What would the picture of points look like? You might have expected that the graph would be sprinkled with points fairly randomly covering the whole of the available space. But as the picture builds up you are in for a surprise. The distribution of the points is anything but random. Great expanses of the picture are completely empty, while most points seem to fill a narrow diagonal band running across the picture.[93]

This remarkable picture enables us to understand the Universe in a broad brush way by drawing just three straight lines. The first is the line that indicates where black holes are located in the diagram and hence which regions describe situations that are trapped within them and so will never be seen.

As we saw on page 147, black holes are regions of the Universe in which such a large amount of material has been compressed into such a small volume that it cannot escape its own gravitational pull after it falls inside a critical spherical 'horizon' surface. The radius of this trapped spherical surface, *R*, is equal to $2GM/c^2$, where G is Newton's constant which measures the strength of gravity, M is the mass inside the sphere of radius R and c is the velocity of light.[94] This is an unusual relation because it describes regions whose size is directly proportional to their mass, quite unlike solid objects of our everyday experience, whose mass is proportional to the cube of their size. The black hole line can be seen on the picture. The whole area in the top left of the diagram is empty because it describes things with $R < 2GM/c^2$, and they lie, unobservable to us, inside black holes.

All the things around us and beneath our feet are made of atoms of various sorts. The same is true of peanuts, planets, and asteroids. All these solid bodies have slightly different material composition, but to a good approximation they are bodies made of similar arrays of packed atoms. This enables us to estimate their densities because the atoms are closely packed side by side and so the density of one of these solid objects is on the average roughly the same as the density of a single atom of the sort that make them up. This density is fixed according to the identity of the atom, and varies by a fairly small amount as we go from one material to another – from water to rock or to metal for example – because it is determined by universal constants of Nature: the masses of electrons and protons, the electric charge on a single electron, the speed of light, and Planck's quantum constant. What this reveals is that, because the density of atomic solids is a constant and density is their mass divided by their volume, and their volumes are proportional to the cube of their size, R, their masses are fairly accurately proportional to the cube of their sizes. On our picture we have drawn this line of constant atomic density. It runs right through the great band of objects from single atoms right up to stars, whose average density is also given by the density of atoms.

We need only one more line to complete our story. When we go down to the micro-world, where quantum effects dominate the structure of matter, we begin to feel the impact of the wavelike quality that all masses possess. This wavelike quality is more akin to a crime wave, or a wave of hysteria, than to the waviness of a water wave. It is a wave of information. If an electron wave passes through your detector, then you are more likely to detect the 'particle' that we call an electron. The wavelength of these particle waves is inversely proportional to the mass of the object and so for large things like you and me the wavelength is tiny, far smaller than our actual size and you can forget about our quantum qualities in everyday life. But when the quantum wavelength becomes larger than an object's physical size it becomes essentially quantum-like in its behaviour. One of the features of this situation is that we cannot define its motion and its position simultaneously with unlimited precision. There is always some uncertainty that is fixed by Planck's constant of Nature. This uncertainty principle, first identified by Werner Heisenberg, ensures that the product of the mass and the size of *observable* small objects must always be larger than this constant. This restriction is the third 'quantum boundary' in our picture. If we tried to observe any particle that lies in the empty region in the bottom left of the picture then the very act of observing it would perturb it to the right and across the quantum boundary.

We see that the distribution of almost everything we see in the Universe is determined by three lines. The exceptions are the galaxies up in the top right. They are not solid objects, but are collections of orbiting stars in which the force required for their circular orbits is supplied by gravity. Still, they are very close to the continuation of the line joining the planets and the stars.[95] Lastly, there are two other objects – the nuclei of atoms and neutron stars – that do not lie on the line of constant atomic density. If you draw a line from the nuclei to the neutron stars and the pulsars they create, you get a new line that is parallel to the atomic density line. Neutron stars have the same density as the nuclei of atoms. They are like great crushes of neutrons crammed together, side by side, so their overall density is about the same as that of a single neutron or an atomic nucleus – 100,000 million times greater than the atomic density that characterises you and me. So, our new parallel line is another line of constant density determined by the constants of Nature, but this time it is nuclear density not atomic density.

Finally, we must mention a strange and mysterious curiosity. Our Universe has three dimensions of space.[96] Suppose that it had more – four, five, six . . . any number you like: what would our size *vs* mass picture of all the things it contains look like? The answer is rather surprising. It would contain none of the familiar things at all: no atoms, no molecules, no planets, no stars, and no galaxies. Only in three-dimensional worlds can the forces of Nature bind things together to make any structure.[97] It is clear that we should not be surprised to find ourselves inhabiting a three-dimensional world – for we could exist in no other.

Schrödinger's cat. The animal trapped in a room together with a Geiger counter and a hamme which, upon discharge of the counter, smashes a flask af prussic acid. The counter contains a tra of radioactive material—just enough that in one hour there is a 50% chance one of the nuclei w decay and therefore an equal chance the cat will be poisoned. At the end of the hour the total wa function for the system will have a form in which the living cat and the dead cat are mixed in equ portions. Schrödinger felt that the wave mechanics that led to this paradox presented an unaccep able description of reality. However, Everett, Wheeler and Graham's interpretation of quantum m chanics pictures the cats as inhabiting two simultaneous, noninteracting, but equally real worl

Schrödinger's Cat Paradox illustrated by Bryce DeWitt and Neill Graham.

THAT CRAZY MIXED-UP CAT

SCHRÖDINGER'S CAT PARADOX

How wonderful that we have met with a paradox. Now we have some hope of making progress.

Niels Bohr[98]

QUANTUM mechanics is a black box. It provides a clear recipe for predicting what we should see if we carry out an experiment involving atoms or sub-atomic particles. It's just that what goes on in-between is something of a mystery. The predictions of the outcomes of quantum experiments are always probabilities: you should see one thing 10 per cent of the time, say, and something else the other 90 per cent. The uncertainty that this element of randomness expresses is not like other types of so-called 'randomness' that we encounter in social science or everyday life. The latter variety of randomness arises because of incomplete information. It just recognises that we have some uncertainty in the state of affairs under consideration but we could reduce it if we worked harder and gathered more data. Quantum randomness is different. It exists even when we have all the information that it is possible to have about the world. It is intrinsic: it arises from the unavoidable effect of being an 'observer' of the world you are also a part of. You can't experiment on the world as if you are a bird-watcher in a perfect hide. In some sense you always shift its state when you observe it. More correctly, we should say that you are always inextricably linked to it.

These shifts should be insignificant when the things being observed are large, but they become crucial once they become very small. As we saw in the previous chapter, when the quantum wavelength of an object is comparable to or bigger than its physical size then quantum fuzziness starts to become noticeable and the world behaves in ways that look nothing like our everyday expectations.

In 1935, Erwin Schrödinger,[99] the Austrian physicist who found the key equation that allows the probabilities for the different observable outcomes of measurements to be calculated, proposed an extraordinary imaginary experiment. It has for ever afterwards been known as 'Schrödinger's cat paradox'.

Suppose a cat is placed in a sealed room next to a Geiger counter and an occasional source of radioactive particles. If the Geiger counter registers one of these (for all practical purposes) completely random quantum radioactive decays within the first hour, it triggers the release of a poisonous gas into the room that immediately kills the cat. If none of the radioactive atoms decay in the hour then the cat survives. The experiment ends on the hour when we look into the room and determine whether the cat is dead or alive.

Schrödinger argued that, according to the standard interpretation of quantum mechanics, before we look into the room the cat must exist in some probabilistic mixture of 'dead' and 'alive' states. Only after being looked at will it exist in one or other of two definite states, 'dead' or 'alive', but with a probability that is determined by solving Schrödinger's equation. But when, where, and how does the crazy mixed-up, neither wholly dead nor alive, cat change into the dead cat or the living cat that you find when you finally open the door? Are you the 'observer' who brings about the definite outcome, or is the Geiger counter, or even the cat itself? Or does quantum mechanics simply not apply to large things like cats?

Schrödinger was trying to make the point that he didn't believe that quantum theory could be a complete description of physical reality. It was our knowledge of the cat, rather than the cat itself, that should be regarded as being in the mixed-up state. Proponents of the standard interpretation of quantum mechanics, including Niels Bohr himself, would have argued that there is no such thing as 'the cat itself'; the only reality is our knowledge of the cat. There is no way to decouple the cat from the observer.

The cat paradox has come to epitomise the problem of the interpretation of quantum mechanics,[100] and seems symptomatic of a fundamental incompleteness in our understanding of what is going on.[101] Perhaps we don't yet have the most illuminating way of expressing quantum mechanics, and are being misled by some odd by-products of our clumsy way of looking at it.[102] No one doubts that at root it is a wonderful approximation to the ultimate truth about the way the micro-world works; its predictions are confirmed with unparalleled experimental precision, and all the world's computers, microelectronic devices, and technical systems are based upon it. Yet, the picture of 'that darned cat' has haunted physicists for more than seventy years and challenges them to solve the problem of the interpretation of quantum mechanics in a way that is convincing and transparent to all. The answer to the conundrum of the cat may well also tell us whether or not it is legitimate to apply quantum theory to entities that are real but cannot be observed – like the Universe itself.

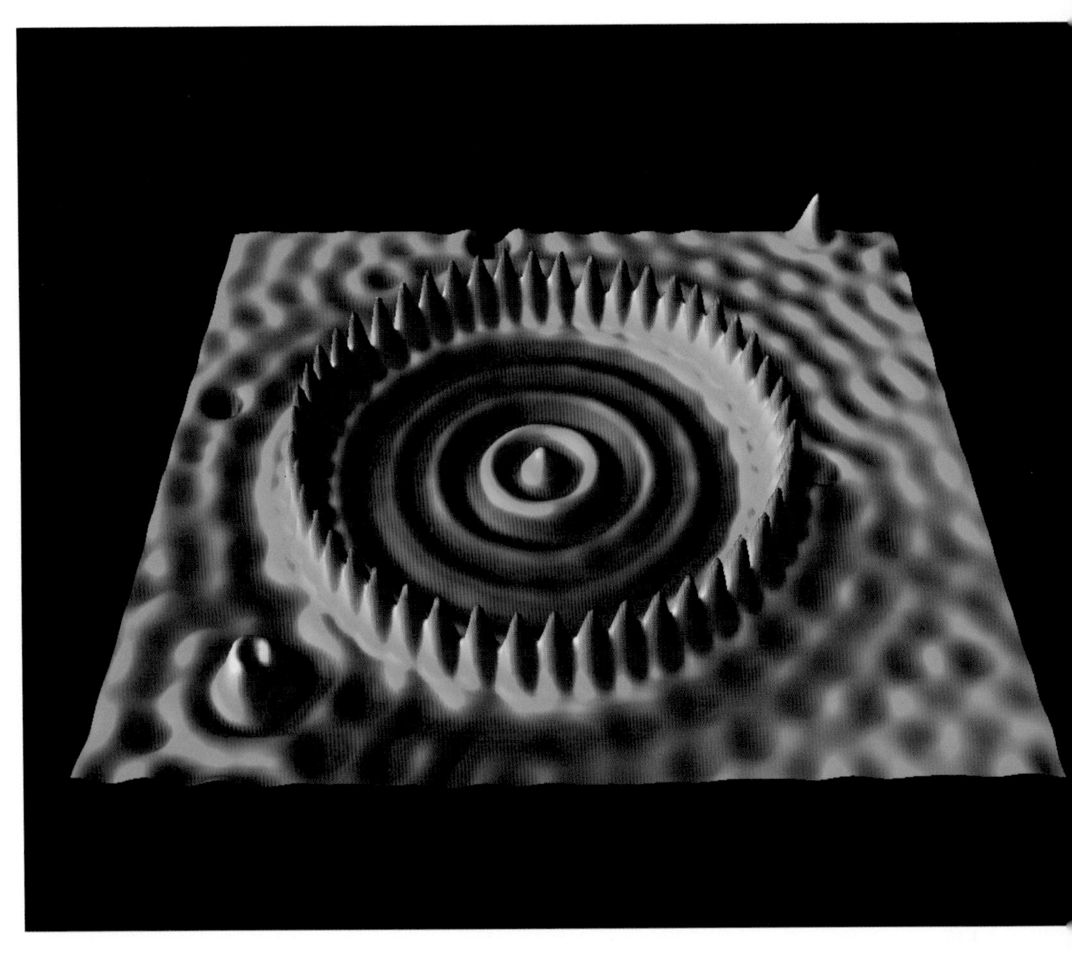

A scanning tunnelling microscope has been used by Crommie, Lutz and Eigler to move 48 atoms of iron into a ‘corr

ROOM AT THE BOTTOM

THE QUANTUM CORRAL

I have seen the future and it is small.

Larry Smarr[103]

THE touchstone of technical progress used to be our ability to engineer in the large: the biggest bridge, the tallest building, the longest aircraft. But things have gradually been turned on their head. Today, it is the skill to engineer in the small that reveals progress at our most important research frontier. It all began long ago with the creation of ever smaller transistor radios, then portable computers, mobile phones, and CD players. Small is beautiful. Meanwhile physicists have been moving in the same direction but in a realm that is too small to be visible to the naked eye. Nanotechnology is the growing science of engineering on the scale of individual atoms, moving them around so as to produce structures and 'machines' not much bigger than a few atoms across. One day, our bodies may be maintained by little nano-machines dredging our arteries clear and monitoring the internal state of our health in real time.

Dramatic advances in the manipulation of single atoms were made possible by the invention of the scanning tunnelling microscope. Don Eigler, Michael Crommie, and Chris Lutz have been responsible for two types of image that have attained iconic status in physics, and also at its interface with art. In 1993, they created[104] an image of what became known as a 'quantum corral' by using the tip of the scanning microscope to shift and confine individual atoms. By building up barriers, like 'fences', using atoms of iron, they could close them up and confine atoms of cobalt. In the picture shown we can see an almost circular 'corral' of forty-eight iron atoms. The most remarkable thing about these structures is their size. Lengths are discussed in nanometres: a nanometre is a billionth of a metre. A typical quantum corral will be between 10 and 20nm across. The diameter of a single human hair is about 200,000nm across and the size of a single atom of silicon is about half a nanometre. A nano-guitar has been made from a single crystal of silicon only 10,000nm long and has six strings, each about 50 nm wide. It is even possible to 'pluck' the strings, although the notes produced have very high frequencies and are inaudible to the human ear.

These structures also give rise to the quantum 'mirage' phenomenon, shown opposite. The electrons and other subatomic particles involved have wavelike qualities which can be made to resonate so that wave peaks add to wave peaks to produce a strong focusing of energy at particular positions. This was first demonstrated by Hari Manoharan, Christopher Lutz, and Don Eigler, in 2000, with an elliptical corral of cobalt atoms on a copper surface. The cobalt atoms reflect the surface electrons around the copper atoms to produce the wave undulations predicted by the equations of quantum mechanics. By making the shape of the corral an ellipse it is possible to place one magnetic cobalt atom at one of the two foci of the ellipse and thereby create the image of another cobalt atom at the other focus – a quantum mirage. The same physical and electronic properties were detected around both the foci of the ellipse even though a physical atom had only been placed at one of them! Their picture appeared on the front cover of *Nature*[105] in February 2000 and attracted massive media attention because of its striking visual appearance: for the first time we seemed to be 'seeing' atoms and herding electrons.

The single cobalt atom is placed at the focal point where the purple peak rises up inside the yellowy-orange elliptical ring of thirty-six cobalt atoms. The shallower purple ridge at the other focus of the ellipse is the mirage – no additional atom was physically placed there.

This experiment exploits an unusual property of the ellipse which is not a consequence of quantum mechanics. If you make an elliptical billiard table and place the cue ball at one focus and the target ball at the other then no matter how you hit the cue ball, it will eventually hit the target ball so long as friction doesn't stop it first (or it goes in one of the pockets). This is a property of the ellipse that we also encounter in whispering galleries. At the University of Virginia, Charlottesville, there is a famous oval reception room originally designed by Thomas Jefferson. Despite its large size, if you stand at one of the focal points you can hear the conversations of people far away at the other focal point as if they are beside you. Something similar occurs in the oval corral with the quantum waves associated with the atoms and electrons. Their quantum 'echoes' converge to oscillate with strong intensity at the second focal point.

In conventional electronic engineering these phenomena would be problematic. Try to make a wire to carry electrons that is too small and the wave-like quality of the electrons will reach outside the wire and become entangled with other parts of the circuitry – with potentially disastrous results. But if you start out with these unusual quantum qualities, you can engineer devices so as to exploit the quantum effects in positive ways. This is the basis of the new field of photonics. It aims to develop computing devices on the nano scale which make use of corralled atoms to store bits of information and process them. Just as atoms bind together to make new molecules and compounds, so it might be possible to join nano devices together like Lego bricks to create complex atomic machines. The quantum corral is not just the pretty face of quantum mechanics: it is the face of things to come.

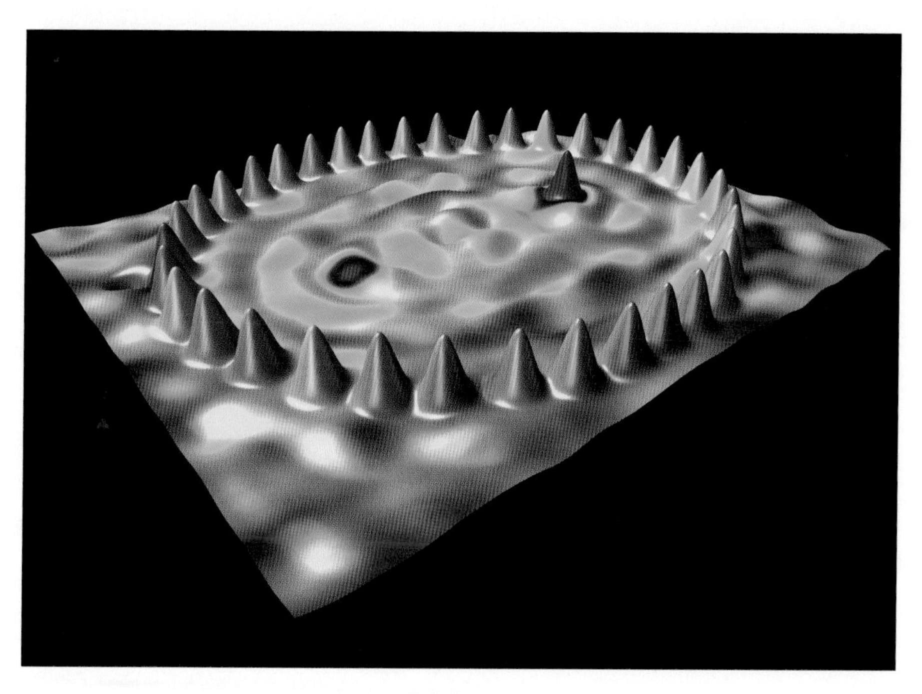

A single cobalt atom (purple peak) located at the focal point of a confining elliptical corral of 36 cobalt atoms creates a quantum mirage of another atom at the other focus of the ellipse (purple spot).

COSMIC IMAGERY JOHN D. BARROW

QUANTUM MIRAGE

ENTANGLED PHOTONS

It is wrong to think that the task of physics is to find out how Nature is. Physics concerns what we can say about Nature.

Niels Bohr

EINSTEIN worried deeply about quantum mechanics. He never accepted it as a complete and consistent description of the world and dissipated a lot of mental energy on trying to devise simple thought-experiments that would reveal quantum mechanics to be inconsistent or incomplete in some unacceptable way. The most famous attempt along these lines was made in a paper co-authored with two collaborators, Nathan Rosen and Boris Podolsky, that appeared in the 15 May issue of the *Physical Review* in 1935.[106] Podolsky and Rosen were research associates at the Institute for Advanced Study in Princeton, where Einstein worked from 1933 until his death in 1955. Their provocative article grew out of ideas that Einstein began to put forward in 1930, and was actually written by Podolsky because he had the greater facility with English (Einstein's earlier great papers had been written in German). Einstein was never very happy with the final version of the paper, because he thought that 'the essential thing . . . was smothered by the formalism' that Podolsky had used to explain it.[107] It has, nonetheless, become one of the most influential papers ever written in physics and defined what became known as the 'EPR paradox'.

Einstein set out to challenge Bohr's widely accepted doctrine that no phenomenon could be said to 'exist' until it was observed: and so physicists are

Opposite: a quantum entanglement made visible by Kwiat and Reck in 1995.

only involved in discovering what we say about reality rather than discovering what it *is*. On the contrary, Einstein believed that the true nature of things existed independently of whether or not they were observed.

The EPR thought-experiment considered the decay of an elementary particle into two photons moving off with the same energies in opposite directions. One feature of this decay is that the photons each have a special quantum property called 'spin'. If one of them is produced with a clockwise spin then the other must have an anticlockwise spin of the same magnitude so that when they are added the result is zero. This reflects the fact that spin is a conserved quantity in Nature, like energy and momentum; if the initial state before the decay has zero spin, then the sum of the spins of all the decay products at the end will also have to be equal and opposite and sum to zero.

Einstein argued that this meant that if you measured the spin of one of the decay products and found it to be clockwise, then the spin of the other one must be anticlockwise whether you choose to measure it or not. No intervention by an observer was needed to determine the spin of the second particle. Therefore, the unmeasured spin is *real* because it is predictable. It is as if you have a pair of red gloves which get separated in the laundry. If you first find the right-hand one, then you know before you ever find it that the other one will be left-handed.

Quantum mechanics tells us that before the spins are measured each photon has an equal chance of being found with either a clockwise or an anticlockwise spin – there is no in-built bias for one direction or the other. But, as soon as one of the spins is determined by a measurement, mere chances become certainties and the other one is definitely determined as well.

The word 'paradox' is associated with this state of sequence of events, because it is mysterious how the second photon comes to 'know' what spin it must have as soon as the first photon's spin is measured. A second measurement to discover it could be pre-programmed to take place so quickly that there isn't time for a light signal to send any information about the result of the first measurement to the venue of the second one. Einstein thought this communication problem required us to believe in some sort of 'spooky action at a distance'. Remarkably, the EPR thought-experiment became a real experiment in 1982, when it was performed by Alain Aspect and his colleagues in France. In effect, they changed the spin of an initial particle every ten billionths of a second before its decay. In the meantime, they made measurements of the equal and oppositely orientated spin of the decay photons when they were separated by four times the greatest distance that light could have travelled in the short time since the initial spin was last altered. The predictions of quantum mechanics were confirmed! The photons did 'know' what spin they should have.

Bohr had never been impressed by the EPR paradox. He simply did not accept Einstein's definition of 'reality' and he stubbornly refused to distinguish between reality and observed reality. Although the two spinning particles do not seem to be able to influence one another, they are entangled, Bohr claimed, by the presence of an observer who decides to learn about the spin of the second particle by making a measurement of the first. The observer is an inseparable part of the experiment. One must not think of the two spins as independent entities until after the intervention of the observer who separates them by the final act of measurement: the results of the first *and* second measurement cannot be predicted before a measurement is performed. The quantum description of events was not an incomplete description of the world, as EPR had claimed in the title of their paper, but merely one that is non-local and far more extensive than they had presumed.

Quantum entanglement of causally separated, or 'non-local', events has developed into a major area of study in recent years. Understanding what it means and how it can be exploited will determine whether we can produce routine and reliable forms of quantum encryption and computation, with all the remarkable benefits that they promise. In 1995, the phenomenon of entanglement was captured on photographic film by Paul Kwiat and Michael Reck, who were then both working at the University of Innsbruck. They photographed a beam of ultra-violet laser light of wavelength 351 nm passing through a crystal of a material called beta barium borate, a material chosen for its transparency and other good optical properties. This process results in about one in 10 billion of the photons being rotated ('polarised') into photons of lower energy, with wavelength 702 nm, which are oriented in two different directions.[108] The total energy is conserved[109] and there is a small chance that the two new photons will be entangled in the way that EPR envisaged.

A photograph, with no lens used to amplify the effects, was able to capture the conditions under which the two different polarisations become entangled. Only about 1 per cent of the one in 10 billion photons that get converted into lower-energy photons end up becoming entangled. But there are so many photons that this still creates a good chance of finding some. In the picture, we see that the light that shines through the crystal forms rings. The two green rings in the picture are formed by photons of the two different polarisations. At each of the two places where the green rings intersect one another, the polarisation of one member of the entangled pair is the opposite to its value at the other point of intersection. This is a photograph of entangled quantum states.

These phenomena have no counterpart in the non-quantum picture of the world. They show up the mismatch between our intuitive common sense about the way the world works and the deep nature of reality. But they promise us huge practical benefits that Einstein and Bohr could never have foreseen. Imagine a message encrypted in one of the two tangled photons. If someone tries to mess with the message then the entangled state will reveal this action via the other photon far away. Quantum encryption was developed in a practical way by Artur Ekert[110] and is a focus of huge worldwide interest by seekers after secrecy. One day it may protect your bank account and computer files in ways that are 100 per cent secure.

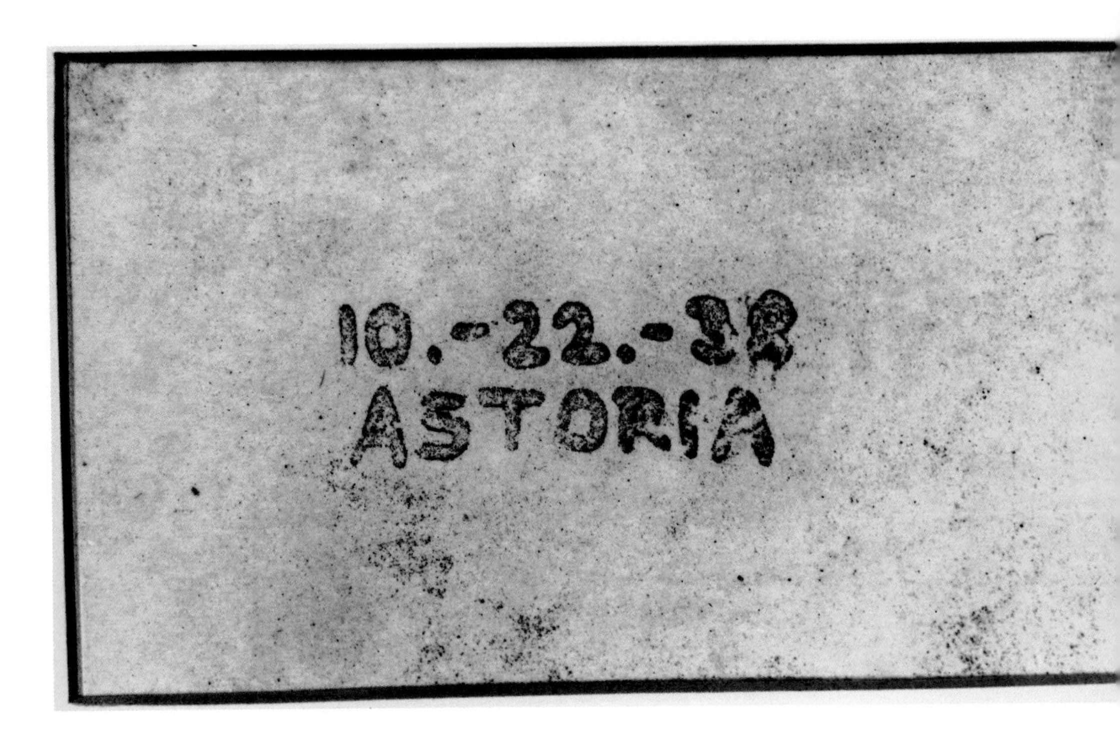

The first photocopy, made by Chester Carlson on 22 October 1938, in Astoria, New York.

DÉJÀ VU ALL OVER AGAIN
XEROGRAPHY

A photograph is not only an image (as a painting is an image), an interpretation of the real; it is also a trace, something directly stencilled off the real, like a footprint or a death mask.

Susan Sontag[111]

MANY schoolteachers, university lecturers, and professors despair that learning has been replaced by photocopying. In the 1970s and 1980s, before cheap microcomputers appeared on the scene, everyone seemed to be photocopying every book, article, and letter they read or were supposed to read. The economies of entire universities seem to rest upon the high charges they made for students to use the photocopying machines in their libraries. Rumour had it that PhD stood for 'Doctor of Photocopying'. So who made the first photocopy and set this juggernaut of paper consumption in motion?

The culprit was an American patent lawyer and amateur inventor named Chester Carlson.[112] Despite graduating in physics from Cal Tech in 1930, Carlson couldn't find a steady job, and his parents were impoverished because of their chronic poor health. The deepening American economic depression hit hard, and Carlson had to settle for any job he could get. The result was a spell at the Mallory battery company, where he found himself in the patent department. Anxious to make the very best of any opportunity, he worked for a law degree at night school and was soon promoted to become manager of the whole department. It was in that job that he began to be frustrated that there were never enough copies of the patent documents available for all the organisations that seemed to need them. All he could do was send them off to be photographed – and that was expensive – or copy them out by hand, an unappealing task because of his failing eyesight and painful arthritis. He had to find a cheap and painless way of making copies.

There wasn't an easy answer. Carlson spent the best part of a year researching messy photographic techniques, before his library searches uncovered the new property of 'photoconductivity' that had been discovered recently by the Hungarian physicist Paul Selenyi. He had found that when light strikes the surfaces of certain materials, the flow of electrons – their electrical conductivity – increases. Carlson realised that if the image of a photograph or a page of text was shone on to a photoconductive surface, then electrical current would flow in the light areas but not in the dark printed areas and an electrical copy would be created. He created a makeshift home electronics lab in the kitchen of his Jackson Heights apartment in the Queens district of New York City and experimented through the night with numerous techniques for duplicating images on paper.[113] After being evicted from the kitchen by his wife, his lab moved to a beauty salon owned by his mother-in-law, in nearby Astoria. The first successful copy that he made, with the help of his assistant, an unemployed German physicist named Otto Kornei, was created in Astoria on 22 October 1938 – and those space and time coordinates were preserved for posterity in what he copied.

Carlson took a zinc plate, coated it with a thin layer of sulphur powder and wrote the date and place '10–22–38 Astoria' in black ink on a microscope slide. Turning the lights down, he charged up the sulphur by rubbing it with a handkerchief (just like we might rub a balloon on a woollen sweater) and then placed the slide on top of the sulphur and put it under a bright light for a few seconds. He carefully removed the slide and covered the sulphured surface with lycopodium fungus powder before blowing it off to reveal the duplicated message. The image was fixed with heated wax paper so that the cooling wax congealed around the fungus powder.

He dubbed his new technique 'electrophotography' and tried hard to sell it to businesses, including IBM and General Electric, because he didn't have any money for further R&D. But none showed the remotest interest. His equipment was awkward and the processing complicated and messy. Anyway, carbon paper worked just fine, everyone said!

It was not until 1944 that the Battelle Research Institute in Columbus, Ohio, approached Carlson and entered into a joint agreement to improve his crude process with a view to commercialisation.[114] Three years later, the Haloid Company, a manufacturer of photographic papers, from Rochester, bought all the rights to Carlson's invention from them and began planning to market his copying devices. Their first change, with his agreement, was to abandon Carlson's cumbersome name for the process. 'Electrophotography' was superseded by 'xerography' thanks to the suggestion of a professor of Classics at Ohio State University. The etymology is Greek, literally 'dry writing'. In 1948, the Haloid Company shortened this to the trademark 'Xerox'. The 'xerox machines' they marketed soon enjoyed commercial success, and in 1958 the company reflected this by changing its name to Haloid Xerox Inc. The new 1961 machine, the Xerox 914, the first to use ordinary paper, was such a big success that they dropped the Haloid completely to become simply the Xerox Corporation in 1961. Revenues ran to $60 million that year, and by 1965 they had grown to a staggering $500 million. From then on, their company received the ultimate accolade: like Hoover, their name became both an adjective, a verb, and a noun to describe the very process and the machines that they (and their frustrated competitors) produced.[115] Carlson became fabulously wealthy but gave away two thirds of his income to charities. Sadly, he collapsed and died unexpectedly, in 1968, aged just sixty-two, while walking down the street after attending a conference in New York.[116] His first photocopy created an unnoticed change in the working practices of the whole world. Information transfer would never be the same: pictures as well as words could now be routinely copied.

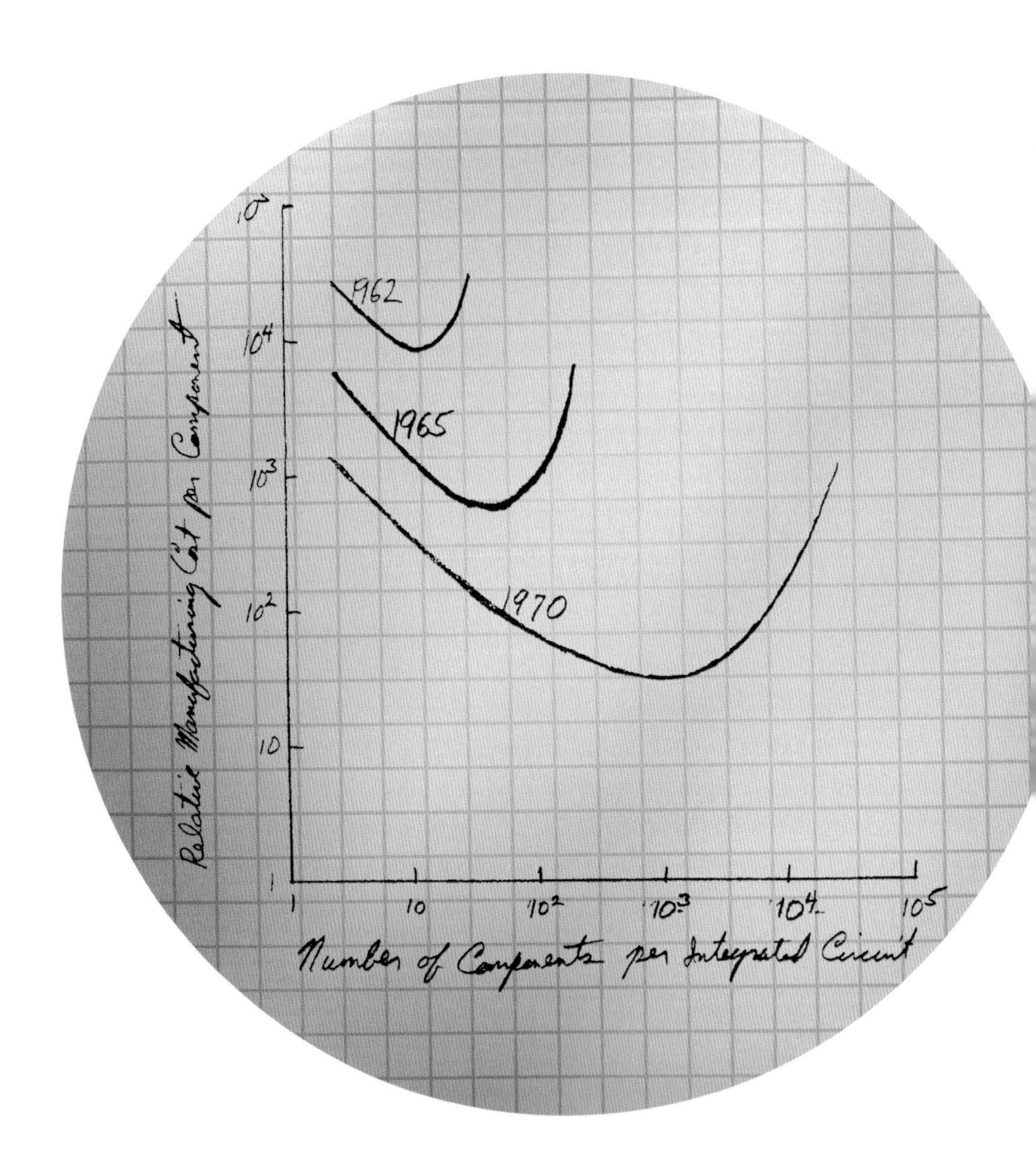

Gordon Moore's original 1965 drawing of the change in the manufacturing cost per computer chip versus the number of transistors per square inch of integrated circuit. The trend displayed became known in the industry as 'Moore's Law'.

ONE THING LEADS TO ANOTHER

MOORE'S LAW

On two occasions I have been asked [by members of parliament], 'Pray, Mr Babbage, if you put into the machine wrong figures, will the right answers come out?' I am not able rightly to apprehend the kind of confusion of ideas that could provoke such a question.

Charles Babbage

COMPUTERS are useless', claimed Pablo Picasso, because 'they only give you answers.' Still, sometimes answers are jolly useful, especially if you can get them quickly, cheaply, and reliably. They may not push the barriers of abstract art but they have produced one of the most dramatic revolutions in the whole of human history in a period of little more than twenty-five years.

Many of us can still remember a world before personal computers, word processors, and mobile phones, but an increasing fraction of the workforce can neither remember that era nor quite imagine what it must have been like. The consequence of this decisive technological development has been an acceleration in the pace of life in advanced societies, and the inexpensive linking of members of developing countries to the global oasis of knowledge. For the ordinary user of technology, this progress seems to be a bit like a runaway train. There is a never-ending series of new products, improvements in speed and memory storage, enhanced interactivity, and wireless capabilities. New companies appear out of nowhere, and grow Google-like into economic forces that surpass the strength of entire countries. Are there any simple patterns behind this headlong race towards the future?

In 1965, long before the appearance of the personal computer, Gordon Moore, the co-founder of the chip-manufacturing company Intel, decided to look for some pattern behind the progress that was occurring at the cutting edge of the computer chip industry. He knew that the progress of his company was rapid, but he was interested in seeing whether that progress was speeding up or slowing down and, being an astute businessman, whether it was predictable. He calculated that the number of transistors that could be miniaturised and fitted on a square inch of integrated circuit was doubling at a predictable interval, and their cost was reducing proportionately.[117] Every two years, the transistor density doubled and the chip cost halved. The reduction in size meant that power consumption decreased and integrated circuits could operate at higher speeds.[118] The picture has been continually updated and acts as a benchmark for the average rate of progress in the microelectronics industry. The title 'Moore's law' for this simple trend was invented by Carver Mead, a professor of computer science at Cal Tech and another microelectronics pioneer.

At the root of this revolution lies the metal-oxide semiconductor, or MOS, that grew from the invention of the first bipolar junction transistor made from germanium at Bell Labs in New Jersey, in 1947. Germanium was superseded by silicon in the 1950s, because its oxidised form, silicon dioxide (just sand or quartz are examples), is a wonderful electrical insulator and is easy to shape and move. Jack Kilby of Texas Instruments invented the integrated circuit in 1958, and the following year it became possible to make thin sheets of silicon on which huge numbers of transistors could be linked. Mass production was now possible and, by 1970, many companies were in the business of producing integrated circuits. The first microcomputer became possible in 1971, when Intel made the first single-chip microprocessor based on a transistor – the Intel 4004. At that stage, a single transistor was 10,000 nm in size and the chip contained about 2300 of them. Today, the latest Intel Itanium-2 processor contains 410 million transistors, 45 nm in size, on a single chip 3 cm^2 in area.

A simple measure of progress in the sheer number of processors might not tell the whole story. It ignores the fact that they might each start to do more

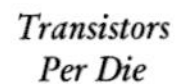

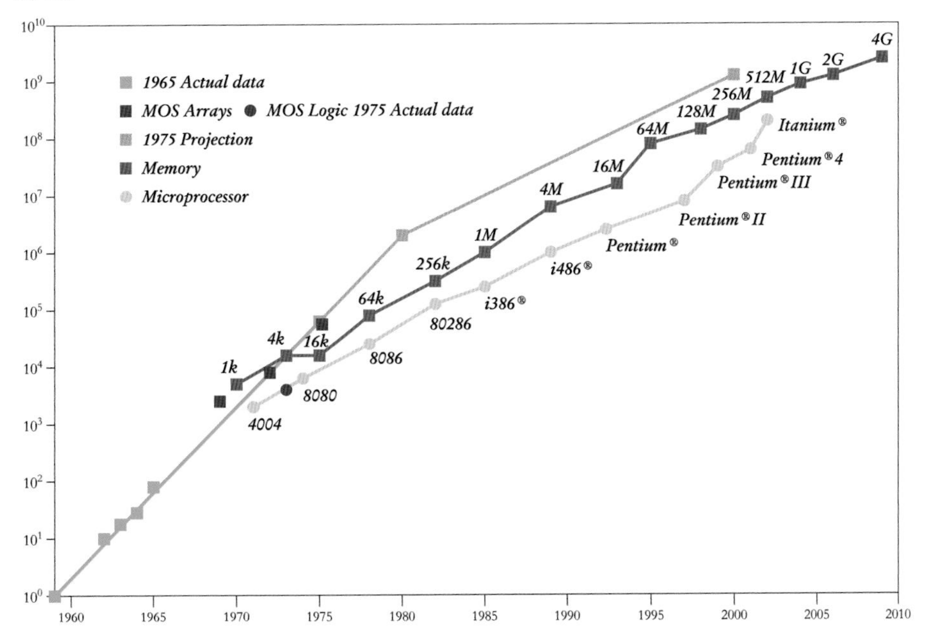

A modern representation of Moore's Law showing the almost-constant rate of increase in the number of transistors on each rectangular wafer of integrated circuit (called a 'die') with time. This level of predictability played a key role in the complementary evolution of hardware and software development in the computer industry. Moore's original law was that the number of transistors per integrated-circuit die (green line) was doubling every two years. It turns out that this also applies to other aspects of computers, such as memory (red line) and microprocessor speed (yellow line).

sophisticated things. Another way to measure progress would have been to chart the reduction in cost to store information on disk, but Moore's measure was the one that the industry regarded as most indicative of what counted. Why did it take on an almost legendary status?

Moore's 'law' did more than simply summarise progress. Of course, it is not a law of Nature, like the law of gravitation, that prescribes how change must occur. However, it played an important role in stimulating progress. It served as a benchmark for what the computer industry knew it should achieve by a given date.[119] Exceptional progress would be defined as progress above and beyond an extrapolation of the normal expectation mapped out by Moore's two-year doubling curve. Most important of all, it enabled parts of the computer industry whose business it was to manufacture new drives and peripherals to keep ahead of the game. If significant hardware developments take several years to research, plan, and put into production, it is no good waiting for new processing power to arrive before you respond. You need to anticipate progress and have your products ready just in time to exploit its first availability. Moore's law played a crucial role in guiding manufacturers to plan product development optimally. In effect, it became a self-fulfilling prophecy about the growth in microchip density.

Will Moore's progression go on for ever? At present, it is expected[120] that the size of transistors will halve every six years, down to 18 nm by 2010, and below 10 nm by 2016. Alas, no; it describes a type of progress that uses the same basic techniques over and over again, with finer quality and more precise robotic dexterity in engineering. Eventually, manufacturing starts to encounter a realm so small that new laws of physics start to affect the way the engineering is done. New structures become possible, smaller versions of old ones impossible. When chip size is reduced, the oxide thickness on the transistor has to be reduced as well. A transistor size of 30 nm (3000 times smaller than the width of a human hair) requires an oxide thickness of about 0.7 nm – just two atoms thick. This presents manufacturing problems and also the need to have sufficient electrical capacitance to control electrical current flows adequately. Very small devices suffer from having larger leakage of current. Materials with superior electrical properties to silicon, such as hafnium, might provide a solution for a while, and keep the Moore curve on track. But once the density of components on a silicon wafer starts to become as crowded as the atoms themselves, we are reaching a fundamental barrier. The distance between individual atoms of silicon is 0.27 nm. In this realm, we enter a new world of nano[121] engineering (see pages 523–525). Here we face the unusual possibilities of intrinsically quantum computing and methods of computing which can exploit the spin of electrons rather than their electric charge, the properties of carbon fullerene nanotubes, and photonic devices. Progress in the direction of pure silicon miniaturisation will eventually level off, but computing capability will then develop in a new direction that may well follow its own pattern of technical progression. Everyone will be looking for a simple trend of the sort that Gordon Moore first projected for the evolution of silicon microelectronics. Perhaps there will be a Moore's 'second law'?

Gabriel Orozco's *Sand on Table*, 1992.

THE SECRET OF THE SANDS

THE SANDPILE

To see a World in a Grain of Sand
And a Heaven in a Wild Flower,
Hold Infinity in the palm of your hand
And Eternity in an hour.
William Blake[122]

NEWTON could tell us deep truths about the world by observing mundane objects. Is that possible any more? It would be easy to believe that fundamental discoveries always require millions of dollars or euros and armies of people, along with huge particle colliders, batteries of computers, gigantic telescopes, or satellites in orbit. Fundamental science has become bigger and bigger science. But there are some beautiful exceptions. One of the most impressive examples grew from careful thought about an observation that we have probably all made at one time or another. It has become a paradigm for the development of forms of complexity that appear to organise themselves out of disorder.

Create a pile of sand by pouring it down on to a flat surface, like a small table top. The force of gravity ensures that the sand falls downwards on to the table and the pile starts to build up. Gradually, the pile steepens. Avalanches of sand continually occur. At first, the infalling grains have effects only very close to where they fall, but as the pile steepens the avalanches become more extensive in their effects. Eventually, something odd happens: the pile gets no steeper. A critical slope is achieved and adding further sand just produces avalanches

A sandpile builds up, producing a pile with a maximum slope. The continued addition of sand precipitates avalanch
sand over all possible sizes, which maintain the pile at its maximum slope. This process, by which a concatenatic
chaotically unpredictable individual events (the falling sand grains) organise themselves to create a large scale c
has been widely studied as a possible paradigm for the development of many forms of complexity. It was first studie
Bak, Tang and Wiesenfeld in 1987.

that maintain the same slope. If the pile is sitting on a table top then eventually the sand will start to flow over the edge of the table at the same rate that it falls in from above.

What has happened is remarkable. Each of the infalling sand grains is following a chaotically sensitive trajectory in the sense that a small deflection on the way down from other grains results in a big change to its subsequent history – maybe falling down the other side of the pile. Yet the net result of all these chaotically falling grains is a highly organised pile with a particular slope.[123] Stranger still, the pile maintains its organised slope by instabilities – avalanches – which occur on all dimensions from the size of a single grain right up to the length of the pile's slope.

This process was given the name 'self-organising criticality' by Per Bak, Chao Tang and Kurt Wiesenfeld, who first recognised its significance in 1987.[124] The adjective 'self-organising' captures the way in which the chaotic input seems to arrange itself into an orderly pile. The attribute of 'criticality' reflects the precarious state of the pile at any time. It is always about to experience an avalanche at some place or other. The sequence of events that maintains its state of order is a slow build-up of sand somewhere on the slope, then a sudden avalanche, slow build-up, sudden avalanche, and so on. When the little local build-up of sand creates a small hill steeper than the critical slope a collapse occurs. Overall stability is maintained by local instability.

What is unexpected about this situation is that the pile continually evolves towards a precariously unstable state, whereas most systems, like a ball rolling into a bucket, seek out the most stable resting place. The sandpile is increasingly susceptible to disturbances of all sizes as it nears its critical state and always exists as a transient orderly state. If it forms on a table top, then sand arrives at the same rate that it falls off the edge of the table; the structure of the pile as a whole persists but is composed of different grains of sand. All that is needed for this type of critical structure to arise is for the frequency of an avalanche to depend only on a mathematical power of the size of the avalanche – that power will be negative as large avalanches are rarer than small ones.[125] This means that there is no preferred size of avalanche. This wouldn't be true if the sand was sticky and tended to form balls of a particular size that just rolled down the side of the pile. Closer examination of the details of the fall of sand has revealed that avalanches of asymmetrically shaped grains, like rice, produce the critical scale-independent behaviour even more accurately because the rice grains always tumble rather than slide.[126] Powders, on the other hand, behave quite differently.

At first, it was hoped by its discoverers that the sandpile might be the paradigm for the development of all types of organised complexity. This was too optimistic. But it turns out to provide clues as to how many types of complexity organise themselves. The avalanches of sand can represent extinctions of species in an ecological balance, the bankruptcies of businesses in an economic system, earthquakes, or volcanic eruptions in a model of the pressure equilibrium of the Earth's crust, and even the formation of ox-bow lakes by a meandering river. Bends in the river make the flow faster there, which erodes the bank, leading to an ox-bow lake forming. After the lake forms, the river is left a little straighter. This process of gradual build-up of curvature followed by sudden ox-bow formation and straightening is how a river on a plain 'organises' its meandering shape.

At first, it seems rather spooky that all these completely different problems should behave like a tumbling pile of sand. A picture of Richard Solé's, showing the dog being taken for a bumpy walk, reveals the connection. If we have a situation where a force is acting (for the sandpile it is gravity, for the dog it is the leash) and there are many possible equilibrium states (valleys for the dog, temporarily stable hills for the sand), then we can see what happens as the leash is pulled. The dog moves slowly uphill and then is pulled swiftly across at the top, begins slowly climbing again, then jumps across again. This staccato movement of slow build-up and sudden change, time and again, is what characterises the sandpile with gradual build-up of sand followed by an avalanche. We can see from the picture that it will be the general pattern of behaviour in any system with very simple ingredients.

The nice feature of these insights is that they show that it is still possible to make important discoveries with the simplest everyday objects simply by asking the right questions and observing closely what happens. It is a modern counterpart to Newton's prism – a picture that threatens to create a new worldview.

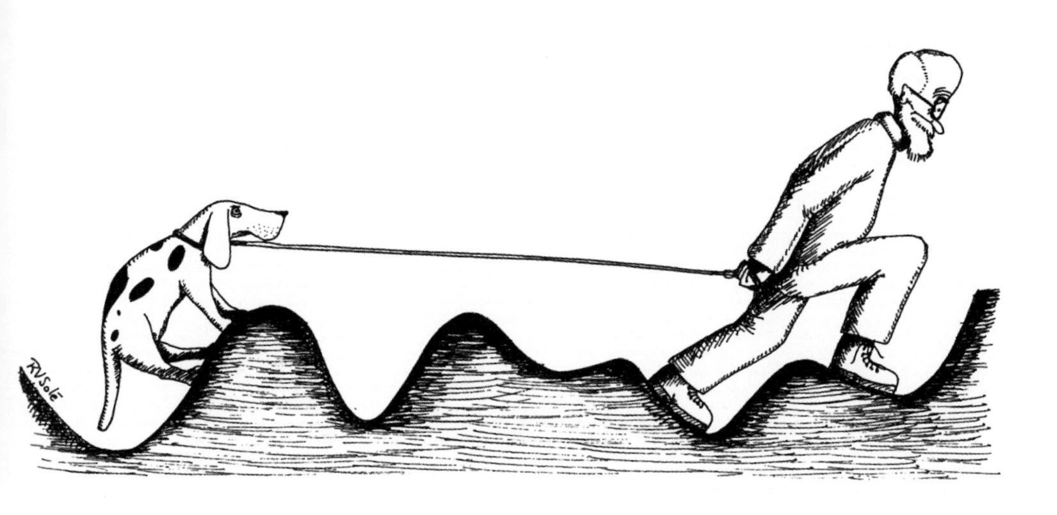

A picture by Richard Solé showing the reason for the general pattern of 'slow change followed by sudden jump' seen in the evolution of systems with many possible equilibrium states, through which they move under the influence of an external force. In this case the force is elastic and the equilibrium points are the valleys in the undulating terrain. The sudden jumps occur every time the dog is pulled over the top of a hill.

NOTES

Several firms have gone so far as to announce that they will not burden their text with footnotes, as if conferring a favour upon their readers. Others have slyly encouraged a writer or two to put up a website for the footnotes that have been refused the hospitality of the book itself. The notion seems to be that this way the scholar can find the 'dull' citations if needed while the general reader can have an uninterrupted good read.

We know this to be nonsense of course. The layperson as well as the scholar enjoys footnotes. They can be charming, an encouragement to read on, worth every penny of the extra expense.

Chuck Zerby, *The Devil's Details*

PART 1 STARS IN YOUR EYES

1. H. Mankell, *Chronicler of the Winds*, Harvill Secker, London (2006), p.63.

2. John Updike, 'The Accelerating Expansion of the Universe', *Physics Today* (April 2005), p. 39.

3. These pictures are widely reproduced and there are no restrictions on their reprinting. If a newspaper editor wants an astronomical image at short notice, then Hubble images are immediately available at no cost and with a minimum of effort. They have become instantly recognisable icons and their periodic release a much-awaited event.

4. Note that in ancient times the Earth was generally believed to lie at the centre of the solar system and the Sun to orbit around it.

5. Aries, Taurus, Gemini, Cancer, Leo, Virgo, Libra, Scorpio, Sagittarius, Capricorn, Aquarius and Pisces.

6. *Julius Caesar* III i 60.

7. For a fuller account of this see J. D. Barrow, *The Artful Universe Expanded,* Oxford UP, Oxford (2005), chapter 4.

8. The 'hole' in the southern sky is difficult to see in Cellarius' map because he has made full use of the space to create a balanced artwork and there have been many additions to the ancient map of constellations.

9. Colures are the two principal celestial circles. One passes through the two poles and the two solstices, the other through the poles and the equinoxes.

10. B. Schaefer, 'The Epoch of the Constellations on the Farnese Atlas and their Origin in Hipparchus's Lost Catalogue', *J. Hist. Astronomy* 36, 167–196 (2005); http://www.phys.lsu.edu/farnese/JHAFarneseProofs.htm

11. The key point about the dating is that it rules out the other prime candidates to have supplied the sky catalogue for the globe. Aratus and Eudoxus are around 275 BC and 355 BC, while Ptolemy's *Almagest* dates from AD 130.

12. C. Swartz, *Le Zodiaque expliqué*, 2nd edn, Migneret et Desenne, Paris, transl. from Swedish (1809). The first edition was published in 1807. The title of the first French edition was *Recherches sur l'origine et la signification des Constellations de la Sphère grecque.* Swartz was a graduate of Uppsala University who then worked as a civil servant.

13. E. W. Maunder, *The Astronomy of the Bible*, Hodder and Stoughton, London (1909). His work is also discussed by A. Crommelin, 'The Ancient Constellation Figures', in *Splendour of the Heavens*, vol. 2 , ed. T. Phillips and W. H. Steavenson, Hutchinson, London (1923).

14. M. W. Ovenden, 'The Origin of the Constellations', *Philosophical Journal* 3, 1–18 (1966). This work was elaborated by Archie Roy, 'The Origin of the Constellations', *Vistas in Astronomy* 27, 171–97 (1984).

15. M. A. Evershed, 'The Origins of the Constellations', *Observatory Magazine* 36, 179–181 (1913); D. R. Dicks, *Early Greek Astronomy to Aristotle*, Cornell UP, Ithaca (1970); E. C. Crupp, 'Night Gallery: The Function, Origin and Evolution of Constellations', *Archaeoastronomy* 15, 43–63 (2000); J. H. Rogers, 'Origins of the Ancient Constellations', *Journal of the British Astronomical Association* 108, 9–27 and 79–89 (1998).

16. Aratus was a native of Cilicia like the apostle Paul, who quotes from his poem in his famous address about the 'unknown god' to the Athenian Court of the Areopagus on Mars' Hill that is recorded in Acts 17. Paul's remark 'for in him we live, and move, and have our being; as certain also of your own poets has said, For we are also his offspring' quotes one of the early lines of Aratus' poem that reads, following its dedication 'To God above':

> In every place our need of Him we feel;
> For we his offspring are

For a translation of the whole poem see G. R. Mair, *Aratus' Phaenomena*, Loeb Classical Library, Heinemann, London (1921).

17. Eudoxus is Greek for 'of good repute', and interestingly Cnidus lies at a latitude of 36 degrees 40 minutes north. He lived from about 409 to 356 BC.

18. For an interesting and detailed series of studies about how the constellations may have been built up over many thousands of years by the creation of marker stars to define equinoxes and solstices that would remain fairly stable over two millennia, see the articles by Alex Gurstein, 'On the Origin of the Zodiacal Constellations', *Vistas in Astronomy* 36, 171–90 (1993); 'Dating the Origin of the Constellations by Precession', *Soviet Physics Doklady* 39, 575–8 (1994); 'Prehistory of Zodiac Dating: Three Strata of Upper Paleolithic Constellations', *Vistas in Astronomy* 39, 347–62 (1995); 'When the Zodiac Climbed into the Sky', *Sky and Telescope* (October 1995), pp. 28–33; 'The Great Pyramids of Egypt as Sanctuaries Commemorating the Origin of the Zodiac: An Analysis of Astronomical Evidence', *Soviet Physics Doklady* 41, 228–32; (1996); 'In Search of the First Constellations', *Sky and Telescope* (June 1997), pp. 46–50; 'The Origins of the Constellations', *American Scientist* 85(3), 264–273 and 500–501 (1997); 'The Evolution of the Zodiac in the Context of Ancient Oriental History', *Vistas in Astronomy* 41, 507–25 (1997). For a fuller account see also J. D. Barrow, *The Artful Universe Expanded,* Oxford UP, Oxford (2005), chapter 4.

19. For example Ovenden and Roy have Hipparchus observing from Rhodes with a latitude of 31 degrees north (which actually passes through Alexandria), whereas the ancient latitude is 36 degrees and the modern one is 36.4 degrees.

20. B. Schaefer, 'The Latitude and Epoch for the Formation of the Southern Greek Constellations', *J. Hist. Astron.* 33, 313–50 (2003).

21. It is only possible to determine a limit because of problems with the visibility of stars near the horizon and the fact that unknown constellations might have extended to the south of the known ones.

22. Schaefer, ibid., p.330.

23. Assuming a 2-degree difference between the most northerly latitude limit and the actual latitude location to allow for missing star groups when observing low on the horizon.

24. Quoted in the *Times Higher Education Supplement*, 26 May 2006, Peep's Diary, p. 15. There is a lot more to this joke. At Glasgow University it takes seventy-six – one to change the light bulb, fifty to fight for the bulb's right not to change and twenty-five to hold a counter-protest. At Strathclyde, it takes five – one to design a nuclear-powered bulb that never needs changing, one to figure out how to power the rest of Scotland using the naked bulb, two to install it and one to write the computer program that controls the wall switch. At St Andrews it also takes five – one to arrange the party, two to co-ordinate the press, one to call the electrician, and one to get daddy to pay. At Napier, it takes only one, but he receives ten course credits for it. At Dundee, it takes ten – one fits the bulb and nine petition for the electrification of Dundee.

25. O. Gingerich, *The Book That Nobody Read*, Walker, New York (2004), gives an engaging account of Copernicus' book and its influence. Gingerich has inspected all the known extant copies of Copernicus' book to determine the extent to which they were read, and by whom.

26. Such an arrangement had been proposed by Aristarchus of Samos in the third century BC. Writing in *The Sand Reckoner*, Archimedes (287–212 BC) describes this work: 'But Aristarchus of Samos brought out a book consisting of certain hypotheses in which the premises lead to the conclusion that the universe is many times greater than that now so called. His hypotheses are that the stars and the sun remain motionless. That the earth revolves about the sun in the circumference of a circle, the sun lying in the middle of the orbit.'

27. G. B. Riccioli, *Almagestum Novum*, Bologna (1651).

28. This general idea is usually known as the anthropic principle and was first stressed in an astronomical context by Brandon Carter, in a talk to mark the 500th anniversary of the birth of Copernicus in Cracow. It can be found discussed most extensively in J. D. Barrow and F. J. Tipler, *The Anthropic Cosmological Principle*, Oxford UP, Oxford (1986). More applications are discussed in J. D. Barrow, *The Constants of Nature,* Jonathan Cape, London (2003).

29. An observed property of the Universe may be an unlikely outcome of a theory in which many possibilities can arise. However, this is not necessarily a good reason for rejecting the theory if the conditional probability of the observed property being an outcome, given that observers must also exist in the universe, is not small. Considerations of this sort play an important role in evaluating the predictions of any cosmological theory which possesses an intrinsically random element – so that there are many possible outcomes for different aspects of the Universe. In any quantum cosmological theory such a random element is inevitable.

30. In those times such objects were referred to as 'nebulae'. Only after about 1930, when the interpretation of galaxies as great islands of billions of stars became clear, were they always referred to as galaxies. The term 'nebulae' was then used to describe the colourful effects of exploding stars on the gas and dust that surround them. The latter are the most photographed astronomical objects and the ones most commonly found on astronomy magazine covers all over the world.

31. Today it can be seen with binoculars in the constellation of Canes Venaciti, close to the Plough (or the Big Dipper, as it is known in the USA).

32. Art historians have been puzzled by the source of Van Gogh's astronomical inspiration. Some believe that the key is to be found in a letter the artist wrote to his brother Theo in June 1889, in which he wrote that 'This morning I saw the country from my window a long time before sunrise, with nothing but the morning star, which looked big.' It is entirely plausible that the morning star

(actually the planet Venus) should inspire Van Gogh to create a painting of the sky, but the final image he created of it is much more than the morning star before sunrise.

33. From a letter to his brother Theo written *c.* 9 July 1888 in Arles, from *The Letters of Vincent van Gogh, 1886–1890*, ed. and transl. by Robert Harrison, Scolar Press (1977), no. 506.

34. The picture labelled 'visible' shows emissions from wavelengths of 3.6 microns (blue), 4.5 microns (green), 5.8 microns (orange) and 8.0 microns (red), which are roughly ten times greater than those that can be seen by the unaided human eye.

35. Ejnar Hertzsprung, *Encyclopaedia Britannica*, 15th edn, University of Chicago Press, Chicago (1995).

36. D. H. DeVorkin, *Henry Norris Russell: Dean of American Astronomers,* Princeton UP, Princeton, NJ (2000).

37. The light from a receding source is received with a lower frequency and longer wavelength than it was emitted with. This is called 'redshifting' of the light because shifting to longer wavelength corresponds to a shift towards the red end of the spectrum of visible light. Light from an approaching source is received with a higher frequency and a shorter wavelength than when it was emitted and this effect is called 'blueshifting'.

38. E. Hertzsprung, 'Zur Strahlung der Sterne', *Zeitschrift für Wissenschaftliche Photographie* (1905).

39. The account of the talk appears in the August 1913 paper by Russell, '"Giant" and "Dwarf" Stars', *Observatory Magazine* 36, 324–9 (1913), but the crucial figure showing his graph was lost and was not published. The version shown here is from H. N. Russell, 'Relations between the Spectra and Other Characteristics of Stars', *Nature* 93, 252 (1914). Russell generously remarks in his talk and article that 'These series were first noticed by Dr Hertzsprung, of Potsdam, and called by him "giant" and "dwarf" stars. All I have done in this diagram is to use more extensive observational material.'

40. M. Lachièze-Rey, J. P. Luminet and J. Lareda (transl.), *Celestial Treasury: From the Music of the Spheres to the Conquest of Space*, Cambridge UP, Cambridge (2001).

41. The four standard varieties are emission, reflection, and planetary nebulae, together with supernova remnants.

42. The word 'plasma' was first used to describe this type of ionised gas by the American chemist Irving Langmuir in 1929, although this new state of matter was first identified by the English physicist William Crookes in 1879. It should not be confused with the medical use of the word 'plasma,' which has no connection with ionised gases at all.

43. See Kessler's contribution to R. W. Smith and D. H. DeVorkin, *The Hubble Space Telescope: Imaging the Universe*, National Geographic, Washington (2004).

44. J. D. Barrow, *The Artful Universe Expanded*, Oxford UP, Oxford (2005).

45. 1 December 2005 blog discussion of new HST picture of the Crab Nebula: http://www.littlegreenfootballs.com/weblog/?entry=18439

46. Jocelyn Bell recalled that Fred Hoyle immediately identified neutron star supernova remnants as fitting the bill for the pulsing times involved when the discovery was first presented to other Cambridge astronomers, while others had suspected pulsating white dwarf stars.

47. Typically, the rotation rate increases as the inverse square of the radius, assuming no mass loss and that no strong magnetic fields are present.

48. http://wilmington.craigslist.org/boa/175710361.html

49. It has a cluster of other smaller companion objects which constitute our Local Group of galaxies.

50. For the history of its rediscovery in Europe during the pre- and post-telescopic eras see http://www.seds.org/messier/m/m031.html

51. For these descriptions of the shapes of galaxies see the section on Hubble's 'Tuning Fork' classification of galaxies (pages 56–59).

52. The oldest humans discovered so far, in Ethiopia, are about 200,000 years old.

53. J. L. Synge, *The Hypercircle in Mathematical Physics*, Cambridge UP, New York (1957).

54. E. P. Hubble, 'Extragalactic Nebulae', *Astrophys. J.* 64, 321–369 (1926).

55. The designation 'EN' is given to an elliptical galaxy by calculating N to the nearest whole or half integer from the ratio of the longest diameter (a) to the shortest diameter (b) of the elliptical image on the photograph by means of the formula $N = 10(a-b)/a$. So for a circular image $a = b$ and $N = 0$, while for the flattest ellipticals seen $N = 7$ and the ratio of the diameters, $b{:}a$, is 3:10.

56. E. P. Hubble, *The Realm of the Nebula*, Yale UP, New Haven (1936).

57. H. Henry, *Virginia Woolf and the Discourse of Science: The Aesthetics of Astronomy*, Cambridge UP, Cambridge (2003).

58. A. C. Doyle, 'The Boscombe Valley Mystery', in *The Adventures of Sherlock Holmes*, George Newnes Ltd, London (1892).

59. H. Arp, *Atlas of Peculiar Galaxies, Astrophys. J. Supplement*, vol. 14 (1966).

60. F. Zwicky, E. Herzog, and P. Wild, *Catalogue of Galaxies and Clusters of Galaxies*, Caltech UP, Pasadena (1960–1968).

61. B. A. Vorontsov-Velyaminov, *The Catalogue of Interacting Galaxies*, Moscow (1959), as made available on the California Institute of Technology website at http://nedwww.ipac.caltech.edu/level5/VV_Cat/frames.html

62. The Large and Small Magellanic Clouds are named after the great Portuguese navigator Magellan Ferdinand (1480–1521), whose crew first discovered them.

63. There had been three spectacular naked-eye supernovae visible from Earth during the previous 1000 years, before the invention of the telescope. The first, observed in China on 4 July 1054, was reportedly brighter than the full Moon and remained visible in the daytime for a month. It was in the constellation of Taurus and its remnant forms the Crab Nebula. The second was observed as an 'extra star' in the constellation of Cassiopeia by the great Danish astronomer Tycho Brahe on 11 November 1572. He reported that it was as bright as Jupiter on the sky. The third was first noticed on 9 October 1604, when it became brighter than any star in the sky and all the planets except for Venus. Kepler began to study it eight days after this. It lies in the Milky Way, in the constellation of Ophiuchus.

64. In 2003, 330 were discovered.

65. Picture in H. Kragh *Cosmology and Controversy*, Princeton UP, Princeton, NJ (1999), p. 18, and the paper is also reprinted in J. Bernstein and G. Feinberg *Cosmological Constants*, p.77, and is recorded as being received 17 January 1929.

66. A. N. Whitehead, *Science and the Modern World*, Mentor, New York (1925).

67. In ordinary 'flat' space we are familiar with the apparent brightness of a source of light a distance *d* away being proportional to its intrinsic brightness but inversely proportional to d^2. Hence, knowing the intrinsic brightness you can determine *d*; or, with two sources of light of the same intrinsic brightness, knowing the ratio of their apparent brightnesses gives the ratio of their distances.

68. What we call 'galaxies'. The light was emitted by stars in local galaxies.

69. The first determination of a slope was actually made by Georges Lemaître in 1927 using some of Hubble's data.

70. M. L. Humason, 'The Large Radial Velocities of N.G.C. 7619', *Proc. Nat. Acad. Sci.* 15, 167–8 (1929).

71. This is generally known as the 'recalibration of the distance scale' in astronomy.

72. W. L. Freedman et al. 'Final Results from the Hubble Space Telescope Key Project to Measure the Hubble Constant', *Astrophys. J.* 553, 47–72 (2001).

73. Hubble first discovered Cepheid variable stars in the Andromeda galaxy in 1924.

74. *Astrophysical Journal* 553, 47–52 (2001).

75. A. Chaikin, 'Are There Other Universes?', *Science*, 5 February 2002; http://www.space.com/scienceastronomy/generalscience/5mysteries_universes_020205-1.html

76. This property of being the same everywhere is called spatial homogeneity. It means that every observer in the Universe would see the same cosmic history. Our Universe cannot be exactly spatially homogeneous (we could not exist if that were the case), but it is almost homogeneous out to very large distances. On the average, the deviations from uniformity are only about one part in 100,000. PTechnically, it is the uniformity in the gravitational potential rather than in the density of matter and radiation that is important.

77. This requires that the density of matter ρ and its pressure *p* have the property that $\rho + 3p/c^2$ is positive, where *c* is the velocity of light in vacuum. In modern theories of the so called 'inflationary universe', it is hypothesised that $\rho + 3p/c^2$ was negative for a brief period in the very early history of the Universe. This is possible because there may exist forms of energy which have negative pressure (i.e. tension) but positive density. Their mutual gravitational interactions can then be repulsive.

78. Note, however, that open universes can also be finite in total extent if their topology is more complicated, like a torus rather than an open plane, as Friedmann himself recognised and stressed in his early writings.

79. Friedmann had interpreted the closed universes as being 'periodic', implying that they could run through a sequence of cycles. This is mathematically permissible but required the universe to experience an infinite density at the moment of bounce unless the behaviour of the contracting universe is changed by new forms of matter which bring the contraction to a halt before it reaches infinite density.

80. The second law of thermodynamics leads to a steady increase in pressure which has a different gravitational effect on the expansion of successive cycles, making them grow larger at the time of maximum expansion.

81. J. D. Barrow and M. Dąbrowski, 'Oscillating Universes', *Monthly Notices of the Royal Astronomical Society* 275, 850–862 (1995).

82. G. Lemaître, 'A Homogeneous Universe of Constant Mass and Increasing Radius', *Mon. Not. R. Astron. Soc.* (1927). This paper was first published in the *Annales de la société scientifique de Bruxelles* 47, 49–56 (1927) but Arthur Eddington had it translated into English so as to appear in the leading international journal of the Royal Astronomical Society. Eddington felt some conscience about this because he had rediscovered one of the results of Lemaître's (the instability of the static universe) and had forgotten about Lemaître's work on this problem even though Lemaître had worked with him as a visiting researcher in Cambridge a few years earlier. During this period Lemaître had worked at the Cambridge University Observatories and been attached to St Edmund's House, the Catholic students' residence that was the precursor of St Edmund's College.

83. M. Lachièze-Rey, J. P Luminet and J. Loreda, op.cit p.155.

84. A. S. Eddington, *The Expanding Universe*, Cambridge UP, Cambridge (1933).

85. The illustrations are credited to B. G. Seielstad.

86. W. Allen, 'Strung Out', *New Yorker*, 28 July 2003, p. 96.

87. This is a long and convoluted story that is told in different degrees of detail in S. Weinberg, *The First Three Minutes*, Basic Books, New York (1977); R. Alpher and R. C. Herman, *Genesis of the Big Bang*, Oxford UP, New York (2001); H. Kragh, *Cosmology and Controversy*, Princeton UP, Princeton, NJ (1999). Robert Dicke, the research group leader, wrote later that 'There is one unfortunate and embarrassing aspect of our work on the fire-ball radiation. We failed to make an adequate literature search and missed the more important papers of Gamow, Alpher and Herman. I must take the major blame for this, for the others in our group were too young to know these old papers. In ancient times I had heard Gamow talk at Princeton but I had remembered his model universe as cold and initially filled only with neutrons': R. H. Dicke, *A Scientific Autobiography* (1975), unpublished, held by the National Academy of Sciences.

88. R. Alpher and R. C. Herman, 'Evolution of the Universe', *Nature* 162, 774 (1948).

89. A. A. Penzias and R. W. Wilson, *Astrophys. J.* 142, 419–42 (1965). The accompanying theoretical paper is R. H. Dicke, P. J. E. Peebles, P. G. Roll, and D. T. Wilkinson, 'Cosmic Black-Body Radiation', *Astrophys. J.* 142, 414–419 (1965).

90. This was first stressed by A. Doroshkevich and I. D. Novikov, 'Mean Density of Radiation in the Metagalaxy and Certain Problems in Relativistic Cosmology', *Soviet Physics Doklady* 9, 111–13 (1964).

91. The term 'black body' for a perfect absorber and radiator of electromagnetic radiation was coined by the German physicist Gustav Kirchoff, in 1862.

92. D. P. Woody and P. L. Richards, 'Spectrum of the Cosmic Background Radiation', *Phys. Rev. Lett.* 42, 925 (1979).

93. This is an orbit around the Sun for which the period of a complete orbit is the same as the Sun's rotation period – the solar analogue of a geostationary orbit around the Earth.

94. J. Mather et al., 'Measurement of the Cosmic Microwave Background Spectrum by the COBE FIRAS Instrument', *Astrophys. J.* 420, 439–444 (1994); Fixsen et al., 'The Cosmic Microwave Background Spectrum from the Full COBE FIRAS Data Sets', *Astrophys. J.* 473, 576–587 (1996).

95. http://lambda.gsfc.nasa.gov/product/cobe/firas_image.cfm

96 The steady-state theory was an attempt to evade this conclusion by hypothesising that the universe expands but matter is created continuously at just the right rate to keep the average density and temperature constant. The steady-state universe has an infinite past and an infinite

future and looks the same at all times and from all places on the average. Unfortunately, this simple picture, proposed first in 1948 by Hermann Bondi, Thomas Gold, and Fred Hoyle, eventually fell foul of many different astronomical observations which showed that the Universe had been quite different in the past, the most notable of which was the discovery of the microwave background radiation and the helium and deuterium abundances produced in the Big Bang model. However, it has many features in common with modern editions of the inflationary universe.

97. C. Hayashi, 'Proton–Neutron Concentration Ratio in the Expanding Universe at the Stages Preceding the Formation of the Elements', *Prog. Theo. Phys.* 5, 22 (1950).

98. R. A. Alpher, J. W. Follin, and R. C. Herman, 'Physical Conditions in the Initial Stages of the Expanding Universe', *Phys. Rev.* 92, 1347–61 (1953).

99. F. Hoyle and R. J. Tayler, 'The Mystery of the Cosmic Helium Abundance', *Nature* 203, 1108 (1964).

100. P. J. E. Peebles, 'Primordial Helium Abundance and the Primordial Fireball' (papers I and II), *Phys. Rev. Lett.* 16, 410 (1966), and *Astrophys. J.* 146, 542–552 (1966).

101. R. Wagoner, W. Fowler, and F. Hoyle, 'On the Synthesis of Elements at Very High Temperatures', *Astrophys. J.* 148, 3 (1967); see also R. Wagoner, W. Fowler and F. Hoyle, 'Primordial Nucleosynthesis Revisited', *Astrophys. J.* 179, 343 (1972).

102. J. B. Rogerson and D. G. York, 'Interstellar Deuterium Abundance in the Direction of Beta Centauri', *Astrophys. J. Lett.* 186, L95–L98 (1973). This was a very important measurement for cosmology. The abundance of deuterium in the Universe had previously been deduced from its abundance in deuterated molecules (like the fraction of heavy water, D_2O, relative to H_2O in sea water). However, deuterium tends to bind into molecules preferentially compared to hydrogen and so the molecular abundances do not accurately reflect the relative abundances of the free deuterium and hydrogen atoms. The *Copernicus* satellite observed the spin-flip lines, analogous to Lyman-alpha in hydrogen, for the deuterium atom in interstellar space and so provided the first true measure of the D to H atomic ratio in space. It acted as a major stimulus for detailed studies of nucleosynthesis and early-universe cosmology. The author began his DPhil thesis research in this area in 1974.

103. This letter was published a few weeks later in a short book by Kepler entitled *Kepler's Conversation with Galileo's Sidereal Messenger,* transl. E. Rosen, Johnson Reprint Corp., New York (1965).

104. We discussed this feature of infinite and finite universes in our earlier look at the idea of the expanding universe. An infinite universe need have no centre and no edge. A finite universe can also avoid having a centre or an edge if it is curved, like the two-dimensional surface of a ball.

105. Quoted by E. R. Harrison, *Darkness at Night*, Harvard UP, Cambridge, Mass. (1987), p. 48.

106. E. Halley, 'On the Infinity of the Sphere of Fix'd Stars', *Phil. Trans. Roy. Soc.* 31, 22 (1720/21), and 'Of the Number, Order and Light of the Fix'd Stars', *Phil. Trans. Roy. Soc.* 31, 24 (1720/21).

107. E. R. Harrison, *Darkness at Night*, Harvard UP, Cambridge, Mass. (1987).

108. A. S. Eddington, *The Expanding Universe*, Cambridge UP, Cambridge (1933), p. 122.

109. C. D. Shane and C. A. Wirtanen, *Astron. J.* 59, 258 (1954).

110. E. J. Groth, P. J. E. Peebles, M. Seldner, and R. M. Soneira, 'The Clustering of Galaxies', *Scientific American* 237, 76 (1977).

111. S. Bhavsar and J. D. Barrow, 'What the Astronomer's Eye Tells the Astronomer's Brain', *Quart. J. Roy. Astr. Soc.* 28, 109–28 (1987).

112. For the results of an interesting workshop in 1991 at Penn State University which brought together experts in galaxy clustering and statistics, see E. D. Feigelson and G. J. Babu (eds), *Statistical Challenges in Modern Astronomy*, Springer-Verlag, New York (1993).

113. M. Geller and J. Huchra, *Science* 246, 897 (1989).

114. Quoted in C. Brownlee, 'Hubble's Guide to the Expanding Universe', National Academy of Sciences classics, online at http://www.pnas.org/misc/classics2.shtml

115. The Hubble Deep Field is located at RA 12h 36m 49.4000s DEC +62° 12' 58.000" (J2000 Equinox) in a small part of Ursa Major about 140 arc seconds across, which is roughly the size of a cricket ball 100 metres away. The field is optimally placed in the Northern Continuous Viewing Zone (CVZ). The CVZ is a special region where the Hubble Space Telescope can view the sky without being blocked by the Earth or have interference from the Sun or Moon. This fixed the declination to be near +62 degrees.

116. http://hubblesite.org/newscenter/newsdesk/archive/releases/1998/41/text/

117. The next step in this spectacular project was the 'Hubble Ultra Deep Field' which captured nearly 10,000 galaxies looking out to a population of about a hundred small red galaxies so distant that their light began its journey towards us when the Universe was only 800 million years old. This image was built from 800 exposures taken during 400 orbits of the Space Telescope taking 11.3 days during the period from 24 September 2003 to 16 January 2004.

118. The idea of time travel was first suggested by H. G. Wells in his novel *The Time Machine*, first published in 1895, but was not shown to be allowed by Einstein's theory of general relativity until Kurt Gödel's paper of 1949, 'An Example of a New Type of Cosmological Solution of Einstein's Field Equations of Gravitation', *Rev. Mod. Phys.* 21, 447–50 (1949).

119. See Note 37.

120. Minkowski was a talented mathematician who made many contributions to pure mathematics. He was a student contemporary and close friend of David Hilbert, the greatest mathematician of the day. Remarkably, it was Minkowski who suggested to Hilbert that he should deliver a lecture at the International Congress of Mathematicians in 1900 that would lay out the greatest unsolved problems of mathematics for the next millennium. His letter suggested to Hilbert that 'What would have the greatest impact would be an attempt to give a preview of the future, i.e. a sketch of the problems with which future mathematicians should occupy themselves. In this way you could perhaps make sure that people would talk about your lecture for decades in the future.' Indeed, they have talked about it for more than a century afterwards. Tragically, Hermann Minkowski died unexpectedly from a burst appendix when just forty-four years old. Hilbert wrote a moving obituary that paid tribute to Minkowski's unfulfilled genius: 'Since my student time Minkowski was the best and most reliable friend who stuck to me with all the fidelity and deepness of his character. Our science that had brought us together and that was at our hearts appeared to us like a flourishing garden. We loved to detect concealed trails and discovered many prospects which seemed to us beautiful. When one of us showed it to the other and when we admired it jointly our enjoyment was complete. He was a present from heaven to me, as it happens to come only rarely, and I can be thankful that I was able to own it for such a long time. Suddenly, death has taken him away from our side. But what death cannot take is his noble image in our hearts and the awareness that his spirit is acting in us.'

121. H. Lorentz, A. Einstein, H. Minkowski, and H. Weyl (eds), transl. W. Perrett and G. B. Jeffrey, *The Principle of Relativity: A Collection of Original Memoirs*, Dover, New York (1952); Minkowski's section is entitled 'Space and Time' (pp. 73–91). This quotation is taken from the beginning of his section, on p. 75.

122. As was assumed in Newton's theory of gravity, in which gravity acted instantaneously 'at a distance'.

123. D. Mellor, *Real Time II*, Routledge, London (1998).

124. G. F. R. Ellis, 'Physics in the Real Universe: Time and Spacetime' (2006), http://arxiv.org/ftp/gr-qc/papers/0605/0605049.pdf

125. J. Brennan, 'Freewill in the Block Universe' (2006), http://www.u.arizona.edu/~brennan/freedomblockuniverse.pdf

126. R. Penrose, 'The Light Cone at Infinity' in *Relativistic Theories of Gravitation*, ed. L. Infeld, Pergamon, London (1964). This is the paper of a talk delivered in Warsaw on 31 July 1962. In between these dates the author published another conformal diagram in his paper 'Asymptotic Properties of Fields and Space–Times', *Phys. Rev. Lett.* 10, 67 (1963).

127. F. Hoyle, 'Cosmological Tests of Gravitation/Theories' *Varenna Lectures*, Corso XX, Academic Press, New York, (1960), p. 141; see also G. F. R. Ellis, 'Relativistic Cosmology', in *Varenna Lectures*, Corso XLVII, Academic Press, New York, (1973), p. 177.

128. President George Bush, on his first meeting with rival presidential candidate John McCain in a TV debate in 2000.

129. Cosmic rays are not 'rays' of radiation like 'light rays' or 'x-rays' but fast-moving particles, such as protons, muons, or very light nuclei, bombarding us from space.

130. S. W. Hawking and G. F. R. Ellis, *The Large Scale Structure of Space–time*, Cambridge UP, Cambridge (1972), and S. W. Hawking and R. Penrose, *The Nature of Space and Time,* Cambridge UP, Cambridge (1995).

131. This beginning does not have to be simultaneous throughout the Universe; nor does it have to be experienced by every past history.

132. *Macbeth* I, iii, 58.

133. See Note 70.

134. In the cosmological theory, without inflation the expansion grows more slowly and it is not possible to 'grow' our whole visible part of the Universe from a region that is small enough for light signals to cross it at an early time in the Universe's history. Different parts of the region that will expand to become our visible Universe therefore end up uncoordinated, with very different density, temperature, and expansion rate.

135. J. D. Barrow, 'Cosmology: A Matter of All or Nothing', *Astronomy and Geophysics* 43, 4.9–4.15 (2002).

136. The latest WMAP technical papers can be found at http://map.gsfc.nasa.gov/m_mm/pub_papers/threeyear.html

137. *Troilus and Cressida* I, iii, 345.

138. It is most probable that our 'bubble' would have inflated by far more than would be necessary to smooth out the region that we can see – if not, there would be a strange and rather un-Copernican coincidence. This means that it is most likely that our encounter with very different

conditions, and even very different physics, in a neighbouring bubble, lies in the far, far future after all the stars have died.

139. Jonathan Swift, *Gulliver's Travels: A Voyage to Lilliput*, Motte, London (1726). The original text is available online with other materials at http://lee.jaffebros.com/gulliver/contents.html

140. C. Misner, K. Thorne, and J. A. Wheeler, *Gravitation*, W. H. Freeman, San Francisco (1973).

141. There is a simple Newtonian analogue of the situation that arises when matter is so compressed that the 'escape speed' needed to escape from its gravitational pull becomes equal to the speed of light. This occurs for an object with a radius R and mass M related by $R = 2GM/c^2$, where G is the gravitation constant and c is the velocity of light in vacuum. However, this Newtonian 'black hole' is not so dramatic as the relativistic one. In relativity the radius $R = 2GM/c^2$ cannot be escaped from: the horizon is a surface of no return for the infalling traveller. However, in the Newtonian edition you can always escape as far as you like beyond this radius; you simply can't get infinitely far from this radius – this is the definition of 'escape speed': the speed necessary to escape the pull of the object completely, i.e. to get infinitely far away from it. The possibility of a Newtonian black hole was first noticed by John Michell in 1783. He imagined a star that was 500 times bigger than the Sun but with the same density. This resulted in an escape speed equal to that of light. He noted that such objects would be unseeable by direct observation but they could be detected indirectly if they were in a binary orbit with another visible star. The motion of the visible member of the pair could betray the presence of the unseen companion. Modern astronomers detect many black holes by looking for them in binary pairs with large visible stars. The best way to detect them is to look for the effects of the material that is steadily captured from the visible star as it swirls down towards the black hole, forming a disc of accreting material around it. The infall heats this material up to millions of degrees and it emits x-rays, which can be observed. The periods of the flickering of the x-rays also reveal the small size of the 'sink' down which the material eventually disappears at the centre because the typical flickering time interval is about R/c in duration.

PART 2 SPATIAL PREJUDICE

1. There has been continual argument about whether Armstrong fluffed his lines or, as NASA claimed, there was static on the sound link which resulted in the rehearsed statement 'one small step for a man, one giant leap for mankind' missing an indefinite article. The transcript and audio file of the mission communications can be read and heard at http://www.hq.nasa.gov/office/pao/History/alsj/a11/a11.step.html

2. W. H. Lambright, *Powering Apollo: James E. Webb of NASA*, Johns Hopkins UP, Baltimore (1995).

3. The scattering of sunlight by the Earth's atmosphere, first explained by Tyndall, results in the largest amount of scattering for the light with the highest frequency. Thus the blue end of the spectrum is scattered most and the red end least. Hence when you look around the sky away from the disk of the Sun you will be seeing the part of the light spectrum that has been scattered most by the molecules of nitrogen and oxygen in the atmosphere and hence appears blue. At sunset you look towards the Sun and see predominantly the part of the spectrum that is scattered least, and hence appears red. When the scattering particles become larger, as in the case of water droplets, the scattering is about the same for all frequencies of light and so clouds look white.

4. Not all these initiatives are generally regarded as beneficial, the elimination of malaria being a benefit that was prevented by the banning of DDT.

5. E. Schumacher, *Small is Beautiful: A Study of Economics As If People Mattered*, Harper and Row, New York (1973).

6. B. Ward and R. Dubos, *Only One Earth: The Care and Maintenance of a Small Planet*, Norton, New York (1972).

7. In response to these images, on Christmas Day 1968 the American poet Archibald MacLeish wrote in the *New York Times*, alongside the photo of the Earth from space, these words to celebrate the accomplishment of the returning *Apollo 8* astronauts: 'To see the Earth as it truly is, small and blue and beautiful in that eternal silence where it floats, is to see ourselves as riders on the earth together, brothers on that bright loveliness in the eternal cold – brothers who know now that they are truly brothers.'

8. The picture shown is created from a number of different pictures each taken when different parts of Earth's surface are in darkness.

9. This energy must have negative pressure (or 'tension'), which is gravitationally repulsive and so accelerates the expansion of the Universe.

10. The most likely candidates are new forms of heavy neutrino-like particles that possess only weak and gravitational interactions. If they took part in electromagnetic and strong interactions like the constituents of atoms then they would create effects on the primordial nucleosynthesis of the lightest elements that we do not see and they would survive in too great a number to be in agreement with what is observed. Particle physicists hope that in the next few years they will detect these weakly interacting dark matter particles in particle accelerator experiments on Earth and then detect the cosmic sea of dark matter particles directly by catching these particles in detectors deep underground as they pass through the Earth.

11. The visible Universe is that region of the entire (possibly infinite) Universe from which light has had time to reach us during the 13.7 billion years that the Universe has been expanding.

12. K. Schroeder, *Permanence*, Tor Books, New York (2002), p. 181.

13. Ozone was first made in the laboratory in 1840 by Christian Friedrich Schöbein.

14. The discovery that ozone absorbs ultra-violet light with wavelengths of less than about 290 nm and is responsible for a cut-off in the solar ultra-violet light getting to the Earth's surface below a wavelength of 293 nm is due to W. N. Hartley, 'On the Absorption Spectrum of Ozone', *J. Chem. Soc.* 39, 57–61 (1881).

15. M. J. Molina and F. S. Rowland, 'Chlorine Atom-Catalysed Destruction of Ozone', *Nature* 249, 810–12 (1974).

16. J. C. Farman, B. G. Gardiner and J. D. Shanklin, 'Large Losses of Ozone in Antarctica Reveal Seasonal ClOx/NOx Interaction', *Nature* 315, 207–10 (1985).

17. For a historical sequence of images showing the change in the maximum extent see http://www.theozonehole.com/ozoneholehistory.htm

18. The manufacture of CFCs was banned by the 1992 revisions to the 1987 UN Protocol.

19. For an overview of the development of the study of atmospheric ozone see the expanded version of Rowland's Nobel Prize acceptance lecture in F. S. Rowland, 'Stratospheric Ozone Depletion', *Phil. Trans. Roy. Soc. B* 361, 769–90 (2006).

20. This is because of the long lifetimes of some of the existing CFC molecules in the atmosphere. The most dramatic effect of legislation was the fall in CH_3CCl_3 concentrations in the atmosphere, because its lifetime against chemical destruction is about five years. Other CFCs fall at a rate of only about 1 or 2 per cent a year.

21. Amos 8: 9. The prophet is writing about Nineveh and the 'day' referred to was the eclipse of 15 June 763 BC. It is mentioned in the Assyrian state archives after being observed at Nineveh.

22. Literally, 'bad star'.

23. On the average the Moon appears a little smaller than the face of the Sun but because of small variations in their distances from the Earth over time there are periods when the Moon appears slightly larger.

24. Berkowski was a local daguerreotypist and his work was used by the Royal Observatory director, Dr A. L. Busch, who, unfortunately, omitted to record Berkowski's Christian name for posterity.

25. Stable hydrogen-burning stars are all of similar sizes and habitable planets will have to lie in a narrow zone around the star where the surface temperature allows water to be liquid over a reasonable surface area.

26. See J. D. Barrow, *The Artful Universe Expanded*, 2nd edn, Oxford UP, Oxford (2005), chapter 4, for a fuller description.

27. Eddington was a Quaker and a pacifist who was not willing to serve as a combatant in the war when he was drafted in 1917. His imprisonment was only avoided by the intervention of the Astronomer Royal, Sir Frank Dyson, who proposed that Eddington's conscription be postponed so that he could lead an eclipse expedition to test the general theory of relativity in 1919. The effort expended on this expedition to Principe Island off the West Coast of Africa, and Eddington's subsequent announcement of the success of Einstein's predictions were remarkable given that Einstein was German. Both Eddington and Einstein played an important role in bringing about a post-war reconciliation between Britain and Germany that was mediated by scientific exchanges. The details of the data analysis process for the Principe expedition and the complementary expedition to Sobral, Northern Brazil, led by Andrew Crommelin, have been subjected to considerable scrutiny. The integrity of the results was questioned in an article by R. Earman and C. Glymour, 'Relativity

and Eclipses: The British Eclipse Expeditions of 1919 and their Predecessors', *Hist. Stud. Phys. Sci.* II, 49 – 85 (1980). However, many of the original Sobral plates survive (the Principe plates do not) and were re-processed using modern automatic plate-measuring machines so as to use position information about the night ascension of the stars (which was ignored in 1919) at the instigation of Andrew Murray at the Royal Greenwich Observatory in 1978. The results were published by Murray's assistant, Geoffrey Harvey, in G. M. Harvey, 'Gravitational Deflection of Light: A Reexamination of the Solar Eclipse of 1919', *The Observatory* 99, 195–198 (1979) and agree fully with the original results; see also C. A. Murray and P. A. Wayman, 'Relativistic Light Deflections', *The Observatory* 109, 189–191 (1989). For a detailed and definitive analysis of the original data analysis by Eddington and Crommelin, which ultimately rejects the arguments of Earman and Glymour, see D. Kennefick, 'Not Only Because of Theory: Dyson, Eddington and the Competing Myths of the 1919 Eclipse Expedition', http://arxiv.org/ abs0709.0685.

28. F. W. Dyson, A. S. Eddington and C. Davidson, 'A Determination of the Deflection of Light by the Sun's gravitational Field, from Observations Made at the Total Eclipse of May 29, 1919', *Phil. Trans. Roy. Soc. A, Containing Papers of a Mathematical or Physical Character* 62, 291–333 (1920), plate on p. 332.

29. Elliptical orbits will not quite close in Einstein's theory and the ellipse precesses very slowly over long periods of time. The effect is very small and could only be seen affecting the orbit of a planet very close to the Sun, where its gravitational pull is strong. In the case of Mercury, the observed precession of the elliptical orbit amounted to 43 seconds of arc per century and is as predicted by Einstein's theory.

30. The end of the opening paragraph of H. G. Wells's *The War of the Worlds* (1898). The book is online at http://www.fourmilab.ch/etexts/www/warworlds/b1c1.html. Wells draws on the observations of Schiaparelli in the opening chapter to set the scene for the Martian invasion: 'Men like Schiaparelli watched the red planet – it is odd, by-the-bye, that for countless centuries Mars has been the star of war – but failed to interpret the fluctuating appearances of the markings they mapped so well. All that time the Martians must have been getting ready.'

31. The term *canali* had been used earlier, in 1858, in connection with Martian surface features by the Italian Jesuit priest and astronomer Pietro Secchi, who was the Director of the Vatican Observatory. However, he was not using the term to describe long artificial channels, just particular features of the surface topography.

32. The broadcast text and audio is available at http://www.members.aol/com/jeff1070/script.html

33. The two light-coloured boulders at the bottom are likely to be meteorites.

34. D. Frost, interview in *The Observer* Colour Magazine Supplement, 5 June 2005, p. 6.

35. *Pioneer 10* was launched on 2 March 1972 and *Pioneer 11* on 5 April 1973. Both used Atlas-Centaur three-stage rockets.

36. The motions of the two *Pioneer* probes has consistently puzzled astronomers because there is a small unexplained deviation from the expected motion that is revealed by an unexpected blue shift from the Doppler tracking of the spacecraft. It amounts to an anomalous acceleration of 8.74 $\pm$ 1.33 x 10^{-8} cm^{-2} towards the Sun. This has become known as the 'Pioneer anomaly' and explanations for it range from gas leaks to new forces of Nature. For a discussion of the possibilities see the article by S. G. Turyshev, M. N. Nieto, and J. D. Anderson, 'Study of the Pioneer Anomaly: A Problem Set', *American J. Phys.* 73, 1033–44 (2005).

37. The plaques are attached to the antenna supports where they are shielded from being eaten away by cosmic dust impacts. The etching on the metal surface is 0.38 mm deep.

38. M. Wolverton, *The Depths of Space: The Story of the Pioneer Planetary Probes*, Joseph Henry Press, Baltimore (2004). This is also available online at http://darwin.nap.edu/books/0309090504/html

39. The pulsar map was designed by Frank Drake. For an interesting study of how much work is necessary (for us) to locate the Earth from the pulsar map, see the article 'Reading the Pioneer/Voyager Pulsar Map', by the astronomer Robert Johnson at http://www.johnstonsarchive.net/astro/pulsarmap.html. There are inevitably small errors in some of the original sky positions, but Johnson manages to locate the Earth to within ten light years and the date of the Pioneer launch as 1969.9 ± 1.3 years. There are however several possible snares. The pulsars will spin more slowly with time, and both they and the solar system are moving relative to one another. Over very long periods of time these motions will need to be folded into the computations if the right pulsars and directions are to be deduced; see also C. Sagan, L. S. Sagan, and F. D. Drake, 'Message from Earth', *Science* 175, 881–4 (1972).

40. This great discovery was made by Henk van der Hulst in 1945 while being guided by his thesis advisor Jan Oort at Leiden Observatory. The typical time for the transition to occur is too long for this to be observed routinely in a laboratory experiment with reasonable likelihood. However, in astronomical environments the number of atoms involved is so huge that many transitions are easily seen by radio telescopes.

41. Actually, an important possibility was missed. The plaque could easily have been constructed to have both its dimensions equal to 21 cm so as to reinforce this point, whereas it was made 22.9 cm long, 15.2 cm wide and 1.27 mm thick.

42. Besides Carl Sagan, the committee consisted of Frank Drake, Ann Druyan, Timothy Ferris, Jon Lomberg and Linda Salzman Sagan. Various other scientific advisors were called upon, and the fuller story of the compilation and intent of the message contents can be found in the book by C. Sagan et al., *Murmurs of Earth: The Voyager Interstellar Record*, Ballantine Books, New York (1984).

43. For the full listing see http://voyager.jpl.nasa.gov/spacecraft/music.html

44. H. C. Oersted, *Anden i Naturen*, Landskrona (1869). The relevant sections are on pp. 151–92. For an English translation see H. C. Oersted, *The Soul in Nature*, transl. L. & J. B. Homer, with supplementary contribution from the Leipzig edn., Henry G. Bohn, London (1852). This work is also discussed in M. J. Crowe, *The Extraterrestrial Life Debate 1750–1900*, Cambridge UP, Cambridge (1986 and 1988), pp. 256–7, 308–9.

45. The best source of information about Neovius' message is an article by the Finnish mathematician Raimo Lehti, entitled 'Edvard Englebert Neovius, an Early Proponent of Interplanetary Communication', which appears in the Proceedings of the 1993 Finnish Interdisciplinary Seminar on SETI, *Thoughts on Life and Mind beyond the Earth*, ed. Jouko Seppänen, Helsinki University of Technology, Dept. of Computer Science and Picaset Oy, Helsinki (2004), pp. 32–42. It is also published in *Acta Astronautica* 42, 727–38 (1998). I am indebted to Jouko Seppänen for providing me with a copy of these articles. My account of Neovius and his message draws heavily upon Lehti's article. I am most grateful to Professor Lehti for his help with these sources.

46. Note that in the second to last line of the message there is an error: the word 'fhb' should read 'fh'.

47. For K. Arnold's first description of an unidentified flying object as a 'saucer', on 24 June 1947, see Kenneth Arnold and Ray Palmer, *The Coming of the Saucers*, Private Publication, Boise, Ida., and Amherst, Wis. (1952).

48. C. S. Lewis, *The Lion, the Witch and the Wardrobe*, Geoffrey Bliss, London (1950), p. 23.

49. We know that the Scandinavian bishop Olaus Magnus wrote about the shapes of snowflakes in 1555, but he believed that they could take on all sorts of suggestive shapes which sound more like clumps of snow rather than single snowflake crystals. The hexagonal character was, in retrospect, first written about by Thomas Hariot in his unpublished notebook of 1591, twenty years before Kepler's work. However, Joseph Needham discovered that the hexagonal character of the snowflake was appreciated by the Chinese at least as early as 135 BC. Han Ying, in his work *Moral Discourses Illustrating the Han Text of the 'Book of Songs'* of that year, writes as if it is common knowledge that 'Flowers of plants and trees are generally five-pointed, but those of snow, which are called *ying*, are always six-pointed.' The twelfth-century philosopher Chu His speculated that six was a perfect Earthly number 'so as snow is water condensed into crystal flowers, these are always six-pointed.' However, the early Chinese scholars accepted the snowflake symmetry as a fact of Nature and there was no attempt to investigate its source or mathematical basis in the way that Kepler did. For further details see R. Temple, *The Genius of China*, André Deutsch, London (2006), p. 175.

50. J. Kepler, *On the Six-Cornered Snowflake*, Prague (1611), ed. and transl. C. Hardie, Oxford UP, Oxford, (1966).

51. Ibid., p. 33.

52. W. A. Bentley and W. J. Humphreys, *Snow Crystals*, Dover, New York (1962); first published 1931 by McGraw Hill.

53. D. C. Blanchard, *The Snowflake Man: A Biography of Wilson A. Bentley*, McDonald & Woodward Publ. Co., Blacksburg, Va. (1998). See also J. Briggs Martin, *Snowflake Bentley*, Houghton Mifflin, Boston (1998).

54. J. Nittmann and H. E. Stanley, 'Non-Deterministic Approach to Anisotropic Growth Patterns with Continuously Tunable Morphology: The Fractal Properties of Some Real Snowflakes', *J. Physics A* 20, L1185 (1987).

55. A. von Humboldt, 'Sur les lignes isothermes', *Annales de chimie et de physique* 5, 102–112 (1817); the figure is between pp. 112 and 113 in a fold-out.

56. Lillian B. Miller (ed.), *The Selected Papers of Charles Willson Peale and His Family*, Yale UP, New Haven, Conn. (1983 and 1989), p.683. Peale was a painter and museum director who acted as Humboldt's host during a visit to Philadelphia and Washington.

57. A. von Humboldt, *Essai sur la géographie des plantes,* Paris, 1805.

58. A. Bierce, *The Devil's Dictionary*, Dover, New York (1958), entry for 'Geology', p. 49; first published by Neale Publ. Co. (1911).

59. M. Du Carla-Boniface, 'Expression des nivellements; ou, Méthode nouvelle pour marquer sur les cartes terrestres et marines les hauteurs et les configurations du terrain', Paris (1782), in F. de Dainville, 'From the Depths to the Heights', transl. A. H. Robinson, *Surveying and Mapping* 30, 389–403 (1970), p. 396.

60. S. Winchester, *The Map That Changed the World: William Smith and the Birth of Modern Geology*, HarperCollins, New York (2001).

61. A. Bierce, *The Devil's Dictionary*, Dover, New York (1958), p. 140.

62. F. Galton, *Meteorographica, or Methods of Mapping the Weather*, Macmillan, London (1863).

63. The creation of meteorology as a predictive science needed three key ingredients: good data about the actual state of the upper atmosphere in many places; an understanding of how the

future predictions of a system's behaviour can be chaotically sensitive to small amounts of ignorance about their current state; and the advent of fast computers to carry out large numbers of computations in something close to real time. In the early years of the twentieth century, one of the pioneers of the subject, the English meteorologist Lewis Fry Richardson, might take six weeks to carry out calculations by hand that would predict the weather for a few hours to the future. Much of the pioneering work to turn meteorology into a real science by using the laws of hydrodynamics and thermodynamics, together with knowledge of the Earth's rotation, to predict how weather systems should change with pressure and temperature, was done by the Norwegian group led by Vilhelm Bjerkenes in Bergen. Bjerkenes' programme was unachievable in the pre-computer era. The fastest progress in the creation of weather maps based on real computations came in the USA after the Second World War, by using the first-generation computers for numerical prediction of weather patterns. The meteorological theory developed by Jule Charney was digitised in 1950 at the Institute for Advanced Study in Princeton and led to the first computer-generated weather maps. Today, the refinement of these forecasts by improving the quality of data inputted into them, and the speed and accuracy of computation, is a major international enterprise. Although we see the output only in the daily TV and newspaper reports, where little detail is provided, the level of detail required by agencies fighting the consequences of catastrophic climatic extremes – farmers, mariners, and airlines – is far more demanding. Getting the weather right can be a matter of life and death for many.

64. The modern study of chaos began with the work of Edward Lorenz, a meteorologist who noticed the sensitivity of his computations to rounding errors in a famous paper, written in 1963, 'Deterministic Non-Periodic Flow', *J. Atmos. Sci.* 20, 130–41 (1963). Lorenz also introduced the proverbial 'butterfly effect' name for chaos in the title of another of his papers: 'Can the Flap of a Butterfly's Wing Stir Up a Tornado in Texas?', which was first presented as a talk to the AAAS meeting in 1972 and appears in E. N. Lorenz, *The Essence of Chaos*, University of Washington Press, Seattle (1996), as an appendix.

65. R. T. Bakker, *Raptor Red*, Bantam, New York (1995).

66. Ironically, Owen was motivated by opposition to Darwin's new theory of evolution in his identification of dinosaurs as a separate taxonomic group.

67. Hawkins describes the exhibits and the geological terrains created to accommodate them in a chronological pattern in the article, B. W. Hawkins, 'On Visual Education as Applied to Geology', *Journal of the Society of Arts* 2, 444–9 (1854).

68. J. O. Farlow and M. K. Brett-Surman (eds), *The Complete Dinosaur*, Indiana UP, Bloomington, Ill. (1997), gives a comprehensive survey of every aspect of dinosaurs.

69. It is believed the Laetoli footprints were made by a species known as *Australopithecus afarensis*. Several hundred specimens have so far been found.

70. N. Agnew and M. Demas, 'Preserving the Laetoli Footprints', *Scientific American* (September 1998), pp. 44–55. Mary D. Leakey and J. M. Harris (eds), *Laetoli: A Pliocene Site in Northern Tanzania*, Clarendon Press, Oxford (1987).

71. For a summary of our understanding of human origins, see R. Leakey, *The Origin of Humankind*, Basic Books, New York (1994) and Phoenix, London (2001); and C. Stringer and P. Andrews, *The Complete World of Human Evolution*, Thames and Hudson, London (2005).

72. L. Fuchs, Preface to *De historia stirpium* (*On the History of Plants*), Basel (1542), quoted in I. Bernard Cohen, *Album of Science: From Leonardo to Lavoisier, 1450–1800*, Scribner's, New York (1980), p. 16.

73. The first botanical work with woodcut illustrations was Konrad von Megenberg's *Das Buch der Natur* (*The Book of Nature*), published in 1475. It places a greater premium on accuracy than on art in its representation of plants. This emphasis had not always been the case. The most elegant and beautiful renderings of plants from Nature are those made by Hans Weiditz, a student of Albrecht Dürer, for Otto Brunfels' *Herbarum vivae eicones* (*Living Portraits of Plants*) in 1530.

74. Vesalius' book was not the first illustrated medical treatise; that was due to Johannes de Ketham, whose *Fasiculo de Medicina*, first printed in 1491, merged six other medieval works.

75. I. B. Cohen, *Album of Science: From Leonardo to Lavoisier, 1450–1800*, Scribner's, New York (1980), p. 207.

76. Copernicus was also a medical student at Padua for some time.

77. The thirty-eighth edition of Henry Gray's *Gray's Anatomy* appeared in 1995, published by Churchill Livingstone, London. The 1918 edition, with 1247 plates, many of which were in colour, is available online at http://www.bartleby.com/107/

78. In the twentieth century, *Gray's Anatomy* has continued to pass through many new editions. It has exploited new standards of graphic design, colour, and multi-toning, to present state-of-the-art diagrams. It has even diversified into Anatomy Colouring Books as revision aids for medical students.

79. For more about Hooke's remarkable career as a scientist and an architect, see L. Jardine, *The Curious Life of Robert Hooke: The Man Who Measured London*, Harper, London (2005); S. Inwood, *The Man Who Knew Too Much: The Inventive Life of Robert Hooke, 1635–1703*, Pan, London (2003); and J. Bennett, *London's Leonardo*, Oxford UP, Oxford (2003).

80. This can still be seen in the National Museum of Health and Medicine in Washington, DC.

81. R. Hooke, *Micrographia: or, Some Physiological Descriptions of Minute Bodies Made by Magnifying Glasses*, J. Martyn and J. Allestry, London (1665). It can be read online at http://digicoll.library.wisc.edu/cgi-bin/HistSciTech/HistSciTech-idx?id=HistSciTech.HookeMicro

82. The modern term for a biological 'cell' emerged from this work because the tiny enclosed substructure of plants reminded him of monks' cells.

83. Joan D. Vinge, *The Snow Queen*, Questar, New York (1980).

84. Later the standard size of drawing paper in the paper trade, 26x34in, became known as 'atlas' because of its use in map collections.

85. A. Wegener, *The Origins of Continents and Oceans* (1915, 1920, 1922 and 1929 editions).

86. Since Wegener's insights geologists have also mapped the network of ridges under the oceans where plates appear to be moving apart. They are the sites of frequent earthquakes, where molten rock is forced up from below the crust and solidifies to form new crust. Where the plates converge and collide, huge mountain ranges like the Himalayas are pushed upwards. If one plate slips underneath the other then deep trenches in the oceans and volcanic activity result. By mapping earthquake activity and the historical changes in magnetisation of rock, it is possible to map the plate boundaries in some detail and determine the slow rate of movement.

87. M. Jenkins, *To Timbuktu*, Morrow, New York (1997).

88. P. Whitfield, *The Image of the World: 20 Centuries of World Maps*, Pomegranate Artbooks, San Francisco (1994).

89. We encountered such a transformation earlier on p. 125–6 in the Carter–Penrose representation of infinite space and time in a finite diagram.

90. Imagine a pair of compasses with one point set at the centre of the Earth while the other can touch the Earth's surface (which we assume to be perfectly spherical). Then the shortest distance on the Earth's surface is given by the shorter of the two arcs of the circumference of the circle that passes through the two points. This is called the great circle route and was first recognised as the shortest distance between two points on the Earth's surface for navigational purposes by the Portuguese navigator Pedro Nunes, in the fifteenth century.

91. This is sometimes called a rhumb line.

92. The 'Vitruvian Man' is a famous drawing completed by Leonardo da Vinci in 1492. It shows a detailed concern for proportion that Leonardo derived from the Roman architect Vitruvius, who died in 25 BC. Vitruvius believed that the human body displayed some harmonious mathematical proportions which reflected those of the Universe as a whole. His accompanying notes reveal that Leonardo uses the prescription that Vitruvius lays down in his work *De Architectura* 3.1.3, where he specifies the proportions of the human body thus: the navel is naturally placed in the centre of the human body, and, if in a man lying with his face upward, and his hands and feet extended, from his navel as the centre, a circle be described, it will touch his fingers and toes. It is not alone by a circle that the human body is thus circumscribed, as may be seen by placing it within a square. For, measuring from the feet to the crown of the head, and then across the arms fully extended, we find the latter measure equal to the former, so that lines at right angles to each other, enclosing the figure, will form a square. Vitruvius' book *De Architectura* was rediscovered in 1414 by a Florentine, Poggio Bracciolini, and Leonardo is reputed to have provided the illustrations for the edition that was published in 1521 under the editorship of Cesare Cesariano; see I. K. McEwen, *Vitruvius: Writing the Body of Architecture*, MIT Press, Boston (2004), and B. Baldwin, 'The Date, Identity, and Career of Vitruvius', *Latomus* 49, 425–34 (1990).

93. J. D. Barrow, *The Artful Universe Expanded*, Oxford UP, Oxford (2005), chapter 3.

94. A. M. Worthington and R. S. Cole: 'Impact with a Liquid Surface, Studied by the Aid of Instantaneous Photography', *Philo. Trans. Roy. Soc. London A* 189 (1897), 138. A. M. Worthington, *A Study of Splashes; with 197 Illustrations from Instantaneous Photographs*, Longmans, Green & Co., London and New York (1908).

95. D'A. W. Thompson, *On Growth and Form*, Cambridge UP, Cambridge (1917), p. 235. See also the discussion by Martin Kemp, entitled 'Stilled Splashes', in M. Kemp, *Visualizations*, Oxford UP, Oxford (2000), pp. 78–9. Kemp points out that one of Worthington's orginal 'splashes' of milk was used for the design of the Milk Marketing Board's logo in the UK during the 1960s.

96. *Seeing the Unseen: Photographs by Harold Edgerton*, online exhibition at http://www.susqu.edu/art_gallery/seeing/seeing.htm, and *Seeing the Unseen: Dr Harold E. Edgerton and the Wonders of Strobe Alley*, exhibition catalogue, Publishing Trust of George Eastman House, Rochester, NY (1994); this was associated with an exhibition of his work at the Lore Degenstein Gallery, Susquehanna University.

PART 3 PAINTING BY NUMBERS

1. Douglas Adams, *The Restaurant at the End of the Universe*, Harmony Books, New York (1982), p. 43.

2. H. Weyl, *Symmetry*, Princeton UP, Princeton (1952).

3. These faces have to be convex so that no straight line drawn between any two points on a surface polygon will pass outside that surface. Roughly speaking it means that they bulge outwards.

4. These are called, interchangeably, the Platonic solids, regular polyhedra, or the regular polytopes.

5. In 1852, Schläfli proved that in four-dimensional spaces there are six Platonic solids, but there are only three in five- and all higher-dimensional spaces; L. Schläfli, 'Theorie der vielfachen Kontinuität', unpublished manuscript (1852), later published in *Denkschriften der schweizerischen naturforschenden Gessel.* 38, 1–237 (1901). The six Platonic solids in four dimensions consist of the 4-simplex, which is made of 5 tetrahedra, 3 meeting at an edge, the hypercube, which is made of 8 cubes, 3 meeting at an edge, the 16-cell, which is made of 16 tetrahedra, 4 meeting at an edge, the 24-cell, which is made of 24 octahedra, 3 meeting at an edge, the 120-cell, which is made of 120 dodecahedra, 3 meeting at an edge, and the 600-cell, which is made of 600 tetrahedra, 5 meeting at an edge.

6. A fascinating series of discoveries have been made by archaeologists who have found hundreds of carved stones, many in regular polyhedral shapes, at Neolithic sites in Scotland. They date back to 2000 BC and are typically about 7 cm in diameter. It is not known what they were used for, and they are not found in burial sites. Sometimes they are made from soft materials like sandstone that are easily carved, but they are also found in granite. Dodecahedral forms are found, but there are no reports of an icosahedron. Could it be that some of the Platonic solids played some practical, decorative, or ceremonial role in human cultures more than 1500 years before Plato lived? See K. Critchlow, *Time Stands Still: New Light on the Megalith Mind*, Gordon Fraser, London (1979), p. 133, and D. N. Marshall, 'Carved Stone Balls', *Proceedings of the Society of Antiquaries of Scotland* 180, 40–72 (1976/77). The original stones are part of the Ashmolean Museum of Art and Technology collection, in Oxford.

7. Geminus according to T. L. Heath, *A History of Greek Mathematics* (2 vols), Oxford (1921).

8. See http://members.aol.com/Polycell/regs.html

9. There are ten in four-dimensional space. In more than four dimensions there are no star polyhedra at all.

10. The classic modern study of polyhedra is by the great Canadian geometer H. S. M. Coxeter, *Regular Complex Polytopes*, Cambridge UP, Cambridge (1974). The study of these polyhedra has also played a key role in the philosophy of science thanks to Imre Lakatos, who wrote a remarkable PhD thesis that was to become a classic in the philosophy of the scientific method. Lakatos's thesis was a dialogue concerning the famous formula which Leonhard Euler discovered, that the sum of the number of vertices and faces of a regular polyhedron is equal to the number of edges plus 2: for example, the octahedron has 8 faces, 6 vertices, and 12 edges, and $8 + 6 = 12 + 2$. See his thesis, which was published after his early death, in 1974, at the age of fifty-one: I. Lakatos, *Proofs and Refutations,* Cambridge UP, Cambridge (1976). Cayley introduced the 'stellated' adjective as a translation of the French 'étoile' which was used by Joseph Bertrand the previous year in J. Bertrand, 'Note sur la théorie des polyèdres réguliers', *Comptes rendus des*

séances de l'Academie des Sciences 46, 79–82 and 117 (1858). Cauchy, in his demonstration that Poinsot has found all these stellated polyhedra, called them 'polyhedra of a higher kind' in A. L. Cauchy, 'Recherches sur les polyèdres', *Journal de l'École Polytechnique*, 16, 68–86 (1813).

11. Star polyhedra are not very common in the arts, but the pinnacle of the sacristry of St Peter's at the Vatican displays a star polyhedron, just beneath the Cross.

12. A collection of star, or non-convex, Archimedean polyhedra also exist. There are three with the symmetry of the cube and octahedron, fourteen with the symmetry of the icosahedron and dodecahedron, and three with the symmetry of the truncated dodecahedron, see http://polyhedra.mathmos.net/entry/stellatedarchimedeans.html

13. The word 'soccer' is an abbreviation derived from the full title, 'Association Football', which originally distinguished the game from Rugby Football – abbreviated 'rugger' – in England.

14. A new design was introduced by FIFA for the 2006 World Cup in Germany.

15. It later turned out that the idea of molecular structures using the truncated icosahedral structure had been suggested on symmetry grounds by a Japanese chemist, E. Osawa, in 1970 and by two Russian chemists, Bochvar and Gal'pern, in 1973, and the truncated icosahedral symmetry had been found in the clathrin protein by two other Japanese scientists, Kamaseki and Kadota in 1969, see http://www.rie.shizuoka.ac.jp/~pmslhome/Fullerene/Osawa8xbE.htm

16. Later investigations by Curl, Kroto, and Smalley and others showed that there were carbon clusters with any even number of constituent atoms between 40 and 80. These could all form closed structures in agreement with a famous result of Leonhard Euler's about the existence of polyhedra: that if n is an even number greater than 22 then at least one polyhedron can be made with faces composed of 12 pentagons and $(n -20)/2$ hexagons. So, as n gets much larger than 20 there are many possible polyhedral structures.

17. J. Baggott, *Perfect Symmetry: The Accidental Discovery of Buckminsterfullerene*, Oxford UP, Oxford (1994); H. Aldersey-Williams, *The Most Beautiful Molecule: An Adventure in Chemistry*, Aurum Press, London (1995).

18. P. Auster, *The Music of Chance*, Viking, New York (1990), pp. 73–4.

19. Suppose they are finite in number so there is a biggest one, then multiply them all together and add one to the result; this new number is either a bigger one or it is divisible by a new prime that is bigger than the assumed biggest prime. Either way, the assumption that there is a biggest prime has led to a contradiction, so the list of primes cannot be finite.

20. He is sometimes known as Nichomachus. A translation of this work appears in the *Britannica Great Books of the Western World* series, vol. 11, University of Chicago Press, Chicago (1980). The description of Eratosthenes' sieve occurs in chapter XIII on p. 818: 'The production of these numbers is called by Eratosthenes the 'sieve', because we take the odd numbers mingled together and indiscriminate and out of them by this method of production separate, as by a kind of instrument or sieve, the prime and incomposite by themselves, and the secondary and composite by themselves, and find the mixed class by themselves.

'The method of the 'sieve' is as follows. I set forth all the odd numbers in order, beginning with 3, in as long a series as possible, and then starting with the first I observe what ones it can measure, and I find that it can measure the terms **two** places apart, as far as we care to proceed. And I find that it measures not as it chances and at random, but that it will measure the first one, that is, the one two places removed, by the quantity of the one that stands first in the series, that is, by its own quantity, for it measures it 3 times; and the one two places from this by the quantity of the second in order, for this it will measure 5 times; and again the one two places farther on by the quantity of the third in order, or 7 times, and the one two places still farther on by the quantity of the fourth in order, or 9 times, and so on *ad infinitum* in the same way.

'Then taking a fresh start I come to the second number and observe what it can measure, and find that it measures all the terms **four** places apart, the first by the quantity of the first in order, or 3 times; the second by that of the second, or 5 times; the third by that of that third, or 7 times; and in this order *ad infinitum*.

'Again, as before, the third term 7, taking over the measuring function, will measure terms **six** places apart, and the first by the quantity of 3, the first in the series; the second by that of 5, for this is the second number; and the third by that of 7, for this has the third position in the series.

'And analogously throughout, this process will go on without interruption, so that the numbers will succeed to the measuring function in accordance with their fixed position in the series; the interval separating terms measured is determined by the orderly progress of the even numbers from 2 to infinity, or by the doubling of the position in the series occupied by the measuring term, and the number of times a term is measured is fixed by the orderly advance of the odd numbers in series from 3.

'Now if you mark the numbers with certain signs, you will find that the terms which succeed one another in the measuring function neither measure all the same number – and sometimes not even two will measure the same one – nor do absolutely all of the numbers set forth submit themselves to a measure, but some entirely avoid being measured by any number whatsoever, some are measured by one only, and some by two or even more.

'Now these that are not measured at all, but avoid it, are primes and incomposites, sifted out as it were by a sieve.' [Ch. XIII, paragraphs 2–8]

21. T. L. Heath, *A History of Greek Mathematics*, 2 vols., Oxford UP, Oxford (1921).

22. Our Modern Pentathlon, an Olympic event, combines running, swimming, fencing, shooting, and riding. The ancient Greek pentathlon was more demanding and combined running (three different distances), jumping, discus throwing, javelin throwing (separately for accuracy and distance), and riding; later boxing or wrestling or a combined martial art, *pankration*, were added. The winner was crowned as the 'Victor Ludorum' of the Ancient Games, so Eratosthenes was quite well thought of.

23. When we hit on a prime number p, all its multiples by numbers smaller than p will already have been crossed out. The first one still standing will be $p \times p$.

24. D. N. Lehmer, *Factor Table for the First Ten Millions Containing the Smallest Factor of Every Number Not Divisible by 2, 3, 5, or 7 between the Limits 0 and 10017000*, Hafner Publishing Co., New York (1956). Reprinted from the Carnegie Institution of Washington Publication No. 105.

25. 'Machine Solves Intricate Tasks of Mathematics', New York *Herald Tribune* (12 July 1931), quoted in the biographical entry on the St Andrews mathematics history site: http://www-history.mcs.standrews.ac.uk/References/Lehmer_Derrick_N.html

26. This rule was first proposed by Adrienne Legendre in 1796.

27. J. Dieudonné, *Mathematics – The Music of Reason*, Springer, Berlin (1987), p. 37.

28. Specifically, Euclid gave twenty-three initial definitions – for example, that of a 'point' as being 'that which has no part'; and five postulates which govern what can be constructed and allowed to 'exist' mathematically – for example, the first postulate says that it is possible 'to draw a straight line from any point to any other point'. Lastly, after the five postulates, there are Euclid's five commonsense truths – for example, that if two things are both equal to a third thing then they must be equal to each other. However, the fifth of these commonsense truths, that the whole is greater

than one of its parts, while sounding obvious, is in fact false. Any infinite set of numbers (such as all the positive whole numbers) has the property that the members of any infinite subset of it (say, all the even numbers) can be put in a one-to-one correspondence with its members. Hence the infinite set and its subset have the same size. See J. D. Barrow, *The Infinite Book*, Jonathan Cape, London (2005).

29. For an account of Pythagoras's discovery, see http://www.maa.org/mathland/mathtrek_11_27_00.html

30. It has also been known as the Peacock's tail, or the Franciscan's cowl, and there have been Greek references to the 'theorem of the married women' in the literature, see D. E. Smith, *History of Mathematics*, Dover, New York (1958), p. 289.

31. The tablet is YBC 7289 in the Yale Babylonian Collection. Details originally published by Otto Neugebauer and Abraham Sachs: see O. Neugebauer, *The Exact Sciences in Antiquity*, Dover, New York (1969).

32. For more on Babylonian arithmetic see G. Ifrah, *The Universal History of Numbers*, Harvill, London, (1998); J. D. Barrow, *Pi in the Sky*, Oxford UP, Oxford (1992); and H. J. Nissen, P. Damerow, and R. K. Englund, *Archaic Bookkeeping*, University of Chicago Press, London (1993).

33. J. Dieudonné; 'The Work of Nicholas Bourbaki', *American Math. Monthly* 77, 134–45 (1970).

34. Euclid, *Elements* VI, 31. It is obviously true if you replace the squares by semi-circles, for example, since every term is just multiplied by $\pi/2$.

35. M. Senechal, 'The Continuing Silence of Bourbaki – An Interview with Pierre Cartier', *The Mathematical Intelligencer*, No. 1, 1998, pp. 22–28.

36. http://www.math.utah.edu/~cherk/mathjokes.html

37. J. Widmann, *Behennde vnnd hubsche Rechenung auff allen Kauffmanschafften* (*Mercantile Arithmetic*) (1489).

38. See F. Cajori, *A History of Mathematical Notations*, Dover, New York (1993), p. 230, part I, section 201; originally published as 2 vols. in 1928 and 1929.

39. It is known that Widmann inspected and annotated this manuscript at the Dresden library and appears to have used the notation in his lecture course at the University of Leipzig in 1486, as a set of notes taken by one of his students still exists in the Leipzig Library (Codex Lips. 1470).

40. J. W. L. Glaisher, 'On the Early History of Signs + and - and on the Early German Arithmeticians, *Messenger of Mathematics* 51, 6 (1921/2).

41. A whetstone was a stone for sharpening knives and the title of Recorde's book a play on the phrase to sharpen one's wits. In fact, in 1525 the German mathematician Christoff Rudolf wrote the first German text on algebra with the title *Coss*. This was a pun on the Latin for 'whetstone', *cos* and the Latin for 'thing', *cosa* (recall *cosa nostra* – 'our thing' in Italian), which was used to describe an unknown quantity in algebra.

42. He also wrote the wonderfully titled work *The Urinal of Physick* in 1547, which the *Dictionary of Scientific Biography*, New York (1970–90), described as 'A traditional medical work on the judgement of urines, full of sensible nursing practice, [but] less modern than his mathematical works and less critical of authority.'

43. F. Cajori, *A History of Mathematical Notations*, part 1, section 262.

44. M. Stifel, *Deutsche Arithmetica*, Nuremberg (1545); see Cajori ibid., p. 250.

45. J. H. Rahn, *Teutsche Algebra*, Zurich (1659).

46. F. Cajori, 'Rahn's Algebraic Symbols', *Amer. Math. Monthly* 31, 65–71 (1924).

47. V. Wing, *Harmonicon coeleste,* London (1651), p. 5.

48. J. M. Barrie, *The Plays of J.M. Barrie: Quality Street*, Scribner's Sons, New York (1948), Act II, p. 113.

49. Pascal's treatise is online at http://www.lib.cam.ac.uk/RareBooks/PascalTraite/. The triangle is on the third page.

50. If a statement $S(n_0)$ is true when applied to some particular number n_0 and the truth of the statement $S(n)$ implies the truth of the statement $S(n + 1)$, then the statement is true for all n equal to or greater than n_0.

51. For example, it has turned out to have a deep and unexpected connection to the Sierpinski carpet fractal, which we will see later in this section of the book. Suppose we just pick on two properties of the numbers in Pascal's triangle – for example, whether they are odd or even. Colour white the slots in the triangle containing even numbers and colour black those containing odd numbers. What does the triangle look like now? The result is striking. We get the nested triangles that we see in fractals like the Sierpinski carpet. If we modify the rules for colouring – say, using black to mark numbers that are divisible by 3 or by 5 or by 9, then we create analogous infinite 'carpets'.

52. Leibniz created a sort of mirror image of the Pascal triangle that is called the 'harmonic triangle'. It can be thought of as building from the bottom upwards rather than top-down. The two diagonal edges of the triangle are composed of the reciprocals of the whole numbers, 1/1, 1/2, 1/3, 1/4, 1/5, . . . etc., which is known as the 'harmonic series' . Each number in the triangle which does not lie on these edges is equal to the difference between the number diagonally above it to the right and the one to the right of it on the same row. So, for example, 1/60 = 1/30 – 1/60.

53. Holbein's famous picture, in the National Gallery in London, shows a carefully positioned symbolic display of objects. On the table, there is the book *Rechnung* written in 1527 by Peter Apian (1495–1552). It was the first arithmetic text to be written in German and on its title page appears Pascal's triangle. The full title of the book is *Ein Newe und wohlgregrundte wunderweysung aller Kauffmans Rechnung* (1527). In the picture it is held open on the page demonstrating division, its location in the picture below the globe thematically highlighting the dividing line of the Treaty of Tordesillas.

54. It was first referred to in print as 'Pascal's triangle' in 1886, in the influential textbook *Algebra* by George Chrystal, although it was referred to as 'Pascal's table for combinations' by Pierre Raymond de Montmort in 1708.

55. There was an Indian tradition beginning with Pingala's book on the metre of Sanskrit poetry that may have appeared around 200 BC. His rule for finding the number of combinations that can be taken two or three at a time from a total of six options created a list, known as the 'Meru Prastaara' ('The Staircase of Mount Meru'), that was put into triangular form by the tenth-century AD. Mount Meru, or Mount Sumeru, is a sacred mountain in Hindu and Buddhist tradition and the ascending nature of the number triangle gave rise to the idea of ascending the stair of numbers to the gods.

56. J. Needham, *Science and Civilisation in China*, III, Cambridge UP, (Cambridge (1959), pp. 133–7; Ho Peng-Yoke, *Yang Hui Dictionary of Scientific Biography* 14, 538–46 (1976); and Chu Shih-chieh, *Dictionary of Scientific Biography* 3, 265–71 (1971). For some comparative discussion see A. W. F. Edwards, *Pascal's Arithmetical Triangle*, Johns Hopkins UP, Baltimore (2002), chapter 5, and the Pascal's triangle website at http://binomial.csuhayward.edu/

57. Jonah 1:7.

58. Some false dice were intriguingly dynamic. If a solid region of a suitable wax was gently warmed by holding the die in the hand, then it would melt and flow along a small interior channel towards one face. Placing metal in the interior so as to respond to hidden magnets is another variant.

59. The book *The Newtonian Casino* by J. Bass tells the story of the attempts by a group of young mathematicians to beat the casinos in Las Vegas by simulating the behaviour of roulette wheels and analysing trends with hidden miniature computers (as a result the state of Nevada introduced a new law banning computers in their casinos). They were particularly interested in detecting 'biased' roulette wheels, because these are the ones that you can hope to win against in the long run. You can never beat a perfectly random wheel. Indeed, a good definition of a random process is one against which there is no long-term winning strategy.

60. D. Bartholemew, *God of Chance*, SCM Press, London (1984) and F. N. David, *Games, Gods and Gambling*, London (1962).

61. I. Hacking, *The Emergence of Probability*, Cambridge UP, Cambridge (1975).

62. B. Russell, 'Vagueness', in *Essays on Language, Mind, and Matter 1919–26*: *The Collected Papers of Bertrand Russell*, ed. J. Slater, Unwin Hyman, London (1923), p. 145. Quoted in O. Lemon and I. Pratt, 'On the Insufficiency of Linear Diagrams for Syllogisms', *Notre Dame Journal of Logic* 39, 573–80 (1998).

63. The first use of the term 'Venn diagram', according to the 2nd edn of the *Oxford English Dictionary*, occurs in the book *A Survey of Symbolic Logic* by Clarence Irving Lewis in 1918.

64. J. Venn, *Philosophical Magazine and Journal of Science* S. 5, Vol. 9, No. 59 (July 1880). For a historical discussion see M. Baron, 'A Note on the Historical Development of Logic Diagrams: Leibniz, Euler and Venn', *The Mathematical Gazette* 53, 113–25 (1969).

65. They can, however, create less than eight; for example, if A, B and C did not intersect at all, then the interior of U would be divided into 4 regions. However, some commentators refer to these as Euler diagrams. In general, a Venn diagram with n curves has $2n$ distinct regions (including the exterior empty region) and no more than two curves can intersect at a single point.

66. L. Euler, *Lettres à une Princesse d'Allemagne*, St Petersburg, l'Académie Imperiale des Sciences (1768).

67. O. Lemon and I. Pratt, 'On the Insufficiency of Linear Diagrams for Syllogisms', *Notre Dame Journal of Logic* 39, 573–80 (1998). The general form of Helly's theorem says that if there are N convex regions in an n-dimensional space, where $N>n$, and any choice of $n+1$ of the collection of convex regions has a non-empty intersection then the collection of N convex regions has a non-empty intersection. We are considering the case $n = 2$ and $N = 4$. Helly's original paper appeared in Jber. Deutsch. Math. Vereinig *32*, 175–6 (1923).

68. The result applies to regions which are convex. A convex region is one where, if you draw a straight line from one point inside it to another, then the line stays inside the region. This is clearly the case for a circle, but would not be the case for a region shaped like the letter S.

69. Besides the later variants of Euler's and Venn's constructions introduced by Lewis Carroll in his book *Symbolic Logic* of 1896, further refinements that allowed more complicated logical relationships to be represented were made by Charles Peirce, twenty years after Venn, as reported in his *Collected Papers*, Harvard UP, Cambridge (1933), and more recently by Sun-Joo Shin, in his

two books *The Logical Status of Diagrams*, Cambridge UP, Cambridge (1994) and *The Iconic Logic of Peirce's Graphs*, MIT Press, Cambridge, Mass. (2003).

70. For a philosopher's approach to the problem of depicting information see C. Peacocke, 'Depiction', *The Philosophical Review* 96, 383–410 (1987).

71. Patents began to be taken out on belts for industrial use in 1949. A collection of pictures from these patent requests can be seen in C. A. Pickover, *The Möbius Strip*, chapter 4.

72. C. A. Pickover, *The Möbius Strip: Dr August Möbius's Marvelous Band in Mathematics, Games, Literature, Art, Technology, and Cosmology.* A short biography can also be found online at http://www-groups.dcs.st- and.ac.uk/~history/Biographies/Mobius.html. See also J. Fauvel, R. Flood, and R. Wilson, *Möbius and His Band: Mathematics and Astronomy in Nineteenth-Century Germany*, Oxford UP, Oxford (1993), p.122.

73. A. F. Möbius, *Werke*, vol. 2 (1858), p. 519.

74. E. Breitenberger, 'Johann Benedict Listing', in I. M. James (ed.), *History of Topology* (Amsterdam, 1999), pp. 909–24.

75. M. C. Escher, *The Graphic Work of M. C. Escher*, transl. J. Brigham, Ballantine, London (1972).

76. J. Thulaseedas and R. J. Krawczyk, 'Möbius Concepts in Architecture', http://www.iit.edu/~krawczyk/jtbrdg03.pdf

77. A. C. Clarke, 'The Wall of Darkness', in *The Collected Stories*, Gollancz, London (2000), pp. 104–19.

78. This can be found in the collection of pieces of mathematically related fiction by C. Fadiman (ed.), *Fantasia Mathematica*, Simon & Schuster, New York (1958), pp. 222–237.

79. Speaking on the issue of whether a moment of silence needed to be formalised into the school day in Utah public schools, quoted in C. Gaither and A. Cavazos-Gaither, *Mathematically Speaking*, IOP, Bristol (1998), p. 356.

80. The etymology is trigonon = triangle made from tri (three), gonia (angles) and metron (measure).

81. The angles are measured in radians where 180 degrees is defined to be equal to π radians.

82. G. G. Joseph, *The Crest of the Peacock: The Non-European Roots of Mathematics*, 2nd edn, Princeton UP, Princeton, NJ (2000).

83. The early use of the word 'sine' or 'sinus' referred to the length of the chord on the circle of which the adjacent and hypotenuse were radii. It was Leonhard Euler who finally made it conventional to use it to describe the ratio of opposite and hypotenuse.

84. J. Gullberg, *Mathematics: From the Birth of Numbers*, Norton, New York (1997), p. 461.

85. This reflects the geometrical property of a tangent to any point on the circumference of a circle which is a line that 'touches' the circle at that point and is perpendicular to a line drawn from the point to the centre of the circle.

86. In his multi-volume work *Geometriae rotundi*, Basle (1583).

87. What mathematicians call a periodic function with period p if the value of the function at x is always the same as its value at $x+p$, so $f(x) = f(x+p)$. The power of the method is that it applies to very general sorts of functions, even those that have discontinuities in them – that is, those where you need to take the pencil off the paper occasionally when you draw their graphs.

88. That is a function f(x) with period 2π can be written as the infinite series $f(x) = a_0 + \sum_{n=1}^{\infty}[a_n \cos(nx) + b_n \sin(nx)]$ where a_0, a_n, and b_n are all constants that are easily determined from the function f(x) that is being expressed by the Fourier series.

89. M. West, *The Wit and Wisdom of Mae West*, Putnam's Sons, New York (1967), p. 35.

90. The origin of coordinates, $x = 0$ and $y = 0$, was taken at the lowest point.

91. Jefferson writes to Paine, 'You hesitate between the catenary, and portion of a circle. I have lately received from Italy a treatise on the equilibrium of arches by the Abbé Mascheroni. It appears to be a very scientifical work. I have not yet had time to engage in it, but I find that the conclusions of his demonstrations are that "every part of the Catenary is in perfect equilibrium."' Paine had earlier referred to it as a 'catenarian arch' but was interested in whether a bridge arch should have this shape or that of the arc of a circle, see http://members.aol.com/jeff570/math word.html

92. The chain is assumed to be ideal in that it has uniform mass per unit length, is perfectly flexible, and has zero thickness. The mathematical function cosh is defined in terms of the exponential function e^x by $\cosh(x) = (e^x + e^{-x})/2$.

93. Although Hooke recognised the correspondence between the arch shape and the hanging chain, he was not able to derive the equation for the catenary. In 1671, in his *Description of Helioscopes*, Hooke says that he has found 'a true mathematical and mechanical form of all manner of Arches for Building', and the solution is:

abcccddeeeeeefggiiiiiiii-illmmmmnnnnnooprrsssttttttuuuuuuuux.

Translated in 1705, this cryptogram reads: '*Ut pendet continuum flexile, sic stabit contiguum rigidum inversum* – As hangs the flexible chain, so inverted stand the touching pieces of an arch.'

94. The arch is 630 ft high and 630 ft wide at the base. Its equation is $y = -127.7\cosh(x/127.7) + 757.7$, where quantities are in feet. Note that $x = 0$ corresponds to the highest point where $y = -127.7 + 757.7 = 630$. Ground level corresponds to $y = 0$.

95. The slope of the cable at any point is just given by the ratio of the weight below it (which is equal to the weight per unit length times x divided by the tension. But this slope is also equal to the derivative dy/dx. Hence equating the two and integrating we have the equation of a parabola with the lowest point located at $x = 0$ and $y = 0$. The beautiful parabolic Tsing Ma suspension bridge in Hong Kong is the sixth-largest in the world. It spans 1377 m and is 206 m high. Its equation is therefore $y = (x/2301.13 \text{ m})^2$ because its two extremities pass through the points where $x = \pm 688.5$ m and $y = 206$ m if the origin of coordinates is located at the lowest point of the cable.

96. *The Saturday Review*, 15 April 1978.

97. J. Wallis, *De sectionibus conicis*, Pars 1, Proposition 1, and also J. Wallis, *Arithmetica infinitorum*, Proposition 91.

98. Aristotle's views about the impossibility of creating a true physical vacuum were tied to his beliefs about the non-existence of actual physical infinities. In a true vacuum, he believed, there would be no resistance to motion and bodies would be able to move with infinite speed. For more about these ideas and their consequences, see J. D. Barrow, *The Book of Nothing*, Jonathan Cape, London (2000), and J. D. Barrow, *The Infinite Book*, Jonathan Cape, London (2005).

99. Alan Lerner, title song of the 1965 musical *On a Clear Day*.

100. See Newton's second letter to Bentley about the infinite, I. B. Cohen, *Isaac Newton's Papers & Letters on Natural Philosophy*, Harvard UP, Cambridge, Mass. (1958), p.295.

101. These one-to-one correspondences between different infinite collections had been noticed long before by Albert of Saxony (1316–90) and by Galileo in his *Dialogues Concerning Two New Sciences* (First Day, sections 78 and 79), but they regarded the situation as a paradox that made infinity an unsatisfactory ingredient in mathematical reasoning. Cantor's masterstroke was to make it the defining characteristic of an infinite collection. For details see J. D. Barrow, *The Infinite Book,* Jonathan Cape, London (2005), chapter 4.

102. The next bigger infinity up the tower can be constructed from a given one by forming all the possible subsets of its members. For a finite set with N members there are 2^N possible subsets; this collection of all subsets of a set is called its 'power set'.

103. Cantor used the ∞ sign at first but abandoned it in 1882, because it was so frequently used by others in the traditional rather undetermined and ambiguous way. He needed to make clear the distinction between the different orders of infinity that he had so precisely distinguished.

104. This proviso is no real penalty because these numbers are rationals and we know that they can be counted. However, it serves to remove the ambiguity created by decimals that end with ever-recurring 9's because 0.2599999999 . . ., say, is the same as 0.2600000000 . . . Hence by eliminating those decimal expansions that end in zeros we uniquely identify the number 26/100 as 0.259999 . . . and not as 0.260000 . . .

105. J. D. Barrow, *The Infinite Book*, Jonathan Cape, London (2005), chapter 4.

106. This is reminiscent of the medievals' celestial agent invoked to avoid the creation of a perfect vacuum, see J. D. Barrow, *The Book of Nothing*, Jonathan Cape, London (2000), for a fuller account of the early arguments about the possibility of creating a true physical vacuum in Nature.

107. W. Playfair, *The Statistical Breviary*, London (1801).

108. H. G. Funkhouser, 'A Note on a Tenth Century Graph', *Osiris* 1, 260–62 (1936). The graph is part of a manuscript that was discovered in 1877 by S. Günther. A more extensive review of graphical methods which includes this find can be found in H. G. Funkhouser, 'Historical Development of the Graphical Representation of Statistical Data', *Osiris* 3, 269–404 (1937).

109. For a detailed analysis see the fascinating and detailed paper by H. P. Lattin, 'The Eleventh Century MS Munich 14436: Its Contribution to the History of Co-ordinates, of Logic, of German Studies in France', *Isis* 38, 205–25 (1948). Lattin disagrees with the interpretations of the graph by Günther and Funkhouser and also points out that the graph appears in a number of other manuscripts of the period.

110. A typical problem of his day was to work out the actual speed of a body at point B if you were given the average speed between two points A and B, together with a constant rate of acceleration.

111. The figure is from Part I of chapter 15, reproduced in J. Murdoch, *Album of Science: Antiquity and the Middle Ages*, Scribner's Sons, New York (1984), p. 156.

112. W. Playfair, *The Commercial and Political Atlas*, Wallis, London (1786), pp. 3–4.

113. J. J. Sylvester, *American J. of Mathematics* 1, 64–128 (1878). This is similar to the use which the word 'graph' is put to in the context of the area of mathematics called 'graph theory', which studies the structure of linkages between points and regions of space. We will look at some aspects of it in the context of the 'four-colour theorem' later.

114. T. L. Hankins, 'Blood, Dirt and Nomograms', *Isis* 90, 50–80 (1999). There is another use of the word graph in mathematics, to denote a connection drawn between different points.

115. W. Playfair, *The Commercial and Political Atlas*, Wallis, London (1786). Plate 20 in the 3rd edn; facsimile edn (ed. Ian Spence), Cambridge UP, Cambridge (2005).

116. I. Spence, 'William Playfair (1759–1823)', *Oxford Dictionary of National Biography*, Oxford UP, Oxford (2004).

117. R. H. Thurston, *A History of the Growth of the Steam Engine*, D. Appleton, New York (1878); see http://www.history.rochester.edu/steam/thurston/1878/

118. These included L. Lalanne, *Ann. de Ponts et Chaussées,* 2nd Ser. 11, 1 (1846). L. Lalanne, *Abaque, ou Compteur universal, donnant à vue à moins de 1/200 près les résultats de tous les calculs d'arithmétique, de géometrie et de mécanique pratique*, Carilan-Goery et Dalmont, Paris (1844).

119. Playfair's *Atlas* had not made much of an impact in England (see H. G. Funkhouser, 'Historical Development of the Graphical Representation of Statistical Data', *Osiris* 3, 269–404 (1937)), but in France it had been enthusiastically received and used by many scientists. Lalanne was one of the greatest enthusiasts for Playfair's new means of representing information.

120. L. Lalanne, 'Mémoire sur les tables graphiques', *Ann. de Ponts et Chaussées*, 2nd Ser. 11, 13 (1846).

121. L. Lalanne, 'Appendice sur la représentation graphique des tableaux météorologiques et des lois naturelles en général', in L. F. Kaemtz (ed.), *Cours complet de météorologie*, transl. and annotated by C. Martins, Paulin (1845), pp. 1–35.

122. A. W. Crosby, *The Measure of Reality*, Cambridge UP, Cambridge (1997), chapter 8.

123. G. Cattin, *Music of the Middle Ages*, vol. 1, Cambridge UP, Cambridge (1984), pp. 48–53.

124. The English translation is: 'So that your servants may, with loosened voices, resound the wonders of your deeds, clean the guilt from our stained lips O Saint John.'

125. Guido also became famous for the mnemonic picture of a hand with which this chapter opened, known as the 'Hand of Guido' or the 'Guidonian Hand', although it appears nowhere in his own writings. Each of the five fingers contained information about sequences of notes and the solemn scale. This device played an influential role in the theory of medieval music and led to the rejection of late non-diatonic music on the grounds that it didn't follow Guido's pattern because *non est in manu* ('it is not in the hand').

126. R. Rastall, *The Notation of Western Music*, St Martin's Press, New York (1982), pp. 136–7.

127. S. Eschenbach, 'The First Pitching Machine', http://www.inventionandtechnology.com/xml/2004/2/it_2004_2_dept_postfix.xml

128. James Hinton even had unusual medical views. He wrote a book entitled *The Mystery of Pain* in which he put forward the theory that 'all that which we feel as painful is really *giving* – something that our fellows are better for, even though we cannot trace it.' His son Charles later tried to create a mathematical formulation of this idea using higher-dimensional geometry and infinite series.

129. C. Hinton, *Dublin University Magazine* (1880). It was reprinted as a pamphlet, with the title 'What is the Fourth Dimension: Ghosts Explained', by Swann Sonnenschein & Co. in 1884. Mr Sonnenschein was a devotee of Hinton's ideas and published nine more of his pamphlets in the next two years. They were then gathered together and published as a two-volume collection entitled *Scientific Romances.* Those that feature extra dimensions are reprinted in C. Hinton, *Speculations on the Fourth Dimension: Selected Writings of Charles Hinton*, ed. R. Rucker, Dover, New York (1980).

130. C. Hinton, 'A Mechanical Pitcher', *Harper's Weekly*, 20 March 1897, pp. 301–2. You can also see a picture of the machine in the issue of *Scientific American* for 26 June 1897 or at the webpage http://www.inventionandtechnology.com/xml/2004/2/it_2004_2_dept_postfix.xml

131. A. Miller, *Einstein, Picasso: Space, Time, and the Beauty that Causes Havoc*, Basic Books, New York (2002). For a detailed study of the other sources of this development see L. D. Henderson, *The Fourth Dimension and Non-Euclidean Geometry in Modern Art*, Princeton UP, Princeton (1983). Henderson relates the important links with nineteenth-century fascinations with mysticism and the occult. A summary can be found in L. D. Henderson, *Leonardo 17*, 205–10 (1984).

132. It derives from *tesser* for 'four' and *aktis*, meaning 'ray'. Whereas a three-dimensional cube has 8 vertices, 12 edges, and 6 square faces, the tesseract has 16 vertices, 32 edges, 24 square faces, and 8 cubes comprising it.

133. It was published first in the February 1941 issue of *Astounding Science Fiction* magazine and then collected in R. A. Heinlein, *The Unpleasant Profession of Jonathan Hoag*, Gnome Press, New York (1959).

134. The American architect (and theosophist) Claude Bragdon was the first to incorporate aspects of four-dimensional geometry into some of his buildings, notably the Chamber of Commerce Building in Rochester, NY. He described some designs in his book *A Primer of Higher Space: The Fourth Dimension*, Rochester (1913).

135. C. Hinton, *A Picture of Our Universe* (1884); see *Speculations on the Fourth Dimension: Selected Writings of Charles Hinton*, ed. R. Rucker, Dover, New York (1980), p. 41.

136. R. Descartes, *Descartes' Conversation with Burman*, transl. J. Cottingham, Oxford UP, Oxford (1976), para. 79.

137. H. von Koch, 'Sur une courbe continue sans tangente, obtenue par une construction géométrique élémentaires, *Arkiv för Matematik* 1, 681–704 (1904); H. von Koch, 'Une méthode géométrique élémentaire pour l'étude de certaines questions de la théorie des courbes planes', *Acta Mathematica* 30, 145–74 (1906).

138. The infinite structure has the unusual property that its perimeter is a continuous curve (you can draw it without removing a pencil from the surface of the paper) that nowhere possesses a tangent because it is too spiky. Koch aimed to supply another example of a curve with this property, first discovered in another case by the German mathematician Karl Weierstrass in 1872.

139. For any closed curve on a plane surface the area enclosed A and the length of the perimeter, p, of the enclosing curve always obey the inequality $p^2 \geq 4\pi A$. The largest area enclosed by a curve of a given perimeter is when the curve is a circle and the equality sign holds in this formula. For a given area A the formula places no limit on how large p can be, if it wiggles about in a sufficiently complicated way as it does in the Koch snowflake.

140. See for example http://mathworld.wolfram.com/KochSnowflake.html

141. Note that the infinite sum inside the [. . .] brackets is a geometric progression with each term 4/9 of its predecessor. Its sum is therefore equal to $1/(1-4/9) = 9/5$.

142. W. Sierpinski, *C. R. Acad. Paris* 160, 302 (1915) and 162, 629–32 (1916).

143. After n steps there will be 8^n black squares in the carpet and the fraction of the total area of the carpet that is black is $(8/9)^n$ which goes to 0 as n goes to infinity.

144. These allowed deformations are changes into curves which are topologically equivalent. These are changes that can be made without cutting or tearing the line or surfaces involved. So

a sphere can be deformed in this way into a cube or an ellipsoid but not into a ring doughnut because the latter has a hole in it, but the doughnut can be deformed into a coffee cup, hence the joke that the definition of a topologist is someone who can't tell a doughnut from a coffee cup.

145. Sign off at http://www.linuxforums.org/forum/linux-networking

146. The first mass-produced Apple computer retailed at $1295 in 1977.

147. The quantities z and c are both complex numbers and so have real and imaginary parts ($z = x + iy$). These two parts can be plotted separately on the two axes of a graph and so specifying z or c fixes a location in two-dimensional space. If the complex constant $c = a + ib$ then the complex transformation $z \rightarrow z^2+c$ corresponds to the two real transformations of the x and y coordinates: $x \rightarrow x^2-y^2+a$ and $y \rightarrow 2xy+b$.

148. There are estimated to be about 10^{80} atoms and 10^{90} photons in the visible universe.

149. In 1991, a Japanese mathematician Mitsuhiro Shishikura proved that the boundary of the Mandelbrot set is connected and has (fractal) dimension 2. This means that although it is a curve on a flat surface it is so wiggly that it needs as much information to specify it as a whole two-dimensional area, even though it is not necessary to fill the whole plane to achieve this – as the example of the Sierpinski carpet shows. So no curve can have greater complexity dimension than the Mandelbrot boundary.

150. B. Mandelbrot, *The Fractal Geometry of Nature*, W. H. Freeman, New York (updated edn, 1983).

151. G. Julia, 'Mémoire sur l'itération des fonctions rationelles', *J. Math. Pure Appl.* 8, 47–245 (1918).

152. P. Fatou, 'Sur les equations fonctionelles', *Bull. Math. France* 47, 161–271 (1919); and 48. *Bull. Math. France* 33–94 and 208–314 (1920).

153. See, for example, R. Penrose, *The Emperor's New Mind*, Oxford UP, Oxford (1989).

154. B. Ernst, *The Eye Beguiled*, Taschen, Cologne (1992), p. 95.

155. L. S. Penrose and R. Penrose, 'Impossible Objects: A Special Type of Visual Illusions', *Brit. J. Psychology* 49, 31 (1958).

156. This subject area was taken up by Richard Gregory, whose famous book *Eye and Brain* attracted a very wide readership and placed the whole subject of visual illusion in the domain of neuropsychology cognitive science. See also R. Gregory, 'Perceptual Illusions and Brain Models', *Phil. Trans. Roy. Soc. B* 171, 278–96 (1968), and R. Gregory, 'Knowledge in Perception and Illusion', *Phil. Trans. Roy. Soc. B* 352, 1121–7 (1997).

157. Although it can be fooled. Craftsmen have constructed physical models which appear, when viewed from a particular angle, to *be* physical manifestations of impossible triangles. However, they are not what they seem and the projection is used to hide some of the construction from our sight.

158. R. Gregory, *Eye and Brain*, 3rd rev. edn (1977), first published 1966; *The Intelligent Eye*, London (1977); 'Perceptual Illusions and Brain Models', *Proc. Roy. Soc. Lond. B* 171, 278–96 (1968).

159. For a discussion of the neurovisual activity that leads to these illusions see A. Fraser and K. J. Wilcox, 'Perception of Illusory Movement', *Nature* 281, 565–6 (1979); J. Faubert and A. M. Herbert, 'The Peripheral Drift Illusion: A Motion Illusion in the Visual Periphery' *Perception* 28, 617–21 (1999); A. Kitaoka and H. Ashida, 'Phenomenal Characteristics of the Peripheral Drift Illusion', *Vision* 15, 261–2 (2003); A. Kitaoka, B. Pinna, and G. Brelstaff, 'New Variations of Spiral Illusions', *Perception* 30 637–46 (2001).

160. G. Tomasi di Lampedusa, *Il Gattopardo* (*The Leopard*), quoted by Andrea Camilleri in *The Shape of Water,* Picador, London (2003), p. 107.

161. R. Berger, 'Undecideability of the Domino Problem', *Memoirs of the Amer. Math. Soc.* 66, 1741 (1966).

162. See M. Gardner, 'Extraordinary Nonperiodic Tiling That Enriches the Theory of Tiles', *Scientific American* January 1977, pp. 110–21. Contributions were also made by the remarkable amateur Robert Ammann and John Conway to the two-tile problem. An excellent review of the development and structure of these tilings can be found in David Austin's article 'Penrose Tiles Talk across Miles' at http://www.ams.org/featurecolumn/archive/penrose.html

163. For his remarkable life story see M. Senechal, 'The Mysterious Mr Ammann', *The Mathematical Intelligencer* 26, 4 (2004).

164. B. Grünbaum and G. C. Shepherd, *Tilings and Patterns*, W. H. Freeman, San Francisco (1987).

165. D. Shechtman, I. Blech, D. Gratious and J. W. Cohn, *Phys. Rev. Lett.* 53, 1951 (1984).

166. C. Jencks, *The Garden of Cosmic Speculation*, Frances Lincoln, London (2003); J. D. Barrow, 'Gardening by Numbers', *Nature* 427, 296 (2004).

167. P. Lu and P. Steinhardt, 'Decagonal and Quasi-Crystalline Tilings in Medieval Islamic Architecture', *Science* 315, 1106–10 (2007); online at http://www.physics.harvard.edu/~plu/publications/Science_315_1106_2007.pdf

168. See for example the summary by the UK Patent Office at http://www.patent.gov.uk/about/ippd/faq/biofaq.htm#5

169. A summary of the case is reported at http://docs.law.gwu.edu/facweb/claw/penrose.htm

170. Quoted by Robin Wilson, *Four Colours Suffice*, Allen Lane (2002).

171. A. Cayley, 'On the Colouring of Maps', *Proc. Roy. Geog. Soc. and Monthly Record of Geog., New Monthly Ser.* 1, 259–61 (1879).

172. A. Kempe, 'On the Geographical Problem of the Four Colours', *Amer. J. Maths.* 2, 193–200 (1879).

173. R. Wilson, *Four Colours Suffice*, Allen Lane (2002).

174. K. Appel, W. Haken, and K. Koch, 'Every Planar Map is Four Colorable', *Illinois J. Maths.* 21, 429–567 (1977); part I on pages 429–90 is by Appel and Haken and the second part by all three authors. See also K. Appel and W. Haken, 'The Solution of the Four Color Map Problem', *Scientific American* 137 (8), 108–21 (1977).

175. Appel and Haken were in something of a race to complete their computer-assisted proof. Two other groups were using similar techniques in a race to be first to establish the result.

176. See for example T. Tymoczko, 'The Four-Color Problem and Its Philosophical Significance', *Journal of Philosophy* 76, 57–83 (1979).

177. A classic example of this sort is the famous proof by Walter Feit and John Thompson that all groups of odd order are soluble. It was 254 pages long and took up the whole issue of *Pacific J. Math.* 13, 775–1029 (1963).

178. The Proceedings appear in issue no. 1835 of the journal *Phil. Trans. R. Soc. A* 363, 2331–461 (2005), and details of the meeting can be found on the website at http://www.royalsoc.ac.uk/event.asp?id=1334

179. R. Tagholm, *Poems Not on the Underground*, ed. 'Straphanger', 5th edn, Windrush Press, Gloucs. (1996), p. 18.

180. K. Garland, 'The Design of the London Underground Diagram', *The Penrose Annual* 62, London (1969).

181. http://www.tfl.gov.uk/tube/maps/realunderground/realunderground.html. This website shows a dynamical transformation of the Underground diagram into the geographical map.

182. The current editions (printable in a variety of formats and languages) are available online at http://www.tfl.gov.uk/tube/maps/

183. O. Green, *Underground Art: London Transport Posters 1908 to the Present*, Studio Vista, London (1989).

184. Quoted in E. R. Tufte, *The Visual Display of Quantitative Information*, 2nd edn, Graphics Press, Cheshire, Conn. (2001), p. 143.

185. A. De Moivre, *The Doctrine of Chances or a Method of Calculating the Probabilities of Events in Play*, London (1738). The 1733 paper was written in Latin. A copy is given in R. C. Archibald, *Isis* 8, 671–83 (1926), and an English translation of the contribution to the 1738 edition of the book is in D. E. Smith, *A Source Book in Mathematics*, McGraw-Hill, New York (1929); see also R. H. Daw and E. S. Pearson, 'Abraham De Moivre's 1733 Derivation of the Normal Curve: A Bibliographical Note', *Biometrika* 59, 677–80 (1972) for a discussion.

186. The first person to draw a graph of a probability distribution appears to have been Johann Lambert (whom we encountered as one of the first to draw graphs of any sort in the eighteenth century), in his book *Photometria sive Mensura et Gradibus Luminis*, Detleffsen, Augsberg (1760), on p. 140, where he draws a possible distribution for the distribution of errors; see A. Hald, *A History of Mathematical Statistics from 1750 to 1930*, Wiley, New York (1998), p. 81.

187. A. De Morgan, *An Essay on Probabilities and on Their Application to Life Contingencies and Insurance Offices*, Longmans, Orme, Brown, Green & Longmans, & John Taylor, London (1838). The pictures can be found on pp. 132, 133, 141.

188. Ibid., p. 143.

189. A. Quetelet, *Lettres sur la théorie des probabilités, appliquée aux sciences morales et politiques*, M. Hayez, Brussels (1846). Quetelet seems to have learned of this mathematical distribution from Laplace, whose understanding of it owed much to Gauss, see S. M. Stigler, *The History of Statistics: The Measurement of Uncertainty before 1900*, Harvard UP, Cambridge (1986).

190. More precisely, any sum of many independent and identically distributed random variables will become more and more closely approximated by the normal distribution as the number in the sum increases without limit. The normal probability distribution for the outcome x is given by $p(x) = (1/\sqrt{2\pi V}) \exp[-(x-\mu)^2/2V]$, where μ is the mean value and V is the variance.

191. Not all physical processes are the sums of independent random variables. For example, many fragmentation processes are the product of many random variables. In that case it is the *logarithm* of the distribution of fragment sizes after many stages of fragmentation that is the sum of independent probabilities and so normal. Hence the distribution of fragment sizes is a lognormal distribution.

192. A. Quetelet, *Research on the Propensity for Crime at Different Ages* (1831), p.3, transl. and introduced by S.F. Sylvester, Anderson, Cincinnati (1984), p.3 P. Beirne, 'Adolphe Quetelet and the Origins of Positivist Criminology', *Amer. J. Sociology* 92, 1140–69 (1987).

193. This determinism was taken by some as a denial of the part played by free will in human actions, and thereby an abnegation of human responsibility for individual actions. An interesting critique was launched in 1902 by the Russian mathematician, Pavel Nekrasov, whose strong Russian Orthodox religious beliefs set him against Quetelet's emphasis. An account is given by E. Seneta, 'Statistical Regularity and Freewill: L. A. J. Quetelet and P. A. Nekrasov', *Int. Statistical Review* 71, 319–34 (2003).

194. Jouffret used the term 'bell surface' ('La surface . . . en forme de cloche') for the two variable normal distributions in 1872, but it does not seem to have caught on. Francis Galton used the English description 'bell-shaped curve' in 1876 in his *Catalogue of the Special Loan Collection of Scientific Apparatus at the South Kensington Museum* (1876). In the twentieth century the abbreviated 'bell curve' became much used. The term 'normal' distribution was first used by the philosopher Charles S. Peirce but has the unfortunate implication that any data not following this distribution are abnormal or peculiar in some way. A chapter of Galton's 1889 book *Natural Inheritance* has the title 'Normal Variability'.

195. Letter to Charles Knight, quoted in I. B. Cohen, *The Triumph of Numbers*, W. W. Norton, New York (2005), p. 153.

196. R. Herrnstein and C. Murray, *The Bell Curve*, Free Press, New York (1994).

197. A useful summary and list of links to the various responses and counter-responses can be found at http://en.wikipedia.org/wiki/The_Bell_Curve#Content

PART 4 MIND OVER MATTER

1. Bertrand Russell, *In Praise of Idleness* (1932), www.zpub.com/notes/idle.html

2. *Nieuwe Rotterdamsche Courant*, 4 July 1921, see M. Janssen et al (eds), *The Collected Papers of Albert Einstein*, vol. 7, Princeton UP, Princeton (2002), section 61, appendix D (see 1).

3. This quotation was part of a supposedly 'popular' explanation of relativity by Einstein entitled 'The History of Field Theory', written in 1929. It can be read online at http://www.rain.org/~karpeles/einsteindis.html

4. W. Wordsworth, *The Prelude.* Wordsworth was a great admirer of Newton and his work. The third book of his autobiographical prose poem *The Prelude* reflects on his early life at Cambridge and the marble statue of Newton in Trinity College chapel. Wordsworth was a student at St John's College, and his rooms overlooked the antechapel of Trinity. The full stanza is:

> And from my pillow, looking forth by light
>
> Of moon or favouring stars, I could behold
>
> The antechapel where the statue stood
>
> Of Newton, with his prism and silent face,
>
> The marble index of a mind for ever
>
> Voyaging through strange seas of Thought, alone.

5. The date is probably misremembered as it is known he was not in Woolesthorpe then; it was the year of the Plague. One contemporary of Newton's recorded that it was 1664 and he purchased it to test Descartes' theory of colours. Other commentators think he meant to write 1666, see D. Gjertsen, *The Newton Handbook*, Routledge & Kegan Paul, London (1986), pp. 507–8.

6. The allusion is to the defeat of Richard III by Henry Tudor at the Battle of Bosworth Field.

7. The idea that rainbows were created by the interaction between sunlight and water droplets, however, was not new. The basic idea had been proposed by Qutb al-Din al-Shirazi (1236–1311) of Shiraz, Iran. A fuller theory was published by Theodoric of Freiburg in 1307 and remained the standard description until the work of Descartes and Newton; see D. C. Lindberg, 'Roger Bacon's Theory of the Rainbow: Progress or Regress?', *Isis* 57, 236 (1966), and *Theories of Vision from al-Kindi to Kepler*, University of Chicago Press, Chicago (1976).

8. J. Keats, *Lamia* (1819):

> Philosophy will clip an Angel's wings,
>
> Conquer all mysteries by rule and line,
>
> Empty the haunted air, and gnomed mine –
>
> Unweave a rainbow.

9. It is even possible to see a 'moonbow' on a bright night, caused by refraction of sunlight reflected by the Moon and refracted by moisture. However, it appears white to our eyes because we lose our colour vision in dim conditions.

10. Secondary or even (very rarely) tertiary rainbows can be formed by light that undergoes two or three reflections around the inside of the water droplet before exiting. The secondary is brightest at an angle of about 52 degrees but forms on the same side of the sky as the Sun so is hard to see.

11. There is a claim that 'orange' was added by Newton between red and yellow in memory of one of his patrons, William III, Duke of Orange, who died two years before *Opticks* was published. The word was not commonly used to describe colour. Orange is a place in southern France. Elsewhere, Newton refers to this colour as 'citrus', or 'yellowish-red' when he describes the same colour formed by interference fringes in 'Newton's rings'. The inclusion of orange in the spectrum may have also had the goal of giving some special status to the political cause associated with William of Orange.

12. Named from the Latin *indicum*, meaning Indian.

13. Indigo was a vegetable dye that was greatly in demand in Newton's day and was imported from India, the Caribbean, and South America, although it had been banned in England for more than a century up to 1685 because it was believed to be toxic and damaging to fabrics, but these effects had been caused by some of the catalysts added to the dyeing process rather than indigo itself. It is used extensively still to dye denim.

14. W. Gilbert, *De Magnete*, London (1600), transl. S. P. Thompson, Chiswick Press, London (1900).

15. This 'dip' was first seen by Robert Norman in 1581.

16. James Clerk Maxwell, *Treatise on Electricity and Magnetism* (1861).

17. G. N. Cantor, *Michael Faraday: Sandemanian and Scientist: A Study of Science and Religion in the Nineteenth Century*, Macmillan, Basingstoke (1991).

18. J. D. Barrow, *The World within the World,* Oxford UP, Oxford (1988).

19. P. M. Harman, *Energy, Force and Matter: The Conceptual Development of Nineteenth-Century Physics,* Cambridge UP, Cambridge (1982).

20. R. Feynman, *QED*, Princeton UP, Princeton, NJ (1985), p. 113.

21. The actual radius r of a circular orbit is linked to the speed v because of the equality required between the centripetal force, mv^2/r, and the electrostatic force of attraction, e^2/r^2 in cgs units, or $e^2/4\pi\varepsilon_0 r^2$ in SI units.

22. N. Bohr, 'On the Constitution of Atoms and Molecules', *Philosophical Magazine* 25, 24 (1913).

23. Or equivalently that the angular momentum is equal to whole numbers of multiples of a quantum of angular momentum.

24. If the electron moves from a distant orbit labelled by quantum number $n = M$ into an orbit closer to the nucleus labelled by quantum number $n = N$, then the frequency of the emitted radiation will be proportional to $(1/N^2 - 1/M^2)$. Johann Balmer, a Swiss mathematics teacher from Basel, actually proposed a formula of this type as a result of juggling with numbers so as to fit the experimental data in 1885. It was refined by Janne Rydberg a few years later. Bohr was guided by the existence of Balmer's formula because he claimed that it showed that there was a simple mathematical rule behind the otherwise perplexing complexity of the spectral frequencies of light from atoms.

25. J. Emsley, *Nature's Building Blocks,* Oxford UP, Oxford (2003), p. 527.

26. A. Lavoisier, *Traité élémentaire de chimie* (1789). This was a major promotional textbook for the new chemistry and played a key part in chemistry education for many years to come.

27. This is reminiscent of modern definitions of an elementary particle. Of course, Lavoisier did not know of the internal structure of atoms: nuclei composed of protons and neutrons (which are, in turn, made of quarks) surrounded by electrons.

28. J. J. Berzelius, 'Essay on the Cause of Chemical Proportions, and on Some Circumstances Relating to Them: Together with a Short and Easy Method of Expressing Them', *Annals of Philosophy* 2, 443–54 (1813); *Annals of Philosophy* 3, 51–2, 93–106, 244–55, 353–64 (1814), reproduced in D. M. Knight (ed.), *Classical Scientific Papers*, American Elsevier, New York (1968).

29. Berzelius learnt of Dalton's work from discussions with him at one of the Royal Institution Lectures Dalton delivered on the subject in London in 1804.

30. Denoting copper sulphate ($CuSO_4$) combining chemically with hydrochloric acid (HCl) to produce sulphuric acid (H_2SO_4) and copper chloride ($CuCl_2$).

31. Most notable perhaps is the case of Julius Meyer, who plotted a graph of atomic volumes of forty-nine elements against their weights in 1868 and found a periodic variation. He prepared a paper for publication which he asked a friend to comment on. Unfortunately the friend was so slow in responding that Mendeleyev's more comprehensive scheme was published before Meyer could submit his version for publication.

32. In 1815, an English chemist, William Prout, had devised a detailed scheme wherein all elements were made from hydrogen – 'Prout's hypothesis', as it was known throughout the nineteenth century.

33. D. Q. Posin, *Mendeleyev: The Story of a Great Scientist*, McGraw-Hill, New York (1948).

34. It can be seen at the Mendeleyev Museum and Archives, St Petersburg State University.

35. Hydrogen is left out because of its unique properties, and the intervening rare gases, such as helium, had not yet been discovered.

36. The Greek *eka* prefix means 'follows'.

37. G. Holton, *Introduction to Concepts and Theories in Physical Science*, 2nd rev. edn with S. Brush, Princeton UP, Princeton, NJ (1985), p. 337

38. Atomic volume is atomic weight divided by density for solids and liquids.

39. It should be remembered that there may well still exist unfound superheavy elements which are unstable and exist for only brief periods of time.

40. Primo Levi's famous book *The Periodic Table* told of a series of remarkable incidents in his extraordinary life, some of which was spent as an industrial chemist, some in Nazi concentration camps. The incidents and characters are described by chapters, each of which bears the name of a chemical element. One well-known Italian astrophysicist (the late Nicolò Dallaporta) appeared as a young teaching assistant in the chapter entitled 'Potassium', which recounts events in Trieste in 1941. After I met Dallaporta in Trieste during the 1980s, where he was the Co-Director of SISSA with my thesis supervisor, Dennis Sciama, I was able to confirm first-hand that he was just as charming and friendly as the young Levi had discovered forty years before. I discovered that Dallaporta was even nicknamed Potassium by some Italians as a result of his leading role in this chapter of Levi's great book. Levi wrote elsewhere that the focus on the periodic table was an important psychological factor in his period of wartime imprisonment and torture. He knew that while his captors might tell him that black was white, and change the ethical standards by which people lived, they could not change the fact of the periodic table of the elements: there was a bedrock of absolute truth that no one could alter – see Primo Levi, *The Periodic Table*, Michael Joseph, London (1985).

41. http://www.ordinarygweilo.com/2004/03/not_a_drop_to_d.html

42. R. M. Roberts, *Serendipidty: Accidental Discoveries in Science*, J. Wiley, New York (1989), pp. 75–81.

43. Ibid. Some historians have been more sceptical about the origins of Kekulé's dreams. See, for example, A. J. Rocke, 'Hypothesis and Experiment in Kekulé's Benzene Theory', *Annals of Science* 42, 355–381 (1985).

44. The best alternative to Kekulé's hexagonal structure was that of a triangular prism (with two triangular faces and three rectangular faces), but it failed to explain all the derivatives of benzene that Kekulé's model succeeded in doing.

45. The routine ('a double unicycle without the seats') can be found at www.passingdb.com/patterns.php?id=34

46. S. Zamenhof, G. Brawerman, and E. Chargaff, 'On the DesoxyPentose Nucleic Acid, from Several Microorganisms', *Biochim. et Biophys. Acta* 9, 402 (1952).

47. J. D. Watson and F. H. C. Crick, 'Molecular Structure of Nucleic Acids', *Nature* 171, 737–8 (1953).

48. J. D. Watson, *The Double Helix: A Personal Account of the Discovery and Structure of DNA*, Penguin, London (1968).

49. H. Mankell, *Before the Frost*, Harvill, London (2004), p. 290.

50. J. C. Kendrew, 'The Three-Dimensional Structure of a Protein Molecule', *Scientific American* 205 (6) (1961), pp. 98–9.

51. The official website of the package can be found at http://www.avatar.se/molscript/

52. P. J. Kraulis, P. J., Domaille, S. L. Campbell-Burk. T. Van Aken and E. D. Laue, 'Solution Structure and Dynamics of Ras p21.GDP Determined by Heteronuclear Three- and Four-Dimensional NMR Spectroscopy', *Biochemistry* 33, 3515–31 (1994).

53. It is an enzyme that hydrolyses GTP (guanosine triphosphate) to GDP (guanosine diphosphate). Its chemistry plays a central role in sending information about growth to cells. Some forms of cancer arise from mutated forms of this protein, when it gets locked into a signalling cycle of perpetual growth instead of acting, as it should, as a regulating molecular-switch that turns growth on and off.

54. Dorothy L. Sayers, *The Unpleasantness at the Bellona Club*, Victor Gollancz, London (1935), chapter 4, p. 29 (first published 1928).

55. The removal of an electron from hydrogen requires 13.6 eV but the energy that binds the two protons and two neutrons into the helium nucleus is 28.3 MeV. The ratio of 2 million between these energies reflects the relative strengths of the electromagnetic forces and strong nuclear forces that bind the electrons and nucleons, respectively.

56. The masses are in units of 1.66×10^{-27} kg. These atomic mass units (amu) are defined so that the carbon-12 nucleus has mass exactly equal to 12.

57. Max Planck was the first physicist to suggest that a mass difference of this sort might exist because of the $E = mc^2$ formula when a bound system was broken into pieces: M. Planck, 'Das Prinzip der Relativitat und die Grundgleichungen der Mechanik', *Verh. Deutsch. Phys. Ges.* 4, 136–141 (1906) and 'Zur Dynamik bewegter Systeme', *Ann. der Phys.* 26, 1 (1908). The first person to apply the idea to the atomic nucleus was Paul Langevin, 'L'inertie de l'énergie et ses conséquences' in *J. de Phys.* (Paris) 3, 553 (1913). The calculations could not be completed correctly until after the neutron was discovered in 1932. Initially, they were developed as proposed tests of Einstein's $E = mc^2$ formula.

58. Note that the binding energy makes a negative contribution to the total energy.

59. F. W. Aston, *Mass Spectra and Isotopes*, Edward Arnold, London (1933), which is based on his earlier *Isotopes*, Edward Arnold, London (1922). In the displayed picture he plots the 'packing fraction', equal to 10,000 x (M – A)/A, where M is the atomic mass, versus the mass number, A. An interesting early history can be found in G. Audi, 'The History of Nuclidic Masses and of their Evaluation', *International Journal of Mass Spectroscopy* 251, 85–94 (2006).

60. Nucleons are the constituents of the nucleus in the form of either protons or neutrons.

61. An MeV is a million electron volts. For comparison with atomic binding energies, the binding energy of the single electron in an atom of hydrogen is 13.6 eV.

62. This opening line of John Donne's sonnet ('Holy Sonnet XIV') inspired J. Robert Oppenheimer to name the first nuclear test site 'Trinity'. It was located at Alamogordo in New Mexico and was the site of the Manhattan Project test on the morning of 16 July 1945. This was followed by the first atomic bomb blast over Hiroshima on 6 August and the second over Nagasaki on 9 August. The Japanese surrendered on 14 August.

63. The 'Little Boy' bomb dropped on Hiroshima was a fission device with 60 kg of uranium-235 and created a blast equivalent to 13 kilotons of TNT, killing approximately 80,000 people immediately; the 'Fat Man' bomb dropped on Nagasaki was a fission bomb with 6.4 kg of plutonium-239, equivalent to the blast from 21 kilotons of TNT. About 70,000 people were killed instantly.

64. M. Light, *100 Suns 1945–1962*, Jonathan Cape, London (2003). See also S. Weart, *Nuclear Fear: A History of Images*, Harvard UP, Cambridge, Mass. (1988).

65. If the explosion is detonated underground or underwater then the characteristic mushroom cloud will not occur in the same way because of the vapourisation of large amounts of solid or liquid in the blast.

66. It was entitled *The Formation of a Blast Wave by a Very Intense Explosion* and prepared for the British Civil Defence Research Committee of the Ministry of Home Security.

67. G.I. Taylor, 'The formation of a Blastwave by a Very Intense Explosion I: Theoretical Discussion', *Proc. Roy. Soc. A* 201, 159–74 (1950) and II: The Atomic Explosion of 1945, *Proc. Roy. Soc. A* 201, 175–186 (1950).

68. This is a beautiful application of what is known as the method of dimensions in physics. Taylor wants to know the radius, R, of a spherical explosion at a time, t, after it occurs (call that detonation time $t=0$). It is assumed to depend on the energy, E, released by the bomb and the initial density, r, of the surrounding air. If there exists a formula $R = kE^{a}r^{b}t^{c}$, where k, a, b, and c are numbers to be determined, then, because the dimensions of energy are $ML^{2}T^{-2}$, and of density are ML^{-3}, where M is mass, L is length and T is time, we must have $a = 1/5$, $b = -1/5$, and $c = 2/5$. The formula is therefore $R = kE^{1/5}r^{-1/5}t^{2/5}$. Assuming that the constant k is fairly close to 1, we see that the unknown energy is given approximately by $E = rR^{5}/t^{2}$.

69. A ton of TNT is a unit of energy equal to 10^{9} cal, which is about 4.2×10^{9} J.

70. This song was written in 1919 by James Kendis, James Brockman, and Nat Vincent, who merged their names into Jaan Kenbrovin because they all had separate contracts with music publishers.

71. A nucleus of the helium atom which contains two protons and two neutrons.

72. The positively charged antiparticle of the electron.

73. P. M. S Blackett, Sir Charles William Wilson, *Biographical Memoirs of the Royal Society* 6, 269–295 (1960).

74. Quoted in J. Gleick, *Genius – the Life and Science of Richard Feynman*, Pantheon, New York (1992), p. 244.

75. For an insightful semi-popular account see R. Feynman, *QED: The Strange Theory of Light and Matter*, Princeton UP, Princeton, NJ (1985).

76. For a history see D. Kaiser, *Drawing Theories Apart: The Dispersion of Feynman Diagrams in Postwar Physics,* University of Chicago Press, Chicago (2005).

77. R. Feynman, 'Space–Time Approach to Quantum Electrodynamics', *Phys. Review* 67, 769–89 (1949).

78. For other examples see R. Penrose, *The Road to Reality*, Jonathan Cape, London (2004), pp. 241–2, and R. Penrose, 'Applications of Negative-Dimensional Tensors', in *Combinatorial Mathematics and Its Applications*, ed. D. J. A. Walsh, Academic Press, London (1971), pp. 221–4. The following description of how the diagrams shown work appears at http://phys.wordpress.com/2006/07/: 'choose a closed polygon to represent the kernel letter of each tensor, and add an upwards leg for each contravariant index, and a downwards one for each covariant index. Index contraction is represented by joining the respective legs. A wiggly horizontal line represents symmetrisation; a straight one anti-symmetrisation. One can cross legs to indicate index shuffling. The metric gets no kernel figure (it's just an arch), so that contractions of indexes in the same tensor are easily depicted, and raising and lowering indexes amounts to twisting the requisite leg up or down. To indicate covariant differentiation, circle the tensor being differentiated and add the corresponding downwards (covariant) leg. And so on and so forth. Note also that commutative and associative laws of tensor multiplication allow your using any two dimensional arrangement of symbols that fits you, which aids in compactifying expressions.'

79. R. Feynman, *Surely You're Joking, Mr Feynman*, Norton, New York (1997).

80. Mark 3:24.

81. See J. D. Barrow, *The Universe That Discovered Itself*, Oxford UP, Oxford (2000) for a fuller discussion of the emergence of the concept of a law of Nature.

82. J. D. Barrow, *The Artful Universe Expanded,* Oxford UP, Oxford (2005).

83. J. D. Barrow, *The Book of Nothing*, Jonathan Cape, London (2000).

84. The uncertainty principle means that we can only measure an energy change ΔE over an interval of time Δt if $\Delta E \times \Delta t > h/2\pi$ where h is Planck's constant, and $h = 6.6 \times 10^{-34}$ *Js*. Processes that obey this inequality are called 'real' because they are directly observable, whereas processes for which $\Delta E \times \Delta t < h/2\pi$ are called 'virtual'. The act of observing them would involve an action that violates the inequality.

85. D. J. Gross and F. Wilczek, 'Ultraviolet Behaviour of Non-Abelian Gauge Theories', *Phys. Rev. Lett.* 30, 1343 (1973), and 'Asymptotically Free Gauge Theories I', *Physical Review D* 8, 3633 (1973); H. D. Politzer, 'Reliable Perturbative Results for Strong Interactions', *Phys. Rev. Lett.* 30, 1346 (1973), and 'Asymptotic Freedom: An Approach to Strong Interactions', *Physics Reports* 14, 129 (1974).

86. H. Georgi and S. Glashow, *Phys. Rev. Lett.* 32, 438 (1974).

87. These new particles will have masses roughly equal to the mass equivalent (via Einstein's famous energy–mass equivalence formula $E = mc^2$) of the energy at which the cross-over in the corresponding interaction strengths occurs. In the case of the electro-weak unification this was a big success. The requirement of unification predicted the existence of two new particles, the so-called W and Z bosons, with masses close to the energy of the unification of the weak and

electromagnetic force strengths. These new particles were subsequently discovered in particle-collider experiments at CERN in 1983, with masses at the expected values predicted by the theory of Glashow, Weinberg and Salam, who shared the 1979 Nobel Prize for physics for their achievement. Strangely, the Nobel Prize for the confirming discovery of the W and Z particles was given to Carlo Rubbia and Simon van der Meer, five years later, in 1984.

88. U. Amaldi, W. de Boer, and H. Fürstenau, 'Comparison of Grand Unified Theories with electroweak and strong coupling constants measured at LEP', *Phys. Letters* B 260, 447–455 (1991).

89. J. C. Pati, 'Topics in Quantum Field Theory and Gauge Theories', *Proceedings VIII International Seminar on Theoretical Physics*, GIFT, Salamanca, June 13–17 1977, ed. J. Azcárraga, *Lecture Notes in Physics* 77, 221–291 (1978).

90. It has been argued that a true intersection only occurs when supersymmetry is included in Nature's symmetry budget. The addition of the extra population of super partners for all the standard particles leads to a small change in the evolution of force strengths with energy and a closer 'triple point' of unification at high energy.

91. The graph plots the logarithms of the mass against the logarithm of the size.

92. D. Brown, *Angels and Demons*, Corgi, London (2001), p. 620.

93. B. J. Carr and M. J. Rees, 'The Anthropic Principle and the Structure of the Physical World', *Nature* 183, 341 (1978). J. D. Barrow and F. J. Tipler, *The Anthropic Cosmological Principle*, Oxford UP, Oxford (1986); V. F. Weisskopf, 'Of Atoms, Mountains, and Stars: A Study in Qualitative Physics', *Science* 187, 605–12 (1975).

94. The velocity needed to escape from the gravitational pull of a mass M of radius R is equal to the square root of $2GM/R$. When this escape velocity is equal to the speed of light the object has $R = 2GM/c^2$.

95. Typically, if the average speed of objects that are clustered by gravity is v, then the size of the aggregate will be roughly $R = 2GM/v^2$ if the total mass is M.

96. We should qualify this to say that it has three large dimensions *in* space. Current superstring theories of high-energy physics actually predict that the Universe has far more than three space dimensions. This is only possible if the 'other' dimensions are extremely small in extent and so don't affect the stability of structures in our three 'large' dimensions. One of the challenges for experimental physics is to find evidence for the possible existence of the other dimensions in high-energy physics experiments, like those planned with the Large Hadron Collider (LHC) at CERN in 2008. Their existence could also be revealed by slow changes in our supposed 'constants' of Nature. See J. D. Barrow, *The Constants of Nature*, Jonathan Cape, London (2002).

97. G. Whitrow, 'Why Physical Space Has Three Dimensions', *Brit. J. Phil. Soc.* 6, 13 (1955); J. D. Barrow, 'Dimensionality', *Phil. Trans. Roy, Soc. A* 310, 337–46 (1983).

98. N. Bohr, in L. Ponomarev, *The Quantum Dice*, IOP, Bristol (1993), p. 75.

99. E. Schrödinger, *Naturwissenschaften Die gegenwartige Situation in des Quantenmechanik* 23, 807–12, 823–8, and 844–9 (1935); English translation by J. D. Trimmer, *Proc. Am. Phil. Soc.* 124, 323 (1980) – the cat paradox appears on p. 338. The English version can also be found in J. A. Wheeler and W. Zurek, *Quantum Theory and Measurement*, Princeton UP, Princeton, NJ (1983) p. 156.

100. B. S. De Witt and N. Graham, *The Many-Worlds Interpretation of Quantum mechanics*, Princeton UP, Princeton, NJ (1973). H. Everett, 'Relative State Formulation of Quantum Theory' *Rev. Mod. Phys.* 29, 454 (1957).

101. J. D. Barrow and F. J. Tipler, *The Anthropic Cosmological Principle*, Oxford UP, Oxford (1986), chapter 7.

102. Another paradox of quantum mechanics was framed by Eugene Wigner and is known as the paradox of 'Wigner's friend'; it was first published in E. P. Wigner, *The Scientist Speculates – An Anthology of Partly-Baked Ideas*, ed. I. J. Good, Basic Books, New York (1962), p. 294, and also reprinted in J. A. Wheeler and W. Zurek, *Quantum Theory and Measurement,* Princeton U P, Princeton, NJ (1983). Einstein, B. Podolsky, and N. Rosen, 'Can Quantum-Mechanical Description of Physical Reality be Considered Complete?', *Phys. Rev.* 47, 777 (1935).

103. http://www.wired.com/wired/archive/11.06/nano_spc.html

104. M. F. Crommie, C. P. Lutz, D. M. Eigler, E. J. Heller, 'Waves on a Metal Surface and Quantum Corrals', *Surface Review and Letters* 2 (1), 127–37 (1995).

105. *Nature* 403, 512 (2000).

106. A. Einstein, B. Podolsky and N. Rosen, 'Can Quantum-Mechanical Description of Physical Reality be Considered Complete?', *Phys. Rev.* 47, 777–80 (1935). See also J. A. Wheeler, and W. H. Zurek, *Quantum Theory and Measurement.*, Princeton University, and *The Universe That Discovered Itself*, Oxford UP, Oxford (2000).

107. Einstein remarked that 'Still, it did not come out as well as I had originally wanted; rather, the essential thing was, so to speak, smothered by the formalism [*Gelehrsamkeit*].' , quoted in A. Fine, *The Shaky Game: Einstein, Realism and the Quantum Theory*, 2nd edn, University of Chicago Press, Chicago (1996), p. 35.

108. Polaroid glass removes one of the polarisations. If you hold two pairs of Polaroid sunglasses one behind the other and rotate one of them, you find an orientation where no light is transmitted because one pair blocks all vertically aligned photons and the other pair blocks all the horizontally aligned photons. A little bit of light usually gets through in practice if there are imperfections in the polarising lens material.

109. Energy is inversely proportional to the wavelength of the light so the 351 nm photon transforms into two photons with half its energy and double its wavelength, 702 nm.

110. See the article 'Cracking Codes II' by A. Ekert in *PLUS* magazine online at http://plus.maths.org/issue35/features/ekert/ for a detailed account of the process that transforms the laser splitting through borate crystals to an unusual and totally secure encryption procedure.

111. S. Sontag, *On Photography*, Farrar Straus & Giroux, New York (1977).

112. There was a long history of mechanical copying of documents before Carlson's innovation, from hand-cranked machines to carbon paper. For an illustrated history of these methods see the article 'Antique Copying Machines' at http://www.officemuseum.com/copy_machines.htm

113. He applied for his first patent in October 1937.

114. Carlson's materials were replaced by superior alternatives. The sulphur was replaced by selenium, which was a much better photoconductor and the lycopodium powder by an iron powder and ammonium salt mixture that created a much clearer final print.

115. D. Owen, 'Making Copies', *Smithsonian* (August 2004), pp. 91–97.

116. For more biographical information see http://members.tripod.com/~earthdude1/xerox/

117. This observation was first made in an article by Moore published in the 19 April 1965 issue

of *Electronics Magazine*. In 2005, Intel were offering $10,000 for an original copy of this magazine issue in mint condition for their company archive after Moore realised that he had long ago lent his own copy to someone who had not returned it. The prize eventually went to David Clark, an English engineer who had a complete set of back numbers of the magazine (see the BBC news story at http://news.bbc.co.uk/1/hi/technology/4472549.stm), but not before librarians in a host of university libraries decided they needed to secure their own copies. The original statement was that 'The complexity for minimum component costs has increased at a rate of roughly a factor of two per year . . . Certainly over the short term this rate can be expected to continue, if not to increase. Over the longer term, the rate of increase is a bit more uncertain, although there is no reason to believe it will not remain nearly constant for at least 10 years. That means by 1975, the number of components per integrated circuit for minimum cost will be 65,000.'

118. G. Moore, 'Progress in Digital Integrated Electronics', *IEDM Tech. Digest* (1975), pp. 11–13.

119. In computational science, for example, when one sees progress in the development of the computational effectiveness of a big project, such as studying the clustering of millions of stars, it is interesting to gauge the researchers' progress by factoring out the pure Moore's law progress. This leaves the real intellectual progress that has resulted from their ingenuity rather than that which has been achieved by the computer industry that they are merely piggy-backing on.

120. H. Wong and H. Iwai, 'The Road to Miniaturization', *Physics World* 18 (9), 40–44 (2005).

121. Nanotechnology takes its name from the nanometre scale of 10^{-9} m.

122. W. Blake, *Auguries of Innocence*, first published by O. G. Rossett in his 1863 edition of Gilchrist's *Life of William Blake.* See *The Complete Poems* (Penguin Classics), Penguin, London (1997).

123. There is nothing deeply fundamental about the angle of the slope. It will be slightly different for different types of sand under different conditions of dampness.

124. P. Bak, C. Tang, and K. Wiesenfeld, 'Self-Organised Criticality: An Explanation of 1/f Noise, *Phys. Rev. Lett.* 59, 381 (1987).

125. Typically the frequency of avalanche occurrence is proportional to 1/size.

126. P. Ball, *The Self-Made Tapestry*, OUP, Oxford (1999). The rice grains are unable to fall by sliding or rolling when their local pile does not exceed the critical slope and they are stopped more effectively when the slope becomes sub-critical after an avalanche.

PICTURE CREDITS

6–7 NASA/Francesco Ferraro (Bologna Conservatory)/ESA; 10 © The British Library Board. All rights reserved (Maps C.6.c.2. pp. 192–3); 15 AKG-images/Erich Lessing; 16 Dr Gerry Picus/courtesy Griffith Observatory; 17 Science Photo Library; 20 Science Photo Library/Royal Astronomical Society; 23 Reproduced from E. Harrison, *Cosmology: The Science of the Universe* (Cambridge University Press, 1981); 26 above reproduced from J. P. Nichol, *Thoughts on some Important Points relating to the System of the World* (Edinburgh, 1846); 26 below NASA/ESA, S. Beckwith (STSd) and The Hubble Heritage Team; 28 photo SCALA, Florence/Digital Image © The Museum of Modern Art, New York; 30 NASA/JPL – Caltech/R. Kennicutt (Univ of Arizona)/DSS; 32 © European Southern Observatory; 35 Reproduced from Owen Gingerich, *Album of Science: The Physical Sciences in the Twentieth Century* (courtesy Charles Schribner's Sons/Macmillan, 1989); 38 Science Photo Library/NASA/STSCI/J. Hester & P. Scowen. ASU; 41 Photo SCALA, Florence/© 2004. Photo Smithsonian American Art Museum/Art Resource; 44 NASA/ESA, J. Hester, A. Lott (ASU)/Davide De Martin (Skyfactory); 47 Science Photo Library/Royal Astronomical Society; 50 Ciel et Espace/Paris Observatory, © B. & S. Fletcher; 53 Science Photo Library/ Royal Astronomical Society; 56 Science Photo Library/Royal Astronomical Society/First published in E. Hubble, *The Realm of the Nebula* (1936)/courtesy of Yale University Press; 60 © Janos Rohan/www.astrojan.ini.hu; 62 Courtesy of the Observatories of the Carnegie Institution of Washington; 64 © Mark Westmoquette (University College London); 68 Anglo-Australian Observatory/ David Malin Images; 70 Science Photo Library/NASA/ESA/STSCI/P. Challis & R. Kirschner, Harvard; 72 First published in E. Hubble, *The Realm of the Nebula* (1936)/courtesy of Yale University Press; 75 First published in E. Hubble, *The Realm of the Nebula* (1936)/courtesy of Yale University Press; 77 Courtesy of the Observatories of the Carnegie Institution of Washington; 80 Science Photo Library; 83 Reproduced from E. Harrison, *Cosmology: The Science of the Universe* (Cambridge University Press, 1981); 85 Archives Lemaître Université Catholique de Louvain Institut d'Astronomie et de

Géophysique G. Lemaître Louvain-la-Neuve, Belgigue/*Celestial Treasury: From the Music of the Spheres to the Conquest of Space* by Marc Lachièze-Rey, Jean-Pierre Luminet and Joe Laredo (Cambridge University Press (16 Jul 2001), p155; 86 Science Photo Library/Take 27 Ltd; 88 M.C. Escher's *Cubic Space Division* © The M.C. Escher Company-Holland. All rights reserved. www.mcescher.com; 90 courtesy of Elizabeth M. Davies; 92 NASA; 97 courtesy of Macau Postal Authority; 98 Science Photo Library, courtesy of Robert V. Wagoner; 102 Science Photo Library, courtesy of Robert V. Wagoner; 104 Science Photo Library/Royal Astronomical Society; 109 Science Photo Library/Royal Astronomical Society; 110 J. P. Huchra & M. Geller/Harvard-Smithsonian Center for Astrophysics; 114 J. P. Huchra & M. Geller/Harvard-Smithsonian Center for Astrophysics; 116 NASA/R. Williams (STScl)/Hubble Deep Field Team; 119 NASA/Sloan Digital Sky Survey Team/NSF/DOE; 120 below courtesy of Dover Publications Inc.; 123 Science Photo Library; 126 reproduced from Roger Penrose, 'The Light Cone at Infinity', from *Relativistic Theories of Gravitation*, ed. L. Infeld (courtesy Pergamon, London, 1964); 128 reproduced from F. Hoyle, 'Cosmological Tests of Gravitation Theories', Varenna Lectures, Corso XX, (courtesy Academic Press, New York, 1960); 131 reproduced of G. F. R. Ellis, *Relativistic Cosmology – Proceedings of the International School of Physics 'Enrico Fermi', Course XLVII* edited by B.K. Sachs, Varenna on Lake Como, Villa Monastero, 20th June – 12th July 1969 (courtesy Academic Press New York and London, 1971); 134 NASA/WMAP Science Team; 137 reproduced from J. D. Barrow, Whitrow Lecture, text published in *Astronomy and Geophysics*, August 2002, Vol 43, issue 4; 138 Science & Society Picture Library; 139 NASA; 140 Reproduced from *Celestial Treasury: From the Music of the Spheres to the Conquest of Space* by Marc Lachièze-Rey, Jean-Pierre Luminet and Joe Laredo (Cambridge University Press 2001) p. 133; 143 Metropolitan Museum of Art/The LeWitt Collection, Chester, CT. Exhibited: Sol LeWitt on the Roof: Splotches, Whirls and Twirls, April 26, 2005 – October 30, 2005, The Iris and B. Gerald Cantor Roof Garden, The Metropolitan Museum of Art. Photograph © 2005 The Metropolitan Museum of Art/© ARS, NY and DACS, London 2007; 144 courtesy of Daniel Marlow at Princeton University; 146 © John Wheeler; 150–1 The Royal Collection/© 2007 Her Majesty Queen Elizabeth II; 154 Science Photo Library/NASA; 158 NASA/GSFC/Craig Mayhew and Robert Simmon; 160 Getty Images; 162

Science Photo Library/NASA; 166 © Sky Publishing Corp; 170 Science Photo Library/Royal Astronomical Society Library; 172 *Illustrated London News*; 174 courtesy of Camille Flammarion Collection; 179 Science Photo Library/PL – Caltech/Cornell/ NASA; 180 NASA; 184 Science Photo Library/NASA; 185 Science Photo Library/NASA; 190 reproduced from R. Lehti, 'Edvard Engelbert Neovius, an early proponent of interplanetary communication' (1998); 191 reproduced from R. Lehti, 'Edvard Engelbert Neovius, an early proponent of interplanetary communication' (1998); 194 Advertising Archive/courtesy artist Alex Schomburg; 198 reproduced from W. A. Bentley and W. J. Humphreys, *Snow Crystals*, 1931; 201 Science Photo Library/Kenneth Libbrecht; 203 Science Photo Library/Kenneth Libbrecht; 204 AKG-images; 208 © The British Library Board. All rights reserved (Maps*1180(20); 212 © The British Library Board. All rights reserved (*The Times*, 31st March 1875); 216 *Illustrated London News*; 220 The Linda Hall Library of Science, Engineering and Technology; 222 Science Photo Library/John Reader; 225 Science Photo Library/John Reader; 226 reproduced from *De Historia Stirpium* by Leonard Fuchs, 1542/Courtesy of the Division of Special Collections, Archives and Rare Books, University of Missouri at Columbia; 229 reproduced from *De Historia Stirpium* by Leonard Fuchs, 1542/Courtesy of the Division of Special Collections, Archives and Rare Books, University of Missouri at Columbia; 232 Science & Society Picture Library; 234 Science & Society Picture Library; 235 Science & Society Picture Library; 236 reproduced from *Gray's Anatomy* (facsimile, Grange Books, 2001); 242 Science Photo Library/UC Regents, NATL Information Service for Earthquake Engineering; 245 courtesy US Geological Survey; 248 Bibliothèque Nationale de France; 252 NASA; 254 photo SCALA, Florence/Courtesy of the Ministero Beni e Att. Culturali; 258 Science & Society Picture Library/NmeM; 261 Science Photo Library/Professor Harold Edgerton; 262–3 © 2003, Andrew D. Burbanks, University of Bristol; 266 Science & Society Picture Library; 269 Science & Society Picture Library; 270 reproduced from Wentzel Jamnitzer, *Perspectiva Corporum Regularium* (Siruela, 2006); 272 Science & Society Picture Library; 274 reproduced from I. & M. Hargittai, *Symmetry: A Unifying Concept* (courtesy of Random House, New York, 1996); 275 courtesy US Patent Office Document No 2682235; 276 School of Architecture Bulletin 1955, University Archives, Department of Special Collections, Washington University Libraries; 277 reprinted by permission of Macmillan

Publishers Ltd/Nature, 14th November 1985; 278 Science Photo Library; 284 Bill Casselman/University of British Columbia; 287 reproduced from D. Guedj, *Numbers: The Universal Language* (courtesy of Thames & Hudson, 1998); 290 reproduced from John Murdoch, *Album of Science, Antiquity and the Middle Ages* (courtesy of Charles Scribner's Sons/Macmillan, 1989); 291 reproduced from John Murdoch *Album of Science, Antiquity and the Middle Ages* (courtesy of Charles Scribner's Sons/Macmillan, 1989); 293 above & below reproduced from John Murdoch *Album of Science, Antiquity and the Middle Ages* (courtesy Charles Scribner's Sons/Macmillan, 1989); 294 courtesy of Dover Publications Inc.; 296 reproduced from Florian Cajori, *A History of Mathematical Notations* (courtesy of Dover, New York, 1993: originally published 1928–9); 298 courtesy of Dover Publications Inc; 302 © Cambridge University Press; 304 Science Photo Library; 307 University of York/Department of Mathematics; 308 AKG-images/Erich Lessing; 310 Science Photo Library; 314 reproduced from J. Venn, 'On the Diagrammatic and Mechanical Representation of Propositions and Reasonings', *Philosophical Magazine and Journal of Science* series 5, vol. 59, no. 59 (July 1880); 318 reproduced from L. Euler, *Lettres à une Princesse d'Allemagne*, St Petersburg, l'Académie Imperiale des Sciences (1798); 319 Science Photo Library; 322 courtesy US Patent Office Document #3991631; 323 M.C. Escher's *Möbius Strip II* © The M.C. Escher Company-Holland. All rights reserved. www.mcescher.com; 325 © Nick Mee; 326 © The British Library Board. All rights reserved (12214.BB.3(2)folio 251); 328 above SICD Universities Stasbourg/informatique@scd-ulp.u-strasbg.fr; 330 reproduced from M. R. Spiegel & J. Liu, *Mathematical Handbook of Formulas and Tables* (courtesy McGraw-Hill Book Co., 1999); 332 David Rosen; 334 Science Photo Library; 335 Alamy/lookGaleria; 336–7 Alamy/Peter Adams Photography; 338 Science Photo Library; 342 Science Photo Library; 346 Science Photo Library; 350 courtesy of E. R. Tufte, *The Visual Display of Quantitative Information*, Cheshire, CT: Graphic Press, 1983; 352 above & below reproduced from John Murdoch, *Album of Science: Antiquity and the Middle Ages* (courtesy Charles Scribner's Sons/Macmillan, 1989) p. 156. Ref MSS CL Slo 2156, fol. 164r: Mu CLM 26838, fol.77r; Ox Bl Can Misc 177, fol.16r; 354 reproduced from William Playfair, *The Commercial and Political Atlas and Statistical Breviary* (courtesy Cambridge, 2005 ed. Howard Wainer & Ian Spence); 356 reproduced from William Playfair, *The Commercial and Political*

Atlas and Statistical Breviary (courtesy Cambridge, 2005 ed. Howard Wainer & Ian Spence); 358 courtesy of The University of Chicago; 360 Bibliothéque de l'Ecole des Mines, Paris; 362 Lebrecht Music & Arts; 364 Lebrecht Music & Arts; 365 © The British Library Board. All rights reserved (CD:E:125/30/HUS 050); 366 above Lebrecht Music & Arts; 366 below courtesy of Dolmetsch online; 367 Lebrecht Music & Arts; 368 reproduced from 1999 Kessinger facsimile of C. Howard Hinton, *The Fourth Dimension* (1904); 371 The Metropolitan Museum of Art Gift of The Chester Dale Collection, 1955 (55.5)/Image © The Metropolitan Museum of Art/© Salvador Dali, Gala-Salvador Dali Foundation, DACS, London 2007; 374 reproduced from Peitgen, Jürgens and Saupe, *Fractals for the Classroom* (Springer-Verlag, 1992) from H. von Koch (1904); 376 reproduced from Peitgen, Jürgens and Saupe, *Fractals for the Classroom* (Springer-Verlag, 1992) from H. von Koch (1904); 377 Science Photo Library; 378 Cliff Pickover/www.pickover.com; 380 Creator Wolfgang Beyer; 384 Creator Wolfgang Beyer; 386 reproduced from Bruno Ernst *Optical Illustions* (Taschen, 1986), courtesy of Bruno Ernst & Jand De Rijk; 388 reproduced from Bruno Ernst *Optical Illustions* (Taschen, 1986); 389 M.C. Escher's *Waterfall* © 2007 The M.C. Escher Company-Holland. All rights reserved. www.mcescher.com; 390 reproduced from Bruno Ernst *Optical Illustions* (Taschen, 1986), courtesy of Bruno Ernst & Jand De Rijk; 392 © Akiyoshi Kitaoka 2004; 396 reprinted courtesy of the American Mathematical Society; 398 reprinted courtesy of Peter Lu, Harvard University; 399 reprinted courtesy of Peter Lu, Harvard University/Topkapi Museum; 401 Image courtesy of Kendall Dudley and Mary Eliff; 404 Science Photo Library; 407 reproduced from A. Kempe, 'On the Geographical Problem of the Four Colours', *American Journal of Mathematics* 2 (1879); 412 London Transport Museum/© Transport for London; 414 London Transport Museum/© Transport for London; courtesy of the Victoria and Albert Museum; 415 London Transport Museum/© Transport for London; 418 reproduced from S. M. Stigler, *The History of Statistics: The Measurement of Uncertainty Before 1900* (courtesy Harvard University Press, 1986); 421 reproduced from Augustus de Morgan, *An Essay on Probabilities* (1838); 426-427 © *R & D Magazine*; 432 Camera Press, London/Yousuf Karsh; 435 Topfoto/World History Archive; 438 © Cambridge University Press; 441 © Cambridge University Press; 442 Science Photo Library; 443 © Cambridge University Press; 444 The Wellcome Trust; 449

Science Photo Library/Pekka Parviainen; 450 reproduced from M. Faraday, *Electricity* (1852); 454 © The Nobel Foundation; 456 © The Nobel Foundation; 457 Science Photo Library; 460 The Linda Hall Library of Science, Engineering & Technology; 463 Science & Society Picture Library; 465 Science Photo Library/Ria Novosti; 471 Edgar Fahs Smith Collection/University of Pensylvania Library; 473 Science Photo Library; 474 reprinted by permission of Macmillan Publishers Ltd/*Nature* 25th April 1953; 476 The Wellcome Trust; 477 Science Photo Library/Driscoll, Youngquist & Baldeschwieler, Caltech; 478 Estate of Irving Geis/Illustration Irving Geis. Image from Irving Geis Collection, Howard Hughes Medical Institute. Rights owned by HHMI. Not to be reproduced without prior permission; 480 Estate of Irving Geis; 484 above Science Photo Library; 484 below reproduced from F.W. Aston *Mass Spectra and Isotopes* (Edward Arnold, London, 1933); 488 Time Life Getty/US Navy National Archives; 491 Science Photo Library/US National Archive; 494 © CERN Geneva; 497 © CERN Geneva; 500 originally published in R. Feynman, *Space Time Approach to Quantum Electrodynamics*, Phys. Rev. 76, 769 (1949), pp.772 & 787 © The American Physical Institute; 505 above originally published in R. Feynman, *Space Time Approach to Quantum Electrodynamics*, Phys. Rev. 76, 769 (1949), pp.772 7 787 © The American Physical Institute; 506 reproduced from J.C. Pati, 'An Introduction to Unification', in *Lecture Notes in Physics – 77 – Topics in Quantum Field Theory and Gauge Theories*, ed. J.A. Azcarrage (courtesy Springer Verlag, 1978); 513 above & below reproduced from Ugo Amaldi, 'LEP, ONE YEAR LATER', in *Relativistic Astrophysics, Cosmology and Fundamental Physics – Annals of the New York Academy of Sciences, Volume 647*, ed. J. D. Barrow, L. M. Mestel and P. Thomas (courtesy New York Academy of Sciences, 1991); 514 reproduced from J.D. Barrow and F. J. Tipler *The Anthropic Cosmological Principle* (courtesy Oxford UP, 1986); 518 © The British Library Board. All rights reserved ((B) PR 50); 522 reprint courtesy of International Business Machines Corporation, © International Business Machines Corporation; 525 reprint courtesy of International Business Machines Corporation, © International Business Machines Corporation; 526 image courtesy of Paul Kwiat and Anton Zeilinger; 532 courtesy of Xerox Corporation 536 © Intel Corporation; 539 Science Photo Library; 542 Courtesy of Marian Goodman Gallery, New York; 547 courtesy of Ricard Solé.

Front cover Photo SCALA, Florence/courtesy of the Ministero Beni e Att. Culturali; Back cover Science Photo Library/NASA.

INDEX

Figures in *italics* refer to illustrations and captions